Magnetism of Surfaces, Interfaces, and Nanoscale Materials

Handbook of Surface Science

Volume 5

Series Editor

Alexei A. Maradudin

ELSEVIER

AMSTERDAM • BOSTON • HEIDELBERG • LONDON • NEW YORK • OXFORD
PARIS • SAN DIEGO • SAN FRANCISCO • SINGAPORE • SYDNEY • TOKYO

Magnetism of Surfaces, Interfaces, and Nanoscale Materials

Volume 5

Edited by

Robert E. Camley
Center for Magnetism and Magnetic Nanostructures,
University of Colorado at Colorado Springs,
Colorado Springs, CO, USA

Zbigniew Celinski
Center for Magnetism and Magnetic Nanostructures,
University of Colorado at Colorado Springs,
Colorado Springs, CO, USA

Robert L. Stamps
University of Glasgow, Glasgow, Scotland, UK

ELSEVIER
AMSTERDAM • BOSTON • HEIDELBERG • LONDON • NEW YORK • OXFORD
PARIS • SAN DIEGO • SAN FRANCISCO • SINGAPORE • SYDNEY • TOKYO

Elsevier
Radarweg 29, PO Box 211, 1000 AE Amsterdam, Netherlands
The Boulevard, Langford Lane, Kidlington, Oxford OX5 1GB, UK
225 Wyman Street, Waltham, MA 02451, USA

Library of Congress Cataloging-in-Publication Data
A catalog record for this book is available from the Library of Congress

British Library Cataloguing in Publication Data
A catalogue record for this book is available from the British Library

For information on all Elsevier publications
visit our website at http://store.elsevier.com/

ISBN: 978-0-444-62634-9

Contents

Contributors

A.O. Adeyeye Information Storage Materials Laboratory, Department of Electrical and Computer Engineering, National University of Singapore, Singapore

N.R. Anderson Center for Magnetism and Magnetic Nanostructures, University of Colorado at Colorado Springs, Colorado Springs, CO, USA

Józef Barnaś Faculty of Physics, Adam Mickiewicz University, and Institute of Molecular Physics, Polish Academy of Sciences, Poznań, Poland

Robert E. Camley Center for Magnetism and Magnetic Nanostructures, University of Colorado at Colorado Springs, Colorado Springs, CO, USA

Mario Carpentieri Department of Ingegneria Elettrica e dell'Informazione, Politecnico di Bari, Bari, Italy

Zbigniew Celinski Center for Magnetism and Magnetic Nanostructures, University of Colorado at Colorado Springs, Colorado Springs, CO, USA

Vitalii K. Dugaev Department of Physics, Rzeszów University of Technology, Rzeszów, Poland, and Departamento de Física and CFIF, Insituto Superior Técnico, Universidade de Lisboa, Lisbon, Portugal

Giovanni Finocchio Department of Electronic Engineering, Industrial Chemistry and Engineering, University of Messina, Messina, Italy

Julie Grollier Unité Mixte de Physique CNRS-Thales, Palaiseau, France

I.R. Harward Center for Magnetism and Magnetic Nanostructures, University of Colorado at Colorado Springs, Colorado Springs, CO, USA

Bret Heinrich Surface Science Lab, Department of Physics, Simon Fraser University, Burnaby, BC, Canada

Karen Livesey Center for Magnetism and Magnetic Materials, Department of Physics, University of Colorado at Colorado Springs, Colorado Springs, CO, USA

Gennadiy Melkov Faculty of Radiophysics, Electronics and Computer Systems, Taras Shevchenko National University of Kyiv, Kyiv, Ukraine

Peter J. Metaxas School of Physics, University of Western Australia, Crawley, WA, Australia

Eric Montoya Surface Science Lab, Department of Physics, Simon Fraser University, Burnaby, BC, Canada

João Sampaio Unité Mixte de Physique CNRS-Thales, Palaiseau, France

Claus M. Schneider Peter Grünberg Institute PGI-6, Research Center Jülich, Jülich, and Faculty of Physics, University Duisburg-Essen and CENIDE, Duisburg, Germany

Helmut Schultheiss Helmholtz-Zentrum Dresden—Rossendorf, Institute of Ion Beam Physics and Materials Research, Dresden, Germany

Thomas Sebastian Helmholtz-Zentrum Dresden—Rossendorf, Institute of Ion Beam Physics and Materials Research, Dresden, Germany

G. Shimon Information Storage Materials Laboratory, Department of Electrical and Computer Engineering, National University of Singapore, Singapore

Andrei Slavin Department of Physics, Oakland University, Rochester, MI, USA

Vasil Tiberkevich Department of Physics, Oakland University, Rochester, MI, USA

Roman Verba Institute of Magnetism, National Academy of Sciences of Ukraine, Kyiv, Ukraine

Preface

The study of magnetic behaviors in confined geometries—thin films, interfaces, and nanostructures—has produced both important practical applications and fundamental discoveries about the physics of magnetic materials. In this book, we have tried to highlight the key topics and future developments in this research area. The authors of all the chapters are experts in their fields and have made special efforts to provide both an introduction to the topic as well as insights into and a description of the current state of the art.

The first few chapters are devoted to issues and procedures involved in creating and characterizing magnetic nano-objects. We start with a chapter devoted to the growth techniques and fundamental measurements. This chapter discusses the most commonly used growth methods, evaporation and sputtering. This is followed by a discussion of many characterization tools, including atomic force microscopy, magnetic force microscopy, MOKE magnetometry, vibrating sample magnetometry, ferromagnetic resonance, and Brillouin light scattering (BLS). The next two chapters give a more detailed look at two important techniques. The chapter by Claus Schneider covers methods to study element-specific static and dynamic behavior in magnetic materials. The following chapter covers the study of magnetization dynamics and includes detailed discussions of BLS, BLS microscopy, and ferromagnetic resonance with special attention to the importance of magnetic linewidth.

The next group of chapters concentrates on theoretical results and approaches to nanomagnetism. It begins with a chapter on magnetic nonlinear behavior in small structures by Karen Livesey. Starting with simple macrospin models, and proceeding through more complex theoretical approaches, she presents a clear and insightful discussion of magnetic chaos, Lyapunov exponents, threshold phenomena, and nonlinear behaviors in nanoelements. The following chapter deals with periodic magnetic structures, specifically the magnetic dynamics of an array of dipole-coupled magnetic nanodots. A general theory for the spin waves in such a structure is created and both square and hexagonal lattices are examined. The authors also deal with two important problems, microwave absorption and the conditions for nonreciprocal propagation. The final chapter in this group examines the kinds of thermal behaviors that can occur in nanostructures. A simple self-consistent local-mean-field model is developed and used to understand the thermal and magnetic phase diagrams of a variety of layered structures and small objects.

The final four chapters concentrate on current and application-oriented topics in the physics of nano-objects. The first of these chapters is devoted to spintronic nano-oscillators, which show promise for applications as on-chip microwave generators and for various nonlinear applications including frequency mixing and frequency modulation. The chapter gives a background to spin-transfer torque, the basic physics supporting these oscillators, and then discusses important practical issues for oscillators such as output power and noise reduction. The next chapter examines both the statics and dynamics of domain walls in nanostructures, where the length scale of the domain wall (a few nanometers up to several hundred nanometers) can be about the same size as the nano-object itself. The chapter finishes with a discussion of a variety of possible domain-wall applications, including domain-wall memristors and logic devices. The next chapter covers the Nobel Prize-winning topic of giant magnetoresistance (GMR). We are particularly pleased to have this contribution as one of the authors (Jozef Barnaś) worked extensively with both Nobel Prize winners, Peter Gruenberg and Albert Fert. The GMR effect is reviewed with both theoretical and experimental details, and the physical origin of the effect is highlighted along with recent applications. The final chapter of this group discusses the use of magnetic thin films for signal processing devices which work in the microwave and millimeter frequency ranges. Because of their small size and because the growth requirements are similar to those found in the electronics industry, these devices are generally compatible with on-chip high-frequency electronics. A variety of devices are explored, including band-stop filters, phase shifters, nonlinear devices, nonreciprocal devices, and electrically tuned devices.

We hope readers will find this book both informative and stimulating.

Robert E. Camley
Zbigniew Celinski
Robert L. Stamps

CHAPTER 1

Growth and Characterization of Magnetic Thin Film and Nanostructures

A.O. Adeyeye, G. Shimon

Information Storage Materials Laboratory, Department of Electrical and Computer Engineering, National University of Singapore, Singapore

1 INTRODUCTION

Advances in the controlled growth and characterization of high-quality magnetic thin films have been the key enablers in establishing the basis of modern magnetic read heads and recording media technology in the past decades. Due to their reduced dimensions, magnetic thin films have many novel and interesting phenomena, which are not present in the bulk parent materials. For instance, in Fe-, Co-, and Ni-based epitaxial thin films (with a thickness in the range of tens to hundreds of nanometers), their intrinsic magnetic properties such as exchange constants, absolute magnetic moments, and magnetic anisotropies can undergo a dramatic change as compared to their bulk values [1–7]. Another example is the induced perpendicular anisotropy and enhanced Co magnetic moments in Co/Pd superlattices due to the interfacial interactions between the Co and Pd thin-film layers [8,9]. All these discoveries were made possible due to advances in thin-film growth technology and various characterization techniques.

In designing functional thin-film structures, multilayer heterostructures that are made by depositing alternate layers of ferromagnetic (FM) and nonmagnetic thin films can be explored. In such heterostructures, many interesting phenomena may occur due to the interfacial interaction, interlayer coupling, and other confinement effects. One key example is the discovery of the giant magnetoresistance (GMR) effect in multilayer structures by Fert and Grünberg in the late 1980s [10,11], which was later awarded the 2007 Nobel prize in physics due to its impact on the hard-disk-drive (HDD) recording industry. The GMR effect arises as the dimensionality of the thin film becomes comparable to characteristic lengths for electrical transport measurements such as the mean free path of the electron, the inelastic scattering length, and the spin-diffusion length.

Magnetism of Surfaces, Interfaces, and Nanoscale Materials. http://dx.doi.org/10.1016/B978-0-444-62634-9.00001-1

This chapter is written with the purpose of providing an overview of the important techniques for the growth and characterization of high-quality magnetic thin films and nanostructures. The chapter will be presented in the following order: Section 2 discusses key thin-film growth techniques which have been widely used to fabricate single- and multilayer magnetic thin films both in academic research and in industry. Section 3 highlights various characterization techniques which can be used to obtain important physical properties of the fabricated films such as surface topography and static and dynamic magnetic properties. Section 4 provides practical examples and brief discussions of the growth and characterization of magnetic thin-film structures such as single- and multilayer film/nanostructures, thickness-modulated and bicomponent thin film/nanostructures. Section 5 summarizes the main points of discussion presented in this chapter.

2 THIN-FILM GROWTH

The growth of a high-quality magnetic thin film requires a precise control of deposition rate, film thickness, and in many cases an exact composition of two or more material components in a single film. The mastery of such precision is a culmination of decades of development in many branches of science and technology, including vacuum physics and design, thermodynamics, kinetics of materials, and surface science. All of these factors were necessary in order to achieve the current state of development in thin-film growth technology.

The method available to grow magnetic thin films can be broadly categorized into three groups based on the range of vacuum pressure in which each deposition method operates as shown in Fig. 1. These are thermal evaporation, sputter deposition, and molecular beam epitaxy (MBE). The range of vacuum pressure for each deposition method also dictates the materials' transport mechanisms during the film growth.

The evaporation method is performed in a vacuum pressure ranging from 10^{-5} to 10^{-10} torr. In this pressure range, materials are transported ballistically during the deposition owing to the high vacuum condition, which makes the collision mean free path longer or similar to the source-to-substrate distance. Normally, evaporation is achieved by thermal heating and therefore is called thermal evaporation. MBE operates at a much lower pressure range (10^{-10} to 10^{-11} torr) and has a lower deposition rate, which enables it to grow epitaxial films on substrates. The sputter deposition method, on the other hand, operates at a much higher pressure (10^{-1} to 10^{-3} torr) because it requires a constant feed of an inert gas (usually Argon) for plasma activation in order to extract (sputter) the materials out from the source. In this pressure range, the material

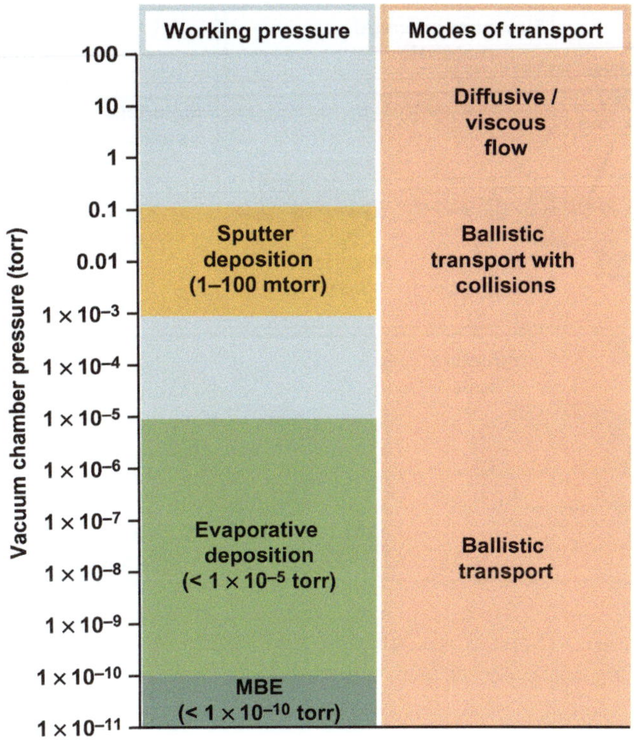

FIGURE 1
A schematic diagram showing the classification of deposition methods to grow magnetic thin film based on the range of working pressure and also the modes of material transport involved.

transport occurs semi-ballistically with frequent collisions due a shorter collision-dominated mean free path, which is comparable to the source-to-substrate distance. In the next section, we will provide a detailed overview of each of the various deposition methods.

2.1 Thermal and Electron Beam Evaporation

In the evaporative process, the source material is placed in a container which is made out of another compatible refractory material. Figure 2 illustrates various methods in which the thermal evaporation can be achieved.

In resistive heating, the source materials are placed on a material "boat" (also called "dimple boat") and heated by the resistive heating electrodes (Fig. 2a). This can control the rate of material evaporation by adjusting the current supply. Besides the simplicity in its setup, there are some disadvantages

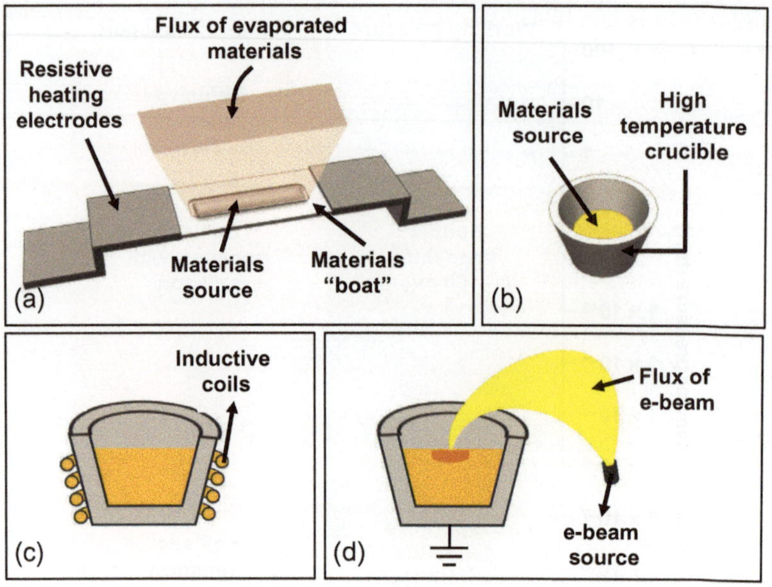

FIGURE 2

Schematic illustration of: (a) resistive heating thermal evaporation, (b) crucible containing source materials, (c) inductive heating, and (d) electron beam heating.

associated with this heating method. First, the contamination from the outgassing or evaporation of the boat and heating circuitry during the evaporation process at high temperature. Second, there is a difficulty in finding a suitable boat material, especially when evaporating refractory materials. Obviously, the boat materials must have a much higher melting temperature than the source materials they carry. In many cases where a nonsuitable boat material is chosen, the boat will melt or break apart due to excessive heating. Third, there can be an alloying and intermixing problem between the source materials and the boat, which can result in poor stoichiometry of the evaporated film.

In order to overcome some of the problems associated with the resistive heating method, the heating can be done inductively. In this method, the source materials are placed in a crucible (Fig. 2b) with water-cooled radiofrequency (RF) inductive coils winding around it to supply the heating power (Fig. 2c). Although the inductive heating method can overcome the heating issue for refractory materials, the crucible itself remains a source of contaminant flux during heating. This persistent problem demands a more ingenious heating method in which local heating of the source materials can be achieved.

The third approach of heating is the electron beam (e-beam) heating method, shown in Fig. 2d. Most commonly known as e-beam evaporation, this technique uses a small flux of electrons produced by an electron gun to locally heat up the source materials while the crucible is water-cooled to prevent evaporative contamination or outgassing. The electron gun is positioned in close proximity to the crucible and is placed at high bias with respect to the center of the crucible via the circuitry underneath. The same circuitry also controls a small-magnitude magnetic field ($\sim$100–200 Oe) to direct the position of the e-beam on the crucible. Although the localized heating overcomes the contamination issue significantly, the evaporation flux has now been modified to a point source rather than an area source. The point source has a detrimental effect on the film uniformity and step coverage. To ensure that different types of materials can be heated uniformly to form an area source, the localized heating can be swept around on the crucible to provide a uniform and large-area material source using a local magnetic field. This is particularly important when heating a nonconductive oxide such as Al_2O_3.

In order to obtain a high-quality magnetic film, the deposition is kept at a low rate (typically at 0.2 Å/s) in a high vacuum chamber and it is accurately monitored by a quartz crystal, which changes its frequency depending on the amount of mass that is deposited on it. The calibration and tooling factor are typically related to the material's physical properties (i.e., density and atomic volume) and the position of the quartz crystal with respect to the substrate, respectively. The tooling factor is used to correct the measured deposition rate by taking into account the relative position between the quartz crystal's location with respect to the substrate's position. The tooling factor depends largely on the specific geometry of the chamber [12].

In an automated process, the detected deposition rate and thickness serve as a feedback loop to control the e-beam power and substrate shutter. When evaporating magnetic thin films onto a patterned resist, particular attention must be paid to the resist step coverage. Due to the nature of the evaporative process, the flux of materials arriving onto the resist pattern has a specific line-of-sight for all the different points on the substrate, as shown in Fig. 3a. As a result, the evaporated film may have a poor conformity to the resist profile due to shadowing, and the coverage at the bottom of the resist pattern will be limited, as shown in Fig. 3b. There are many approaches which can be adopted in order to improve the film conformity. These include rotating the substrate during deposition as depicted in Fig. 3c, the use of multiple sources as illustrated in Fig. 3d, the use of a planetary-type substrate mount shown in Fig. 3e, substrate heating and combination of these. Shown in Fig. 3f is an example of an evaporated film with good surface conformity.

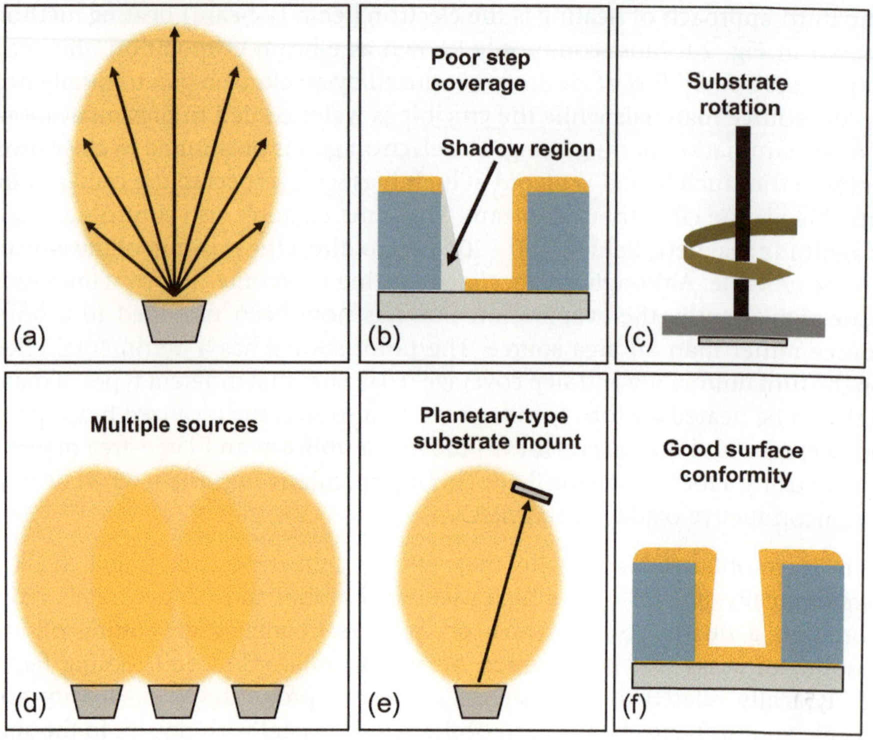

FIGURE 3

Schematic diagram showing: (a) anisotropic evaporation flux of small area source ($\cos^n\theta$ distribution), (b) poor step coverage of evaporated film, (c) substrate rotation, (d) evaporative deposition using multiple sources, (e) planetary-type substrate mount to achieve uniform film deposition, and (f) good surface conformity of evaporated film on patterned resist profile.

2.2 Sputter Deposition

The sputter deposition technique is the method of choice in depositing magnetic recording media and read heads in the HDD industry [13,14] due to its versatility and various advantages over the evaporative techniques. In a typical sputter deposition chamber, a high-purity Ar gas source at a certain calibrated pressure (typically ranging from a few millitorr to tens of millitorr) is maintained to achieve the desired deposition and film grain structures. The plasma is activated by supplying either a direct current or alternating current (RF) voltage source across a cathode and anode. The cathode is a negatively charged electrode at which the target source material is mounted, while the anode is a positively charged electrode at which the substrate is placed.

The process of plasma activation can be understood as follows. When a free electron is discharged near the cathode, it is immediately accelerated toward the positive anode. During this process, electrons will have collisions with Ar atoms such that there is a possibility that they will have gained enough kinetic energy to ionize the Ar atom into an Ar^+ ion. The production of Ar^+ ion in turn generates another free electron, which will be accelerated toward the anode and collide with Ar atoms on its way. Conversely, the generated Ar^+ ions will be accelerated toward the negative cathode. A fraction of Ar^+ ions with high enough kinetic energy may sputter off the source material near the surface. The sputtered material will then travel via semi-ballistic transport toward the substrate. At the same time, a fraction of Ar^+ ions may generate secondary electrons from the surface of the cathode to contribute to the plasma's self-sustaining process while another fraction of highly energetic Ar^+ ions may become implanted into a source target. During the sputtering process, distinct glow discharge regions are observed across the cathode–anode space in which various processes occur [15,16].

Some of the key advantages of sputtering deposition, compared to evaporative deposition methods, are the following: (1) the ability to deposit a wide range of materials (metals, insulators, compounds, and mixtures) with accurate composition and stoichiometry reproduction; (2) the possibility of bias etching processes for *in situ* substrate preconditioning or cleaning; (3) a better step coverage and thickness uniformity; (4) a better control of grain size and its distribution; and (5) support of high-throughput batch processing.

In determining the rate of materials growth during sputter deposition, there are a few parameters that are very important, namely, the sputtering yield, the ionizing gas pressure, and the sputtering power. First, the sputter yield measures the ratio between the number of sputtered atoms/molecules and the number of incident ions arriving at the source target. This depends largely on the relative atomic mass between bombarding ions and target materials and the incident ion energy. A typical sputtering yield is in the range of 0.01–3.5 [15]. Second, the choice of ionizing gas pressure will affect the sputtering yield, deposition rate, and the type of grains of the deposited films. For instance, increasing ionizing gas pressure to a certain optimum value may lead to an increase in sputtering yield and deposition rate. As discussed earlier, the typical range for the ionizing gas pressure is from a few to tens of millitorr. Various types of film grain structures (e.g., columnar with void, columnar with metallurgical grain boundaries, or equiaxed grains) [17] can be formed depending on the gas pressure and substrate temperature. Finally, the sputtering power determines the energy of the ion bombardment and particle–ion collision rates. The typical range of sputtering power is from tens of watts to a few hundred watts. Practically, increasing sputtering power results in increasing film deposition rate. The deposition rate is affected by all these key parameters and it is normally calibrated at a set condition. In the recording media research, the range of achievable grain size is around 4–7 nm [13].

The sputter deposition can be broadly categorized depending on the type of configuration used in the plasma activation and materials extraction process. Three main categories will be discussed: DC, RF, and magnetron sputter deposition. A short paragraph on reactive sputter deposition will also be given at the end of this section.

2.2.1 DC Sputter Deposition

In DC sputter deposition, a negative bias voltage is applied to the source target in order to direct the bombardment of Ar^+ ions in the plasma toward the source and eject the materials from the source. A stream of ejected materials then traverses across the cathode–anode distance and is deposited onto the substrate. In practice, only $-V$ (DC) is applied to the source target while the substrate is grounded, as shown in Fig. 4.

A presputtering of the source target is usually performed in order to clean the target materials from contaminant/oxides on the surface. This is also useful when sputtering a compound or mixture. In a compound or mixture, the initial sputtering rate may differ between different atomic species or components; therefore, the presputtering step helps in reaching the steady-state sputtering composition in order to replicate the source target composition. For the presputtering step, a shutter is placed on top of the source to prevent deposition onto the substrate until the presputtering is completed.

A sputtering system also has the ability to perform *in situ* substrate bias etching. For bias etching, a mixture of oxidizing gas (e.g., O_2) and ionizing gas (e.g., Ar)

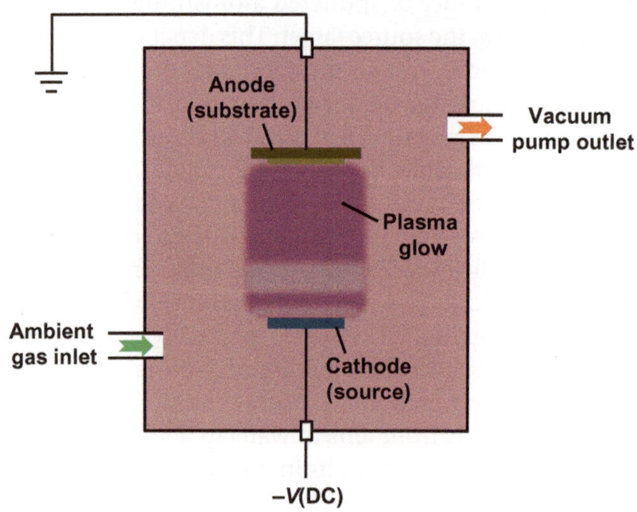

FIGURE 4

A schematic diagram of DC sputter deposition chamber.

in fixed proportions (typically 20% O_2 and 80% Ar) is fed into the chamber for plasma activation. At the same time, a negative bias is applied to the anode where the substrate is placed in order to direct the Ar^+ ions toward the substrate and bombard-etch it, hence the term bias etching. The oxidizing gas is added in order to help remove organic materials or contaminant on the substrate.

DC sputter deposition, despite being the simplest of all the sputtering techniques, suffers from a low deposition rate and a difficulty in depositing insulating materials [18]. The low deposition rate is caused by the low impact-ionization rate of secondary electrons just passing through the plasma without forming Ar^+ ions. This means that there are no new ions introduced in the process and therefore limits the sputtering rate. In addition, at higher power, the situation becomes worse because the secondary electrons' kinetic energy increases significantly, resulting in lesser impact ionization with the gas. Thus, instead of increasing the deposition rate, increasing the power actually decreases the sputtering rate. Moreover, during the sputtering of insulating materials, the accumulation of Ar^+ ions (excess positive charge) on the source compensates the cathode's negative bias and thus reduces the sputtering rate over time. In order to circumvent the problems associated with DC sputter deposition, an RF sputter deposition is used, as discussed in the next section.

2.2.2 RF Sputter Deposition

In RF sputter deposition, a network-matched current source at a standard frequency of 13.56 MHz is usually applied across the cathode–anode space. The RF power source prevents the accumulation of charges on top of the insulating materials by constantly reversing the type of charged species (positive ions or negative electrons) arriving on it. Furthermore, RF power can selectively enhance the oscillatory motion of an electron (due to its smaller mass vs. the heavier ions in the plasma) across the chamber. This improves the impact-ionization rate and thus results in higher plasma density. RF sputter deposition is typically used to achieve faster sputtering rates and to sputter-deposit insulating materials. In RF sputtering, the anode is usually connected to the chamber walls in order to increase the effective area of the anode. A larger effective area for the anode implies that a larger voltage drop will occur at the cathode (source) and therefore most of the sputtering will occur on the source.

2.2.3 Magnetron Sputter Deposition

In addition to using RF sputtering, another enhancement to sputter deposition can be made by adding a set of permanent magnets (e.g., NdFeB) to create a localized magnetic field near the surface of the material's source target in the sputtering gun assembly. For example, the most common setup involves placing a set of radially arranged permanent magnets, as shown in Fig. 5a. In this

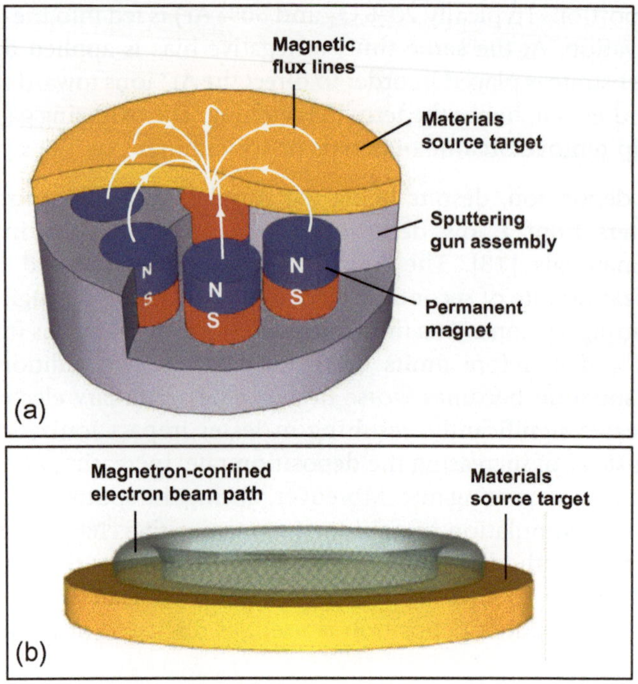

FIGURE 5

A schematic diagram of: (a) magnetron sputtering target with a set of radially arranged permanent magnets showing the magnetic flux lines created, and (b) the path traveled by electron beam due to magnetic field confinement created by the magnetron near the surface of target.

case, a closed magnetic flux is created resembling a half-toroidal shape. These magnetic field regions will cause free electrons to travel in a circular motion close to the surface of the target due to electron–magnetic field interaction (Fig. 5b).

This enhancement effect gives rise to a higher plasma density and a correspondingly higher material sputtering and deposition rate. However, due to the higher sputtering rate, around the half-toroidal area, the target will thin down slightly faster, creating a circular dent on the target. There is another type of magnetron arrangement in which bar magnets are used to create a rectangular electron beam path instead of radial.

Another class of sputter deposition is called reactive sputter deposition. In this method, chemical reactions occur between the species of materials arriving at the surface of a substrate, namely, reaction between different species of sputtered materials and reaction between sputtered materials and reactive gases. The latter method is usually adopted in the deposition process forming oxides or nitride films. For example, Al_2O_3 thin film can be deposited using reactive sputter deposition by sputtering an Al target in the presence of O_2 gas.

Another deposition technique which is not covered in detail in this chapter is the MBE deposition method [19]. MBE operates in an ultrahigh vacuum chamber with a base pressure ranging between 10^{-10} and 10^{-11} torr. As the base pressure is very low, the vaporized materials are transported ballistically. The deposition is capable of growing high-quality epitaxial films. The material source, typically solid, is evaporated in a deep crucible within a cooled shroud called effusion source cells or Knudsen cells. The MBE system is often equipped with *in situ* monitoring tools such as reflection high-energy electron diffraction spectroscopy, low-energy electron diffraction, X-ray photoelectron spectroscopy, Auger electron spectroscopy, and mass spectrometry to monitor the quality of the films.

3 CHARACTERIZATION TECHNIQUES

In this section, a variety of common characterization techniques used in measuring the properties of magnetic thin films and nanostructures will be discussed. The section is divided into three subsections, namely, surface characterization and static and dynamic magnetic characterizations.

3.1 Surface Characterization

The surface characterization of a magnetic thin film can be classified based on the type of information extracted from the surface of the sample. In this section, two techniques will be discussed, namely, atomic force microscopy (AFM) and magnetic force microscopy (MFM), both of which come under the broader term scanning probe microscopy (SPM). SPM uses a scanning probe tip to characterize a surface property, such as topography and magnetic contrast.

3.1.1 Atomic Force Microscopy

Developed in 1986 by Binnig and Gerber (IBM) and Quate (Stanford), the AFM brings a probe into close proximity with the surface. The force is detected by the deflection of a spring, usually a cantilever. The forces between the probe tip and the sample are sensed to control the distance between the tip and the sample. At short (large) probe–sample distances, the forces are repulsive (attractive). The AFM cantilever can be used to measure both attractive force and repulsive force modes. The scanning probe's tip used in AFM is usually made of doped Si and has a typical radius between 25 and 50 nm. The radius of curvature of the tip limits the resolution of an SPM. An Si cantilever extended from a tip carrier, as shown in Fig. 6a, carries the tip. The tip carrier is loaded onto a scanner head, which also includes a photodetector feedback loop. The feedback loop uses a

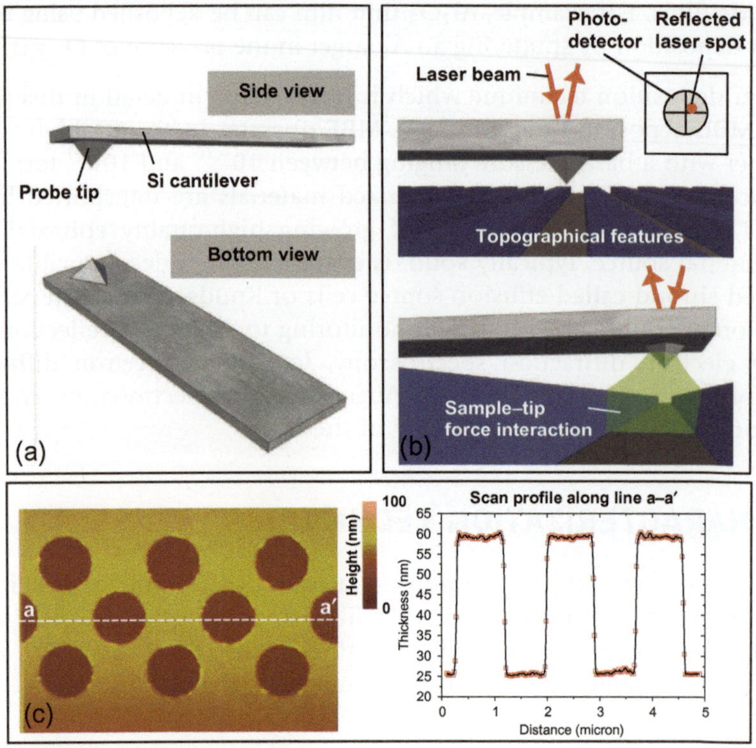

FIGURE 6

Schematic illustrations showing: (a) SPM probe tip carried by Si cantilever, (b) the sample-to-tip force interaction and photodetector detection process, (c) left: AFM plot showing an array of 25-nm NiFe circular disk embedded in 60-nm BARC layer; right: the scan height profile along line a–a′.

laser source, which is positioned on the back surface of the cantilever such that the laser is reflected onto the center position of the photodetector. As the tip scans along a surface, the photodetector detects the change in the position of the laser reflecting off the Si cantilever's back surface due to sample–tip interactions, as shown in Fig. 6b. This information is used by the feedback loop to map the topological features on the surface of the samples. The image is formed by scanning and using the feedback loop to maintain either constant force or constant oscillation amplitude during scanning.

An example of the topographical features on the surface of the sample mapped with an AFM is shown in Fig. 6c. This is an AFM image of an array of 25-nm-thick disks, each with a diameter of 800 nm embedded inside a bottom antireflection coating (BARC), which is 60 nm thick. The scan profile along line a–a′ shows that the thickness changes as the tip scans from left to right, detecting the BARC layer at 60 nm and the disks at 25 nm. This method can be used to determine the film thickness.

Two main modes of imaging are used in a typical AFM system, namely, contact mode and tapping mode. The contact mode, where the tip scans the sample in close contact with the surface, is the most common mode used in force microscopy. The force on the tip is repulsive and is set by pushing the cantilever against the sample surface with a piezoelectric positioning element. The deflection of the cantilever is sensed and compared in a DC feedback amplifier to some desired value of deflection. The feedback amplifier applies a voltage to the piezo to raise or lower the sample relative to the cantilever to maintain the desired value of deflection. In contrast, the tapping mode technique allows high-resolution topographic imaging of sample surfaces that are easily damaged, loosely bound to their substrates, or difficult to image by other AFM techniques. The tapping mode approach can be used to overcome problems associated with friction, adhesion, electrostatic forces, and other difficulties in conventional AFM scanning methods by "tapping" the surface to obtain the surface information but avoid dragging the tip across the surface.

3.1.2 Magnetic Force Microscopy

The MFM uses a magnetized tip to scan the surface of the magnetic sample. Tip–sample magnetic interactions are detected and used to reconstruct the spatial variation of magnetic forces on a sample surface. The MFM can sense the deflection of the cantilever resulting in a force image (static mode) or from the resonance frequency change of the cantilever, which will result in a force gradient image. The MFM signal contains information from both the surface topography and from the magnetic state generated by Van der Waals and magnetic forces, respectively. In order to separate the magnetic contrast information from the topographical information, two successive scans are performed. In the first scan, the tip measures the topography of the sample. In the second scan, the tip–sample distance (Δz) is increased and the biased tip is scanned again along the same line scan, which is now only affected by the magnetic force. This is made possible as the magnetic force scales with Δz following $1/(\Delta z)^3$ relation, while the Van der Waals force scales with a $1/(\Delta z)^6$ relation. The former is more sensitive at larger Δz values during MFM scanning. By subtracting the signal of the first scan from that of the second scan, magnetic contrast information from the sample can be extracted.

3.2 Static Magnetic Characterization

In order to understand the magnetization reversal mechanism in thin films and nanostructures, it is important to characterize the static magnetic properties. Among the wide spectrum of static characterization techniques, this section will focus mainly on two, namely, magneto-optic Kerr effect (MOKE) spectroscopy and vibrating sample magnetometry (VSM).

3.2.1 MOKE Spectroscopy

The Faraday effect describes the rotation of light polarization as it passes through a magnetic medium. The sense of rotation was found to reverse as the magnetization in the medium is switched. In 1877, Kerr observed a similar phenomenon to the Faraday effect when a linearly polarized light is reflected from a magnetic sample, later known as MOKE. By plotting the relative Kerr intensity versus the applied magnetic field, it is possible to characterize the magnetization reversal process of thin films.

In a MOKE system, an incidental light beam, with linear polarization, interacts with a magnetic sample and gains an additional magnetization-sensitive component *k* to produce elliptically polarized light. The Kerr coefficient is typically much smaller than the coefficient of reflection. The magnetization of the sample breaks the spatial and time-reversal symmetries and gives rise to the *k* component. The magnetization-sensitive component is composed of a complex Kerr rotation θ_K and ellipticity ε_K, which is related to the reflectivity of light r according to the expression $\frac{k}{r} = \theta_K + i\varepsilon_K$ (for $k \ll r$). The signal from MOKE is usually modulated using a photoelastic modulator (PEM) to improve the signal/noise ratio. The PEM also introduces a phase retardation (e.g., 0.25λ) in order to phase-separate the Kerr rotation θ_K and ellipticity ε_K components of the MOKE signal.

MOKE can be categorized by the direction of the magnetization vector with respect to the reflecting surface and the plane of incidence, as shown in Fig. 7. In the *longitudinal effect*, the magnetization vector is parallel to both the reflection surface and the plane of incidence. The longitudinal setup involves light reflected at an angle from the reflection surface and not normal to it as in the case of polar MOKE. The change in polarization is directly proportional to the component of magnetization that is parallel to the reflection surface and parallel to the plane of incidence. When the magnetization is perpendicular to the plane of incidence and parallel to the surface, it is said to be in

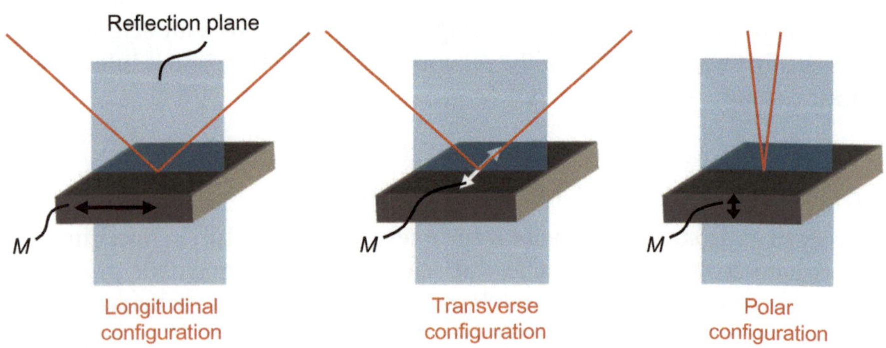

FIGURE 7

Schematic diagrams of various MOKE geometries.

the *transverse* configuration. In this case, the change in reflectivity is proportional to the component of magnetization that is perpendicular to the plane of incidence and parallel to the surface, as above. When the magnetization vector is perpendicular to the reflection surface, the effect is called the *polar Kerr effect*. To simplify the analysis, near normal incidence is usually employed when doing experiments in the polar geometry. It is worth noting that the polarization (electric field vector) of the incident beam can be either perpendicular (s-polarized) or parallel (p-polarized) to the plane of the incident of light. If the incident of light is neither s nor p, then the MOKE signal will become much less noticeable [20–22].

It is important to note that MOKE is a local probe. It measures the magnetization at the position of the laser spot. As such, it may be used to check for variations in properties across the sample. As we will see, this is in contrast to the results from the vibrating sample magnetometer. We will see some examples of MOKE results in the nanostructure section.

3.2.2 Vibrating Sample Magnetometry

VSM is a versatile technique for measuring the magnetic moment of a sample when it is vibrated *perpendicularly* to a uniform magnetizing field. Changes as small as 10^{-5} to 10^{-6} emu can be detected with this method. The VSM technique can be used to obtain the magnetic moment information of samples based on Faraday's law of magnetic induction [23]. Unlike MOKE, VSM measures the magnetic moment of an entire sample. In this technique, a magnetic sample is mounted on the sample holder and placed between the electromagnet poles, typically oriented horizontally (see Fig. 8a). There are a few types of sample holder, as shown in Fig. 8b; depending on the types of measurement to be performed, there are both in-plane and/or out-of-plane sample holders.

The optimum position of the sample is called the "saddle point," which is determined by positioning calibration prior to the measurements. During measurement, the sample is subjected to a constant vibration vertically at a fixed frequency ω. The changing magnetic flux, which is proportional to the magnetic moment in the sample, causes an AC voltage induction to be detected by a set of pickup coils placed near the electromagnetic poles. A flux change is detected when the applied magnetic field, coil position, or sample position is changed. The induced voltage is then fed into a lock-in amplifier to obtain the magnetic moment information of the sample. Figure 8c shows a typical magnetization curve measured using VSM, which gives the magnetic moment of the sample as a function of applied magnetic field as well as other information such as coercivity, switching fields, and their distribution. Detailed discussion on the VSM measurement technique is available in a review paper by Foner [23]. There are other types of measurement techniques that can be used to measure the magnetic moment information from magnetic samples such as a Superconducting Quantum Interference Device [24,25] and BH loopers [26,27].

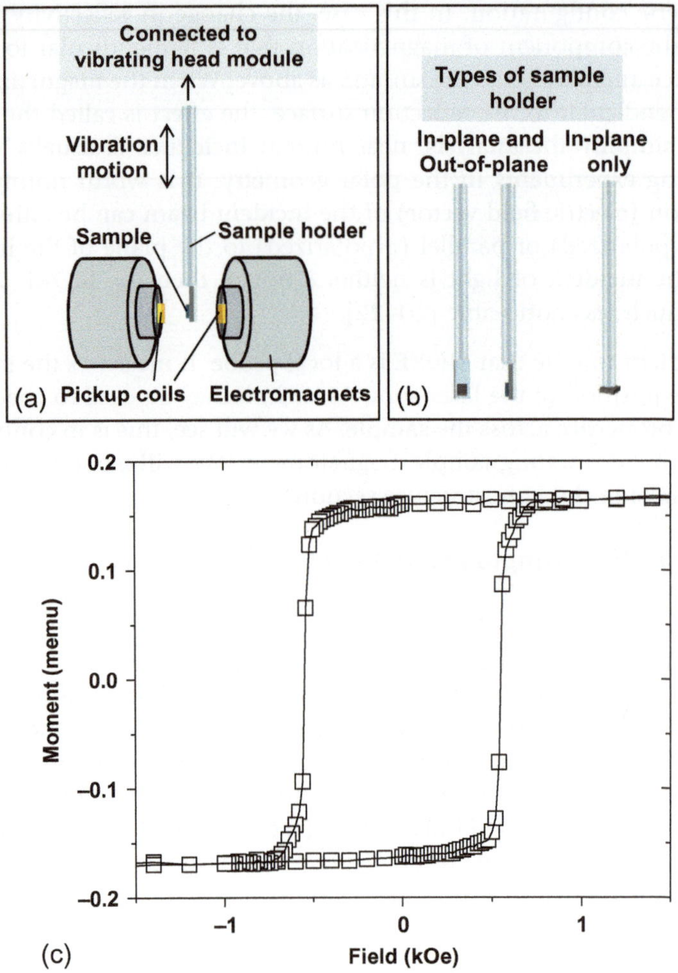

FIGURE 8

Schematic diagram of: (a) part of the VSM setup and (b) types of sample holder used in VSM. (c) Typical magnetization curve which can be obtained using VSM showing the out-of-plane magnetic moment versus applied magnetic field for $[Co(0.5\ nm)/Pd(3\ nm)]_{10}$ film deposited using sputter deposition.

3.3 Dynamic Magnetic Characterization

The characterization of the dynamic behavior of magnetic thin films and nanostructures is very important from both a fundamental viewpoint and for device applications. In this section, two commonly used techniques to probe dynamic magnetization in thin films and nanostructures will be discussed, namely, ferromagnetic resonance (FMR) spectroscopy and Brillouin light scattering (BLS) spectroscopy.

3.3.1 FMR Spectroscopy

FMR spectroscopy is widely used to characterize the dynamic behaviors in magnetic thin films and nanostructures. The phenomenon was first discovered by Griffiths [28] in 1946. A resonant absorption by the magnetic dipoles takes place in the FM film when a high-frequency alternating field is applied perpendicular to the static magnetic field. This was originally described by Griffiths as the anomalous high-frequency resistance of FM metals in which a large increase in the product of magnetic permeability (μ) and electrical resistivity (ρ) was observed a certain magnetic field that depended on the excitation frequency generated by a cavity resonator. As later explained by Kittel [29], when a FM sample is subjected to an externally applied field (H_{app}) and an RF alternating field (h_{rf}), the magnetization M will undergo precession (see Fig. 9) according to the equation of motion:

$$\partial J/\partial t = [M \times H_{eff}] \tag{1}$$

where J is the angular momentum density, t is time, H_{eff} is the effective field which includes the applied fields, H_{app} and h_{rf}, the demagnetizing field, H_d, and the anisotropy field, H_{ani}.

There generally is an applied field along the equilibrium direction of the magnetization. In FMR experiments, h_{rf} is often spatially uniform and couples strongest to spatially uniform precession modes. In many cases, h_{rf} will not couple to modes which have a short wavelength.

When the relation between M and J is given by $M = \gamma J$ where γ is the electron gyromagnetic ratio (and a negative number), Eq. (1) becomes:

$$\partial M/\partial t = \gamma [M \times H_{eff}] \tag{2}$$

The FMR phenomenon can occur when the frequency of h_{rf} matches a natural resonance frequency in the magnetic sample. In such a case, the microwave signal will be strongly absorbed at this resonance frequency. For the theory of FMR

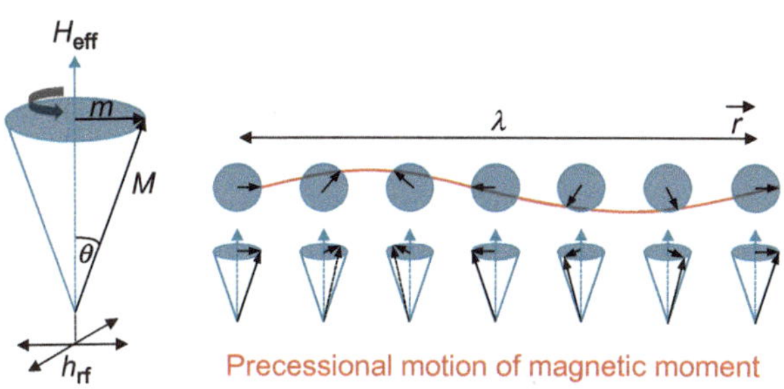

FIGURE 9
Schematic diagram showing magnetization precession in a spin wave with wavelength λ.

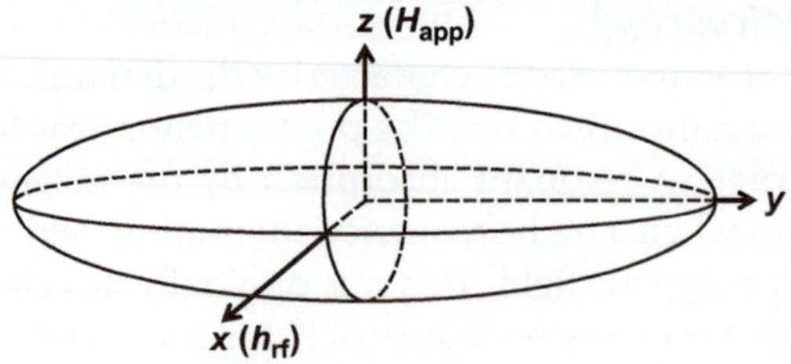

FIGURE 10
A sketch of a general ellipsoid with three principal axes.

absorption, Kittel [30] considered a general ellipsoid having three principal axes x, y, and z with $H_{app}=H_z$ and $h_{rf}=H_x$, as shown in Fig. 10.

Kittel took into account the demagnetizing fields using the relation $H_{di} = -\mathcal{N}_{ij}M_j$, where $\mathcal{N}_{ij}$ is the demagnetizing tensor and H_{di} is the ith component of the demagnetizing field. In a general ellipsoid, $\mathcal{N}_{ij}$ only has diagonal components $\mathcal{N}_x$, $\mathcal{N}_y$, and $\mathcal{N}_z$. Solving Eq. (2) by incorporating a time-varying $M_i = M_0 e^{-i\omega t}$ and $H_{di} = -\mathcal{N}_{ij}M_j$ for all three Cartesian axes yields the famous Kittel formula (Eq. 3) for the resonance frequency ($f_{res} = \omega_{res}/2\pi$, ω_{res} is the angular velocity):

$$\omega_{res} = \gamma\sqrt{[H_z + (\mathcal{N}_x - \mathcal{N}_z)M_z][H_z + (\mathcal{N}_y - \mathcal{N}_z)M_z]} \tag{3}$$

The Kittel formula can be used to predict the FMR frequency of various FM samples at a given H_{app}.

FMR experiments can be performed in a number of configurations. The most common schemes involve either sweeping the frequency at a fixed field, or sweeping the field at a fixed frequency and determining either the resonance frequency (for the first case) or the resonance field for the second case. In either case, one uses the resonance condition, such as Eq. (3), to find various parameters of the material. This is treated in detail in Chapter 3 in this handbook.

We concentrate on a particular example. In a broadband FMR experiment, the Vector Network Analyzer (VNA) can be connected to a coplanar waveguide (CPW) having a characteristic impedance of 50 Ω using coaxial cables and microwave probes. The VNA compares the input and output signals on the coplanar waveguide with respect to their amplitude and phase, allowing measurements of the absorption signal as a function of frequency. Compared with the classical FMR, frequency and magnetic field can both be varied continuously, thereby enabling systematic characterization of the magnetic properties of thin films and reversal mechanisms in magnetic nanostructures, for example [31].

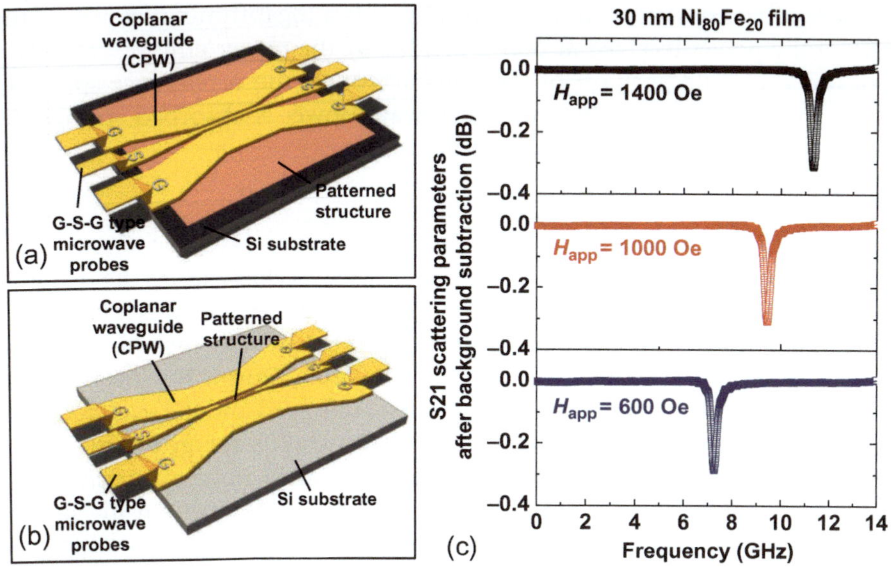

FIGURE 11

(a, b) Schematic diagrams of FMR measurement using G-S-G type microwave probe on a CPW. (c) Plots showing measured S21 scattering (transmission) parameters at various H_{app} obtained from a 30-nm $Ni_{80}Fe_{20}$ film using VNA-FMR after background subtraction.

Figure 11 shows the two approaches in which the CPW can be used to measure the FMR signal. In the first method, the CPW is fabricated on top of an existing magnetic sample using photolithography followed by the deposition of electrode materials and then a photoresist lift-off step, as shown in Fig. 11a. An insulating layer (usually an oxide, e.g., Al_2O_3) is first deposited followed by the deposition of conducting electrode material (typical thickness of 200 nm Au) [32]. The insulating layer ensures that the signal and ground lines are not shunted via the magnetic film/nanostructures underneath.

In the second method, the CPW is first deposited on a blank substrate followed by the fabrication of the nanostructures on top of the central signal line of the CPW, as shown in Fig. 11b. In this method, the insulating layer between the CPW and the sample is no longer necessary because the ground and signal lines are well isolated from each other [33].

The FMR signal is obtained using the VNA by comparing the relative microwave power between the emitting port (port 1) and the receiving port (port 2) through the CPW on the test sample. The transmission characteristic is recorded as a function of the excitation frequency at different values of H_{app}. Typical FMR spectra at different H_{app} values are shown in Fig. 11c. The S21 parameter is a measure of transmission. More specifically, it is a logarithmic measure of the ratio of the power received at port 2 divided by the power emitted from port 1 of the VNA. The position of the transmission dip in a reference spectrum shows the frequency at which the microwave power is absorbed strongly by the sample due to the resonance condition.

3.3.2 BLS Spectroscopy

BLS spectroscopy is another technique that is used to characterize the dynamic magnetization process in magnetic thin films and nanostructures. The underlying principle of BLS is based on the interaction between an incoming photon (a laser source) and spin waves in a magnetic sample. During the inelastic scattering process, the total energy of the system is conserved and the scattered photon may experience either a decrease or an increase in its energy by emitting or absorbing a spin wave, respectively. The former is called Stokes scattering while the latter is called anti-Stokes scattering. The schematic diagram in Fig. 12a illustrates such events. There is a similar conservation law for the momentum of the system parallel to the surface. Momentum conservation is not required for the direction perpendicular to the surface, because there is no translational symmetry in that direction.

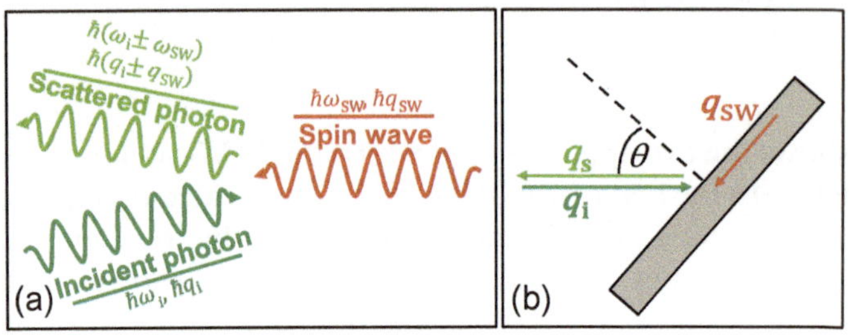

FIGURE 12
Schematic illustrations showing photon and spin wave interaction in BLS.

Consider an incoming photon having an energy $\hbar\omega_i$ and a momentum $\hbar q_i$, which interacts with a spin wave having an energy $\hbar\omega_{SW}$ and a momentum $\hbar q_{SW}$. If the spin wave is absorbed (i.e., anti-Stokes scattering occurs), the energy and parallel momentum of the scattered photon becomes $\hbar\omega_s = \hbar(\omega_i + \omega_{SW})$ and $\hbar q_{s-||} = \hbar(q_{i-||} + q_{SW-||})$, respectively. Conversely, if the spin wave is emitted, the energy and momentum of the scattered photon become $\hbar\omega_s = \hbar(\omega_i - \omega_{SW})$ and $\hbar q_{s-||} = \hbar(q_{i-||} - q_{SW-||})$. Based on the relative orientation of magnetization direction M and the in-plane wave vector q_{SW}, different types of spin wave modes in magnetic films can be detected using BLS and these (bulk modes, standing wave modes, and surface modes) have been described in detailed in Ref. [34].

Practically, the BLS measurement leverages the presence of Stokes and anti-Stokes scattering to detect the characteristic spin wave frequencies of the sample at resonance conditions, i.e., by the amount of the frequency shift between the incident and the scattered photon due to energy and momentum transfer during photon–spin wave interaction.

Furthermore, using BLS spectroscopy, the spin wave dispersion characteristic (i.e., frequency vs. parallel k-vector) can be constructed by varying the incidence angle (θ) of the probing laser light. This is in contrast to the FMR measurement, which characterizes only the uniform spin wave frequency with $k = 0$. By varying θ, the wave vector of the spin wave can be derived as $q_{SW-||} = -2q_i/sin\theta$ by solving the parallel momentum conservation with the detection condition $q_s = -q_i$, (see Fig. 12b). This detection condition applies to BLS spectroscopy performed in the backward scattering geometry, as shown in Fig. 13a. Another BLS spectroscopy geometry, i.e., forward scattering, shown in Fig. 13b is preferred when characterizing transparent samples such as yttrium iron garnet (YIG).

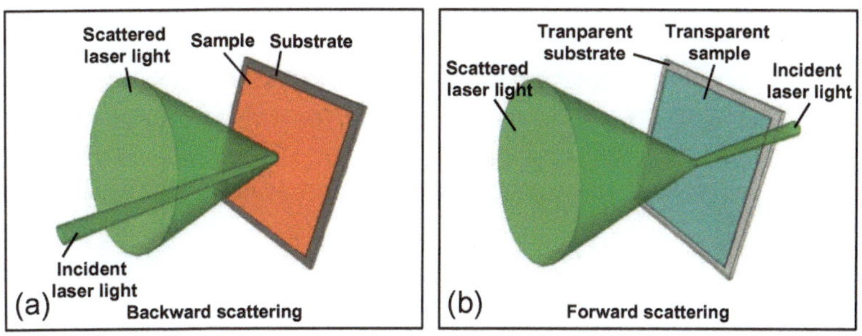

FIGURE 13
Schematic diagram of BLS measurement: (a) backward scattering and (b) forward scattering.

In detecting the shift in laser frequency, an interferometer is usually used to scan the light intensity at different wavelengths (or frequency) by varying the mirror spacing in the interferometer construction using a piezo-motor. The use of an interferometer allows precise and low-noise acquisition of the scattered light. An example of a six-pass tandem Fabry–Perot interferometer and its detailed working principle can be found in Ref. [35]. A typical BLS spectrum showing Stokes and anti-Stokes scatterings peaks is shown in Fig. 14.

In principle, BLS offers some advantages over FMR, namely: (1) the possibility to investigate spin waves with different wave vectors; (2) both thermally excited spin waves (low amplitude) and microwave-excited spin waves (high amplitude) can be detected; (3) a time-resolved BLS measurement is possible by combining the use of a pulse generator and high-speed data acquisition module; and (4) using micro-focused BLS, it is possible to probe the microwave

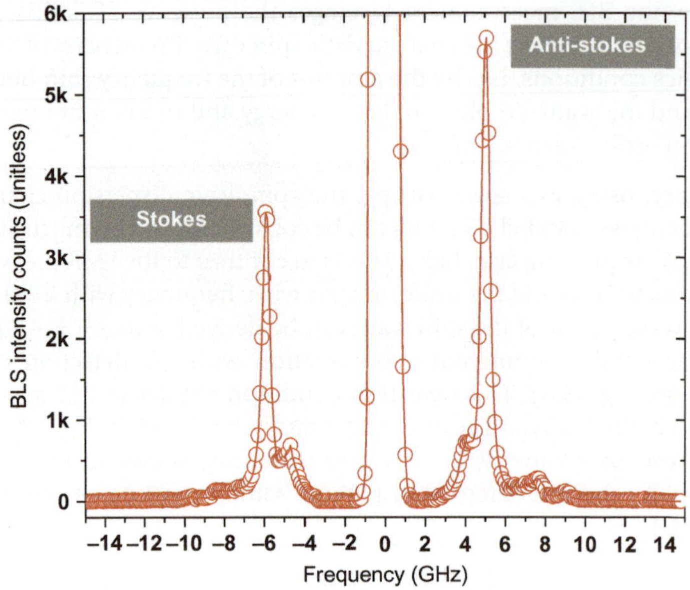

FIGURE 14

A typical BLS spectrum showing Stokes and anti-Stokes scattering peaks obtained from exciting a single 1-μm-diameter $Ni_{80}Fe_{20}$ disk at remanence using microwave power of 10 dBm.

excitation of individual nanomagnets with high spatial resolution detection of ~250 nm [36].

4 MAGNETIC NANOSTRUCTURES

In the last decade, magnetic nanostructures have attracted a lot of interest from researchers around the world both from a fundamental viewpoint and also due to their potential in a wide range of emerging applications such as ultra-high-density storage, magnetic random access memory, spin logic, and magnonics. From a fundamental viewpoint, nanomagnets possess static and dynamic behaviors that are drastically different from their parent bulk materials due to the size confinement effect.

The main challenge in nanomagnetism research is the development of a low-cost, high-resolution technique for fabricating high-quality magnetic nanostructures. The ability to synthesize arrays of magnetic nanostructures with precise dimensionality control is crucial for reproducible magnetic switching processes. Key issues to be considered in the development of any nanofabrication techniques are critical dimension control, resolution, size and shape homogeneity, patterned area, and alignment accuracy. There are two main

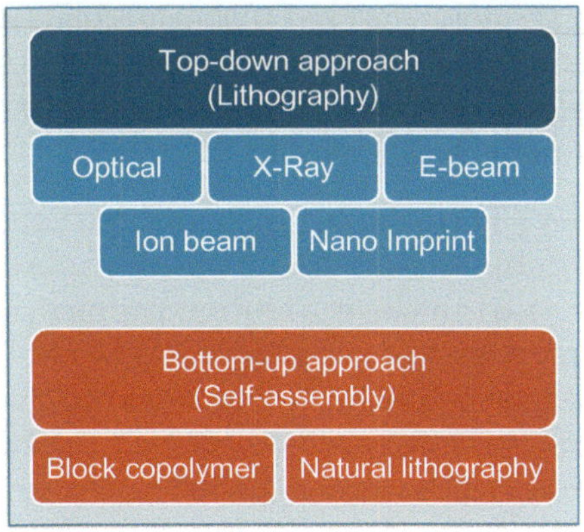

FIGURE 15
Schematic diagrams showing top-down and bottom-up approach in nanostructure lithography.

approaches that can be used to fabricate magnetic nanostructures, namely, the "top-down" and the "bottom-up" approaches, as shown in Fig. 15.

The most common type of top-down approach is lithography, which is a process of patterning fine-scale structures onto a substrate. Generally, a lithographic system uses radiation to expose the recording medium selectively. The recording medium is known as resist. There are two types of resist, namely, positive and negative resists. For positive resist, the exposed parts of the resist breakdown when exposed to light, become softened, and are dissolved off in the developer solution. For negative resists, however, the exposed parts of the resist become cross-linked (polymerized due to a photochemical reaction), harden, and remain on the substrate after the development process. Depending on the type of exposing radiation, there are various types of lithographic techniques such as optical, e-beam, X-ray, and ion beam.

In optical lithography, a substrate is coated with a light-sensitive resist material. The substrate and resist are then exposed to light to capture a predefined pattern (from a mask) in the resist. The resist is then developed so that a pattern from the mask remains on the substrate. The light source used for exposure will depend on the resist sensitivity and more importantly on the minimum dimensions that are required for patterning.

Electron beam lithography (EBL) is a high-resolution patterning method for fabricating magnetic nanostructures using a focused beam of electrons for writing of the desired structures on a thin resist layer. This method has been used to fabricate a wide range of magnetic nanostructures [37–43]. There are limitations, however, with the use of EBL in fabricating large-area magnetic nanostructures. This is largely due to the serial writing process in EBL which is inherently very slow, thus making large-area fabrication impractical. It is also very difficult to fabricate closely packed high-aspect-ratio nanostructure arrays due to proximity effects. The minimum achievable pitch size using EBL which has been reported in the literature ranges between ~10 and 20 nm, with the smallest isolated features ~2–2.5 nm [44,45].

Interference lithography (IL) is a mask-less nanofabrication technique, which has been used widely to synthesize large-area magnetic nanostructures [46–51]. The basic principle of IL is based on Lloyd's mirror experiment exposing a photoresist layer using an interference pattern generated from two obliquely incident beam paths (i.e., direct and reflected beams). The patterned area and resolution are related to the beam source diameter and wavelength, respectively. The IL method allows multiple exposure steps to form a specific pattern on a resist. For example, in order to fabricate dot arrays, a second exposure after rotating the substrate through 90° is needed. The main drawback of the IL method is the limitation in fabricating only periodic patterns. The pitch resolution is linearly proportional to half the wavelength of the light and inversely proportional to the sine of the relative angle between the two beam paths. The typical size dimension for IL is ~200 nm for a laser light source [47,48]. Using an X-ray source, the period can be reduced to ~70–80 nm [50]. Using an extreme ultraviolet light source of 13 nm wavelength gives a theoretical limit of the resolution of 6.5 nm.

Nanoimprint lithography (NIL) is a low-cost and high-throughput process for replicating micro- and nanoscale patterns. NIL has been used to fabricate large-area magnetic nanostructures of different materials [52–54]. NIL patterns a resist by deforming the resist shape through embossing (with a mold), rather than by altering resist chemical structures through light radiation. After imprinting the resist, anisotropic etching is used to remove the residual resist in the compressed area to expose the substrate underneath. Some of the limitations of NIL include the fact that heating cycles may cause a thermal expansion mismatch of mold to substrate, there is a difficulty in aligning to a nontransparent mold, making layer-to-layer alignment problematic, and the large applied pressures required to displace high-viscosity polymers may cause mechanical distortion.

Ion beam lithography (IBL) is a technique for surface modification and fabrication of nanostructures. IBL commonly uses light ions (protons, helium ions) for the exposure of polymeric resists. The use of heavier ions makes it possible to dope the substrate or create thin layers of new chemical compounds. The use of ion irradiation for patterning of a continuous magnetic film without a significant modification of the surface topography was first proposed by Chappert *et al.* [55]. Ion irradiation can induce adverse modifications to the sample, including chemical disordering [56] and intermixing at the interface [55,57].

In the bottom-up approach, chemical or physical forces operating at the nanoscale are used to assemble basic units into larger structures. At the nanoscale, bottom-up approaches provide an important complement to top-down techniques. For example, the ability of block copolymers, which are macromolecular materials comprising two or more chemically different polymeric segments, to self-assemble into a range of morphologies makes them suitable candidates for bottom-up lithographic processes. By using phase separation of the blocks, periodic nanoscale areas with a uniform distribution of size and shape can be formed. This has been used as a template for synthesizing large-area arrays (typically 4 mm × 4 mm) of nanomagnets [58–62]. Nanosphere lithography is another bottom-up nanofabrication technique. This approach is inexpensive and simple to implement. This is a high-throughput form of a self-assembly method in which spherical colloidal particles are used to define a large-area lithographic mask. A monolayer of colloidal particles is deposited in either a random or ordered array over the entire surface of a macroscopic substrate. For a general review of the various techniques for fabricating ordered magnetic nanostructures listed above, the reader is referred to Martín *et al.* [58] and Lodder [63].

Section 4.1 will focus on the use of deep ultraviolet lithography (DUV) to fabricate large-area arrays of magnetic nanostructures with lateral dimensions below the conventional resolution limit of optical lithographic tools using various resolution enhancement techniques. One unique advantage of this technique is the fact that, unlike e-beam lithography, thicker resists can be used to make high-aspect-ratio nanostructures [64].

4.1 Lithography

Image formation in optical lithography can be achieved using contact printing, proximity printing, or by optical projection. In contact printing, the structures to be printed are obtained by illuminating a mask, which is in direct contact with a substrate coated with a resist layer. Contact printing gives good resolution but suffers from mask damage and contamination during mask–substrate contact. For proximity printing, there is a small gap between the mask and the substrate which helps to alleviate the mask damage issue experienced with contact printing, but at a cost of poorer resolution due to light diffraction. The most commonly used lithographic method in the industry is projection lithography. In projection lithography, the mask image is projected on the resist-coated wafer. The projection tool itself can be in the form of a reduction stepper or a scanner. A stepper exposes the entire mask region at once while a scanner exposes the mask through a scanning process.

In optical lithography, the resolution (R) of a given system is determined by the Rayleigh criterion as shown in Eq. (4):

$$R = K_1 \frac{\lambda}{\mathrm{NA}} \tag{4}$$

where λ is the wavelength of the light source, NA is numerical aperture, and K_1 is a process-dependent parameter. The ultimate resolution of the patterned structures can be improved by using a shorter λ. Earlier optical lithographic techniques have used a mercury lamp (broadband 300 nm $\le \lambda \le$ 450 nm), g-line (λ=436 nm), and i-line (λ=365 nm) as the illumination source. DUV lithography uses an illumination source with an even shorter wavelength compared to conventional optical lithography, such as a KrF laser with λ=248 nm or an ArF laser with λ=193 nm. NA is proportional to the refractive index of the medium (n) where the lens is operating, and it can be modified using immersion lithography. Compared to conventional lithography, where the medium between the lens and the resist is air (n=1), immersion lithography may incorporate other media of higher n, such as water (n=1.44) or oil (n=1.52) to improve the NA value and consequently improve the resolution.

Another factor determining the resolution is the value of K_1, which is usually about 0.6 for a conventional system and has a theoretical limit of 0.25 [65]. To further improve the resolution, the K_1 value can be improved by using various resolution enhancement techniques such as off-axis projection, phase shift mask, proximity correction, etc. [66]. For example, the resolution of DUV lithography can be improved by using an alternating phase shift mask (ALT-PSM) instead of a conventional photomask. The key difference lies in the 180° out-of-phase difference between adjacent apertures in the photomask, as illustrated in Fig. 16. Using a conventional photomask at a very small pitch

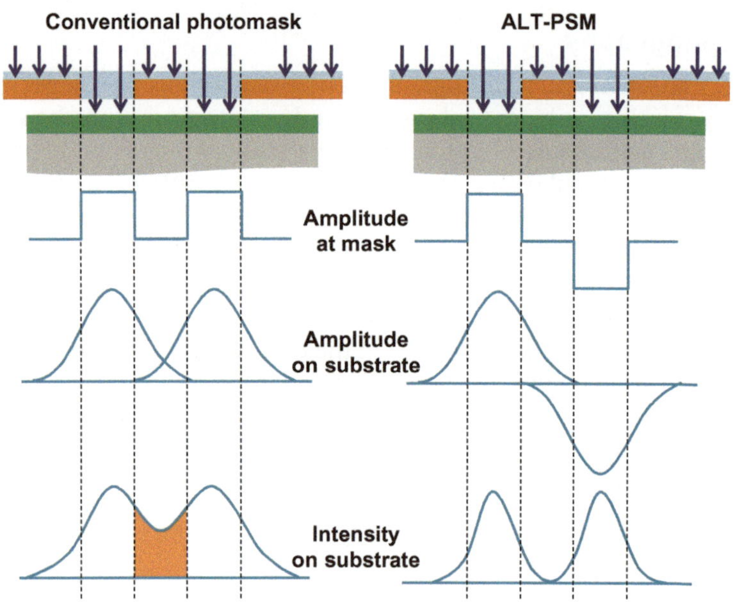

FIGURE 16
Schematic diagram showing the improvement in resolution when using ALT-PSM compared to conventional photomask.

may cause exposure error due to the overlapping intensity tails between adjacent features (red highlighted region in Fig. 16). In ALT-PSM, adjacent features on the mask are phase-shifted 180°. The 180° phase shift can be achieved by etching the quartz layer by one order of λ, calculated based on the refractive index of quartz ~1.5 at the exposure wavelength. Another example of a resolution enhancement technique in DUV lithography is the use of a BARC layer placed at the bottom of the resist layer in order to minimize the standing wave pattern on the resulting resist profile due to substrate reflection.

Despite the improved resolution capability, this technique poses challenges such as precisely controlling the etch depth and the difficulty in applying this method to random patterns. Nevertheless, this technique has shown remarkable versatility when used to produce resist features with periodic patterns [64,67]. The minimum achievable feature size for DUV at $\lambda=193$ nm is ~120 nm. For a more detailed discussion of DUV lithography and how to achieve high-resolution resist patterning, readers are referred to a review article in Ref. [65].

4.2 Pattern Transfer

In any fabrication process, there are two methods that can be used to transfer lithographic patterns unto the underlying substrates, namely, additive and subtractive process. The typical subtractive patterning process (Fig. 17) involves three steps: (a) a uniform layer of the material to be patterned is deposited on the substrate; (b) a lithographic process is then performed such that the areas to be etched are left unprotected by the resist; and (c) etching is performed either using chemicals such as acids (wet chemical etching), or more commonly a plasma environment (dry etching).

The resist "withstands" the etchant and protects the material underneath the resist. The etching process involves the removal of materials that are not protected by the photoresist. In a wet etching process, etchant solutions (such as acids) are used, which can attack the underlying film and leave the resist intact. Wet etching is isotropic leading to undercutting as the film is etched from underneath the resist. If anisotropic etching is desired, directionality must be induced into the etching process using dry etching such as in the reactive ion beam etching. Etch selectivity is an important resist property which is dependent on both the resist material properties and the nature of the etch process for a specific film. It is worth noting that the conventional fabrication technologies used in the microelectronic industry are not always compatible with magnetic nanostructure fabrication because the process involves high temperature, which will degrade the quality of the FM films. It is also very challenging to use reactive ion beam etching to pattern magnetic films because it is difficult for the reactive gases to form volatile compounds when in contact with magnetic materials.

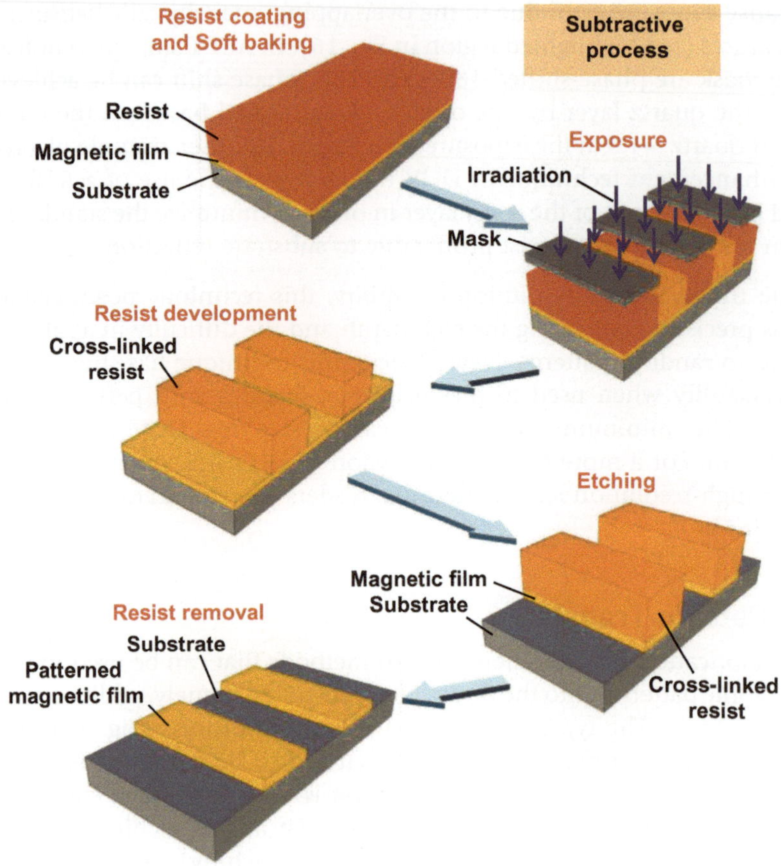

FIGURE 17
A schematic diagram of the subtractive pattern transfer process.

In the additive process, which is also known as the "lift-off" method, shown in Fig. 18, the order of the process is changed as follows: (a) lithography is used to create the negative of the pattern that is desired; (b) material is deposited into the areas not protected by the lithographically produced pattern and on top of the resist; and (c) stripping off the resist, which leaves the new material in a negative region of the patterned resist.

In the conventional lift-off process, a film is blanket-deposited all over the substrate, covering both the photoresist and areas in which the photoresist has been cleared (developed). During the actual lifting-off, the remaining photoresist under the film is removed with solvent, taking the film with it and leaving only the film, which was deposited earlier, directly on the substrate. As a general rule, for successful lift-off, the total resist thickness should be three times the thickness of the film being lifted off. Also, as shown in Fig. 18, negative sloping

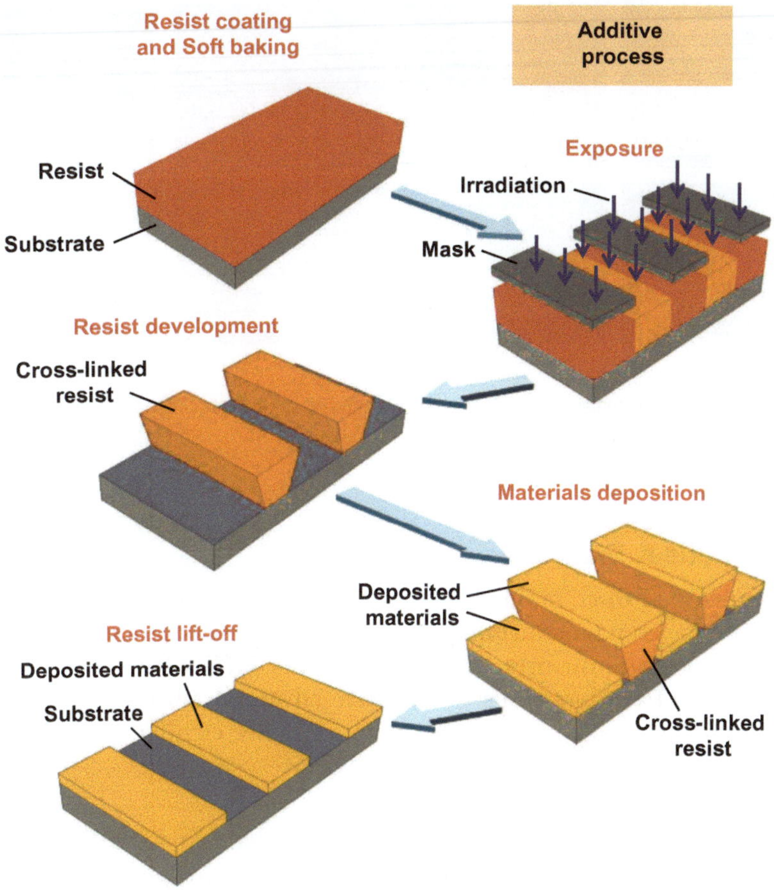

FIGURE 18
A schematic diagram of the additive pattern transfer process.

walls (undercut profile) also help the lift-off process. The resist profile is critical in the lift-off process.

A broad range of magnetic nanostructures can be patterned using DUV lithography. Figure 19 shows some examples of high-quality resist profiles and the corresponding nanostructures after the lift-off process. In addition to the disks, nanodots, and rings seen here and fabricated using DUV lithography, other nanostructures such as stripes, wires, and nanomagnets of various shapes can also be fabricated using this method.

In the next section, a few practical examples of basic and advanced fabrication of magnetic nanostructures and their magnetic behaviors will be presented. These discussions will showcase the state of the art for advanced magnetic nanostructures and their novel magnetic properties.

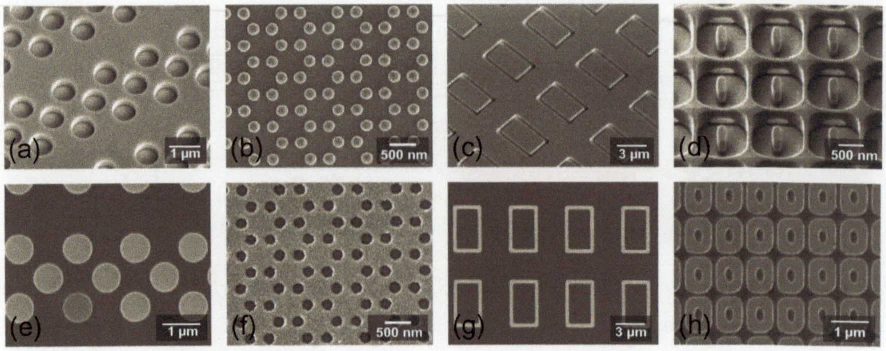

FIGURE 19

SEM images showing examples of: (a–d) resist profiles and (e–h) the resulting nanostructures patterned using DUV lithography. *Taken from Refs. [68–70].*

4.3 Practical Examples

We have selected three main categories of magnetic nanostructures, namely, single-/multilayer nanostructures, thickness-modulated nanostructures, and bicomponent nanostructures, to illustrate some of the methods we have discussed earlier. The discussions will include advanced fabrication techniques such as self-aligned shadow deposition and selective etching processes.

4.3.1 Planar Magnetic Nanostructures

Two examples of magnetic nanostructures fabricated using DUV lithography, namely, antidots and nanowires, will be highlighted in this section. A more comprehensive review of various magnetic nanostructures and their behavior can be found in Ref. [65]. The presence of hole lattices in antidot nanostructures has been found to modify the intrinsic magnetic anisotropy of the film locally [69]. Furthermore, depending on the lattice configuration of the holes in the antidot structures, different directions of easy and hard magnetization axes can be induced. Figure 20a–c shows SEM images of antidot nanostructures having different lattice configurations. The analyses of the magnetic anisotropy of the antidot nanostructures were extracted using the remanent magnetization from VSM measurement as a function of the field angle θ.

When the holes are arranged in a square lattice, a well-defined fourfold anisotropy is observed with the easy axis (EA) and hard axis (HA) oriented at $\frac{\pi}{4}+\frac{n\pi}{2}$ and $\frac{n\pi}{2}$, respectively, for $n=(0, 1, 2, 3, \ldots)$, as shown in Fig. 20d. However, when the holes are arranged in a honeycomb structure, a sixfold anisotropy is observed with the EA and HA oriented at $\frac{\pi}{6}+\frac{n\pi}{3}$ and $\frac{n\pi}{3}$, respectively, for

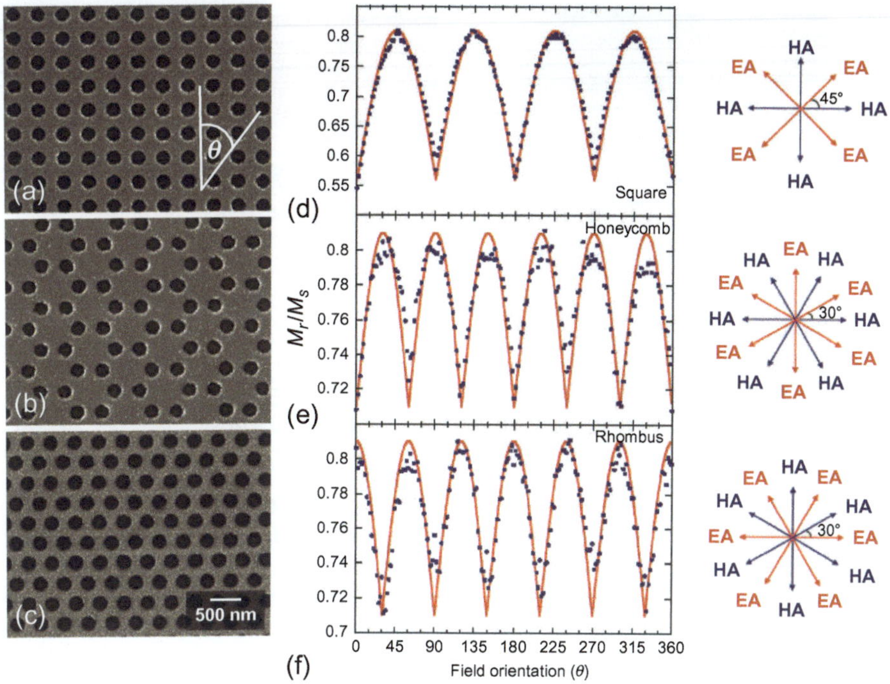

FIGURE 20

SEM images showing antidot nanostructures of different lattice configurations: (a) square, (b) honeycomb, and (c) rhombus lattice. (d–f) The corresponding M_r/M_s plots measured using VSM versus applied field orientations. *Taken from Ref. [69].*

$n = (0, 1, 2, 3, \ldots)$, as shown in Fig. 20e. Interestingly, in the rhomboid lattice the EA and HA anisotropy is now switched while maintaining the same sixfold anisotropy as shown in Fig. 20f. For multilayer antidot nanostructures, as described by Wang *et al.* [71], it was observed that the magnetization response of Co(30 nm)/Cu(t)/NiFe(30 nm) multilayer antidots is largely dependent not only on the field angle θ but also on the thickness of the Cu spacer layer (t_{Cu}).

Another example of magnetic nanostructures that we have fabricated with DUV lithography is nanowires. We have conducted extensive investigations of the magnetization reversal processes in $Ni_{80}Fe_{20}$ nanowires fabricated using DUV lithography. Goolaup *et al.* showed a marked modification in the magnetization process when the magnetic field was applied parallel and perpendicular to the wires' axis [72]. Furthermore, when nanowires of different width are placed close together in a dense array as shown in Fig. 21a and b, a controlled ordering of nanowire ground state can be achieved by systematic magnetic field cycling [73]. The magnetic ordering of the nanowires can be set either parallel

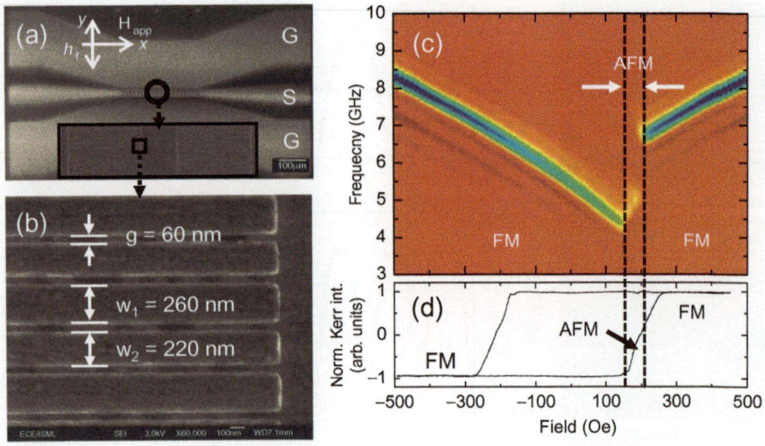

FIGURE 21

SEM images of: (a) EBL fabricated nanowires array on top of CPW patterned using UV lithography, and (b) magnified view of inset (a) showing densely packed nanowire array. (c) 2D-FMR absorption spectra measured using the VNA-FMR technique, where the color shows the S21 parameter at each frequency and field, and (d) normalized Kerr intensity plot versus applied field showing FM and AFM state. *Taken from Ref. [73].*

(FM ordering) or anti-parallel (antiferromagnetic (AFM) ordering). This is evident from the measured FMR signal, where a sharp transition is observed between the FM and AFM states, as shown in Fig. 21c. A corresponding MOKE signal plotted in Fig. 21d shows the different regions where FM and AFM ordering can be distinctively observed.

4.3.2 Thickness-Modulated Magnetic Nanostructures

The previous section highlighted how the behaviors of magnetic nanostructures can be strongly modified by their shape, lateral dimensions, and degree of neighboring interactions. These structures, however, are limited to planar nanostructures, where the film thickness is homogeneous. Thickness modification is often associated with a large alteration in magnetic properties of magnetic nanostructures due to the changing energy landscape. Notably, a stable vortex magnetization state in magnetic nanodisks is favored in a thicker film due to lower exchange energy associated with its vortex core. The thickness modulation will dictate a selective modification of magnetization behaviors in different parts of the magnetic nanostructure.

Thickness-modulated nanostructures can be made using complex multilevel lithography alignment and pattern transfer processes. However, a simple self-aligned lithographic process is desirable. In this section, a fabrication technique that circumvents the need for multilevel lithography and pattern transfer processes will be introduced. The key advantage of this nanofabrication process is that the successive deposition steps are self-aligned. In addition, these depositions are carried out without breaking the vacuum between the steps.

Figure 22 illustrates a typical self-aligned process, which can be used to fabricate thickness-modulated nanostructures.

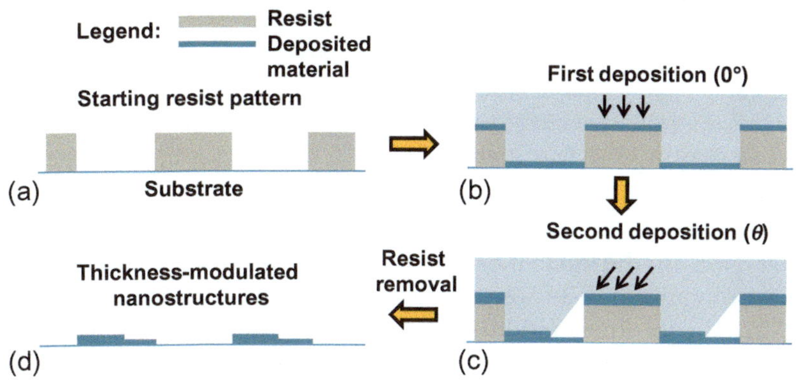

FIGURE 22

Schematic diagram showing the self-aligned fabrication steps for making thickness-modulated nanostructures.

From the starting resist pattern shown in Fig. 22a, successive deposition steps can first be done at normal incident (0°) (Fig. 22b) and a second deposition is initiated at a certain angle (θ) (Fig. 22c). The first step will deposit a magnetic material with uniform thickness while the second step will produce a step modulation on the first deposited layer. The choice of θ will determine the amount of coverage due to the shadowing effect by the resist profile, as marked by the white triangular region in Fig. 22c, where the flux of incoming materials is blocked by the resist. The final lift-off step will produce the desired thickness-modulated nanostructures, as shown in Fig. 22d. Two examples of thickness-modulated magnetic nanostructures fabricated using this technique, i.e., thickness-modulated nanowires and disks, will be presented below.

First, we will discuss the fabrication of a thickness-modulated nanowire (MNW) array [31]. The nanowires have a step modulation along their width, with different width coverages, as highlighted in Fig. 23a. The dynamic properties were characterized using BLS, as shown in Fig. 23b. As described in Ref. [31], the two lowest frequency modes measured using BLS for the MNW have their frequency shifted upward with respect to the homogeneous thickness nanowires (HNWs) of the same width due to the presence of a thickness step; compare modes I and II for HNW and MNW in Fig. 23b. The HNW has a thickness of 25 nm while the MNW has an additional step thickness of 25 nm.

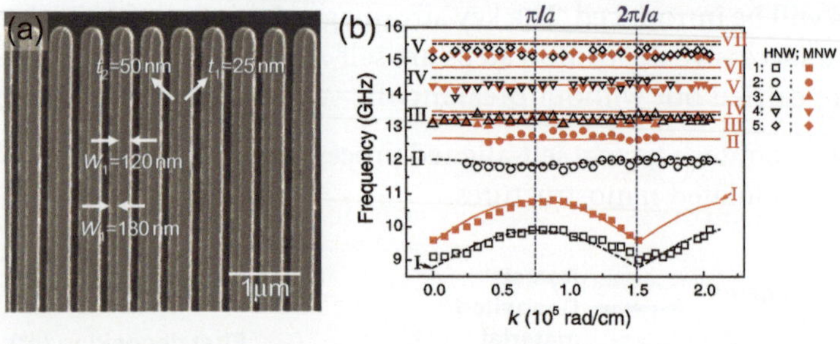

FIGURE 23

(a) SEM image of a thickness-modulated nanowire array and (b) measured (points) and calculated (lines) frequency dispersion for MNWs (filled symbols, solid curves) and HNWs (open symbols, dashed lines). *Taken from Ref. [31].*

Another type of thickness-modulated disk is described in Ref. [74]. The thickness-modulated disks were fabricated using the technique shown earlier in Fig. 22. The thickness modulation is in the form of a lens shape on top of a disk as a result of 45° shadow deposition, as shown in Fig. 24a.

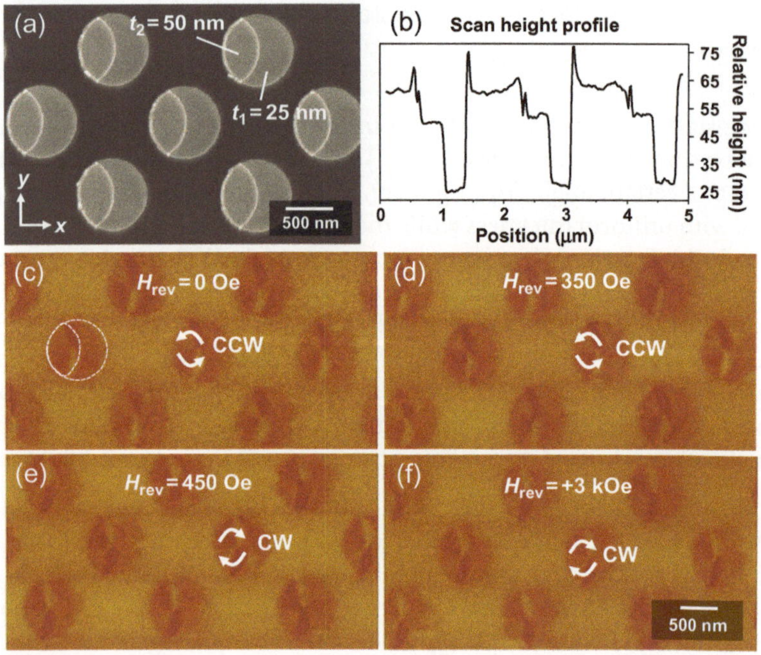

FIGURE 24

(a) SEM image of thickness-modulated disks and (b) an AFM scan height profile showing the thickness modulation. (c–f) MFM images showing vortex core location and chirality at remanence after applying various reversal fields along the *y*-direction. *Taken from Ref. [74].*

The AFM scan height profile shown in Fig. 24b presents a clear thickness modulation of 25 nm. It should be noted that the thickness-modulated disks were embedded within a 60-nm-thick BARC layer for the purpose of achieving resolution enhancement during the resist patterning process (see Section 4.1). The thickness modulation provides an additional shape anisotropy, which defines the vortex chirality in the disk depending on the magnetization reversal history. This is demonstrated in Fig. 24c–f, whereby the MFM images show how the vortex chirality can be systematically switched from counterclockwise to clockwise rotation.

4.3.3 Bicomponent Magnetic Nanostructures

The method described in the previous section may be adapted to fabricate a new form of self-aligned bicomponent magnetic nanostructures. The first variant (method A) involves two successive angle-deposition steps followed by a lift-off process, as depicted in Fig. 25a. It is important to note that the choice of deposition angle in each step is crucial, as it will determine the extent of the

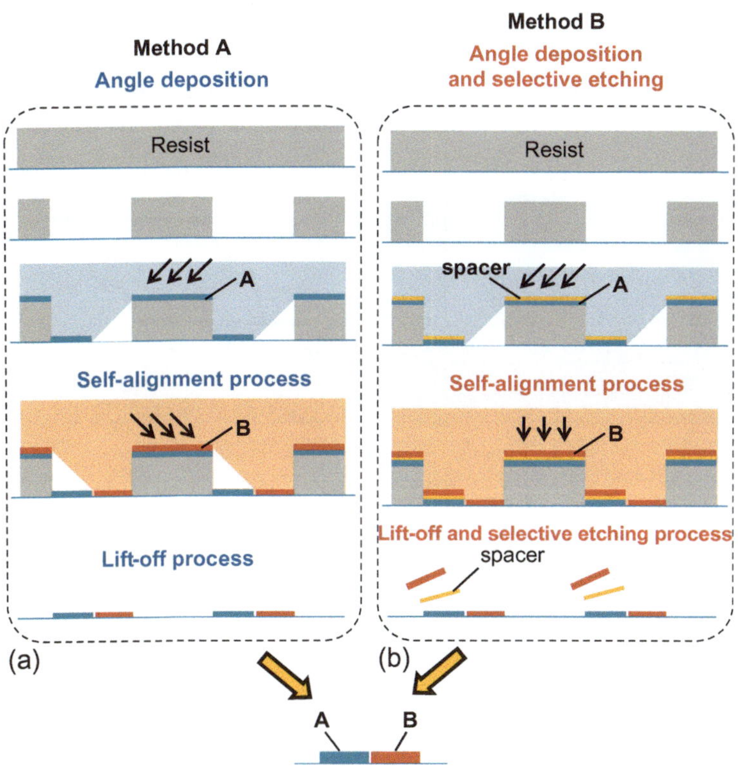

FIGURE 25

Schematic diagrams of self-aligned methods to fabricate bicomponent nanostructures.

overlapping region between the two adjacent materials, or the gap size if they are not overlapping. Consequently, a range of bicomponent nanostructures including magnetostatically coupled, exchanged coupled, or even overlapping nanostructures can be fabricated.

The second method (method B), shown in Fig. 25b, combines angle deposition and a selective etching process. The first angle deposition is done to deposit material A and the subsequent spacer layer. The second deposition is done at normal incident to deposit the second material B. Subsequently, a lift-off process is performed to remove the resist and is then followed by a selective etching of spacer layer, resulting in a magnetostatically coupled bicomponent nanostructure. Compared to method A, this method allows one to maintain the shape conformity of the resulting bicomponent nanostructure to the original patterned structure. Furthermore, by tuning the angle in the first deposition, one can systematically define the gap size between the adjacent materials within the structure.

These methods have been used to fabricate binary nanomagnets, as described in Ref. [75]. Figure 26a shows an SEM image of an array of isolated binary nanomagnets. The binary nanomagnets are made up of NiFe and Ni, which are magnetostatically coupled and separated by a small gap, as is evident in Fig. 26b. Figure 26c shows a plot of a longitudinal MOKE loop showing a clear two-step

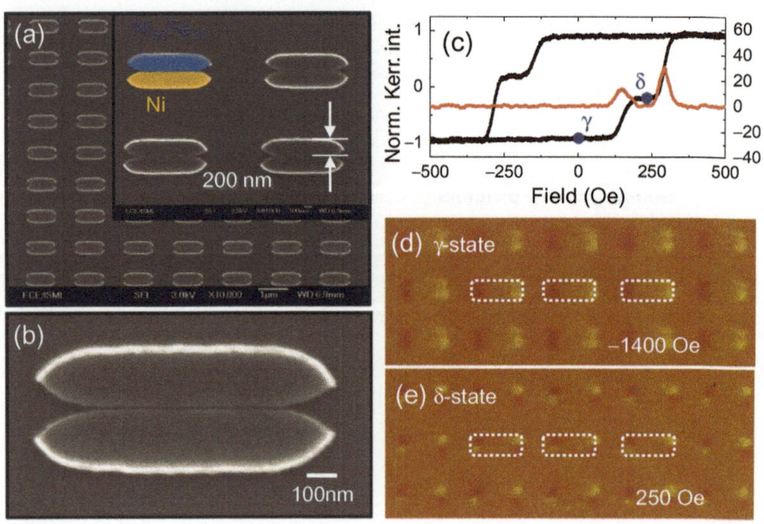

FIGURE 26

(a, b) SEM images of binary nanomagnets. (c) Plot of normalized Kerr intensity of the binary nanomagnets. (d, e) MFM images showing the corresponding magnetic states γ and δ shown in (c). *Taken from Ref. [75].*

reversal, which behaves significantly differently from the reversal of a single-component nanomagnet of the same dimensions involving vortex nucleation and annihilation. Furthermore, probing the magnetic states γ and δ as labeled in Fig. 26c using MFM reveals that the γ-state (δ-state) corresponds to the parallel (antiparallel) alignment of the adjacent components (cf. Fig. 26d and e).

Another example is the bicomponent disks described in Ref. [68]. Figure 27a shows the SEM image of bicomponent disks made up of an NiFe lens and an Fe crescent. The two adjacent materials are magnetostatically coupled together with a gap of 20–30 nm. The longitudinal MOKE loop in Fig. 27b shows a distinct three-step reversal, which is markedly different compared to the behavior of a single-component disk. The simulated spin configuration shows that the three switching steps correspond to the single domain reversal of NiFe lens (Fig. 27c–e), and vortex core nucleation (Fig. 27e and f) and subsequent annihilation (Fig. 27g and h) in the Fe crescent. Other bicomponent magnetic nanostructures fabricated using this method can be found in Ref. [68].

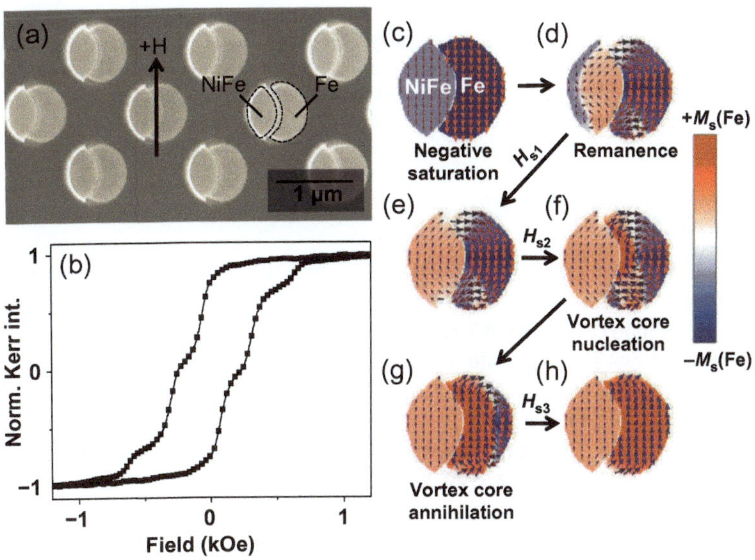

FIGURE 27

(a) SEM image of bicomponent disks array and (b) the corresponding MOKE loop of the array of bicomponent disks. (c–h) The simulated spin configuration during the magnetization reversal of bicomponent disk. *Taken from Ref. [68].*

5 CONCLUSION

In this chapter, we have presented some of the most common techniques for the growth and characterization of magnetic thin films and nanostructures.

The first part of the chapter discusses the various magnetic thin-film growth techniques covering mainly physical vapor deposition, which includes thermal and electron beam evaporation, sputter deposition, and MBE. The second part provides an overview of common characterization techniques used in measuring the properties of magnetic thin films and nanostructures, i.e., surface, static, and dynamic magnetic characterizations. The final part of the chapter covers the basics of lithography and pattern transfer with a particular emphasis on DUV lithography used in the fabrication of magnetic nanostructures. This part ends with some practical examples of basic and advanced fabrication techniques for synthesizing high-quality patterned magnetic nanostructures. We hope that this chapter can provide useful resources and reference materials for researchers and students working on the topic of low-dimensional magnetism, advanced fabrication, and characterization techniques.

Acknowledgments

The authors would like to thank Dr. N. Singh, Dr. S. Goolaup, Dr. S. Jain, Dr. D. Tripathy, Dr. C. C. Wang, Dr. J. Ding, Dr. X. M. Liu, and Y. Huang for their contributions to the experimental work. This work was supported by the National Research Foundation, Prime Minister's Office, Singapore under its Competitive Research Programme (CRP Award No. NRF-CRP 10-2012-03), SMF-NUS New Horizon Awards, and Ministry of Education, Singapore AcRF Tier 2 (Grant No. R-263-000-A19-112).

References

[1] Hicken RJ, Eley DEP, Gester M, Gray SJ, Daboo C, Ives AJR, et al. Brillouin light scattering studies of magnetic anisotropy in epitaxial Fe/GaAs films. J Magn Magn Mater 1995;145:278–92.

[2] Bland JAC, Daboo C, Heinrich B, Celinski Z, Bateson RD. Enhanced magnetic moments in bcc Fe films. Phys Rev B 1995;51:258–72.

[3] Ney A, Poulopoulos P, Baberschke K. Surface and interface magnetic moments of Co/Cu(001). Europhys Lett 2001;54:820.

[4] Gutjahr-Löser T, Sander D, Kirschner J. Magnetoelastic coupling in Ni and Fe monolayers on Cu(001). J Appl Phys 2000;87:5920–2.

[5] Kowalewski M, Schneider CM, Heinrich B. Thickness and temperature dependence of magnetic anisotropies in ultrathin fcc Co(001) structures. Phys Rev B 1993;47:8748–53.

[6] Song O, Ballentine CA, O'Handley RC. Giant surface magnetostriction in polycrystalline Ni and NiFe films. Appl Phys Lett 1994;64:2593–5.

[7] Szymczak H. Surface magnetoelastic behavior of magnetic multilayers (invited). J Appl Phys 1997;81:5411–5.

[8] Stinson DG, Shin SC. Magnetization and anisotropy of Co/Pd multilayer thin films. J Appl Phys 1990;67:4459–61.

[9] Draaisma HJG, den Broeder FJA, de Jonge WJM. Perpendicular anisotropy in Pd/Co multilayers. J Appl Phys 1988;63:3479–81.

[10] Binasch G, Grünberg P, Saurenbach F, Zinn W. Enhanced magnetoresistance in layered magnetic structures with antiferromagnetic interlayer exchange. Phys Rev B 1989;39:4828.

[11] Baibich MN, Broto JM, Fert A, Van Dau FN, Petroff F, Etienne P, et al. Giant magnetoresistance of (001)Fe/(001)Cr magnetic superlattices. Phys Rev Lett 1988;61:2472.

[12] Maiti N, Karmakar P, Barve UD, Bapat AV. An evaporation system for film deposition using electron beam sources. J Phys Conf Ser 2008;114:012049.

[13] Piramanayagam SN. Perpendicular recording media for hard disk drives. J Appl Phys 2007;102:011301.

[14] Weller D, Parker G, Mosendz O, Champion E, Stipe B, Xiaobin W, et al. A HAMR media technology roadmap to an areal density of 4 Tb/in^2. IEEE Trans Magn 2014;50:1–8.

[15] Ohring M. Materials science of thin films. 2nd ed. San Diego: Academic Press; 2002.

[16] Plummer JD, Deal MD, Griffin PB. Silicon VLSI technology: fundamentals, practice and modeling. New Jersey: Prentice Hall Inc.; 2000

[17] Thornton JA. Structure-zone models of thin films, In: Proc SPIE 0821, Modeling of optical thin films; 1988. p. 95–105.

[18] Seshan K. Handbook of thin-film deposition processes and techniques: principles, methods, equipment and applications. 2nd ed. New York: Noyes Publications; 2002.

[19] Cho AY, Arthur JR. Molecular beam epitaxy. Prog Solid State Chem 1975;10(Pt. 3):157–91.

[20] Kerr J. On rotation of the plane of polarization by reflection from the pole of a magnet. Philos Mag Ser 5 1877;3:321–43, 1877/05/01.

[21] Schlenker M, Souche Y. Magneto-optical effects. In: de Lacheisserie ÉDT, et al., editors. Magnetism: fundamentals. 1st ed. Boston: Springer; 2005.

[22] Qiu ZQ, Bader SD. Surface magneto-optic Kerr effect. Rev Sci Instrum 2000;71:1243–55.

[23] Foner S. Versatile and sensitive vibrating-sample magnetometer. Rev Sci Instrum 1959;30:548–57.

[24] Fagaly RL. Superconducting quantum interference device instruments and applications. Rev Sci Instrum 2006;77:101101.

[25] Kleiner R, Koelle D, Ludwig F, Clarke J. Superconducting quantum interference devices: state of the art and applications. Proc IEEE 2004;92:1534–48.

[26] Slade SB, Kassabian G, Berkowitz AE. Combination high sensitivity alternating current susceptometer and high frequency B-H looper. Rev Sci Instrum 1996;67:2871–6.

[27] Crittenden EC, Smith CS, Olsen LO. B-H meter for samples of small cross-sectional area. Rev Sci Instrum 1946;17:372–4.

[28] Griffiths JHE. Anomalous high-frequency resistance of ferromagnetic metals. Nature 1946;158:670–1.

[29] Kittel C. Interpretation of anomalous larmor frequencies in ferromagnetic resonance experiment. Phys Rev 1947;71:270–1.

[30] Kittel C. On the theory of ferromagnetic resonance absorption. Phys Rev 1948;73:155–61.

[31] Gubbiotti G, Kostylev M, Tacchi S, Madami M, Carlotti G, Ding J, et al. Collective spin waves on a nanowire array with step-modulated thickness. J Phys D Appl Phys 2014;47:105003.

[32] Ding J, Kostylev M, Adeyeye AO. Broadband ferromagnetic resonance spectroscopy of permalloy triangular nanorings. Appl Phys Lett 2012;100:062401.

[33] Ding J, Kostylev M, Adeyeye AO. Magnetic hysteresis of dynamic response of one-dimensional magnonic crystals consisting of homogenous and alternating width nanowires observed with broadband ferromagnetic resonance. Phys Rev B 2011;84:054425.

[34] Demokritov SO, Hillebrands B, Slavin AN. Brillouin light scattering studies of confined spin waves: linear and nonlinear confinement. Phys Rep 2001;348:441–89.

[35] Sandercock JR. Tandem Fabry-Perot interferometer TFP-1 operator manual, Available at: http://tablestable.com/uploads/ckeditor/TFP-1/Manual_TFP_1.pdf; 2014.

[36] Demidov VE, Demokritov SO, Hillebrands B, Laufenberg M, Freitas PP. Radiation of spin waves by a single micrometer-sized magnetic element. Appl Phys Lett 2004;85:2866–8.

[37] Chou SY, Krauss PR, Wei MS, Fischer PB. Nanolithographically defined magnetic structures. Scr Met Mater 1995;33:1537–44.

[38] Cowburn RP, Adeyeye AO, Weiland ME. Configurational anisotropy in nanomagnets. Phys Rev Lett 1998;81:5414–7.

[39] Cowburn RP, Koltsov DK, Adeyeye AO, Welland ME, Tricker DM. Single-domain circular nanomagnets. Phys Rev Lett 1999;83:1042–5.

[40] Kirk KJ, Chapman JN, McVitie S, Aitchison PR, Wilkinson CDW. Switching of nanoscale magnetic elements. Appl Phys Lett 1999;75:3683–5.

[41] Adeyeye AO, Welland ME. Magnetization process of a single chain of nanomagnets. Appl Phys Lett 2002;80:2344–6.

[42] Castaño FJ, Ross CA, Eilez A. Magnetization reversal in elliptical-ring nanomagnets. J Phys D Appl Phys 2003;36:2031.

[43] Adeyeye AO, Bland JAC, Daboo C, Hasko DG, Ahmed H. Optimized process for the fabrication of mesoscopic magnetic structures. J Appl Phys 1997;82:469–73.

[44] Joel KWY, Yunjie C, Tianli H, Huigao D, Naganivetha T, Hui Kim H, et al. Fabrication and characterization of bit-patterned media beyond 1.5 Tbit/in^2. Nanotechnology 2011;22:385301.

[45] Manfrinato VR, Zhang L, Su D, Duan H, Hobbs RG, Stach EA, et al. Resolution limits of electron-beam lithography toward the atomic scale. Nano Lett 2013;13:1555–8.

[46] Wassermann EF, Thielen M, Kirsch S, Pollmann A, Weinforth H, Carl A. Fabrication of large scale periodic magnetic nanostructures. J Appl Phys 1998;83:1753–7.

[47] Farhoud M, Ferrera J, Lochtefeld AJ, Murphy TE, Schattenburg ML, Carter J, et al. Fabrication of 200 nm period nanomagnet arrays using interference lithography and a negative resist, In: 43rd International conference on electron, ion, and photon beam technology and nanofabrication, Marco Island, Florida (USA); 1999. p. 3182–5.

[48] Zheng M, Yu M, Liu Y, Skomski R, Liou SH, Sellmyer DJ, et al. Magnetic nanodot arrays produced by direct laser interference lithography. Appl Phys Lett 2001;79:2606–8.

[49] Vogeli B, Smith HI, Castano FJ, Haratani S, Hao Y, Ross CA. Patterning processes for fabricating sub-100 nm pseudo-spin valve structures, In: The 45th international conference on electron, ion, and photon beam technology and nanofabrication, Washington, DC (USA); 2001. p. 2753–6.

[50] Heyderman LJ, Solak HH, David C, Atkinson D, Cowburn RP, Nolting F. Arrays of nanoscale magnetic dots: fabrication by x-ray interference lithography and characterization. Appl Phys Lett 2004;85:4989–91.

[51] Moralejo S, Castaño FJ, Redondo C, Ji R, Nielsch K, Ross CA, et al. Fabrication and magnetic properties of hexagonal arrays of NiFe elongated nanomagnets. J Magn Magn Mater 2007;316: e44–7.

[52] Natali M, Lebib A, Cambril E, Chen Y, Prejbeanu IL, Ounadjela K. Nanoimprint lithography of high-density cobalt dot patterns for fine tuning of dipole interactions, In: The 45th international conference on electron, ion, and photon beam technology and nanofabrication, Washington, DC (USA); 2001. p. 2779–83.

[53] Moritz J, Landis S, Toussaint JC, Bayle-Guillemaud R, Rodmacq B, Casali G, et al. Patterned media made from pre-etched wafers: a promising route toward ultrahigh-density magnetic recording. IEEE Trans Magn 2002;38:1731–6.

[54] Hu W, Wilson RJ, Xu L, Han S-J, Wang SX. Patterning of high density magnetic nanodot arrays by nanoimprint lithography. J Vac Sci Technol 2007;A25:1294–7.

[55] Chappert C, Bernas H, Ferré J, Kottler V, Jamet J-P, Chen Y, et al. Planar patterned magnetic media obtained by ion irradiation. Science 1998;280:1919–22.

[56] Fassbender J, Ravelosona D, Samson Y. Tailoring magnetism by light-ion irradiation. J Phys D Appl Phys 2004;37:R179.

[57] Devolder T, Chappert C, Chen Y, Cambril E, Launois H, Bernas H, et al. Patterning of planar magnetic nanostructures by ion irradiation. J Vac Sci Technol B 1999;17:3177–81.

[58] Martín JI, Nogués J, Liu K, Vicent JL, Schuller IK. Ordered magnetic nanostructures: fabrication and properties. J Magn Magn Mater 2003;256:449–501.

[59] Asakawa K, Hiraoka T, Hieda H, Sakurai M, Kamata Y, Naito K. Nano-patterning for patterned media using block-copolymer. J Photopolym Sci Technol 2002;15:465–70.

[60] Cheng JY, Ross CA, Thomas EL, Smith HI, Vancso GJ. Fabrication of nanostructures with long-range order using block copolymer lithography. Appl Phys Lett 2002;81:3657–9.

[61] Cheng JY, Mayes AM, Ross CA. Nanostructure engineering by templated self-assembly of block copolymers. Nat Mater 2004;3:823–8.

[62] Kubo T, Parker JS, Hillmyer MA, Leighton C. Characterization of pattern transfer in the fabrication of magnetic nanostructure arrays by block copolymer lithography. Appl Phys Lett 2007;90:233113.

[63] Lodder JC. Methods for preparing patterned media for high-density recording. J Magn Magn Mater 2004;272-27(Pt. 3):1692–7.

[64] Singh N, Goolaup S, Adeyeye AO. Fabrication of large area nanomagnets. Nanotechnology 2004;15:1539–44.

[65] Adeyeye AO, Singh N. Large area patterned magnetic nanostructures. J Phys D Appl Phys 2008;41:153001.

[66] Brunner TA. Why optical lithography will live forever. J Vac Sci Technol B 2003;21:2632–7.

[67] Adeyeye AO, Goolaup S, Singh N, Jun W, Wang CC, Jain S, et al. Reversal mechanisms in ferromagnetic nanostructures. IEEE Trans Magn 2008;44:1935–40.

[68] Shimon G, Adeyeye AO, Ross CA. Reversal mechanisms of coupled bi-component magnetic nanostructures. Appl Phys Lett 2012;101:083112.

[69] Wang CC, Adeyeye AO, Singh N. Magnetic antidot nanostructures: effect of lattice geometry. Nanotechnology 2006;17:1629.

[70] Adeyeye AO, Goolaup S, Singh N, Wang CC, Gao XS, Ross CA, et al. Magnetostatic coupling in arrays of elongated Ni80Fe 20 rings. J Phys D Appl Phys 2007;40:6479–83.

[71] Wang CC, Adeyeye AO, Singh N. Magnetic and transport properties of multilayer nanoscale antidot arrays. Appl Phys Lett 2006;88:222506.

[72] Goolaup S, Singh N, Adeyeye AO, Ng V, Jalil MBA. Transition from coherent rotation to curling mode reversal process in ferromagnetic nanowires. Eur Phys J B 2005;44:259–64.

[73] Ding J, Kostylev M, Adeyeye AO. Magnonic crystal as a medium with tunable disorder on a periodical lattice. Phys Rev Lett 2011;107:047205.

[74] Shimon G, Adeyeye AO, Ross CA. Magnetic vortex dynamics in thickness-modulated $Ni_{80}Fe_{20}$ disks. Phys Rev B 2013;87:214422.

[75] Ding J, Adeyeye AO. Binary ferromagnetic nanostructures: fabrication, static and dynamic properties. Adv Funct Mater 2013;23:1684–91.

CHAPTER 2

Element-Specific Probes of Magnetism

Claus M. Schneider
Peter Grünberg Institute PGI-6, Research Center Jülich, Jülich, Germany
Faculty of Physics, University Duisburg-Essen and CENIDE, Duisburg, Germany

1 INTRODUCTION

The progress in modern magnetism and spintronics is closely connected to the development of complex material systems with tailored magnetic and spin-transport properties. The magnetic behavior and spin-dependent functionalities can be tuned independently by choosing the composition of constituents in a compound or the sequence in a layer stack. In particular, in a thin-film stack, each layer often provides a specific function. For an in-depth understanding of the intricate magnetic interplay in such a stack, at least a layer-resolved or even an element-selective characterization is mandatory.

In order to distinguish different elements from each other or discriminate different chemical states of the same element, one needs to take a closer look at the electronic structure of an atom. The binding energies of the strongly bound core electrons are characteristic for a given element. The binding energy of a core electron is determined by the electrostatic interaction of the electron with the nucleus, which may be partially screened by other electrons in the system. A chemical bond with an unlike atom causes a charge transfer in the valence electrons and alters the screening [1]. As a consequence, the binding energy of the core state will be changed. For instance, an oxidation process causes a reduction of the valence electron charge at the atom, thereby lowering the local electrostatic screening and leading to an increase of the binding energy (Fig. 1). Analyzing these binding energy shifts will thus yield detailed information on the chemical state and the bonding situation. The core levels may be probed by a variety of photon- and electron-based spectroscopies, which then provide data on the chemical composition and chemical environment of a chosen species. The potential of this approach in the case of photoelectron spectroscopy was first recognized by Siegbahn, who developed this technique as electron spectroscopy for chemical analysis [2]. Photoelectron spectroscopy usually

Magnetism of Surfaces, Interfaces, and Nanoscale Materials. http://dx.doi.org/10.1016/B978-0-444-62634-9.00002-3

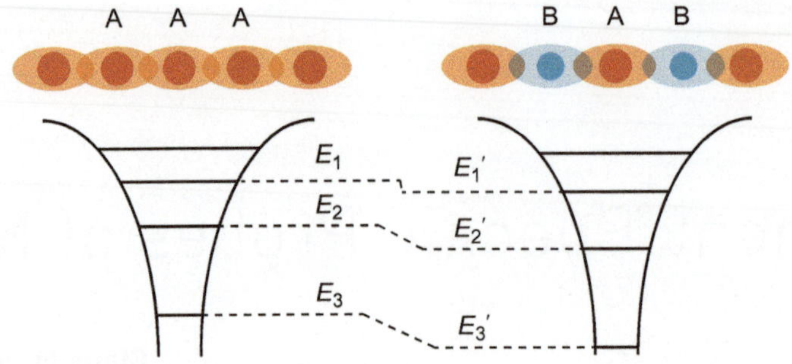

FIGURE 1

Core-level states and chemical sensitivity. The binding energy (E_1, E_2, E_3) of the bound states in the atomic Coulomb potential is determined by the electrostatic interactions and screening. If the chemical environment of the atom "A" is changed, for example, by chemical bonding to an atom "B," the screening is increased or reduced, leading to a shift of the energy levels in the potential well. The case depicted corresponds to an oxidation, resulting in an increase of the binding energy values to E_1', E_2', E_3'.

involves a fixed photon energy. Complementary element-selective information can be obtained from X-ray absorption spectroscopies (XASs) by scanning the photon energy [3].

Magnetism is a cooperative phenomenon which is characterized by the magnetic moments or electron spins orienting or aligning along a well-defined spin quantization axis [4–6]. This long-range spin ordering reduces the total energy of the system and forms a stable state below a critical temperature. We must distinguish two limiting cases taking into account the different types of magnetic order. In a *ferromagnet*, we have a nonzero magnetization vector **M** and an exchange–split band structure, which gives rise to a nonzero spin polarization **P** of the valence electrons. The long-range spin ordering breaks down at the Curie temperature T_C. As was first experimentally demonstrated in 1969 [7], the photoelectrons excited from these valence bands carry a spin-polarization vector **P** which can be measured by appropriate spin polarimeters. We will see in more detail below that even the photoelectrons excited from the core levels are spin-polarized, thereby already combining magnetic sensitivity and chemical specificity.

In an *antiferromagnet*—representing the other limiting case—we still have a spin alignment axis, but the spin polarization averages to zero over a unit cell, because the magnetic moments on neighboring lattice sites compensate each other [8]. The long-range order persists up to the Néel temperature T_N. In antiferromagnets, we need a different approach to unambiguously probe the magnetic state. This is provided by the spin–orbit coupling, which is the quantum mechanical mechanism behind a large class of magneto-optical and magneto-dichroic effects ranging from visible light to the hard X-ray regime. A precondition for the exploitation of magneto-optical effects is a meticulous control of the polarization state of the incident light. It was already predicted in 1975 by Erskine and Stern [9] that core-level absorption spectra taken with polarized light should show a magnetic dichroism, which is related to the magnetic state of the sample. Since this magnetic dichroism is strongest for the core-level excitation, we have again a useful combination of magnetic sensitivity and chemical selectivity. The magnetodichroic response in the soft X-ray regime was experimentally proven for antiferromagnets in 1986 by van der Laan *et al.* [10] and for ferromagnets in 1988 by Schütz *et al.* [11].

The success of element-selective studies in magnetism was promoted through the development of dedicated third-generation synchrotron radiation light sources. These provide a high photon flux, a wide range of tunability of the photon energy, and a well-defined polarization state, i.e., linear (circular) polarization for the studies on antiferromagnets (ferromagnets). Synchrotron radiation has two further interesting properties, which are very useful for nanomagnetism studies. First, the light has well-defined time structure, i.e., it consists of a regular sequence of short pulses each being some 10 ps wide [12]. This "stroboscopic" illumination has been very extensively employed to investigate the magnetization dynamics in the nano- and picosecond timescales. Second, the light from high-brilliance synchrotron sources has also a considerable degree of coherence which enabled pioneering holography experiments. In the free electron laser (FEL) sources which are currently being developed, the degree of coherence will be even higher and the time structure involves light pulses in the femtosecond regime [13,14]. This allows for new experimental approaches which are particularly addressing the ultrafast demagnetization and spin dynamics [15,16].

In this chapter, we will focus on photon-based techniques for the element-selective study of magnetism and nanomagnetism, covering the regime from soft (~100 eV) to hard X-rays (~15 keV). In this regime, we find most of the *photon-in/photon-out* and *photon-in/electron-out* techniques which are currently in use. In order to describe the effects on a common ground, we will start out with a short description of the interaction of photons with a spin-polarized electronic system. We will restrict ourselves to the salient features only, leaving out the many intricacies, as the emphasis should lie on the application to nanomagnetism. We then move on to the various experimental approaches starting with X-ray photoemission and moving to X-ray absorption, reflection, and emission experiments.

2 FUNDAMENTAL ASPECTS OF OPTICAL EXCITATION

The interaction of photons with a solid is a complex process which involves numerous excitation and decay channels (Fig. 2). We can distinguish photon

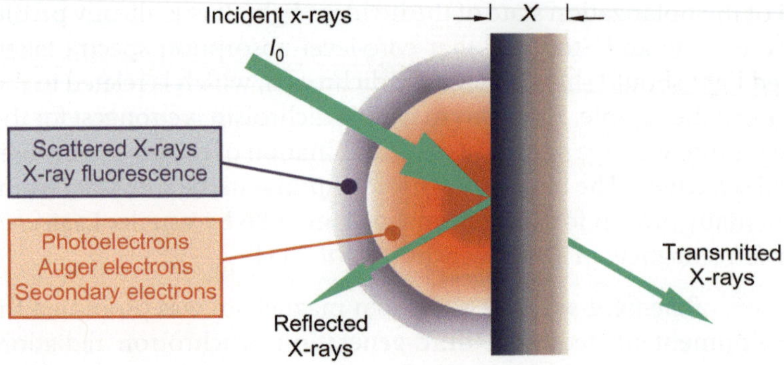

FIGURE 2
Excitation channels in the interaction of X-rays with a solid of thickness *x*. We can distinguish a photon—photon (transmission, absorption, reflection, diffraction, scattering, and fluorescence) and a photon—electron family (photoemission, Auger electron emission, secondary electron yield).

(reflection, scattering, fluorescence) and electron channels (photoelectrons, Auger electrons, secondary electrons), all of which can be exploited in dedicated spectroscopic techniques. In order to introduce the fundamental aspects of the spectroscopic approaches involved in an element-selective study of magnetism, we discuss these aspects first for the case of nonmagnetic systems.

2.1 Electronic States in Solids

The electronic structure of a magnetic solid is often subdivided into two systems: the valence bands and the core states. The valence electronic states are involved in the chemical bonding and are often strongly delocalized. They form a band structure which reflects the crystal symmetries and the crystal potential. The states are located around the Fermi level E_F and range down to a binding energy of $E_B \sim 10$–15 eV. This band structure is responsible for most of the physical properties of an element or material. The valence electrons are also subject to many-particle Coulomb interactions leading to particular collective phenomena, such as superconductivity or magnetism. A fine balance of exchange and correlation mechanisms determines the collective behavior and thus the ground state of a material, for example, the type of magnetic and spin ordering [17,18]. This spin ordering caused by the exchange interaction Δ_{ex} has a profound consequence on the electronic structure. As Δ_{ex} lowers the symmetry of the respective Hamiltonian, the degeneracy of electronic states is removed, causing an energetic splitting. This is particularly pronounced in a ferromagnet, where the valence bands split into majority (spin-up) and minority (spin-down) spin states, resulting in a spin-polarized density of states around the Fermi level E_F.

The core states are found at higher binding energies $E_B > 50$ eV and are localized in the Coulomb potential of the individual atom. The core-level binding energies are characteristic for a specific element. As mentioned above, however, the chemical bonding of the atom to its neighbors causes a shift of the binding energy scale as a function of the chemical state of the atom. A spectroscopic characterization of the core levels thus reveals not only the chemical composition of a material but also the charge state of the individual ion [19,20].

The separation in valence and core states may be convenient, but already the above discussion of the charge transfer shows it to be only an approximation. In fact, all electrons in an atom interact with both the crystal field and atomic Coulomb potential. For a magnetic system, this has the profound consequence that also the core levels are affected by the spin ordering in the valence states and may experience an energetic splitting according to their magnetic quantum numbers m_J [21]. This energy splitting together with a spin polarization in the valence states are key to an element-selective study of magnetic systems.

2.2 Spectroscopy of Core States

This section is based on a lecture given in the 2010 IFF Spring School [22] and is touched on only briefly, as there is a wealth of existing literature on core-level spectroscopies. For a detailed reading we refer, for example, to the books of Hüfner [19], Schattke and van Hove [23], and de Groot and Kotani [20]. The emerging field of hard X-ray photoemission is covered in some review articles [24,25] and a recent feature issue in the *Journal of Electron Spectroscopy and Related Phenomena* [26].

The basis for any element-selective spectroscopic investigation with photons is the photoexcitation process of an electron in a core-level state into an empty state above the Fermi level or above the band gap of an insulator. The photoexcitation event is a quantum mechanical process, the essence of which is captured in Fermi's Golden Rule, which describes the transition probability $P_{i\to f}$ between the initial $|i\rangle$ and final electronic level $|f\rangle$ with binding energies E_i and E_f, respectively:

$$P_{i\to f} = \frac{2\pi}{\hbar}|\langle f|\mathcal{O}|i\rangle|^2\delta(E_f - E_i - h\nu). \qquad (1)$$

In the simplest approximation, the two levels $|i\rangle$ and $|f\rangle$ may refer to the single-electron ground-state electronic structure of the solid; however, this is at the expense of neglecting the role of electronic correlations in the excitation process. Particularly in core-level photoexcitation, the latter may be important, as one creates a hole in the core shell which strongly interacts with the other electrons, thereby often causing a binding energy shift of the levels. The most important quantity in Eq. (1) is the transition matrix element:

$$M_{fi} = \langle f|\mathcal{O}|i\rangle, \tag{2}$$

which essentially depends on the symmetries of the electronic wave functions and the photonic operator $\mathcal{O}$,whereas the delta function in Eq. (1) ensures energy conservation in the excitation process. For low photon flux densities the operator $\mathcal{O}$ can be treated within linear response theory and takes the form:

$$M_{fi} = \frac{-e}{mc}\left\langle f\middle|\vec{A}\left(\vec{r}\right)\cdot\vec{p}\middle|i\right\rangle, \tag{3}$$

with $\vec{A}\left(\vec{r}\right)$ the vector potential of the electromagnetic field and $\vec{p}$ the momentum operator. It is usually assumed that the wavelength of the electromagnetic field is large compared to interatomic distances, i.e., $\vec{A}\left(\vec{r}\right)$ varies only marginally in the spatial region contributing to the transition matrix element.[1] This view is commonly known as *dipole approximation* and simplifies the transition matrix element to

$$M_{fi} = \frac{-ie}{\hbar c}A_0\left(E_f - E_i\right)\left\langle\psi_f\middle|\vec{e}\cdot\vec{r}\middle|\psi_i\right\rangle, \tag{4}$$

with the complex amplitude of the vector potential A_0, its polarization vector $\vec{e}$, and the wave functions of the final and initial states ψ_f and ψ_i, respectively. This form of the transition matrix element is extremely valuable, as the quantity $\left\langle\psi_f\middle|\vec{e}\cdot\vec{r}\middle|\psi_i\right\rangle$ can be evaluated for selected symmetries of the wave functions and yields dipole selection rules, which are very useful for a qualitative interpretation of photoemission spectra. This can be most easily seen for atomic levels, the wave functions of which can be expressed in terms of a radial part and a part containing spherical harmonics $Y_{l,m}$. As the operator $\vec{e}\cdot\vec{r}$ can also be represented in terms of spherical harmonics (e.g., $Y_{1,0}$ for linearly polarized light, or $Y_{1,\pm m}$ for circularly polarized light), the matrix element $\left\langle\psi_f\middle|\vec{e}\cdot\vec{r}\middle|\psi_i\right\rangle$ can be fully calculated by evaluating products of spherical harmonics. The specific mathematics of spherical harmonics yields a nonzero matrix element only for particular relations between l_f, l_i, m_f, m_i, which is the basis of the selection rules. Although being strictly valid only for atomic systems, this approach has been successfully extended to approximately describe the behavior of electronic states at high symmetry points in solids [27]. A more general treatment of dipole selection rules in solids, in particular with respect to crystalline symmetries, has been developed on the basis of group theory [19]. This is important for the interpretation of mapping results of the valence electronic structure.

The general type of spectroscopy is finally determined by the nature of the final state the electron is excited into. Depending on whether the electron ends up in a state below or above the vacuum level, we may distinguish between XAS and

[1]This assumption should be revisited if photoexcitation with hard X-rays is considered.

X-ray photoemission spectroscopy (XPS). A specific case is resonant photoemission, in which photoabsorption and photoemission processes interfere. The main detection schemes for the investigation of magnetic systems differ considerably and their principles will be discussed in the following.

2.3 X-Ray Photoemission Spectroscopy

In photoemission spectroscopy, the excited electron leaves the sample and moves toward the detector with a kinetic energy $E_{\mathrm{kin}} = \hbar^2 k_f^2/2m_e$ related to the wave vector k_f of the electron in the final state. The kinetic energy is determined by the photon energy $\hbar\omega$, the binding energy of the initial electronic state E_B, and the work function Φ_0 of the sample surface according to (Fig. 3)

$$E_{\mathrm{kin}} = \hbar\omega - E_B - \Phi_0. \tag{5}$$

This equation can be used to determine the binding energy of the initial electronic state as long as electronic correlation effects can be neglected. At low binding energies close to the Fermi level E_F, one usually finds rather broad spectral features related to the valence band states, whereas at higher binding energies often narrow peaks are observed, caused by the excitation of core levels at

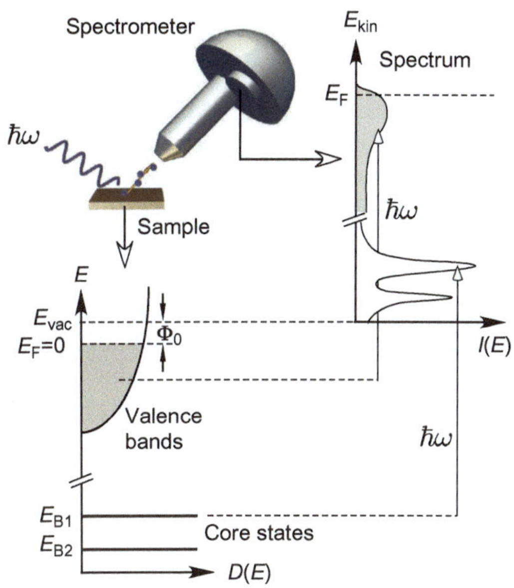

FIGURE 3
Optical excitation in a photoemission experiment performed at a fixed photon energy $\hbar\omega$. Only electrons excited above the vacuum level E_{vac} can leave the crystal and will contribute to the photocurrent. Photoemission from the valence bands contains information about the band structure of a solid, whereas the narrow photoemission lines from the localized core levels at E_{B1} and E_{B2} yield chemical information.

well-defined energies E_{B1}, E_{B2}, etc. From the energy position of these peaks, we can experimentally determine the binding energies, and from this information, we can then deduce the specific elements contained in the sample and their chemical state. In addition to the energy conservation (Eq. 5), there is also a conservation law for the photoelectron wave vector component parallel to the surface $\mathbf{k}_{f,||}$ [19,23]:

$$\mathbf{k}_{f,\parallel} - \mathbf{k}_{i,\parallel} + \mathbf{G}_{\parallel}, \tag{6}$$

i.e., $\mathbf{k}_{||}$ is conserved modulo a reciprocal lattice vector $\mathbf{G}_{||}$. The wave vector component perpendicular to the surface $\mathbf{k}_{\perp}$ is not conserved, because on its way to the detector the photoelectron has to pass through the surface potential barrier, the height and shape of which is generally not known [23]. The wave vector conservation is crucial for band structure investigations, where $\mathbf{G}_{||}$ may complicate the interpretation by causing so-called Umklapp processes [19]. To a first approximation, we may neglect it for core-level photoemission, which will be our main focus. Note that in Eq. (6) we also neglect the momentum of the photon $\mathbf{q}$. This approximation is no longer valid if the experiments are carried out with very high energy and angular resolution or involve hard X-ray excitations [28].

Before the photoelectrons can leave the sample, however, they have to travel from the location of excitation to the surface. Due to strong inelastic scattering of the energetic electrons in the sample, the effect of which is commonly described by the inelastic mean free path (IMFP) λ_{in}, the photoelectrons arriving at the spectrometer without energy loss usually originate from the near-surface region [29]. As a consequence, the information depth for a conventional X-ray photoelectron spectroscopy (XPS) using photon energies up to 1500 eV is only of the order of $\lambda_{in} \simeq 1$–2 nm, which is close to the minimum of the "universal curve" describing the energy dependence of the IMFP [30]. Working in this minimum is certainly an advantage for surface-sensitive experiments, but can be a drawback if true bulk-related information is needed. Numerous experiments during recent years have demonstrated that in complex materials, for instance in oxides, the surface region may significantly differ both electronically and chemically from the bulk [31]. This situation can be improved by exciting the photoelectrons with hard X-rays, resulting in higher kinetic energies [32]. This approach is now known as hard X-ray photoelectron spectroscopy or HAXPES [24,33].

A typical X-ray photoelectron spectrometer employs a hemispherical capacitor as energy filter which is operated at a fixed pass energy to ensure a constant energy resolution thoughout the spectrum. An entrance lens system defines the angular acceptance of the instrument and accelerates or retards the electrons to the pass energy of the analyzer. Behind the energy filter the photoelectrons are received by a two-dimensional detection system which records a certain window in kinetic energy and emission angle in one shot [34]. This is the standard mode of operation. Such a hemispherical spectrometer may also be retrofitted with a spin polarimeter to determine the spin polarization of the emitted electrons from ferromagnets [35,36] (cf. Section 3).

An overview spectrum of the photoemission from a clean iron surface is given in Fig. 4. The photon energy is 1253.4 eV, corresponding to the Mg K_α emission line produced by a standard laboratory X-ray source. The main spectral features result from core-level photoemission from the 2s, 2p, 3s, 3p states and the 3d valence bands (order in decreasing binding energy). As is apparent from this data, the shallow 3s, 3p core levels have a low photoexcitation cross-section at this photon energy and therefore contribute only with a low spectral weight. The same is true for the valence band states. In addition, we find spectral signatures caused by Auger transitions of the LMM, LMV, and LVV type, which are caused by a nonradiative decay of the core hole created in the 2p level (L-shell).

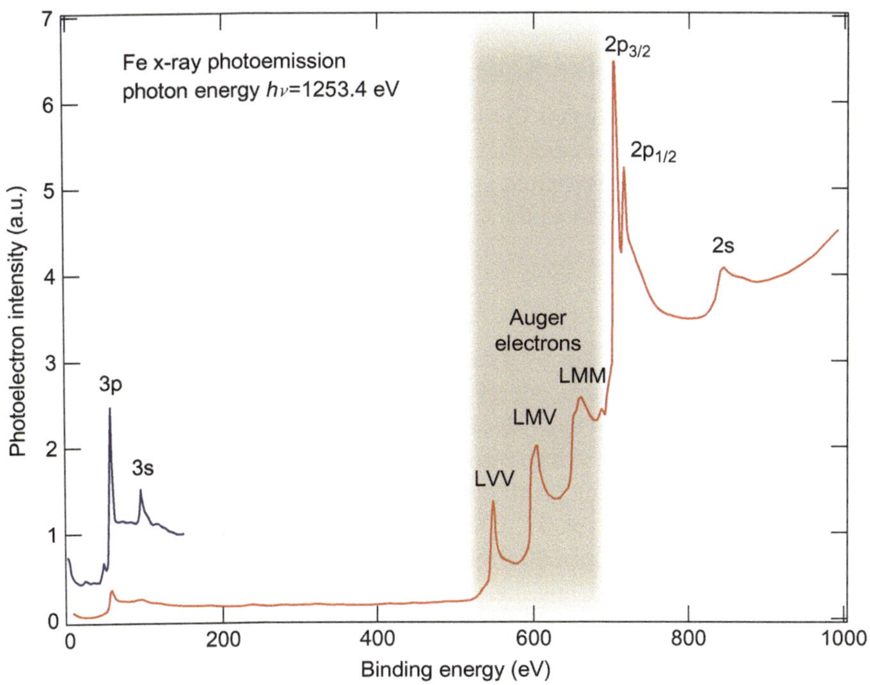

FIGURE 4

Overview spectrum of core-level photoemission from clean iron using a photon energy of 1253.4 keV (Mg K_α). In addition to the core-level photoemission lines (3p, 3s, 2p, 2s), the Auger electron emission lines (LMM, LMV, LVV) are also indicated. Note that Auger electrons have a fixed kinetic energy independent of the photon energy. *After Ref. [37].*

In this three-electron process, the core hole is filled by an electron from the 3p level (M) or the valence band (V) and the difference in binding energies is transferred to another electron in a 3p (LMM) or valence band state (LMV, LVV) [38]. The Auger electrons have a specific kinetic energy—their position on the binding energy scale in an XPS spectrum therefore shifts with the photon energy. They can also be employed for element-selective studies.

The conventional XPS systems are not optimized for high lateral resolution. Employing dedicated X-ray optical and/or electron optical systems, however, X-ray photoelectron spectroscopies can be performed with submicrometer resolution. In the *scanning* approach, a finely focused X-ray beam is rastered across the surface and the photoelectrons are collected in a standard spectrometer in a point-by-point fashion [39]. In the *full-field* approach an immersion lens system is employed to create a magnified image of a micron-sized surface area and pass it through an energy analyzer. This method is known as energy-filtered photoemission microscopy and has been successfully applied in the soft [40–42] and hard [43] X-ray regimes.

2.4 X-Ray Absorption Spectroscopy

The major difference between the X-ray photoemission and photoabsorption processes arises from the states that the photoelectron originating from a level with binding energy E_{B1} is excited into. In XAS, these are the empty states in the region between the Fermi and the vacuum level, i.e., the photon energy fulfills the relation

$$E_{B1} \leq \hbar\omega < E_{B1} + \Phi_0. \tag{7}$$

Consequently, the photoelectron excited from E_{B1} does not leave the crystal.[2] However, the same photon energy will generate photoemission from levels with binding energy $E_{B2} < E_{B1} - \Phi_0$. In the case of X-ray absorption, the individual optical transitions are still governed by dipole selection rules, but there is no wave vector selection of the photoelectron in the final state, as the electron will never leave the crystal. Instead the XAS signal represents an integration over all allowed transitions. The essential physics of the absorption process is usually cast into a Beer–Lambert law, i.e., the intensity of light transmitted through a sample $I(\hbar\omega)$ exponentially decays with the path length x of the photons in the material:

$$I(\hbar\omega) = I_0 \exp(-\mu(\hbar\omega)x), \tag{8}$$

[2]This is true only for the directly excited photoelectrons, whereas electrons excited during inelastic scattering cascade processes may be excited above the vacuum level.

with I_0 denoting the intensity of the incident light. The quantity determined in the experiment is the photon energy- and polarization-dependent absorption cross-section or attenuation coefficient $\mu(\hbar\omega)$. The cross-section shows a resonant enhancement if the photon energy reaches an absorption edge. Depending on the energy range around the edge which is investigated, one may distinguish different approaches. In near-edge X-ray absorption fine structure (NEXAFS) experiments, information is drawn from the chemical state of an atom or also from the orientation of molecular orbitals [44]. Structural information on the local atomic environment, i.e., the nature of the nearest neighbors, can be obtained from extended X-ray absorption fine structure (EXAFS).

In a one-electron approximation, the absorption coefficient in the vicinity of an absorption edge is proportional to an integration over all optical transitions starting from the core level $|\psi_i\rangle$ giving rise to the edge

$$\mu(\hbar\omega) \propto \sum_f \left| \left\langle \psi_f | O | \psi_i \right\rangle \right|^2 \delta(E_f - E_i - \hbar\omega). \tag{9}$$

In the case of the transition metal L-edges, we only need to consider dipole transitions, whereas for K-edge absorption and rare-earth systems quadrupole transitions must also be taken into account.

Measuring $\mu(\hbar\omega)$ can be essentially performed in two specific modes, which yield different information about the sample under investigation. In the *transmission* mode, the photons passing through the entire sample are analyzed as a function of photon energy. The resulting absorption spectrum reflects the bulk properties of the sample. In the *electron yield* mode, the secondary electrons emitted from the sample are recorded as a function of photon energy. These secondary electrons are generated by a two-step process after the absorption of the photon, which generates a core hole. A nonradiative decay of the core hole creates highly energetic Auger electrons. While propagating through the sample the Auger electron suffers multiple inelastic scattering events, thereby creating a cascade of low-energy secondary electrons, a fraction of which is leaving the sample. This total electron yield (TEY) has been shown to be proportional to the absorption signal [44]. Due to the strong electron–electron interaction, however, the information depth in the TEY mode is limited to a few nanometers, i.e., the near-surface region. In the case where a radiative decay of the core hole—X-ray fluorescence or X-ray emission—is involved, the fluorescence signal can also be employed as a measure of the absorption [45].

The spectral shape of the X-ray absorption lines depends strongly on the element and the core level which is excited. In Fig. 5, we show an example of absorption spectroscopy in the soft X-ray regime involving 3d transition metals. With photon energies in the range of 600–900 eV, one addresses the $L_{2,3}$ absorption edges of the ferromagnets. The photoelectrons are excited from

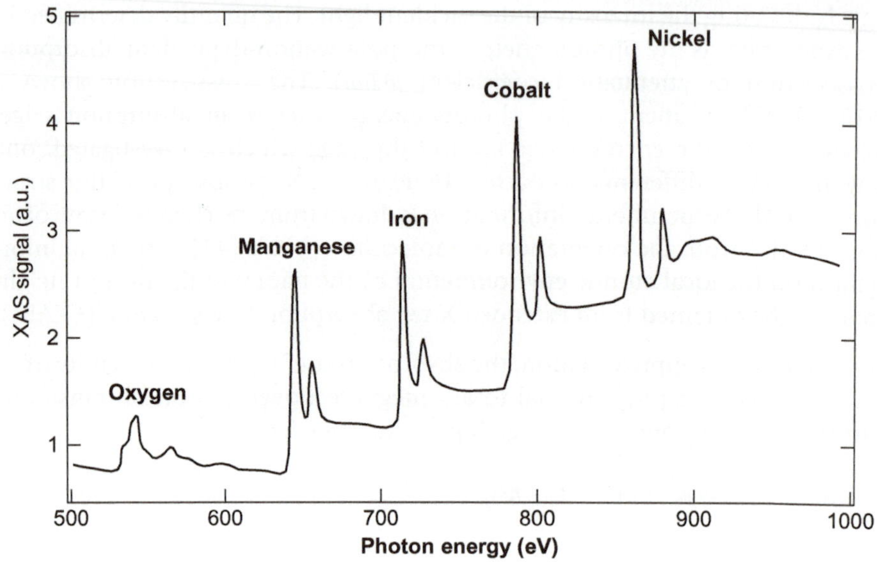

FIGURE 5

X-ray absorption spectroscopy of the $L_{2,3}$-edges of the magnetic transition metals as compared to the oxygen K-edge. The spectra are measured in total electron yield. *Taken from Ref. [46].*

the 2p levels into the empty density of states below the vacuum level. The resulting spectral features are very pronounced, because the 3d transition metals provide a high density of unoccupied states just above the Fermi level. They are often called "white lines." The characteristic doublet structure in the spectra of Mn, Fe, Co, and Ni shown in Fig. 5 is caused by the spin–orbit splitting of the 2p levels. In contrast, the much weaker absorption line at around 550 eV which stems from oxygen is caused by K-edge absorption.

The details of an XAS spectrum depend on the chemical environment of the respective atom. From photoemission we know already that a charge transfer caused by chemical bonding may lead to a change in the binding energy of a core level. Likewise, a change in the crystal field around the atom may result in a splitting of electronic levels. These effects can also be seen in XAS. Figure 6 depicts a set of XAS spectra of the Fe $L_{2,3}$-edges for different material systems containing iron. The top spectrum shows two relatively broad and smooth absorption lines, which are characteristic for clean Fe metal. The bottom spectrum for Fe_2O_3, however, exhibits a pronounced fine structure on both the L_2 and L_3 absorption edges. In Fe_2O_3, the valency of iron is Fe^{3+} and the ions occupy both octahedral and tetrahedral lattice sites with respect to the surrounding oxygen atoms. These two sites create different crystal fields at the position of the Fe^{3+} ions resulting in the multiplet splitting in the spectrum. Between these two extreme cases, we find a spectrum for an iron-telluride

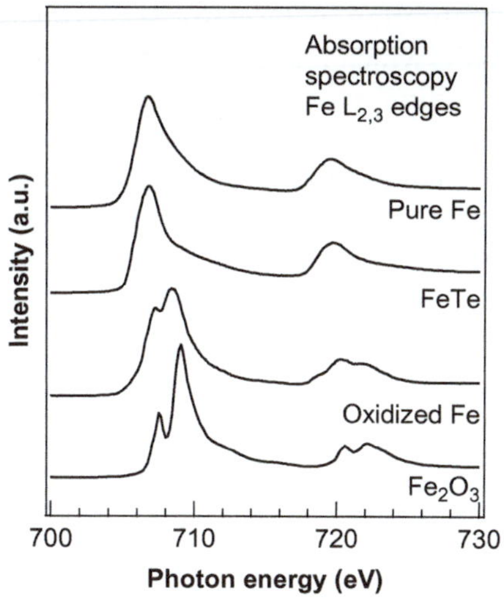

FIGURE 6
Chemical variations in iron-based systems observed in XAS. Clean metallic iron (top) is compared to an iron-telluride compound, a partially oxidized iron surface, and to an iron oxide, namely Fe_2O_3 (bottom). *After Ref. [47].*

compound. In comparison to pure iron the FeTe spectrum exhibits a significant change on the high-energy flank of the L_3 peak. The shoulder in the spectrum is attributed to covalent bonding between the Te 5p and Fe 3d states [48]. The bonding causes small changes in the electronic structure around the Fe atom and in the itinerancy of the 3d electrons, creating extra spectral contributions. The most complicated example spectrum is the one for the oxidized iron. Qualitatively it can be understood as a superposition of a metallic iron spectrum (which gives two smooth absorption lines) and different iron oxides (which contribute with multiplet features). Depending on the amount and composition of the oxide layer, the spectral contribution from these multiplet features will strongly vary.

2.5 X-Ray Reflection Spectroscopy

The X-rays scattered back from a sample surface may contain information of quite different physical properties of the sample, depending on the specific scattering. In general, one may distinguish between elastic and inelastic scattering processes. In inelastic X-ray scattering, energy and momentum are transferred between the photon and elementary excitations in the crystal. If the photon energy is tuned to an absorption edge one also talks about resonant inelastic X-ray scattering (RIXS), which is widely used to study electronic and magnetic excitations in strongly correlated electron systems.

The most prominent application in the elastic scattering regime is X-ray diffraction [49]. For wavelengths which match characteristic length scales in the sample, the resulting diffraction patterns contain structural information, e.g., atomic arrangements and interatomic distances. In the soft X-ray regime, the structural information relates rather to layer thicknesses and is an important tool to verify the structure of thin-film stacks and multilayers. These experiments are usually performed at a fixed wavelength (photon energy) and map the angular variation of the diffracted intensity.

A different mode of operation is employed in X-ray reflection spectroscopy. In this case, the scattering geometry is fixed and the reflected light is analyzed as a function of the photon energy. The variation of the reflectivity contains information of the electronic structure of the sample. In particular, in the vicinity of an absorption edge one obtains element selectivity, which becomes very important for magnetic studies.

The example shown in Fig. 7 maps the reflectivity of an HfO_2 thin film in the energy range between 500 and 550 eV photon energy. This is the range in which the oxygen K-edge is located. The 5-nm-thick HfO_2 film has been grown on a SiO_2/Si template [50,51]. The reflectivity spectrum at an angle of incidence of $\theta = 30°$ (Fig. 7a) reveals strong features at $h\nu = 540$ and above 550 eV. These are Kiessig fringes related to the interference of light reflected from the surface of the film and the interface to the substrate. The small modulation around $h\nu = 532$ eV is related to the excitation of the O K-edge. By moving the experiment geometry to glancing angles (Fig. 7b), the reflectivity signal becomes more sensitive to the film itself and the Kiessig fringes are suppressed. The

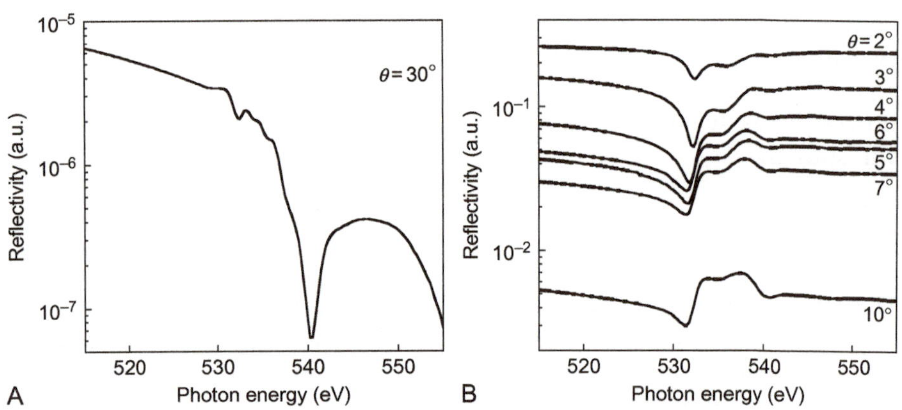

FIGURE 7

X-ray reflectivity spectrum of HfO_2 thin films in the energy range of the oxygen K-edge. (a) Kiessig fringes seen at an angle of incidence of $\theta = 30°$. (b) The anomalous reflectivity at the O K-edge leads to the spectral feature at around at $h\nu = 532$ eV photon energy. *Taken from Ref. [50].*

spectral feature related to the O K-edge becomes much more pronounced. The variation of the signal strength and the details of the spectral shape with the angle of incidence are related to multiple scattering events in the layered structure and can be recovered by optical simulations [50].

3 SPIN-SENSITIVE PHOTOEMISSION

In this chapter, we will focus on magnetic spectroscopies related to photoelectrons leaving the sample. As we have discussed above (cf. Section 2.3), photoemission can to a first approximation be regarded as a single-electron excitation and emission process. The photon operator does not act directly on the spin-dependent part of the wave function and thus the spin of the photoelectron is preserved during the excitation step. Therefore, if the initial electronic state is spin-split due to exchange interactions and magnetic order, the photoelectron does not only carry information about its binding energy and the chemical state, but also about the magnetic state of the sample. Moreover, due to the small IMFP of low-energy photoelectrons the information depth can be of the order of 5 Å. This method is therefore perfectly suited to explore surface magnetic phenomena.

3.1 Core-Level Photoemission from Ferromagnets

Early core-level photoemission investigations of ferromagnets showed unusually broad line widths of the 3s photoemission signals. This was suggested to be caused by a spin-splitting of the core levels due to exchange interactions between the core and valence electrons. Direct experimental proof, however, must resolve the spin character of the split states. The experimental challenges in the case of the 3s core-level photoemission are twofold. First, the photoemission cross-section is small and the spectral feature sits on a strong background caused by inelastic scattering of the 3p photoelectrons. Second, the spin detection of photoelectrons is generally a very inefficient process yielding small signals. We will briefly allude to the function of a spin-polarization detector or polarimeter, before we turn to the core-level photoemission results.

3.1.1 Spin Polarimeter Schemes

The spin polarization carried by the photoelectrons is described by a three-dimensional vector **P** [52]. A spin polarimeter is a device which determines the magnitude of **P** and its orientation in space. Basically all spin detection schemes currently in use employ a spin-dependent scattering of monochromatic electrons from a nonmagnetic or ferromagnetic target [53]. The spin-dependent scattering arises either through spin–orbit coupling or exchange interaction, and the scattering geometry (angle between incoming and outgoing beams) defines the spin-sensitive axis **t** of the detector (Fig. 8). Depending

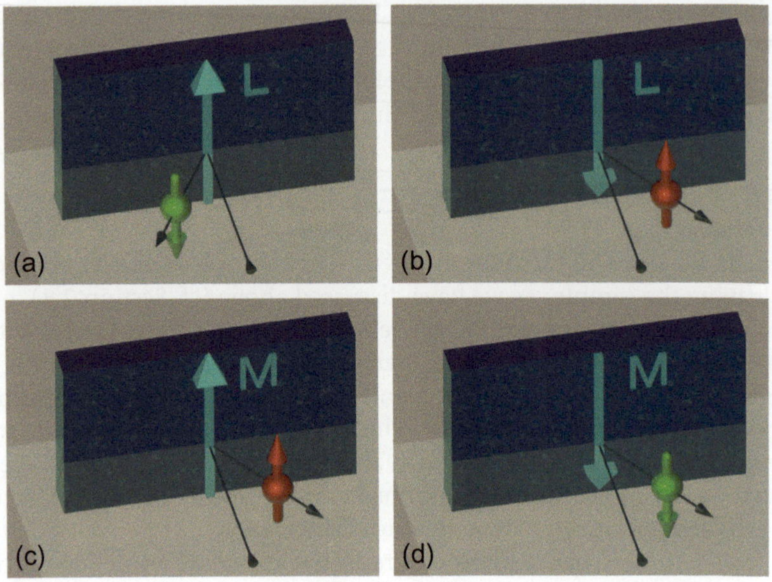

FIGURE 8

Principal mechanisms of spin-polarization detection for an electron beam at normal incidence. (a) The spin–orbit coupling causes a "left–right" asymmetry in the scattering process, i.e., spin-up (spin-down) electrons will be scattered predominantly to the right (left). (b) The orientation of **L** is defined by the scattering geometry. (c) The exchange-based scattering in turn causes an "up-down" asymmetry, i.e., in a fixed scattering geometry, the orientation of the detector magnetization **M** decides about which spin character is predominantly scattered. (d) A reversal of **M** therefore causes the opposite spin character to be scattered more strongly.

on the underlying physical mechanism, the scattering energy may range from very low energies of a few electron-volts for exchange scattering [54] up to 100 keV in the case of Mott scattering [52]. The scattered intensity I_S is usually several orders of magnitude lower than the incident one I_0, i.e., $I_S/I_0 \sim 10^{-4}\ldots10^{-3}$. Therefore, spin polarimetry is generally a slow and time-consuming process. The spin dependence of the scattering process generates a modulation S of the scattered intensity. The spin sensitivity S is normalized to a polarization $P=1$ and takes typical values of $S=0.2\ldots0.3$ depending on the type of polarimeter. For example, if we have an electron beam which is fully polarized along the detector axis (**L** or **M** in Fig. 8) and we turn the incident spin polarization perpendicular to the detector axis, we will observe an intensity change of 20% for $S=0.2$. From experiments one usually determines the spin-related intensity changes from which the spin polarization P is calculated.

In simple cases, we may be able to define the spin quantization axis in the sample by a clever design of the experiment. This is often possible in ferromagnets by choosing the proper magnetization direction **M**. In such cases we may have to measure only one spin-polarization component, namely the one along **M**. In the general case, however, the full spin-polarization vector, i.e., $\mathbf{P}=(P_x, P_y, P_z)$, has to be determined. This requires in principle three independent measurements with the spin-sensitive axis t of the detector oriented at orthogonal directions in space. Some spin polarimeter concepts intrinsically provide two orthogonal spin-sensitive axes (t_x, t_y) along which the data can be collected simultaneously. This simplifies the procedure into two independent measurements at orthogonal geometries, whereby one spin-polarization component is determined redundantly [55].

3.1.2 3s XPS

As a first example, we will discuss spin-resolved photoemission from the 3s core level. This case will also serve to illustrate some general aspects in the interpretation of spin-polarization data. The 3s states have no angular momentum ($L=0$) and therefore we can safely neglect effects from spin–orbit coupling. According to Hund's rule the two electrons residing in the 3s level should have opposite spin. The first spin-resolved 3s photoemission spectra were reported for iron in 1990 [56,57]. The analysis is complicated by the fact that the signal itself is quite weak and sits on a strong background arising from the inelastic scattering of photoelectrons with higher kinetic energy. Moreover, this background of secondary electrons is also spin-polarized. Therefore, a highly efficient spin polarimeter is necessary to resolve the spin-split states [56].

The spectrum in Fig. 9 shows the 3s photoelectron distribution from a single-crystalline Fe layer [56]. The sample was a 5-nm-thick Fe film epitaxially grown on Ag(001). Already from the intensity distribution $I(E)$ it becomes clear that the spectrum consists of two distinct spectral lines with positions at the binding energies marked by "A" and "B" (Fig. 9a), being separated by about 4.5 eV in binding energy. The spin-polarization analysis reveals a strong feature in the polarization $P(E)$ at the position of the 3s photoemission line (Fig. 9b). The spin-polarization vector **P** is aligned with the magnetization vector **M** which lies in the film plane. Along this quantization axis we can treat the spin polarization as a scalar quantity P, its value ranging between +1 (100% spin up) and −1 (100% spin down). The positive and negative values of the spin-polarization spectrum $P(E)$ in Fig. 9b therefore directly map the majority (↑) and minority spin (↓) electrons in the ferromagnet. From the intensity and spin-polarization distributions one may calculate the partial spin spectra displayed in Fig. 9c according to Ref. [52] as

$$I^{\uparrow(\downarrow)}(E)=\frac{I(E)}{2}(1\pm P(E)). \tag{10}$$

They distinguish the spin-up ($I^{\uparrow}(E)$) and spin-down ($I^{\downarrow}(E)$) contributions to the total spectrum and clearly show one distinct feature in each spin channel, whereby the feature "A" is related to spin-down electrons, whereas the feature

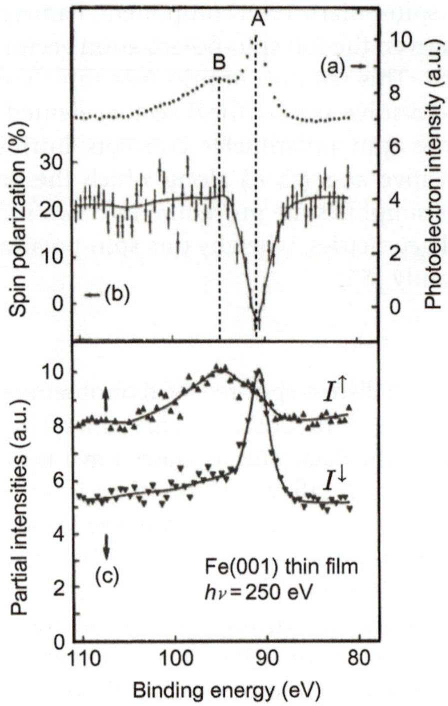

FIGURE 9

Spin-resolved core-level photoemission from the Fe 3s levels. (a) Total photoemission intensity distribution using synchrotron radiation of $h\nu=250$ eV. (b) Experimentally determined photoelectron spin polarization with statistical error bars. (c) Corresponding partial spin spectra revealing the spectral contributions with spin-up (▲, ↑) and spin-down (▼, ↓) character. *From Ref. [56].*

"B" stems from spin-up electrons. This proves that the peak splitting in the intensity spectrum is of magnetic origin.

We should point out that the analysis is somewhat complicated by the fact that the inelastic scattering of energetic electrons in a magnetic solid is strongly spin-dependent [58]. At the photon energy chosen in the experiment ($h\nu=250$ eV) there are spin-polarized photoelectrons with high kinetic energies excited from the valence bands and the Fe 3p core states which undergo inelastic scattering and contribute to the background in the spectrum. In a ferromagnet such as Fe, Co, and Ni, the inelastic scattering of spin-down electrons is usually more efficient, as there are more empty spin-down than spin-up states above the Fermi level to scatter into. As a consequence, the inelastic scattering acts as a kind of spin filter for spin-up electrons, rendering the secondary electrons leaving the crystal to be spin-polarized. This mechanism explains the consistent difference between the partial spectra $I^{\uparrow}(E)$ and $I^{\downarrow}(E)$ appearing in Fig. 9c.

The origin of the spin-splitting in the core levels is still not fully resolved [20]. One mechanism suggested invokes an exchange interaction between the core hole left behind after the photoexcitation process and the spin-polarized valence band electrons. The splitting would thus be caused by the photoexcitation process and cannot be considered a ground-state property. Another mechanism treats the core-level splitting as a ground-state property and is related to an interaction of the core-level electrons with the hyperfine field [21].

3.1.3 2p XPS

The situation in the photoemission from p levels is more involved as these states show an additional splitting due to spin–orbit interaction ($L=1$). In the 2p levels, the spin–orbit split states are well separated into the $J=3/2$ and $J=1/2$ levels, giving rise to two well-defined spectral features. In the presence of exchange interaction each of these J-levels will split into m_J sublevels. In the $J=3/2$ photoemission line, we therefore have four m_J sublevels contributing to the spin-polarization spectrum, whereas in the $J=1/2$ photoemission line there will be only two m_J sublevels [21]. Each sublevel can be associated with a specific spin character. Because of the intrinsic line width in photoemission, the energy splitting of the sublevels is generally not resolved in experiments, so the spin-polarization signal rather represents a weighted average of the individual sublevel contributions.

In Fig. 10, we show results for the photoemission from the Co 2p states obtained with a laboratory X-ray source at a photon energy of 1253.6 eV. The $2p_{1/2}$ and $2p_{3/2}$ levels are separated by about 15 eV in binding energy;

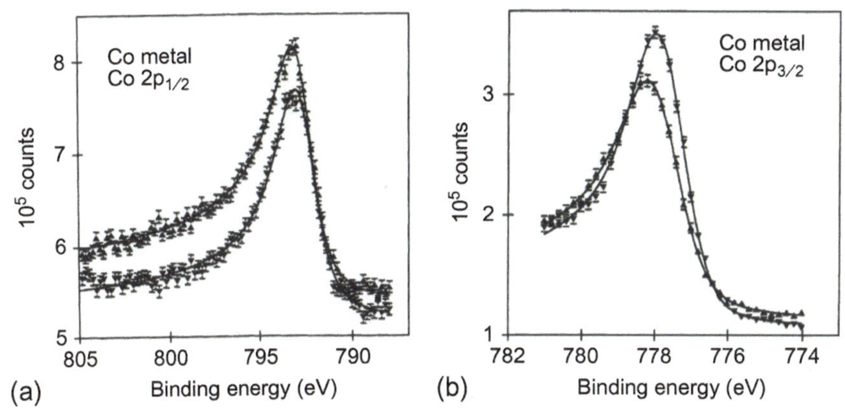

FIGURE 10

Spin-resolved core-level photoemission from the Co 2p levels at a photon energy of 1253.6 eV. (a) Partial spin spectra of the Co $2p_{1/2}$ photoemission line for majority (▲) and minority spins (▼). (b) Partial spin spectra for the Co $2p_{3/2}$ photoemission line. The lines represent a fit using a Doniach–Sunjic line shape convoluted with a Gaussian. *From Ref. [59].*

therefore, it is sufficient to confine the spin-resolved measurements to the photoemission lines only [59]. Note that in spin-integrated photoemission, we observe only a single photoemission line at each of the $2p_{1/2}$ and $2p_{3/2}$ levels. By contrast, the spin-resolved data exhibit a clear energy difference between the partial spectra for spin-up and spin-down contributions at both lines, with the minority spin contribution always found at the lower binding energy side of the peaks, consistent with theoretical treatments of interatomic exchange interactions [21,60]. In both cases, the intensity and the peak position clearly depend on the spin character. This is quite similar to the situation observed in the Fe 3s photoemission. The energy separation of the $I^{\uparrow}(E)$ and $I^{\downarrow}(E)$ contributions at the $2p_{3/2}$ level is of the order of $\Delta E_B \approx 0.5$ eV, whereas the overall energy resolution of the experiment (photoelectron spectrometer and light source) amounts to $\Delta E \approx 1.6$ eV [59]. The spin-polarization analysis thus acts as an effective improvement of the energy resolution.

For reasons of completeness, we also include a photoemission result from nickel metal. Nickel represents a particular case, because due to the narrow bandwidth of the valence states it shows pronounced electronic correlation effects, which complicate the interpretation of photoemission spectra [19]. A famous example is the "6 eV satellite" (Fig. 11). This satellite feature was found to be highly spin-polarized already in early photoemission experiments [62]. In Fig. 11, we reproduce a result obtained by See and Klebanoff [61] using a Mg K_α laboratory X-ray source. In analogy to Fe and Co, the 2p photoemission from Ni should only result in two well-defined

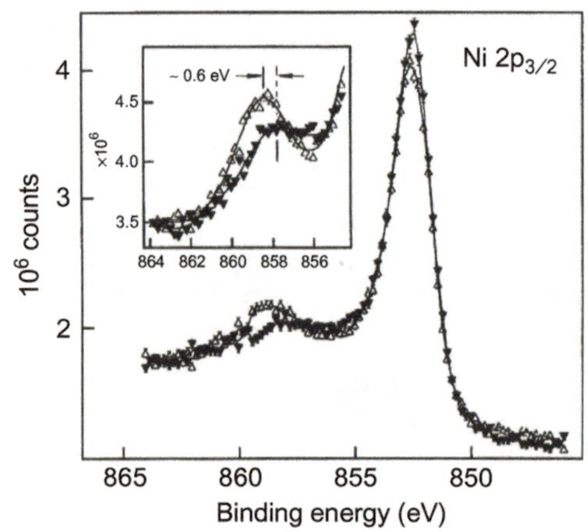

FIGURE 11

Spin-resolved core-level photoemission from the Ni $2p_{3/2}$ levels at a photon energy of 1253.6 eV. Partial spectra for spin-up (Δ) and spin-down (▼) with respect to the sample magnetization. Inset: Magnified part of the spectrum covering the highly spin-polarized structure corresponding to the "6 eV satellite." The spin splitting in the satellite amounts to about 0.6 eV. *From Ref. [61].*

spectral features corresponding to the Ni $2p_{1/2}$ and Ni $2p_{3/2}$ levels and being separated in binding energy by $\Delta E_B \approx 17$ eV. By contrast, the spectra show an additional feature located at about 6 eV higher binding energy than the main photoemission line from the Ni $2p_{3/2}$ level. Moreover, this feature is clearly spin-polarized with majority spin character. A similar result is obtained for the Ni $2p_{1/2}$ photoemission [61].

The reason for the appearance of this satellite is the strong interaction of the core hole in the 2p shell with the remaining $N-1$ electrons in the system. The core hole is screened by valence electrons from the neighboring Ni atoms (extra-atomic screening) [63]. Both 3d and 4s electrons are involved in the screening process, which causes a redistribution of the charge in the photoemitting atom. In particular, a part of the d-density of states is pushed below the Fermi energy. This changes the d-electron count as compared to the atomic ground state of Nickel in a solid ($4s3d^9$). The final state in the photoemission (taking into account the screening) can then be described as a mixture of $3d^9$ and $3d^{10}$ electronic configurations. In a complex process, these configurations lead to two different spectral contributions, representing the main line and the satellite [64]. This example illustrates that magnetic spectroscopy from correlated materials may become a rather involved issue.

3.2 Magnetic Dichroism in Photoemission

So far, we have neglected spin–orbit interaction in the analysis of spin-polarization effects in core level photoemission from magnetic materials. In particular, if the energy splittings induced by spin–orbit coupling are comparable or larger than those caused by the exchange interaction, this will give rise to new phenomena in photoemission. In particular, not only the spin-polarization spectrum $P(E)$ but also the intensity distribution $I(E)$ itself will depend on the magnetization direction and the light polarization. Such a spectral effect is commonly called *magnetic dichroism* and, in this specific case, magnetic dichroism in photoemission.

3.2.1 Magnetic Circular Dichroism in Photoemission

The first observation of a polarization dependence in core-level photoemission from ferromagnets was performed on an Fe(100) surface. The experiment was performed with circularly polarized light [65] incident at 10° with respect to the surface plane and the magnetization direction. In this way, the magnetization vector **M** and the helicity vector $\vec{\sigma} = \sigma^{\pm} \cdot \mathbf{q}/|\mathbf{q}|$ (**q**: photon wave vector) are almost collinear. The main result is reproduced in Fig. 12. The Fe 2p photoemission spectrum taken at a photon energy of $h\nu = 850$ eV shows a clear change in line shape, resulting in a shift of the peak centers on the binding energy scale as a function of magnetization direction. The shift has different magnitudes and goes in opposite directions for the $2p_{3/2}$ and $2p_{1/2}$ photoemission lines. This

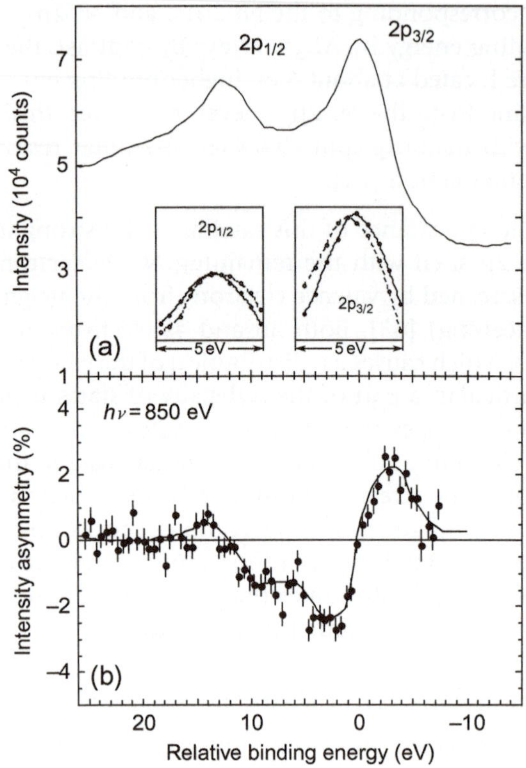

FIGURE 12

Magnetic circular dichroism in core-level photoemission from the Fe 2p levels at a photon energy of 850 eV. (a) Photoemission intensity distribution averaged over spectra taken at opposite magnetization directions. The relative binding energy is measured with respect to the position of the $2p_{3/2}$ line. The insets show the spectra for the individual magnetization directions at the $2p_{3/2}$ and $2p_{1/2}$ lines. (b) Intensity asymmetry revealing the characteristic plus–minus features at the two photoemission lines. *After Ref. [65].*

behavior can be quantified by the intensity asymmetry, which represents the normalized point-by-point difference of two photoemission spectra $I^+(E)$ and $I^-(E)$ recorded for opposite light helicities ($\sigma^+ = +1$, $\sigma^- = -1$) or magnetization directions ($M^+ = +1$, $M^- = -1$):

$$A(E) = \frac{I^+(E) - I^-(E)}{I^+(E) + I^-(E)}. \tag{11}$$

The asymmetry spectrum taken upon magnetization reversal (see Fig. 12b) clearly shows a response of the two photoemission lines in opposite directions. Since the magnetic state of the sample shows up directly in the intensity distribution in a photoemission experiment, this phenomenon is termed magnetic circular dichroism in photoemission. In order to understand this particular behavior, we have to briefly refer to spin-polarization effects in the photoemission from nonmagnetic materials.

In the case of a nonmagnetic system, i.e., in the absence of exchange splitting, the excitation with circularly polarized light is known to produce spin-polarized photoelectrons. This phenomenon is also called the "Fano effect" [66,67] or optical spin orientation [68]. Its quantum mechanical origin lies in a spin dependence of the dipole transition matrix elements. This spin dependence is brought about by spin–orbit interaction, which couples the spin to the orbital part of the wave function for a given sublevel. It is important to note that the light operator acts only on the orbital part of the wave function. In the case of circularly polarized light, the dipole selection rules require $\Delta J = \pm 1$ as well as $\Delta_{m_J} = +1$ and $\Delta_{m_J} = -1$ for light with positive (σ^+) and negative helicity (σ^-), respectively (Fig. 13).

Taking the example of p levels, the first selection rule states that transitions of the type p⇒s and p⇒d will occur, the latter channel being more pronounced because of the larger number of final states available. For reasons of simplicity

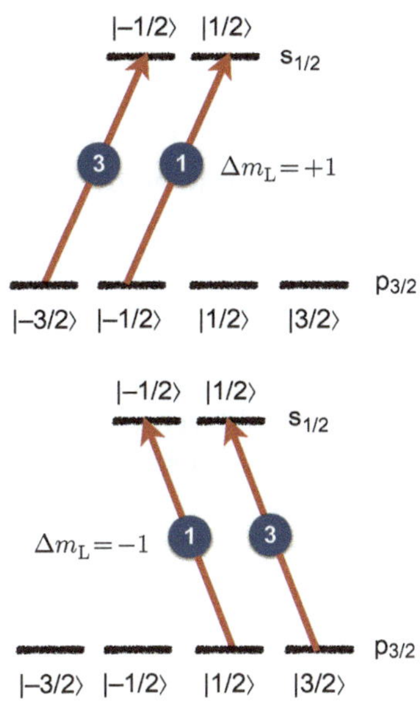

FIGURE 13

Principle of optical spin orientation or Fano effect in an atomic-like picture for transitions $p_{3/2} \Rightarrow s_{1/2}$ with circularly polarized light. The transition strength is determined from the Clebsch–Gordan coefficients and indicated by the encircled numbers. The sum of all transitions for a given helicity into the spin-up ($|+1/2\rangle$) and spin-down ($|-1/2\rangle$) final states yields an absolute spin-polarization value of 50%. A change of the light helicity ($\Delta m_L = +1$ to $\Delta m_L = -1$) reverses the sign of the spin polarization, but keeps the same magnitude.

let us focus in the further discussion on transitions from the $p_{3/2}$ into $s_{1/2}$ levels. For light of positive helicity, the second dipole selection rule requires the transitions to take place into $m_J = -1/2$ and $m_J = 1/2$ final states, each of which corresponds to a pure spin-down or spin-up state, respectively. The m_J sublevels are energetically degenerate. The resulting spin polarization in the free electron-like final state is thus obtained by weighting with the strength of the transition matrix elements and adding up. In our example, we obtain a spin polarization P of −50% (+50%) for $\Delta m_L = -1$ ($\Delta m_L = +1$). A similar analysis shows a finite spin polarization also for the $p \Rightarrow d$ transitions. The quantization axis for the spin polarization is defined by the angular momentum of the photon, i.e., it is oriented parallel or antiparallel to the direction of light incidence. A reversal of the light helicity ($\sigma^+ \Rightarrow \sigma^-$) reverses the photoelectron spin polarization ($P^+ \Rightarrow -P$), but leaves the shape of the spectrum unchanged. The Fano effect has been confirmed by spin-resolved core-level photoemission experiments from copper and tungsten [69].

In a ferromagnetic system the m_J sublevels in the initial and final states are no longer degenerate, but spin-split due to the exchange interaction. As a consequence, a thorough theoretical description must treat spin–orbit and exchange coupling on an equal footing. In the 2p core-level photoemission, we can adopt a simpler picture for a qualitative discussion. To a first approximation, we neglect the exchange splitting in the initial states, because it is small compared to the spin–orbit-induced level splitting. Likewise, the spin–orbit coupling in the final states is small, and we take into account only the exchange-induced spin splitting. In this picture, the final states take the role of an "internal" spin detector which probes the spin polarization of the excited electrons generated by the Fano effect. Depending on the spin character of the photoelectron, the transition strength is modulated and the electron reaches a slightly different final state. The photoemission spectrum is then obtained as an integration over the spectral lines of the individual sublevel contributions [21]. Due to the spin and helicity dependence of the sublevel spectral lines, the shape of the total photoemission spectrum depends on the magnetization state and the light helicity. This is called *magnetic dichroism in photoemission*. The above discussion is only a very simplified one as yet, as it does not take into account the angular dependence of the photoemission process and other issues like photoelectron diffraction [70].

3.2.2 Role of Angular Selection in Photoemission

Up to now we have neglected the wave vector **k** of the photoelectrons in our interpretation of the above dichroism. In fact, both experiments [71–73] and theoretical considerations on the basis of symmetry arguments [74,75] indicated that the dichroic signal may vary strongly with the emission angle of the outgoing photoelectron. Therefore, the effect should be more precisely called *magnetic dichroism in the angular distribution of photoelectrons (MDAD)*. In addition, the symmetry considerations also show that the observation of a magnetodichroic signal is not bound to the use of circularly polarized light. A magnetic dichroism will appear also with linearly and even unpolarized light. The microscopic origin of the magnetodichroic signal is traced back to the interference of different excitation channels of the outgoing photoelectron. This is the reason why MDAD can also be described in the framework of photoelectron diffraction [73,76].

In order to see this interference character without going into details, we may look into some of the symmetry-based derivations made by Venus in Ref. [74]. The general form of MDAD may be written as

$$D^{\pm\sigma}(\mathbf{k},\mathbf{q},\mathbf{M},\mathbf{n}) = \left|\left\langle \psi_f(\mathbf{k},\mathbf{n},\mathbf{M})|O^+(\mathbf{q})|\psi_i(\mathbf{k}_\parallel,\mathbf{n},\mathbf{M})\right\rangle\right|^2 - \left|\left\langle \psi_f(\mathbf{k},\mathbf{n},\mathbf{M})|O^-(\mathbf{q})|\psi_i(\mathbf{k}_\parallel,\mathbf{n},\mathbf{M})\right\rangle\right|^2, \tag{12}$$

whereby ψ_i and ψ_f denote the initial and final state wave functions, respectively, being a function of the wave vector **k**, magnetization **M**, and the surface normal **n**. The quantity O^+ (O^-) denotes the operator for circularly polarized light of positive (negative) helicity and depends on the photon wave vector **q**. If we assume that **k**, **n**, and **M** are coplanar and lie in a mirror plane of the surface, and we allow for elliptically polarized light[3] at an arbitrary angle of incidence, the previous equation can be expressed as

$$D^{\pm\sigma}(\mathbf{k},\mathbf{q},\mathbf{M}) = -2\mathrm{Im}\left[\langle\mathbf{M}|O^1(\mathbf{q})|\mathbf{k}\rangle\,\langle\mathbf{k}|O^2_\perp(\mathbf{q})|\mathbf{M}\rangle + \langle\mathbf{M}|O^1(\mathbf{q})|\mathbf{k}\rangle\,\langle\mathbf{k}|O^2_\parallel(\mathbf{q})|\mathbf{M}\rangle\right]. \tag{13}$$

In this equation, the elliptically polarized light is represented by a coherent superposition of two orthogonal linearly polarized operators O^1 and O^2. Whereas the linear polarization of O^1 can be chosen parallel to the mirror plane containing **k**, **M**, and **n**, the light operator O^2 has components perpendicular ($O^2_\perp$) and parallel to the mirror plane $\left(O^2_\parallel\right)$. The structure of the terms in Eq. (13) clearly reflects the interference of matrix elements related to cross-terms of the dipole operators O^1 and O^2.

As a further result of these considerations, one finds that for an arbitrary angle of incidence, a reversal of the magnetization direction ($M^+ \Rightarrow M^-$) is generally not equivalent to a reversal of the light helicity ($\sigma^+ \Rightarrow \sigma^-$). It is important to note that this equivalence is only obtained for selected experimental geometries where all four vectors (**k**, **q**, **M**, **n**) are coplanar. Therefore, these coplanar

[3] In a realistic experiment, the light inside the crystal will always have some ellipticity, i.e., some linearly polarized contribution, which can be calculated from Fresnel's equations.

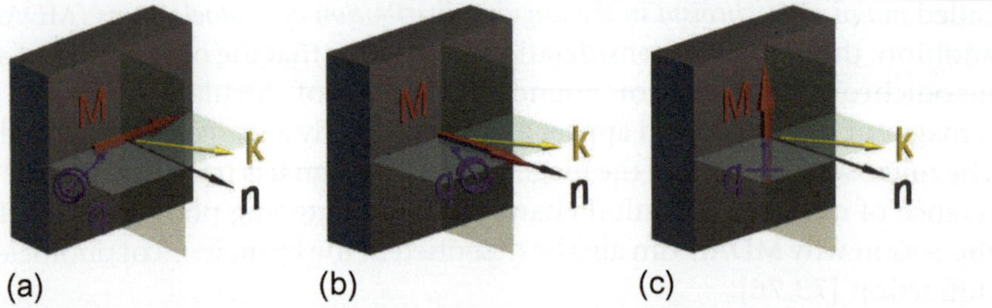

FIGURE 14

Sketch of preferable geometries for MDAD experiments probing different magnetization directions derived from the discussions in Ref. [74]. (a) MDAD probes an in-plane magnetization for grazing incidence of circularly polarized light. (b) A perpendicular magnetization may be probed by MDAD with c.p. light at near-normal incidence. (c) An in-plane magnetization perpendicular to the reaction plane can be probed by linearly polarized light with an electric field vector parallel to the reaction plane (TMOKE geometry).

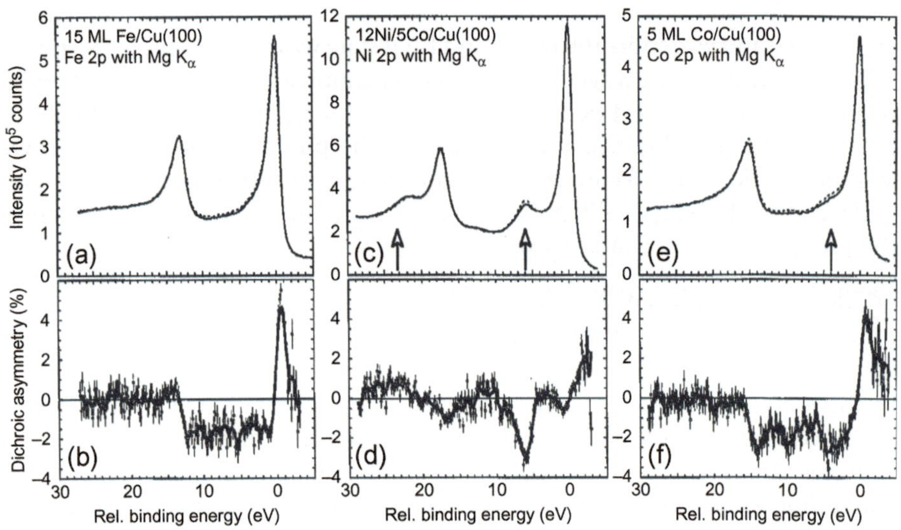

FIGURE 15

Photoelectron intensity distribution curves (a) and magnetic linear dichroism signal upon magnetization reversal (b) for the excitation of the Fe 2p core levels with unpolarized Mg K_α radiation. Panels (c) and (d) give the corresponding results for Ni 2p, and panels (e) and (f) those for Co 2p (arrows mark the position of correlation-induced satellites). *From Ref. [78].*

geometries are preferred in experiments and can be optimized for a given magnetization direction and light polarization (Fig. 14).

As has been mentioned above, magnetodichroic effects in photoemission can also be observed with linearly polarized light [77]. It should be noted that unpolarized light can be seen as an incoherent superposition of s- and p-polarized contributions, but only the p-polarized component of the incident light creates a magnetodichroic signal (see also Fig. 16). As a consequence, a linear magnetic dichroism in photoemission also appears with unpolarized light sources—in contrast to the situation in an absorption experiment (Section 4). The amplitude of the magnetic dichroism with unpolarized light, however, is at most 50% of that measured with linearly polarized light. In Fig. 15, we compile the magnetic dichroism spectra obtained with unpolarized light (Mg K_α radiation) upon magnetization reversal ($M^{\Uparrow} \rightleftharpoons M^{\Downarrow}$) from Fe, Co, and Ni films [78]. In each case, we observe a sizable magnetodichroic asymmetry upon magnetization reversal. The result for Fe agrees closely with that obtained with circularly polarized light. For Ni, we clearly observe the presence of the correlation satellites in the intensity spectra (Fig. 15d). The magnetic dichroism spectrum, however, exhibits the strongest contribution at the position of the 6 eV satellites and only rather small dichroism at the main spectral lines. This finding agrees nicely with multiple electron calculations [64] and is caused by the spin dependence in the screening processes (cf. Section 3.1.3). For Co, there has also been a discussion of correlation-induced satellites, although their effect is expected to be considerably weaker. The data in Fig. 15e exhibit a small shoulder on the high binding energy side of the $2p_{3/2}$ photoemission line, which is connected to some modulation of the dichroic asymmetry in Fig. 15f. This may be an indication of the presence of a correlation-induced satellite also in Co metal.

For the above case, O[1] the transition metal 2p edges, the level splittings due to spin–orbit coupling are much larger than the respective m_J splittings caused by the exchange coupling, which can then be considered as a perturbation. In the transition metal 3p levels, the energetic situation is more complicated as the level splittings induced by spin–orbit coupling and exchange interactions

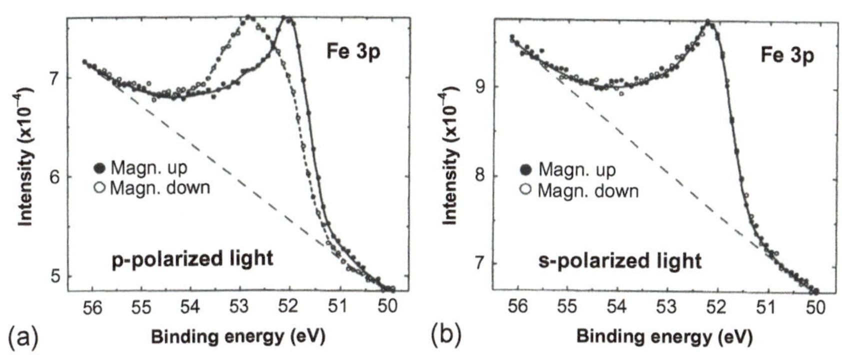

FIGURE 16
Magnetic linear dichroism measured at the Fe 3p levels at a photon energy of 1253.6 eV. The comparison shows that the dichroism disappears for s-polarized light. *From Ref. [77].*

are of the same order of magnitude. The interpretation of magnetodichroic phenomena in photoemission is therefore more involved [79]. Nevertheless, the experiments produce clear MDAD signatures as evidenced, for instance, in Fe 3p photoemission (Fig. 16) [77]. These spectra taken with p-polarized light exhibit a clear magnetodichroic signal upon magnetization reversal, which disappears for s-polarized light (Fig. 16b). The latter observation corroborates our discussion on the use of unpolarized light above.

Magnetic systems containing 4f electrons represent a particularly interesting case for spectroscopic studies. Although the binding energy of the 4f levels in rare earth elements is of the order ~10 eV, their interaction with the valence bands is weak. The 4f states are therefore localized and behave spectroscopically like core levels. Nevertheless, the magnetism in rare earth elements is carried by the 4f states, which means that both spin–orbit and exchange coupling in these states may be strong. As a consequence, we should expect strong magnetodichroic signals in photoemission. This is indeed observed for the 4f photoemission with circularly polarized light from gadolinium in Fig. 17. The sample

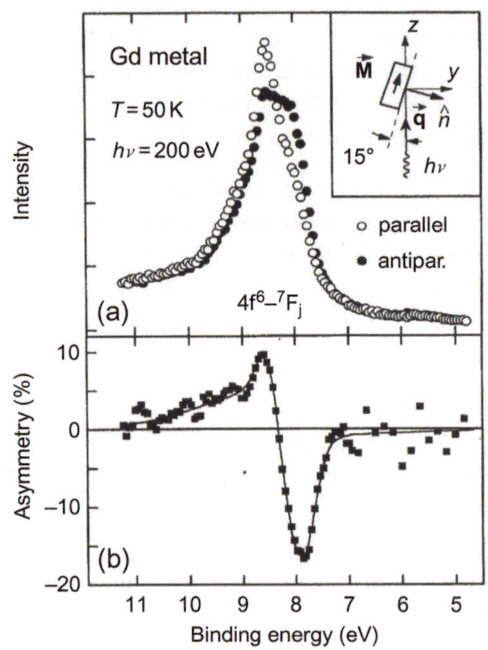

FIGURE 17

MCDAD at the 4f levels of Gd metal upon magnetization reversal with respect to the photon wave vector **q**. The light of $h\nu = 200$ eV photon energy has a degree of circularity of $pC = 55\%$. (a) Intensity spectra for magnetization parallel and antiparallel to the light incidence. (b) Intensity asymmetry calculated from the spectra in (a). Inset: Experimental geometry. *Taken from Ref. [80].*

consisted of an 8-nm-thick Gd(0001) film epitaxially grown on W(110) [80]. Because the Curie temperature of bulk Gd is only $T_C = 292$ K, the experiment has been carried out at $T_C = 50$ K, which ensured a stable in-plane magnetization. The circular magnetic dichroism is determined upon magnetization reversal; however, the degree of circularity was only $p_C = 55\%$. The two intensity spectra $I^+(E)$ and $I^-(E)$ differ markedly from each other, but also from the paramagnetic case [81]. Despite the limited circularity in the experiment, the intensity asymmetry reaches values of up to −20%. The photoemission spectrum from Gd can be described within a multiplet approach, assuming a $4f^6 - {}^7F_J$ final-state multiplet [80]. In this model, the transition probabilities for the $J_f = 0\ldots6$ sublevels can be calculated. For parallel orientation of **M** and **q**, the transition probability successively increases up to the $J_f = 6$ sublevel, whereas for the antiparallel orientation ($\mathbf{M} \upharpoonleft\downharpoonright \mathbf{q}$) the transition probability peaks for the $J_f = 3$ sublevel. As a consequence, in the $\mathbf{M} \upharpoonleft\upharpoonright \mathbf{q}$ ($\mathbf{M} \upharpoonleft\downharpoonright \mathbf{q}$) the total spectral weight is shifted to the high (low) binding energy flank of the photoemission spectrum, in agreement with the experimental findings of a plus–minus feature in the intensity asymmetry.

The MDAD experiments discussed so far provide only magnetic information about the near-surface regime. This comes about by the small electron IMFP at low and moderate photoelectron kinetic energies related to the excitation in the soft X-ray regime. In order to address bulk properties or buried layers, a larger electron IMFP is needed. As already discussed in Section 2.3, this can be achieved by employing hard X-ray excitation, but the photoexcitation cross-sections drop by two to three orders of magnitude when moving from soft to hard X-ray radiation [82]. Spin-polarized photoemission with the present type of spin polarimeters is not possible under these circumstances. Therefore, the question arises whether or not sizable magnetodichroic effects will appear in the hard X-ray regime.

Experiments prove that magnetodichroic phenomena can indeed be observed in hard X-ray photoemission [83]. An example for MDAD at the Co 2p levels is shown in Fig. 18. The sample consists of a complex magnetic layer stack in which the topmost magnetic layer consists of 3 nm of CoFe, which is capped by 2 nm of MgO and 1 nm of AlO_x. The photoelectrons from the Co 2p levels must therefore travel at least 3 nm before they can leave the sample. The data were recorded with circularly polarized light at a photon energy of $h\nu = 7.94$ keV, impinging on the surface under a grazing angle of $\theta = 2°$. We find the typical spectral features at the $2p_{3/2}$ and $2p_{1/2}$ photoemission lines. In remarkable contrast to the soft X-ray experiments, the absolute number of the dichroic asymmetry reaches large values in excess of 40%. Similar results have been obtained for the Fe 2p and shallow 3p levels for both elements Co and Fe [83]. An interesting detail in the data in Fig. 18 is the shoulder at

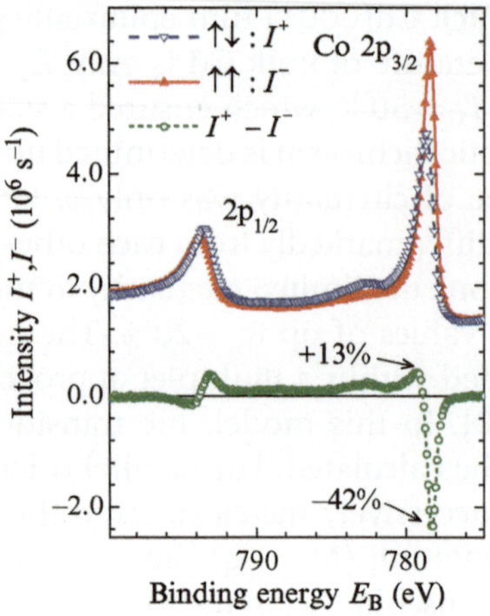

FIGURE 18

Magnetization-dependent photoelectron spectra of the Co 2p core levels from CoFe on top of an IrMn exchange-biasing layer taken with circularly hard X-rays at a photon energy of 7.94 keV. The bottom curve depicts the plain difference of the two intensity spectra taken for **M** 1 ↾ **q** (I^-) and **M** 1 ⇂ **q** (I^+). Extreme values for the intensity asymmetry are marked at selected energies. *From Ref. [83].*

a binding energy of $E_B \approx 783$ eV, which also exhibits a clear magnetic dichroism. This may be another indication of a correlation-induced satellite as mentioned in the context of Fig. 15. These findings demonstrate that even in the hard X-ray photoemission (HAXPES) regime the magnetic dichroism in the angular distribution can be conveniently employed to obtain element-selective information on the magnetic state of a sample. This approach opens up new opportunities in the element-selective study of buried magnetic layers.

4 MAGNETIC DICHROISM IN X-RAY ABSORPTION

Following the notation from Section 3.2, we also call the intensity variation in an absorption spectrum upon magnetization or light polarization a magnetic dichroism, namely X-ray magnetic circular or linear dichroism (XMCD or XMLD). Magnetodichroic phenomena in XAS have found widespread use in the study of magnetism during the last 20 years [84]. In addition to element specificity they also provide the unique advantages of being able to address not only ferromagnetically but also antiferromagnetically ordered systems, and enable studies with high magnetic sensitivity. The latter is particularly important when investigating nanostructures with a very small amount of magnetic material.

4.1 Magnetic X-Ray Circular Dichroism

The first XMCD (magnetic X-ray circular dichroism) studies were reported for the K-edges of iron [11]. In this case, we need circularly polarized hard X-rays of 7.5 keV photon energy to observe the effect. In contrast to the photoemission case, the magnetic dichroism is expressed as the *difference* of two absorption spectra obtained with left- and right-hand circularly polarized light or opposite magnetization directions, i.e.,

$$I_{\mathrm{MCD}}(\hbar\omega) = I_{\sigma^+}(\hbar\omega) - I_{\sigma^-}(\hbar\omega) = I_{M^+}(\hbar\omega) - I_{M^-}(\hbar\omega). \tag{14}$$

This can also be written as the difference in the absorption cross-sections for parallel and antiparallel orientations of light helicity σ and magnetization **M**:

$$\Delta\mu(\hbar\omega) = \mu_{\downarrow\uparrow}(\hbar\omega) - \mu_{\uparrow\uparrow}(\hbar\omega) \tag{15}$$

The result of such an experiment on an Fe–Si alloy sample is shown in Fig. 19 [85]. The Fe K-edge gives rise to a step-like spectrum (upper panel) with some fine structure. The absorption process effectively integrates over all photoelectron wave vectors inside the sample; therefore in most cases only, the magnetization **M** and the light helicity σ play a role in angular variations of XMCD. In order to optimize the dichroic signal the vectors **M** and σ (respectively **q**) should be aligned, as

$$I_{\mathrm{MCD}} \sim \sigma \cdot \mathbf{M} \cdot (\mathbf{q}/|\mathbf{q}|). \tag{16}$$

The resulting magnetic dichroism spectrum shows a clear plus–minus feature at the energy position of the edge, which can be related to the ferromagnetic state of the sample. The absolute value of the K-edge dichroism, however, is only of the order of $\Delta\mu(\hbar\omega) \sim 10^{-4}$. The reason for this will be discussed below in the context of the L-edge magnetic circular dichroism. Because of the large penetration depth of the hard X-rays the K-edge XMCD is very useful to study bulk samples and deeply buried structures, in particular in combination with a theoretical description of the spectra (solid lines in Fig. 19).

A few years after the discovery of magnetic circular dichroism at the Fe K-edges, circularly polarized soft X-rays also became available. In a key experiment on the Ni $L_{2,3}$-edges using TEY detection, a significant XMCD in absorption was verified [86]. In fact, this XMCD signal exceeded the K-edge results by more than two orders of magnitude, reaching magnetization-induced changes of >10% in the spectra. An example for the L-edge XMCD from cobalt is shown in Fig. 20 [87]. The sample was a 15-nm-thick $Co_{50}Pd_{50}$ film with perpendicular anisotropy. The characteristic "white lines" in the Co L-edge absorption spectrum exhibit a strong dependence on the magnetization direction for a fixed light helicity. In particular, the changes at the L_3- and L_2-edges have the opposite sign which gives rise to

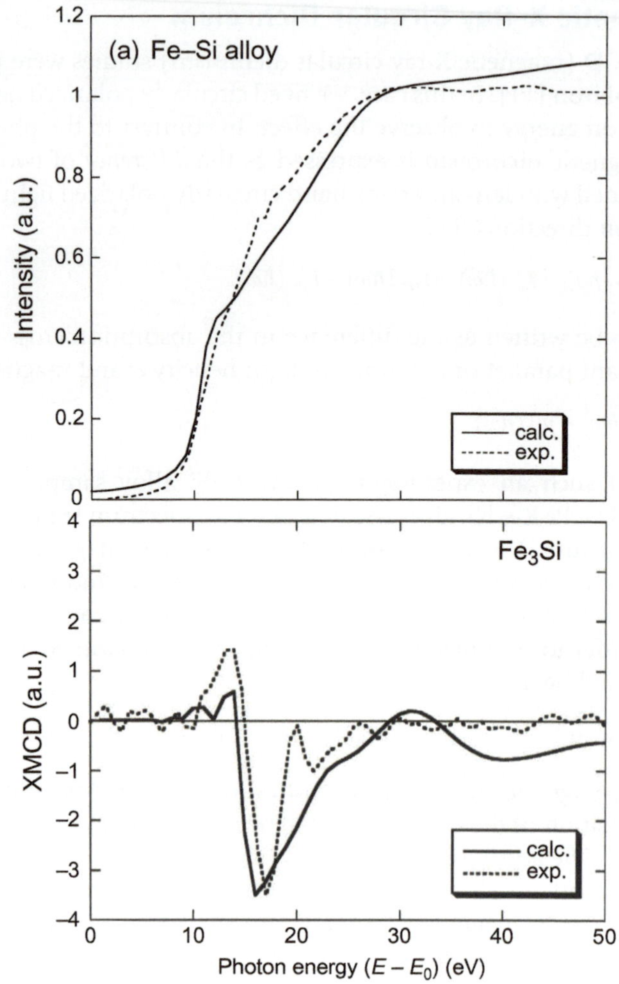

FIGURE 19

Magnetic circular dichroism at the Fe K-edge. The reference photon energy is $E_0 = 7130$ eV. Top: Intensity spectrum. Bottom: XMCD spectrum. The experiment (dotted line) is compared to a theoretical calculation (solid line). *From Ref. [85].*

a characteristic shape of the magnetic dichroism spectrum exhibiting a negative (positive) excursion at the L_3 (L_2)-edge. The opposite sign for the dichroism at the L_3 ($2p_{3/2}$)- and L_2 ($2p_{1/2}$)-edges is a consequence of the spin–orbit coupling in the core states.

In contrast to the dichroism in photoemission discussed in Section 3.2, the magnetic dichroism in X-ray absorption has an important advantage.

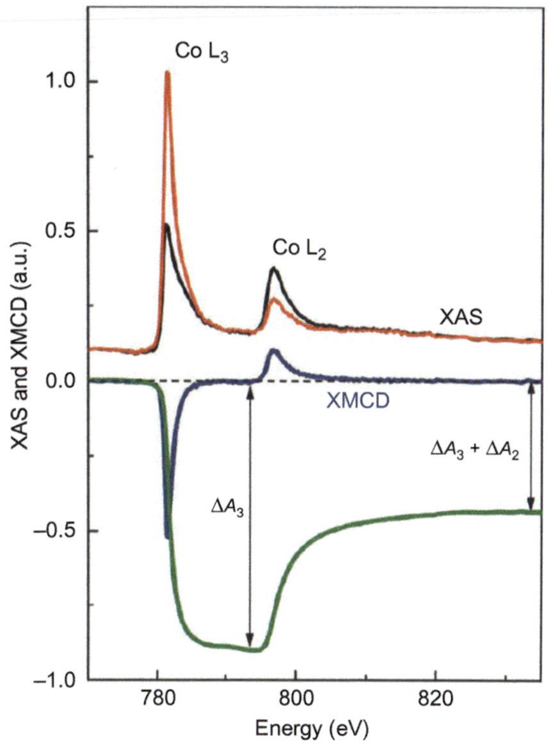

FIGURE 20
X-ray absorption spectra of the Co $L_{2,3}$-edges in a CoPd film using circularly polarized light. The spectra for opposite magnetization directions (top panel) reveal a strong difference, which is plotted separately in the bottom panel. This magnetic X-ray dichroism spectrum shows characteristic structures at the position of the Co L_3- and L_2-edges. The green curve in the bottom panel shows the result of a moving integration of the dichroic signal and defines the quantities ΔA_2 and ΔA_3 required for the evaluation of the sum rules. *After Ref. [87].*

The data can be analyzed in a relatively simple approach to yield quantitative values for the atomic magnetic moment using "sum rules" [88–90]—at least within certain limits [91]. For the transition metal $L_{2,3}$-edges, the sum rules can be expressed for the orbital magnetic moment μ_L (in units of a Bohr magneton μ_B) as

$$\mu_L[\mu_B] = -\frac{2N_h}{N}\int\left(\Delta\mu_{L_3}(E) + \Delta\mu_{L_2}(E)\right)dE \equiv -\frac{2N_h}{N}(\Delta A_3 + \Delta A_2) \tag{17}$$

and for the spin magnetic moment μ_S as

$$\mu_S[\mu_B] = -\frac{3N_h}{N}\int\left(\Delta\mu_{L_3}(E) - 2\Delta\mu_{L_2}(E)\right)dE + 7\langle T_z\rangle \equiv -\frac{2N_h}{N}(\Delta A_3 - 2\Delta A_2) + 7\langle T_z\rangle, \tag{18}$$

with N being

$$N=\int_{L_3+L_2}(\mu_+(E)+\mu_-(E)+\mu_0)\mathrm{d}E, \tag{19}$$

a normalization to the integrated spectrum obtained with unpolarized light. The absorption cross-section $\mu(E)$ should not be confused with the magnetic moments μ_L and μ_S. The quantities ΔA_3 and ΔA_2 denote the dichroic signal integrated over the L_3- and L_2-edges, respectively. The green line in Fig. 20 marks the development of the integration across the edges for the quantity which is proportional to the orbital moment μ_L. The photoexcitation from the core-level electron takes place into the unoccupied density of states—more precisely the unoccupied d states—below the Fermi level. Therefore, the strength of the transition is directly determined by the number of empty d states (holes) N_h, which consequently enters Eq. (18). The quantity $\langle T_z \rangle$ describes an asphericity of the spin magnetic moment. A critical issue is the definition of the integration limit between the L_3- and L_2-edges. In the case of the CoPd alloy, the spin magnetic moment for Co is found to be $\mu_{S,Co}=0.78\mu_B$, whereas the orbital moment takes a value of $\mu_{L,Co}=0.24\mu_B$ [87]. These values differ considerably from bulk cobalt, where values of $\mu_{S,Co}=1.55\mu_B$ and $\mu_{L,Co}=0.15\mu_B$ have been found by the same method [92]. The large orbital moment in comparison to cubic systems is due to the uniaxial magnetic anisotropy in hexagonal bulk cobalt. In the CoPd film, it is the strong perpendicular magnetic anisotropy (PMA) which leads to the large orbital moment.

The microscopic picture behind the XMCD at the $L_{2,3}$-edges is illustrated in Fig. 21, analyzing the dipole-allowed transitions between the 2p levels and the 3d states. For reasons of simplicity, we assume the direction of light propagation and the magnetization vector to be collinear. The most important ingredient leading to a large magnetic dichroism is the spin–orbit splitting in the 2p states. The excitation of the $2p_{3/2}$ and $2p_{1/2}$ states with circularly polarized light generates spin-polarized photoelectrons due to optical spin orientation. The quantization axis of the spin polarization P is defined by the helicity vector (parallel or antiparallel to the propagation direction) of the light. The sign of the spin polarization is opposite for the two spin–orbit split states, and the magnitude of P differs. The spin-polarized photoelectrons are excited into the unoccupied density of states which is spin-split due to the exchange interaction in the ferromagnet. If we take a case similar to nickel, the majority spin d states are completely occupied and there are only empty d states of minority spin character. The probability for a transition from a 2p sublevel, i.e., the strength of the dipole matrix element, will thus be strongly spin-dependent and thus also helicity-dependent, as can be seen by the encircled numbers in Fig. 21. Transitions which do not take place due to the wrong spin character are marked by a red "X." Adding up the sublevel contributions for the

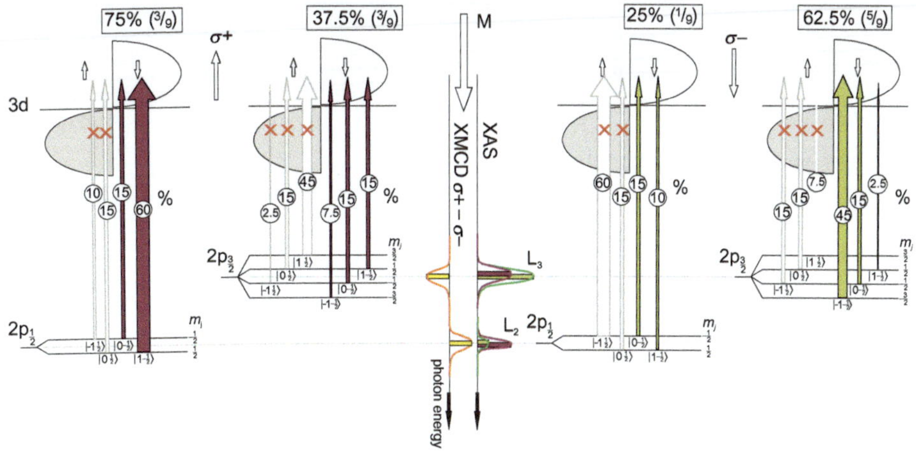

FIGURE 21

Origin of the XMCD for the transition metal L-edges. We assume a nickel-like case (only empty spin-down d states) and a collinear alignment of light propagation direction and magnetization vector. The left- and right-hand panels depict the possible dipole-allowed transitions and their respective strengths (encircled numbers) for σ^+ and σ^- radiation. Transitions marked by a red "X" do not take place due to spin selection by the empty states. The contributions of all sublevels are added up and represented by Gaussians to yield the schematic XAS spectra along the vertical center line. The difference of the spectra for opposite helicity reflects the magnetic dichroism. Although the m_J sublevels of the core states have been drawn split due to the exchange interaction, this splitting is not necessary to understand the salient features of XMCD.

L_3- and L_2-edges reveals quite different transition probabilities for opposite helicities (L_3: 75% vs. 25%; L_2: 37.5% vs. 62.5%), which directly lead to magnetic dichroism.

As observed in experiments the magnitude of the dichroism at the L_2-edge is smaller and changes sign with respect to the L_3-edge value. For reasons of consistency, we have drawn the sublevels of the core states in a spin-split fashion; however, this splitting is not required to understand the occurrence of XMCD. This is different for magnetic X-ray linear dichroism, as we will see below.

The combination of chemical and magnetic information is demonstrated in Fig. 22, which shows the XMCD spectrum from an Fe_3O_4 thin-film sample. The intensity distribution reveals a clear multiplet structure at both the L_2 and L_3 absorption edges and clear differences upon change of the light helicity. The resulting magnetic dichroism spectrum is characterized by a rich substructure with a pronounced "W"-shaped feature at the L_3-edge. This "W" feature is a characteristic fingerprint of magnetite and can be used to identify the type of iron oxide in a sample. The origin of the spectral structure is caused by the

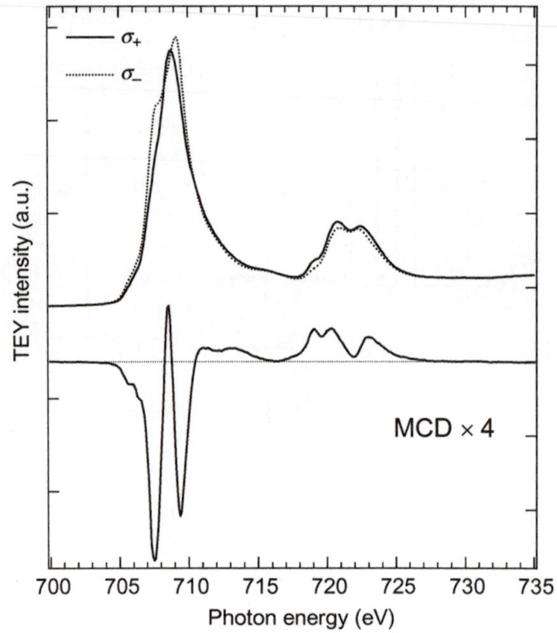

FIGURE 22

X-ray absorption spectra for a magnetized Fe_3O_4 thin film recorded at opposite light helicities (σ^+ σ^-, top) and resulting XMCD spectrum (bottom). The data have been acquired in the total electron yield (TEY) mode. *From Ref. [93].*

different Fe ions in magnetite. We find Fe^{2+} and Fe^{3+} in an octahedral environment, but also Fe^{3+} in a tetrahedral environment. Each type of ion contributes in a characteristic manner to the magnetic dichroism spectrum, as is depicted in Fig. 23. It compiles calculated XMCD spectra for the individual ions and compares them to the spectrum from an Fe_3O_4 bulk crystal. As we can see, the two dips in the dichroism at the L_3-edge are caused by the iron ions in octahedral configuration, whereas the peak stems from the Fe^{3+} ion in a tetrahedral configuration.

The ultimate challenge in nanomagnetism is the characterization of magnetic elements approaching the size of clusters or even individual atoms. Due to the high sensitivity of the resonant X-ray excitation, magnetic dichroism in X-ray absorption is the appropriate experimental approach. It enables the determination of magnetic coupling phenomena and magnetic moments from monoatomic layers and one-dimensional atomic chains [95] down to the single molecule [96] and atom level [97]. In order to be able to study the electronic states in atoms they must be supported by a template. Therefore, the electronic interaction between atom and template must be considered and should be minimized by a clever choice of template. An example for XMCD studies from single atoms of Fe, Co, and Ni is given in Fig. 24. The atoms have been

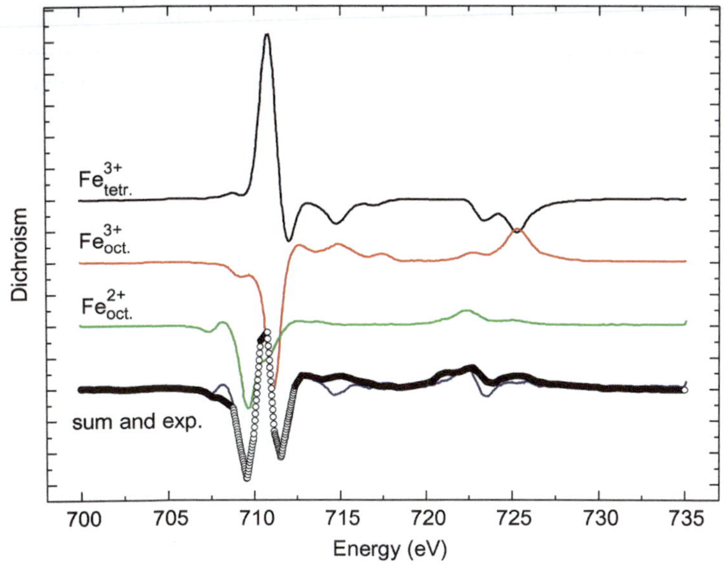

FIGURE 23
Experimental XMCD spectrum from an Fe_3O_4 bulk crystal compared to spectra calculated for Fe^{2+} and Fe^{3+} ions in tetrahedral and octahedral bonding configurations. *From Ref. [94].*

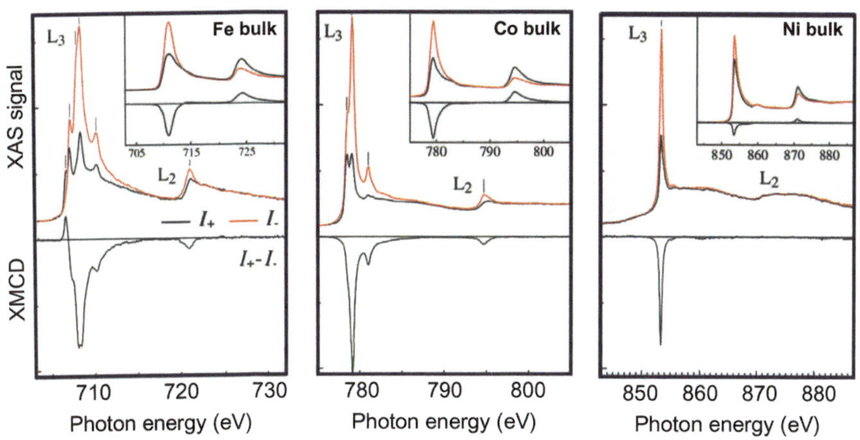

FIGURE 24
XAS and XMCD spectra at the $L_{2,3}$-edges recorded for isolated Fe, Co, and Ni atoms on a potassium surface (corresponding coverages less than 0.02 monoatomic layers). For comparison, the insets reproduce the characteristic smooth XMCD spectra for metallic bulk Fe, Co, and Ni. *From Ref. [97].*

deposited on potassium surfaces and the spatial orientation of the magnetic moments has been obtained by a high magnetic field which was applied during the measurements. A comparison with the metallic bulk data (see insets) reveals considerable differences in the X-ray absorption spectra, which are caused by very pronounced atomic multiplet structures. These multiplet structures indicate that there is only a weak electronic interaction between the orbitals of the 3d metal adatoms and the free-electron-like potassium template, and that the atomic wave functions are localized. From the spectra one can determine both the valence state and the magnetic moment of the individual atoms. One important finding is a very high orbital magnetic moment reaching values of up to $\mu_{Co,L} \sim 0.8\mu_B$ in the case of Co atoms [97]. This is a consequence of the electronic localization of the individual atom. Increasing the coverage also increases the electronic interaction between the magnetic atoms and leads to a subsequent quenching of the orbital magnetic moment in the magnetic clusters.

Molecular structures carrying a magnetic moment present very interesting physical properties and may be useful for future applications in molecular spintronics or quantum information processing. A single molecular magnet often consists of one or several metal atom centers which carry the magnetic moment and are encaged by a ligand structure. A simple case is shown in Fig. 25, which compiles XAS and XMCD results for a molecule with a single transition metal ion as magnetic center [96]. In the depicted porphyrin molecule, an Fe atom is surrounded by a planar arrangement of ligands with a fourfold symmetry. For sterical reasons, the interaction of such a molecule with its environment will take place mostly through the ligands. With respect to nanomagnetism an important question concerns the exchange coupling of such a magnetic molecule to a ferromagnetic element or electrode. In the experiment by Wende *et al.* [96], the Fe-porphyrin molecules were deposited onto an Ni film with a remanent perpendicular magnetization. On the Fe absorption edges which are related to the Fe-carrying molecule a clear XMCD signal is observed even at room temperature. By comparison to the Ni XMCD, the coupling character can be easily determined and is shown to be of ferromagnetic type. Detailed theoretical analyses are needed to disentangle the contributions of the different molecular orbitals to the exchange coupling. As a result, one finds the electronic hybridization of both ligand and central iron atom to be involved in the coupling [96]. The strength and the character of the exchange coupling depend sensitively on the molecular magnet and the details of the surface. Subsequent studies have shown that the insertion of oxygen atoms between molecular magnet and surface can reverse the coupling character to an antiferromagnetic type [98].

4.2 Magnetic X-Ray Linear Dichroism

In the simplest case of an antiferromagnet, the magnetic moments on neighboring lattice sites are oriented antiparallel to each other. Therefore, the crystal

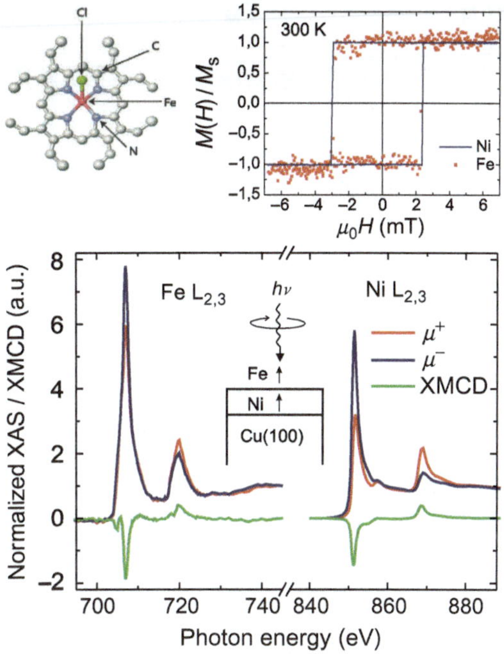

FIGURE 25

Top left: Schematic illustration of the Fe-OEP-chloride molecule. Top right: Element-specific hysteresis loops of the Fe center (filled squares) and Ni film (full line) obtained by following the L_3-edge XMCD maxima of Fe-OEP on Ni/Cu(100) at 300 K. Bottom: Normalized X-ray absorption spectra for right and left circularly polarized light $\mu^+(E)$ (red) and $\mu^-(E)$ (blue), and resulting XMCD (green) at the $L_{2,3}$-edges of Fe and Ni (300 K, 10 mT). The inset depicts the experimental geometry. The arrows for the Fe center and the Ni film show the alignment of the respective magnetic moments. *After Ref. [96].*

does not have a net magnetization **M** and circular magnetic dichroism will disappear. Nevertheless, there is still a well-defined quantization axis of the local magnetic moment which can be used for spectroscopic purposes. The element-selective magnetic spectroscopy of antiferromagnets involves linearly polarized light and was demonstrated first by van der Laan *et al.* [10].

In an XMLD experiment, one compares the absorption spectra obtained at excitations with light of orthogonal linear polarization. A schematic picture of the microscopic mechanisms behind the X-ray linear dichroism at the transition metal $L_{2,3}$-edges are given in Fig. 26. We compare the transition probabilities for the electrical field vector **E** of the linearly polarized light parallel and perpendicular to the spin quantization or alignment axis **m**. It is important to note that adding up the transition probabilities for a given edge, for example the L_2-edge, yields exactly the same values for $\mathbf{E} \parallel \mathbf{m}$ and $\mathbf{E} \perp \mathbf{m}$. So in contrast to XMCD, there is no dichroism in the total transition probability! The linear dichroism appears only if an exchange splitting of the core states is taken into account. This causes the transitions from the sublevels to take place at slightly

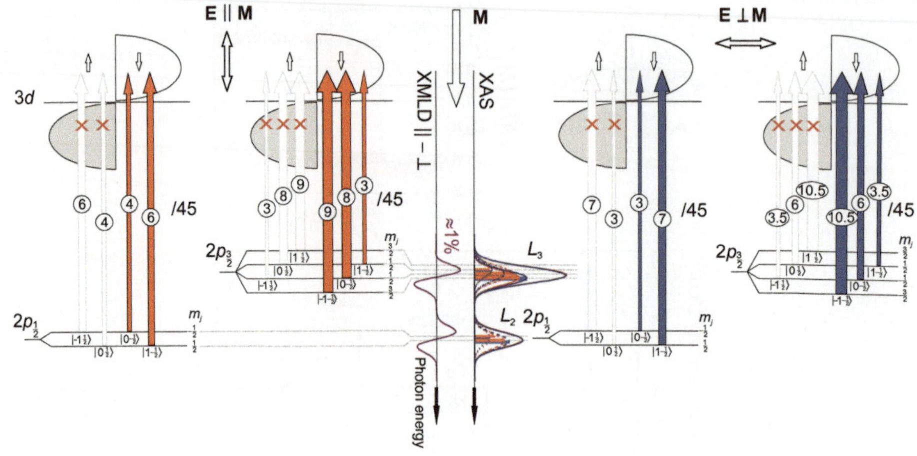

FIGURE 26

Origin of the XMLD for the transition metal L-edges. Again we assume a nickel-like case (only empty spin-down d states). The left- and right-hand panels depict the possible dipole-allowed transitions and their respective strengths (encircled numbers) for the orientation of the electric field vector **E** of the linearly polarized light **E** || **m** and **E** ⊥ **m** (note that **m** indicates the antiferromagnetic spin alignment axis in this case). Transitions marked by a red "X" do not take place due to spin selection by the empty states. The contributions of all sublevels are added up and represented by Gaussians to yield the schematic XAS spectra along the vertical center line. The difference of the spectra for orthogonal orientations of **E** reflects the magnetic dichroism. Note that the spin splitting of the m_J sublevels of the core states is now crucial to allow for a nonzero XMLD.

different photon energies. As a consequence, there is a small change of the absorption line profiles as a function of the light polarization, which results in characteristic bipolar features in the difference or magnetic dichroism spectrum. Compared to the magnetic circular dichroism at the same edges, the XMLD signal almost looks like a derivative of the XMCD spectrum [99]. We should note that an XMLD signal can be observed from *both* ferromagnets and antiferromagnets, because it requires only a spin *alignment* along an axis, but no spin *orientation* along a direction in space.

The angular variation of the XMLD signal with respect to the angle of incidence θ is slightly more complicated and takes the form:

$$\Delta I_{\text{XMLD}} \sim \langle m^2 \rangle \left(3\cos^2\theta - 1\right) \qquad (20)$$

under the assumption of a cylindrical asymmetry in the experiment [100]. The expression $\langle m^2 \rangle$ denotes the square of the local magnetic moment along the spin alignment axis averaged over antiferromagnetic domains. This approach is suitable to describe experiments in polycrystalline samples, in which we average over all possible orientations of **m**. As we will discuss below, this assumption may break down in experiments with single crystals and particularly when individual domains are concerned, because by definition an antiferromagnetic domain has only one spin alignment axis **m**. This singular axis breaks the cylindrical symmetry of the experiment.

One of the classical antiferromagnets is NiO in which superexchange across the oxygen ions leads to an antiferromagnetic spin arrangement of neighboring Ni atoms. In fact, NiO exhibits a layered antiferromagnetic order along the crystallographic $\langle 111 \rangle$ direction, i.e., a stacking of lattice planes with in-plane ferromagnetic order, but alternating spin orientation from layer to layer. As an interesting property of NiO the strongest XMLD signal appears in the atomic multiplet structure at the L_2-edge [100]. The data in Fig. 27a have been recorded in the TEY mode and reveal a strong dependence of the second peak at the Ni L_2-edge on the orientation of the light polarization, which is varied from parallel ($\theta = 0°$) to perpendicular to the surface normal ($\theta = 90°$). The spectra are normalized to the intensity of feature A; the variation of feature B with increasing angle θ marks the XMLD signal.

The magnetic origin of these spectral changes can be proven via temperature-dependent measurements. In Fig. 27b, the geometry has been fixed at normal incidence ($\theta = 90°$), for which the spectral feature B shows the lowest intensity. Upon raising the temperature, the intensity of feature B steadily grows until the

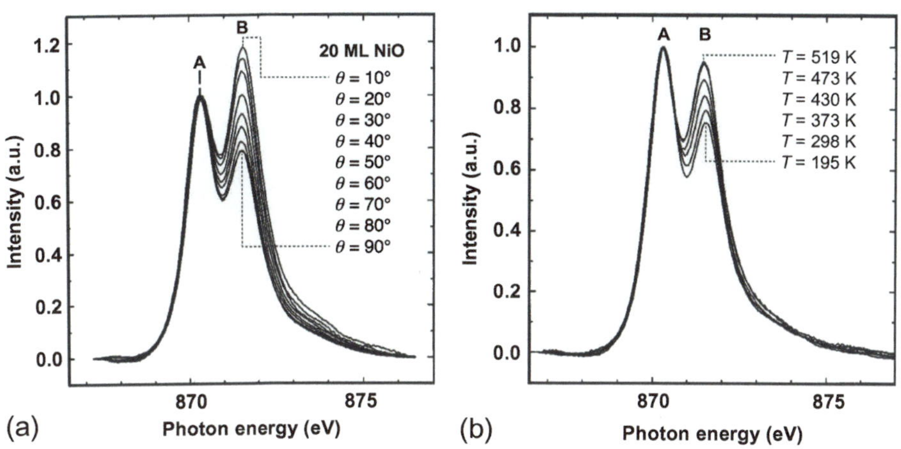

FIGURE 27

(a) Magnetic linear X-ray dichroism (XMLD) at the Ni L_2-edge. The sample is a 20-monolayer NiO(001) film grown on MgO(001). θ denotes the angle between the linear polarization direction and the surface normal. All spectra are normalized to feature "A." With increasing θ the intensity of the feature "B" drops.

(b) Temperature dependence of the XMLD at the Ni L_2-edge. The temperature is varied across the Néel temperature of the NiO film ($T_N \approx 470$ K). *Adapted from Ref. [101].*

Néel temperature of the NiO film ($T_N \approx 470$ K) is reached. Above the Néel temperature, there is no further increase of the intensity of feature B, demonstrating that this behavior is associated with the loss of the antiferromagnetic order in the crystal.

4.3 Hybrid Systems

X-ray magnetodichroic phenomena are uniquely suited to investigate complex magnetic structures, for example, combinations of ferro- and antiferromagnets. In this way, exchange coupling effects such as exchange biasing [102,103] can be studied in great detail. In the following we will illustrate the wealth of information that can be obtained from a magnetic hybrid system by a specific example. It concerns a single-crystalline magnetic oxide stack consisting of NiO films grown on Fe_3O_4 surfaces [104,105].

The data in Fig. 28 have been obtained by X-ray photoemission microscopy and reveal the magnetic domain structure in the NiO/Fe_3O_4 (110) system. The

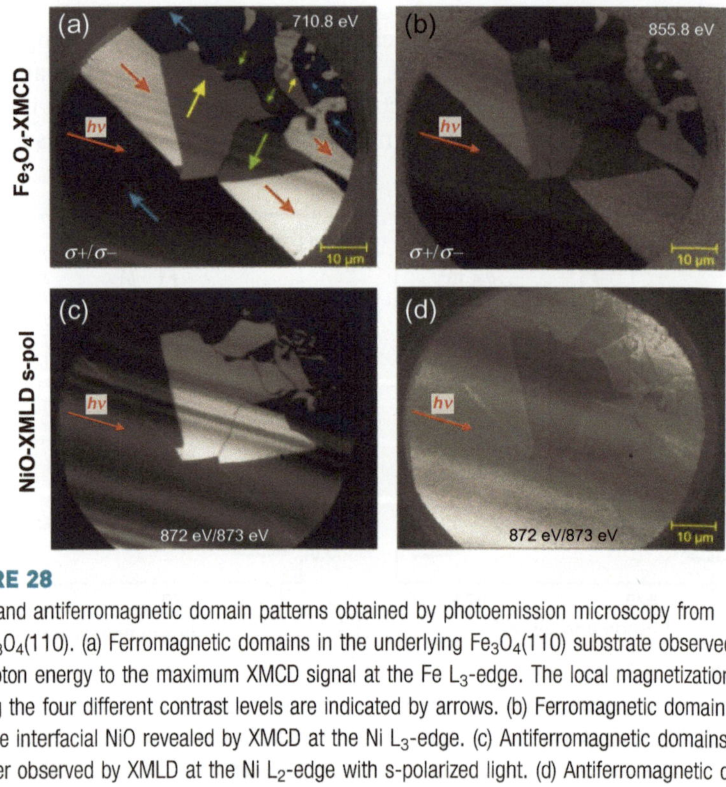

FIGURE 28

Ferro- and antiferromagnetic domain patterns obtained by photoemission microscopy from NiO/Fe_3O_4(110). (a) Ferromagnetic domains in the underlying Fe_3O_4(110) substrate observed by tuning the photon energy to the maximum XMCD signal at the Fe L_3-edge. The local magnetization directions causing the four different contrast levels are indicated by arrows. (b) Ferromagnetic domain contrast from the interfacial NiO revealed by XMCD at the Ni L_3-edge. (c) Antiferromagnetic domains in the NiO top layer observed by XMLD at the Ni L_2-edge with s-polarized light. (d) Antiferromagnetic domains in the NiO top layer observed by exploiting XMLD at the Ni L_2-edge with p-polarized light. The arrow marked "*hν*" indicates the direction of light incidence projected onto the surface plane and relevant photon energies are given in each case. *After Ref. [104].*

ferrimagnet magnetite can be conveniently studied by XMCD at the Fe L_3-edge (see also Fig. 23) and exhibits a clear domain structure with four different gray levels. According to the angular dependence of the XMCD (cf. Eq. 14) these gray levels can be translated in local spin orientations, which are marked by arrows. These are distinguished by their gray level in Fig. 28a.

The domain structure in the antiferromagnet NiO (film thickness 35 monolayers) is made visible by exploiting the XMLD (see also Fig. 27) with s-polarized light at the Ni L_2-edge (Fig. 28c). The XMLD is determined from the intensity difference of the features "A" and "B" discussed in the context of Fig. 27. In this pattern, we can see a replica of the domain structure in $Fe_3O_4(110)$; however, there are only two different contrast levels. From this result we can conclude that there must be a strong exchange coupling between NiO and Fe_3O_4. Two oppositely magnetized domains in Fe_3O_4 lead to antiferromagnetic domains in NiO with the same contrast level. This is due to the fact that XMLD can only distinguish two spin axes in the orthogonal, but not the antiparallel arrangement. The ferromagnetic domain walls in Fe_3O_4 show up as narrow white and dark lines separating the domains in NiO. Note that at this point the orientation of the spin quantization axis in the AFM is not yet known. The angle of light incidence in the PEEM is around 20° with respect to the sample surface. Therefore, an XMLD should also be visible with p-polarized light (cf. Fig. 28d).

A surprising result is reproduced in Fig. 28b. As NiO is an antiferromagnet there should be no magnetic contrast in XMCD experiments at the Ni L-edge. However, the data show a clear replica of the domain structure in Fe_3O_4. This signal stems from the NiO layer at the interface, which is ferromagnetically coupled to the interface. A quantitative analysis of the XMCD signal shows that this transition zone is about one monolayer thick [94].

The question that remains concerns the orientation of the spin quantization axis in the antiferromagnet with respect to the ferromagnet. In theory, there have been two models proposed: (i) a "spin-flip" model, in which the spins in the FM and AFM align collinearly, and (ii) a "spin-flop" model, in which the spins in the FM and AFM assume an angle of 90° [102]. This information can be obtained by investigating the angular variation of the XMLD from NiO in detail. The simple angular dependence given above (Eq. 20) is only valid for the assumption of a cylindrical symmetry, which requires the sample to be polycrystalline. In a single-crystalline sample, this assumption breaks down and the angular dependence becomes more complicated. At the end of an extended analysis, we find the spin-flip coupling for the $NiO/Fe_3O_4(110)$ system (Fig. 29), i.e., the spin quantization axes in NiO and $Fe_3O_4(110)$ are indeed collinear.

Further studies in this material system along other crystalline orientations $\langle 001 \rangle$ and $\langle 111 \rangle$ have shown that the character of the exchange coupling

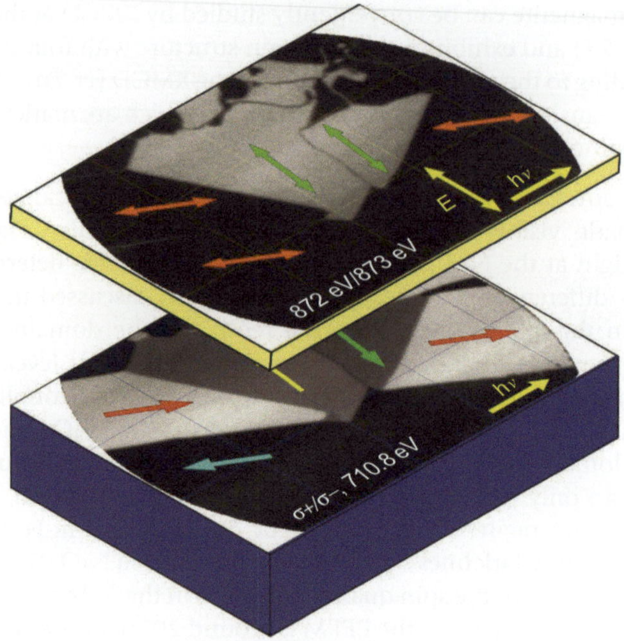

FIGURE 29

Assignment of the spin quantization axes in the antiferromagnet NiO with respect to the underlying ferrimagnet $Fe_3O_4(110)$. In each domain, the local spin alignment (NiO) and spin orientation ($Fe_3O_4(110)$) is indicated by a double- and single-headed arrow, respectively. The arrow marked "*hν*" indicates the direction of light incidence projected onto the surface plane. From this result a spin-flip-type coupling between the two materials can be inferred. *After Ref. [104].*

depends crucially on the bonding situation and even the lattice strain at the interface. For the ⟨001⟩-oriented samples, a spin-flop coupling has been found, whereas for the ⟨111⟩-oriented interface a spin-flip coupling prevails [105].

5 MAGNETIC RESONANT X-RAY SCATTERING

5.1 Basic Aspects

The interaction of X-rays with matter provides a further channel for magnetic studies, employing the scattered or reflected light from a sample. In conventional X-ray scattering, the photon energy should be different from the energy of characteristic absorption edges. By tuning the photon energy to a characteristic optical transition, however, the scattering may be performed in a resonant mode to achieve element selectivity [106]. If we adopt the scattering picture for the moment, in nonmagnetic materials the scattering is taking place solely on the charge distribution in the sample. Similar to XMCD or XMLD phenomena, the scattering process in magnetic materials also contains spin-dependent components. The atomic scattering factor $f(E)$ in a system with magnetization **M** can be given by the expression [107,108]:

$$f(\omega) = \left[(\hat{e}'* \cdot \hat{e})\ (f_0 + f'(E) - if''(E))\right] - \left[i(\hat{e}'* \times \hat{e})\mathbf{M}(m'(E) - im''(E))\right] \tag{21}$$

$$= \underbrace{(\hat{e}' \cdot \hat{e})F^c(\omega)}_{\text{charge}} \underbrace{-i(\hat{e}' \times \hat{e})mF^{m1}(\omega)}_{\text{circular dichroism}} \underbrace{+ (\hat{e}' \cdot m)\ (\hat{e} \cdot m)F^{m2}(\omega)}_{\text{linear dichroism}}. \tag{22}$$

The first term in (21) describes the charge scattering contributions in the non-resonant (f_0) and resonant (f', f'') cases(f', f''). The quantities $\hat{e}$ and $\hat{e}'$ describe the (complex) polarization vector of the incident and reflected light, respectively. The second term contains the magnetic scattering contributions which appear only in the resonant case (m', m''). Reordering the contributions in Eq. (22) shows the terms corresponding to circular and linear dichroism with the transition probabilities $F^i(\omega)$. Thus, the scattering pattern will be changed if magnetic order in the system is present. A simple example of this magnetic scattering is given in Fig. 30. The sample is a polycrystalline Co/Cu multilayer structure exhibiting strong interlayer exchange coupling [110]. The Cu interlayer thickness has been chosen to be about 2 nm, resulting in an antiferromagnetic

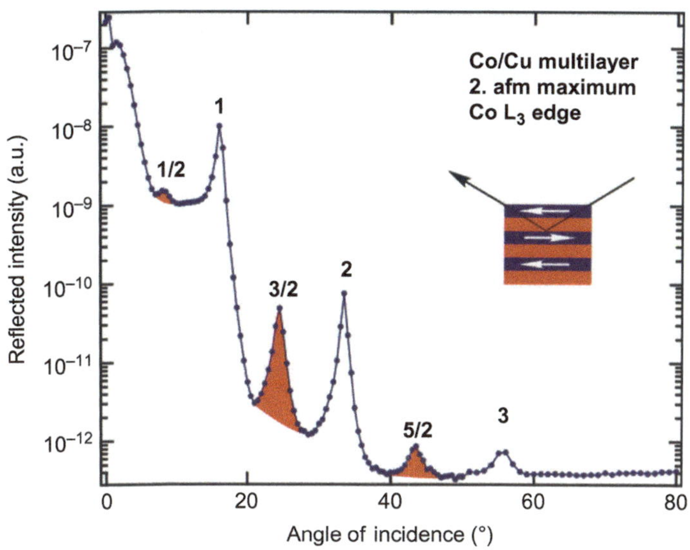

FIGURE 30

X-ray resonant magnetic scattering from a Co/Cu multilayer with p-polarized light at the Co L_3-edge ($h\nu = 776$ eV). The periodicity of the multilayer is [Co(0.9 nm) + Cu(2 nm)]. The scattering peaks induced by the magnetic superstructure (1/2, 3/2, 5/2) are shaded in order to distinguish them from the structural peaks. The inset depicts the magnetic structure in the multilayer which is caused by the interlayer exchange coupling, and corresponds to twice the structural period. *See also Ref. [109].*

coupling scheme, i.e., the magnetization direction between adjacent Co layers changes by 180°. As a consequence, the magnetic period in the multilayer is twice as large as the structural one.

For a resonant excitation at the Co L_3-edge, the wavelength of the light corresponds to $\lambda_{Co,L_3} = 1.6\,nm$ and is thus of the order of the individual layer thickness in the multilayer. In an elastic scattering experiment ($\theta - 2\theta$ geometry), we will therefore expect a diffraction on the multilayer periodicity resulting in a strong modulation of the scattered intensity as a function of the angle of incidence θ. As we can see in Fig. 30, this is indeed the case. We find strong diffraction peaks at around θ values of 15°, 30°, and 50° which arise from the first-, second-, and third-order structural Bragg peak of the Co/Cu multilayer structure. In addition, there appears a set of diffraction peaks at smaller scattering angles, i.e., which can be related to the larger periodicity of the magnetic ordering. These are half-order integer magnetic Bragg peaks. These "magnetic" peaks disappear if the experiment is repeated under off-resonance conditions, or if a sufficiently high magnetic field is employed to overcome the interlayer exchange coupling and align the magnetization vectors in neighboring Co layers in the same direction.

The inherent advantages of the XRMS approach stem from the fact that it is a true photon-in/photon-out technique. The information depth is entirely determined by the propagation and absorption of the light in the sample and reaches values of several tens of nanometers in the case of soft X-ray radiation. As such it is significantly higher than electron-based methods and provides better access to the magnetic properties of buried layers. Second, XRMS studies can be performed in the presence of rather high applied magnetic fields of several teslas and thus permit the investigation of field-induced magnetization reversal processes.

5.2 Complex Layered Structures

In the previous experiment, we have only made use of the intrinsic periodicity of the magnetic structure of the sample. In more complex magnetic film systems, for instance, magnetoresistive sensors or magnetic tunneling junctions, such periodicity is usually not present. Moreover, one is often interested in the response of the system to an external magnetic field. In Fig. 31, we show results obtained from a spin-valve structure. The spin-dependent transport functionality of a spin valve is given by a trilayer stack consisting of a magnetically harder layer (CoFe) which is separated by an interlayer (Cu) from a magnetically soft layer block (FeNi/CoFe). The hard CoFe layer is additionally stabilized by exchange coupling to an antiferromagnet, IrMn in this case. The bottom Ta layer improves the growth and crystallinity of the subsequent layers, whereas the Ta/Cu layers serve as protection against oxidation. The

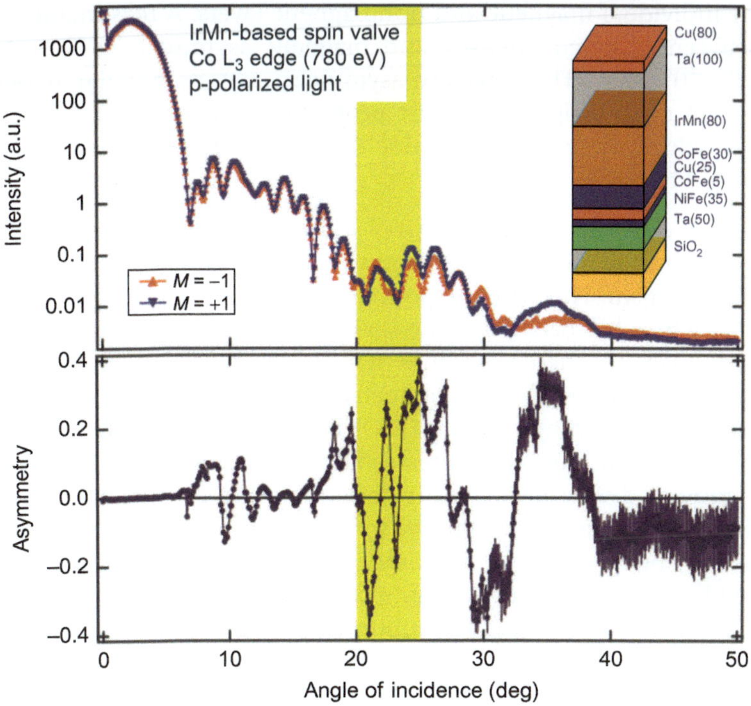

FIGURE 31
X-ray resonant magnetic scattering from a spin-valve system (structure see inset, layer thicknesses in Å) comprising an exchange-biased magnetically hard Co layer. Top: Reflectivities measured with p-polarized light at the Co L_3-edge for opposite magnetization directions. Bottom: The magnetic reflectivity is represented as an intensity asymmetry of spectra taken at $M=+1$ and $M=-1$ and is plotted with error bars.

experimental geometry is a transversal one, i.e., the incident light is p-polarized with the magnetization **M** being oriented perpendicular to the scattering plane.

The resonant scattering experiment yields a complicated diffraction pattern which can be understood in principle on the basis of multiple scattering formalisms, taking into account the resonant and nonresonant scattering contributions at the individual layers and interfaces, and from the substrate. The change of the scattered intensity over several orders of magnitude with the scattering angle is a typical feature of the XRMS experiment. We observe significant changes in the diffraction pattern when reversing the magnetization direction in the layer stack by applying the external magnetic field. These differences are related to the magnetization reversal in the Co film and have been plotted as an intensity asymmetry similar to Eq. (11). The asymmetry exhibits a pronounced structure over an extended range of scattering angles starting at around $\theta=8°$. This structure is caused by the interference of the reflected contributions from the individual magnetic and nonmagnetic layers. A reasonable compromise between scattered intensity and magnetic asymmetry is found in the region $\theta = 20°...25°$. The intensity asymmetry and its angular dependence can be used to investigate magnetic properties and processes with element and layer selectivity and to determine magnetization profiles in a thin-film structure [111].

In the above experiment, the soft X-ray scattering has been used to define an optimum spectroscopic "working point" for the element-selective investigations of magnetism. This working point is given by the angle θ and the photon energy $\hbar\omega$, which has been set to the Co L_3 resonance in the case of the studies presented in Fig. 31. These properties make XRMS a true spectroscopic tool. In fact, with respect to the reflection geometry it can be seen as a high-energy counterpart of the magneto-optical Kerr (MOKE) effect in the visible range of light. Therefore, we can also use the expression X-ray transverse MOKE (TMOKE) for the experiment discussed above. In fact, there are further magneto-optical effects in the visible range, which have soft X-ray counterparts. The most important one may be the longitudinal MOKE (LMOKE), which employs circularly polarized light. The soft X-ray variant of LMOKE has a very close relationship to XMCD, as the magnetic contribution to the absorption coefficient $\Delta\mu(E)$ (Eq. 15) is directly related to the imaginary part of the magnetic scattering factor (quantity $m''(E)$ in Eq. 21). Because of the resonant nature of the X-ray LMOKE the magnetic signal shows direction as an intensity difference [109,112], whereas phenomena such as the X-ray Voigt effect [113] or other quadratic magneto-optical effects [114] need an additional polarization analysis of the reflected light.

So far we have concentrated on studies in the soft X-ray regime. Resonant magnetic reflectivity experiments, however, can also be performed very efficiently in the vacuum ultraviolet (VUV) regime below 100 eV. In this region, we find the M-edges of the transition metals. Because of the larger wavelength and smaller penetration depth of the VUV radiation, the resonant reflectivity signal is only weakly modulated by interference effects in a layer stack. An example of a resonant magnetic reflectivity experiment on Permalloy ($Fe_{20}Ni_{80}$ alloy) is depicted in Fig. 32. The magnetic asymmetry upon magnetization reversal is measured in a transverse MOKE geometry with p-polarized light and displayed in a color code for an extended range of incident angle and photon energy. We observe clear signatures related to the Fe (~54 eV) and Ni M-edges (~66 eV). The maximum asymmetry values are found around the Brewster angle (45°) and reach values of +30% and −10%, respectively. This demonstrates that element-selective studies of magnetism can also be performed in the VUV regime. A quantitative interpretation of the magnetic reflectivity signals, however, must be based on quantum mechanical calculations, because at the transition metal M-edges the energy splittings due to exchange and

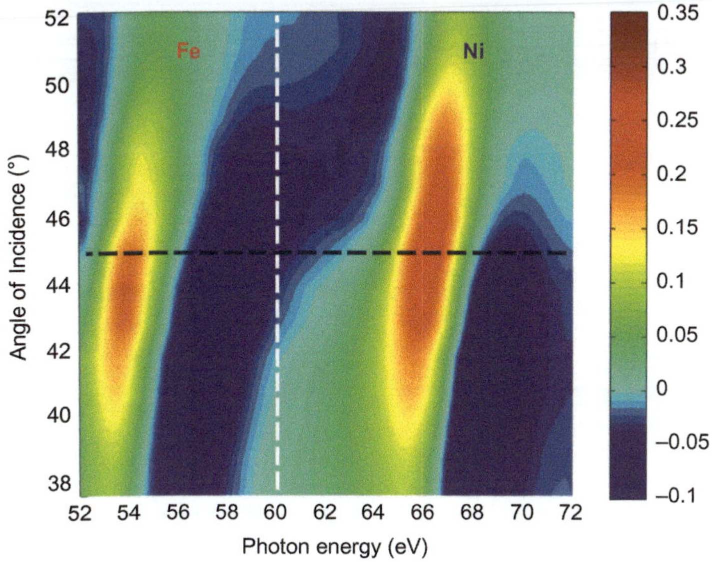

FIGURE 32
TMOKE reflection asymmetry from a Permalloy ($Fe_{20}Ni_{80}$) alloy film at the Fe and Ni M-edges using p-polarized extreme ultraviolet (EUV) synchrotron radiation. The value of the asymmetry is color-coded within the range of −0.1 and +0.35 on a scale of −1 to 1. *After Ref. [115].*

spin–orbit coupling are of the same order of magnitude. This results in a complex hybridization of the $3p_{1/2}$ and $3p_{3/2}$ sublevels.

5.3 Antiferromagnets

In the resonant magnetic scattering approach, we are not limited to the study of ferromagnets. Similar to XMLD in absorption discussed in Section 4.2, there is also a linear magnetic dichroism. This is demonstrated in Fig. 33 for an NiO thin film [116]. The NiO was grown on a Co layer and was subjected to a field-cooling procedure before the experiment. In this procedure, the NiO film is annealed above the Néel temperature and cooled down in the presence of a magnetic field. The field aligns the magnetization in the Co film and due to the exchange coupling also the spin quantization axis m in the antiferromagnet NiO is aligned in space. In this way, we create a single antiferromagnetic domain in the sample.

The experimental result shows first of all a pronounced feature in the reflectance R at the position of the Ni L_3-edge. In the same energy range we also find a clear magneto-optic variation of the reflectivity $\Delta R(E)$ upon rotating the sample around its normal by 90°. The linear dichroism in the reflectivity is expressed as

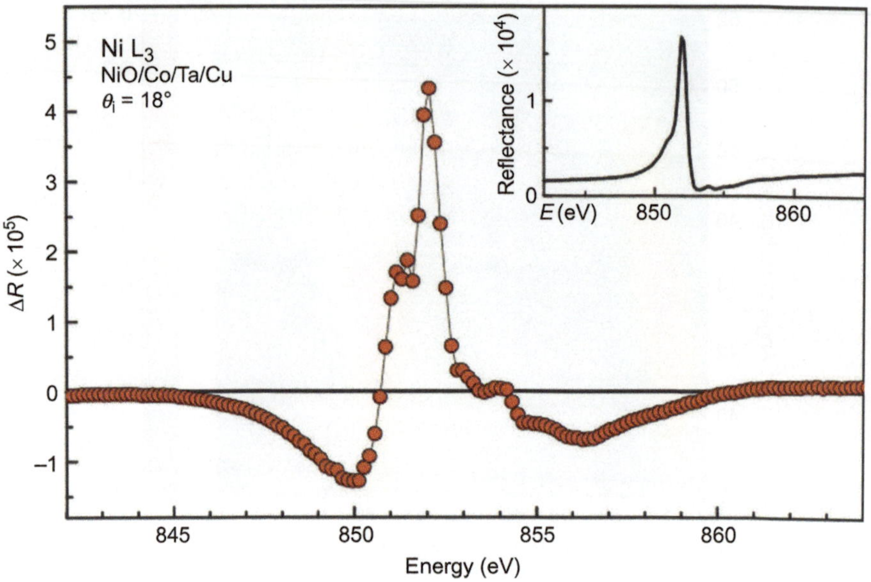

FIGURE 33
The X-ray magneto-optical reflection spectrum ΔR measured for $\theta_i = 18°$ at the Ni L_3-edge of an exchange-biased NiO/Co/Ta/Cu multilayer stack. The antiferromagnetic NiO layer is buried beneath the ferromagnetic layer. Inset: Reflectance spectrum $R_{||}$ at the Ni L_3 absorption edge. *From Ref. [116].*

$$\Delta R(E) = R_{\perp}(E) - R_{||}(E), \tag{23}$$

where $R_{||}$ denotes the reflectivity measured with the spin alignment axis m parallel and $R_{\perp}$ the reflectivity with m perpendicular to the film-projected component of the light polarization vector **E**. The fine structure in the spectrum is caused by the multiplet splitting in the oxide.

In addition to providing an element-selective photon-based access to antiferromagnets, the wider importance of the above linear magnetic dichroic effect becomes clear when considering it within a magneto-optical theory [116]. It can be shown that the linear dichroism addresses specific components of the dielectric tensor $\underline{\underline{\epsilon}}$. In particular, $\Delta R(E)$ probes directly the fundamental magneto-optical quantity $\Delta\epsilon = \epsilon_{||} - \epsilon_{\perp}$, which describes the difference of the tensor components parallel and perpendicular to the spin alignment axis m. For the L_3-edge, the imaginary part of $\Delta\epsilon$ is directly related to the local unoccupied spin-split density of states $D^{\uparrow\downarrow}(E)$, i.e.,

$$\text{Im}\,(\Delta\epsilon) \propto \Delta_{\text{ex}} \frac{\partial}{\partial E}\left[D^{\uparrow}(E) - D^{\downarrow}(E)\right]. \tag{24}$$

In other words, from measurements of the XMLD in a reflection geometry, we can derive information on the on-site d-spin magnetic moment.

A closer inspection of Eq. (24) highlights three further important aspects with respect to the spectroscopic study of antiferromagnets. The first one is the appearance of the quantity Δ_{ex}, which describes the exchange splitting of the $2p_{3/2}$ core states. This corroborates the findings in Sections 3.2 and 4.2, which already showed this exchange splitting to be crucial for the understanding of magnetodichroic phenomena in core-level photoemission and absorption. The second issue is the energy derivative of the density of states in Eq. (24). In fact, a similar magneto-optical derivation for XMCD produces [117]:

$$\mathrm{Im}(\Delta\epsilon) \propto \Delta_{ex}\left[D^{\uparrow}(E) - D^{\downarrow}(E)\right]. \tag{25}$$

Therefore, in simple cases the XMLD spectrum should look like an energy derivative of the respective XMCD spectrum in accordance with our discussion in Section 4.2. The third aspect concerns the antiferromagnetic ordering. The atomic magnetic moment m and thus the sign of Δ_{ex} reverses between neighboring lattice sites, but so does the sign of the local spin density of states $D^{\uparrow}(E) - D^{\downarrow}(E)$. This explains why the XMLD is even in the magnetic moment m in antiferromagnets and even in the magnetization M in ferromagnets.

The above discussion demonstrates that the XMLD in reflection is sensitive to details of the electronic structure represented by the structure of the unoccupied density of states. This sensitivity has been used to explore the role of anisotropies in single-crystalline ferromagnets [118]. In this case, the angular dependence of the XMLD spectra also exhibits a strong anisotropy, which can be related to crystal-field-induced splitting of the density of states.

6 ADDRESSING PICOSECOND MAGNETIZATION DYNAMICS

In the previous chapters, we have focused on the analysis of quasistatic magnetic properties. With some modifications, the same techniques in X-ray photoemission, absorption, or reflection can also be used to study the dynamic behavior of magnetic systems. The general experimental scheme involves either a "resonant" or a "pump-probe" approach. In the resonant mode, the magnetization is excited by a continuous microwave signal. In the pump-probe approach, the magnetization dynamics are initiated by a pulsed stimulus and the transient state at a later time is probed by the X-ray pulse. The nature of the stimulus depends on the temporal regime and the microscopic processes of interest.

In this chapter, we will discuss experiments on magnetization dynamics in the nano- and picosecond regimes. This temporal region covers processes such as domain wall nucleation and motion, magnetization rotation, and precession, which are at the heart of fast magnetic switching processes employed in data storage and spintronics. Of major importance is the dynamic response of small micrometer- and submicrometer-sized magnetic elements. The main stimulus used to initiate and drive magnetization dynamics is an external magnetic field pulse. The resulting temporal evolution of the magnetization distribution in an element can be described in detail by the Landau–Lifshitz–Gilbert (LLG) equation [5].

The magnetic field pulse is generated by passing short current pulses through microcoils (Fig. 34a) or coplanar wave guides (Fig. 34b and c). In order to obtain very short and well-defined pulses and to avoid unwanted reflections, the microcoils and waveguides must be impedance-matched to the electrical circuit. Current pulses down to 100 ps pulse length may be supplied by adapted electrical pulse generators. Even shorter pulses far below 100 ps can be generated by using optical switches which are controlled by femtosecond laser pulses [119]. The micrometer-sized magnetic element is usually placed either within the microcoil (perpendicular magnetic field) or on top of the coplanar wave guide, resulting in an in-plane magnetic field direction.

In order to obtain a pump-probe scheme, the magnetic field pulse must be synchronized with the pulse pattern of the synchrotron radiation. Down to the picosecond regime this can be achieved by electronic delay generators, which are controlled by a microwave signal derived from the RF system driving the cavities in the storage ring.

The microcoil setup is often employed in domain imaging experiments on the basis of X-ray transmission microscopy [120]. The main reason is the specific geometry in XTM, which in most cases requires a normal or nearly normal light incidence on the sample [121,122]. With MXCD as a magnetic contrast mechanism this yields the highest dichroic signal for a magnetization component perpendicular to the sample surface. For this reason, X-ray transmission

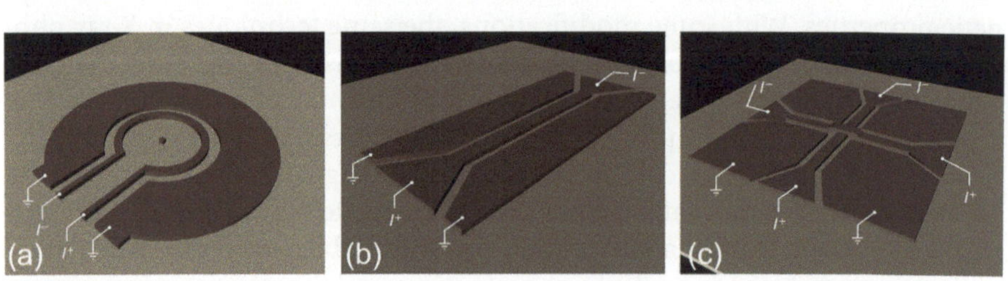

FIGURE 34

Geometries for short field-pulse experiments. (a) Microcoil for generating magnetic fields perpendicular to the film plane. (b) Coplanar waveguide or microstripline for in-plane magnetic fields. Sample is placed on top of the microstripline. (c) Microstripline system for variable in-plane magnetic field pulses. Sample is placed on top of the crossing point of both striplines.

microscopy was successfully applied to investigate the gyration of magnetic vortices and the dynamics of the magnetic vortex core in square- and circle-shaped magnetic elements [123–125].

The coplanar waveguide has become the standard geometry in time-resolved photoemission microscopy experiments [126]. The magnetic contrast mechanism employed in PEEM is XMCD in the TEY. The type of information that can be obtained with this approach is illustrated in Fig. 35. It displays the temporal evolution of the magnetization distribution in rectangular Permalloy microstructures in response to a magnetic field pulse of about 10 ns duration. In the ground state, i.e., before the onset of the field pulse, the magnetic domain pattern is a characteristic Landau pattern determined by stray field minimization principles [128]. It leads to a characteristic sequence of triangular and quadratic domains. Once the field pulse **H**(t) acts on the magnetization, several processes on different timescales set in. These are domain wall displacement, coherent and incoherent magnetization rotation [127]. The magnetization rotation is strongest in those domains with a magnetization vector perpendicular to the external field direction (dark and bright domains). Domains with **H**(t) $\upuparrows$ **M** grow by domain wall displacement into the **H**(t) $\perp$ **M** domains, whereas domains with **H**(t) $\uparrow\downarrow$ **M** are subject to incoherent rotation processes, leading to complex stripe patterns. The contrast in these stripe patterns indicates the local magnetization vector in the individual stripe to be oriented roughly perpendicular to the applied field, whereas neighboring stripes are magnetized almost antiparallel. This configuration can only be established

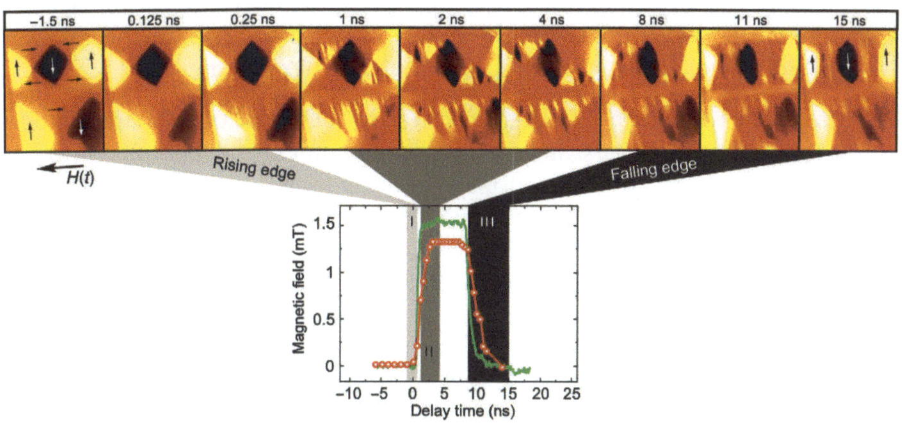

FIGURE 35

Magnetization dynamics in Permalloy microstructures driven by a magnetic field pulse H(t). The current pulse fed into the coplanar wave guide (green) causes a field pulse on the wave guide surface, the shape of which is displayed by the red curve. This field pulse drives the dynamic response in the microstructures which is captured in snapshots at different points in time along the pulse profile. *After Ref. [127].*

by the nucleation of a large number of partial domain walls. On the plateau of the field pulse, which is a few nanoseconds wide, the system settles into a new equilibrium state in the presence of the external magnetic field. This is reflected by a quasistationary domain pattern. Significant changes occur again when the magnetic field decays. For the quasistationary state, this is equivalent to applying a magnetic field in the opposite direction. This explains why the incoherent rotation suddenly appears also in the domains with $\mathbf{H}(t) \upuparrows \mathbf{M}$ and the stripe pattern expands into these regions. A relaxation of these stripe patterns requires that the system drives the domain walls out again, which must happen via the rather slow process of domain wall displacement. It therefore takes several tens of nanoseconds for the Permalloy structure to reach the ground state, i.e., the pattern observed before the field pulse, again. The images have been accumulated over 10^8 repetitions of the pulse sequence and therefore show only the reversible processes in the time evolution of the magnetization. Irreversible and stochastic processes do at best not show up in the images, or they impair the image quality, if they appear more frequently. This is a general limitation of the pump-probe approach, which may be exemplified for the domain wall motion. In pump-probe imaging experiments, the width of the domain walls often appears wider than in quasistatic imaging. The reason must be sought in the fact that during the 10^8 pump-probe cycles the domain wall does not always take the exact same position in each cycle. In the integrated data set the wall therefore appears more or less "smeared out."

Instead of using a laterally resolved domain imaging approach, laterally integrating studies of the magnetization dynamics can also be performed in a similar manner. They are particularly interesting in the case of pulsed or continuous microwave excitations, or if the magnetization dynamics are driven by spin-polarized currents in the spin-torque transfer scheme [129].

Ferromagnetic resonance (FMR) is a well-established method in magnetism to study magnetic excitations, magnetization dynamics, and damping mechanisms [130]. It exists in two variations working with frequencies either in a narrow or in a broad band. In the classical narrowband FMR, the sample is placed in a microwave cavity tuned to a single well-defined excitation frequency. In the more modern broadband approach, one either works in the frequency domain analyzing the microwave spectrum by a vector network analyzer [131,132], or one works in the time domain employing pump-probe schemes [133]. Exploiting the high magnetic sensitivity of XMCD, the classical FMR technique may be augmented by element selectivity. This approach is sometimes called XFMR and has been pioneered by Bailey *et al.* [134]. The authors used a similar pump-probe scheme with coplanar waveguides as described above and drove the system by pulses with a short fall time. In this way, they could measure the free precession of the sublattice magnetization in Permalloy in an applied static field. Within the experimental error, the precession frequencies and phase relations of Ni and Fe magnetic moments were found to be the same.

In Fig. 36, we show an example of a narrowband continuous wave XFMR experiment on a gadolinium-doped yttrium–iron garnet ($Gd_1Y_2Fe_5O_{12}$, Gd:YIG) [135]. The sample is placed on a $\lambda/2$ microstrip resonator and driven at a frequency of 2.21 GHz. The Gd:YIG exhibits a complex ferrimagnetic ordering: Fe ions in tetrahedral sites and Gd ions are coupling antiferromagnetically to the Fe ions in octahedral sites. In summary, this can be seen as an effective two-sublattice (Fe and Gd) system where the sublattice magnetizations m_{Gd} and m_{Fe} are oriented antiparallel to each other. The resonance condition is tuned by an external magnetic field B_0 and the results are compared to the standard FMR signal shown in Fig. 36a. The XFMR signal is proportional to the opening of the precessional cone of the sublattice magnetization, which is measured by the M_z component of the magnetization. In Fig. 36b one clearly sees the sharp increase of the precessional amplitude in the Fe XMCD signal when approaching the resonance condition at $B_0 = 165$ mT. In the same conditions the Gd XMCD signal also shows a sharp resonance (Fig. 36c), indicating that the sublattice magnetizations m_{Gd} and m_{Fe} precess with a fixed phase relation. The opposite sign of the Fe and Gd XFMR signals proves that the strong exchange coupling between the sublattices keeps an antiferromagnetic alignment between m_{Gd} and m_{Fe} also during the precessional motion.

7 ULTRAFAST DEMAGNETIZATION DYNAMICS

In the final section, we will address the element-selective investigation of ultrafast spin dynamics in the sub-picosecond regime. This rapidly growing field was initiated by a pioneering experiment by Beaurepaire *et al.* [136], who showed that a strong infrared light pulse from a femtosecond pulse laser can quench the magnetization in Ni on a timescale of about 100 fs. The nature of this ultrafast demagnetization process raises many questions concerning the energy and angular momentum transfer between the electron, spin, and lattice subsystems in a sample. In a nutshell, the phenomenon involves a complex sequence of microscopic processes on different timescales starting with the excitation of spin-polarized valence electrons above the Fermi level by the infrared light pulse. Subsequently, these highly energetic electrons undergo spin-dependent scattering and dissipate energy, at the same time creating disorder in the spin system. In the last step, the excess energy in the spin system is transferred via spin–lattice relaxation to phonons on a timescale of picoseconds. This interpretation is also called the three-temperature model [136].

Most of the subsequent studies were carried out with standard magneto-optical approaches in visible light, thereby lacking element selectivity. The reason was the lack of suitable X-ray sources which could provide sub-picosecond light pulses. Due to an ongoing development in laser physics and in accelerator-based light sources, such light sources are now becoming available [137]. The scheme of higher harmonic generation (HHG) in gases [138,139] or metal

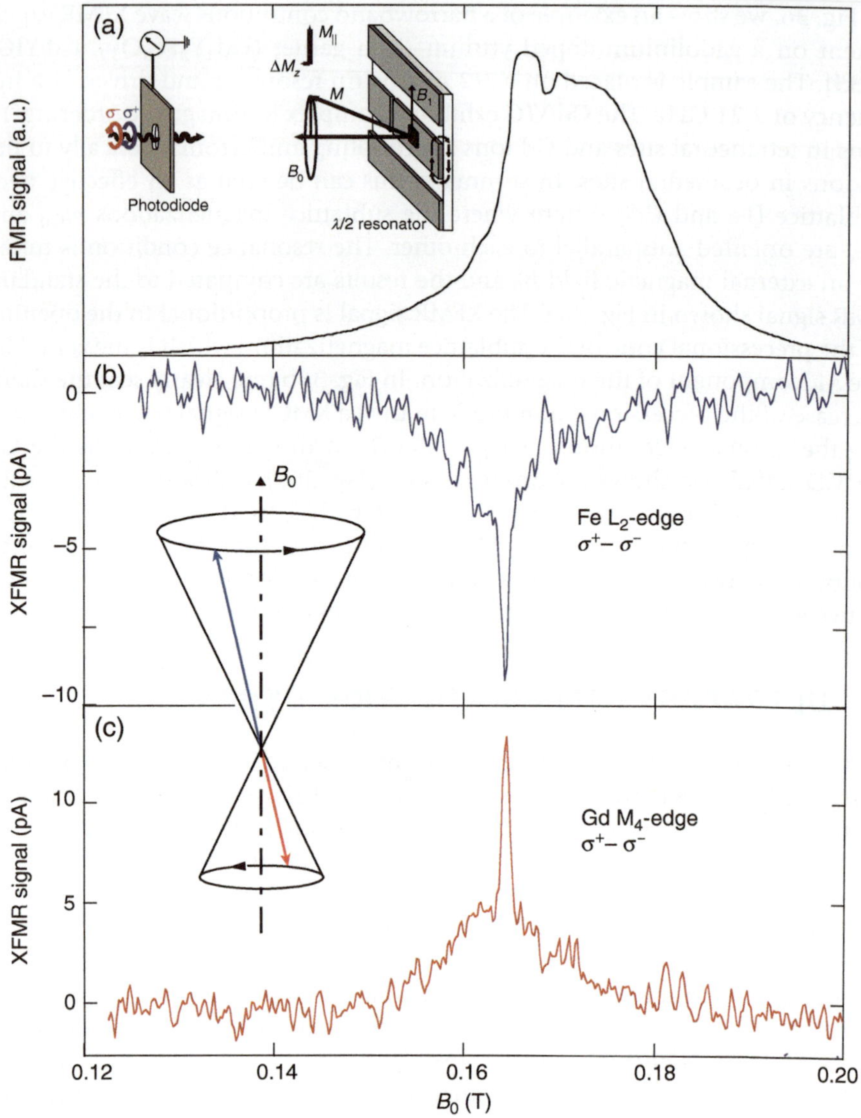

FIGURE 36

(a) FMR spectrum of Gd:YIG measured simultaneously with the XFMR data. (b) XFMR ($\sigma^+ - \sigma^-$) intensity measured at the L_2-edge of Fe (723.8 eV) and (c) at the M_4-edge of Gd (1222 eV). The microwave power is 31 dBm. The data are averaged over 40 sweeps of B_0 in the positive direction, with a sweep time of 80 s and lock-in time constant of 100 ms. *Taken from Ref. [135].*

vapors [140] provides femtosecond light pulses in the VUV regime using table-top laser systems. Storage ring-based solutions involve the femtosecond slicing approach [141] or different harmonic generation schemes reaching up into the soft X-ray regime [142,143]. Extremely intense femtosecond X-ray pulses are now becoming available through FELs [14,144].

The first element-selective studies of ultrafast spin dynamics have been performed with the femtoslicing technique. In femtoslicing, the electron package in an undulator is segmented by a strong laser pulse into microbunches. The radiation pulses from these microbunches are of the order of 100 fs in pulse width and are separated from the normal undulator radiation by apertures because of their particular angular distribution. Although the intensity of the slicing radiation is several orders of magnitude lower than the regular synchrotron radiation, Stamm *et al.* [145] succeeded in mapping the signature of the ultrafast demagnetization in Ni. For this they split off a part of the infrared laser beam to excite the sample. This ensures a good time resolution in the femtosecond regime as both light pulses arriving at the sample are derived from the same source and have a fixed phase relation. In order to follow the evolution of a signal in the time domain, the phase relation can be changed in a very controlled manner by placing an optical delay in the path of the IR laser beam pumping the sample.

Using X-ray absorption at the Ni L_3-edge they were able to probe the ultrafast response of the electron and spin systems. This is shown in Fig. 37a, reproducing the temporal variation of the XAS signal at the initial slope of the Ni L_3-edge measured with linearly polarized light. The curve reflects a transient change in the electron distribution in Ni on a sub-picosecond timescale due to the electron–hole excitations around the Fermi level. The understanding of this result is very important, because the nonequilibrium distribution of valence electrons may affect the validity of XMCD sum rules. A subsequent theoretical treatment, however, proved the sum rules to hold even for this nonequilibrium case, i.e., the XMCD signal can serve as a measure of the transient magnetization [146].

The results of the corresponding XMCD study at the maximum of the Ni L_3-edge absorption are depicted in Fig. 37b and exhibit a reduction of the magnetic dichroism by more than a factor of 2 within about 120 fs after the pump pulse. This is consistent with the early observations in Ref. [136]. Exploiting sum rule arguments, the authors interpret this ultrafast change in the magnetization in terms of an ultrafast spin–lattice relaxation process and thereby rule out a sizable transfer of spin S to the angular momentum L [145]. The orbital magnetic moment of Ni, however, is small and could not be determined independently. In addition, as we will discuss below, ultrafast spin–lattice

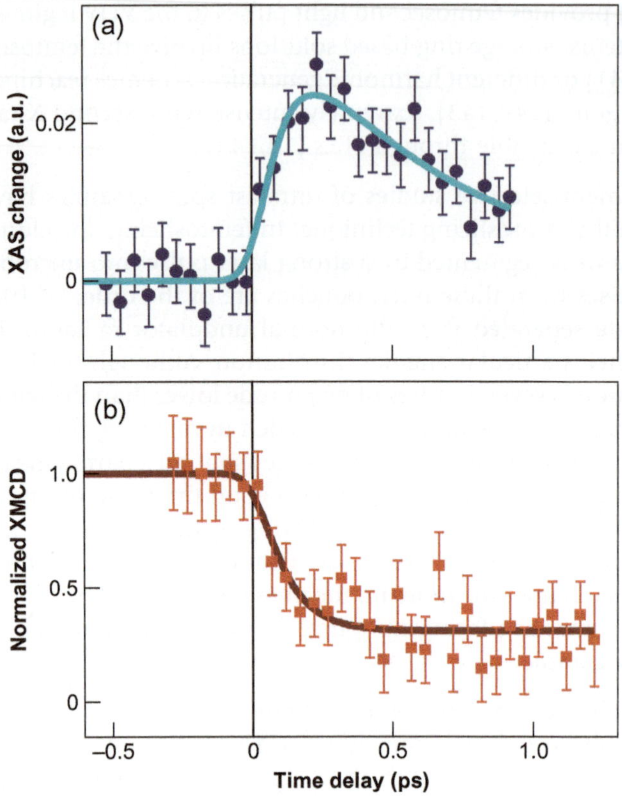

FIGURE 37

Response of the electronic and magnetic structure of Ni to a femtosecond IR pump pulse. (a) Transient change of XAS intensity as a function of pump-probe time delay. (b) Time-resolved XMCD signal as a function of pump-probe time delay. The symbols are experimental data and lines are fits of the three-temperature model. Error bars indicate the standard deviation of the Poisson-distributed single-photon events. *From Ref. [145].*

relaxation is only one of several competing microscopic mechanisms which can drive ultrafast magnetization changes [147].

Subsequent experiments with this slicing approach went a step further and tried to separate the dynamics of spin and orbital magnetic moments [87]. For this purpose, a magnetic material with a strong PMA, namely a CoPd alloy, was chosen. Such a system is characterized by a large orbital magnetic moment in the equilibrium state (cf. Section 4.1) of $\mu_{L,Co} = 0.24\mu_B$. The time-resolved studies of the spin and orbital contributions to the magnetic moment, $\mu_S(t)$ and $\mu_L(t)$, upon excitation reveal a small but distinct difference in the decay times. These take values of $\tau_{\mu_S} \simeq 280$fs and $\tau_{\mu_L} \simeq 220$ fs for the spin and orbital magnetic moments, respectively. The PMA in the CoPd film is a consequence of fine details in the equilibrium electronic structure around the Fermi level which are related to spin–orbit coupling [148]. The faster response of the orbital magnetic moment may be interpreted as a change of the PMA which is caused by driving the electronic structure into nonequilibrium [87].

Together with the PMA, the orbital magnetic moment is also reduced. This finding indicates that ultrafast spin dynamics may also be caused by changes in the strength and orientation of the magnetic anisotropy.

A complementary approach to femtosecond spin dynamics involves the high-harmonic generation scheme on the basis of table-top laser systems. In capillary cells filled with noble gases one achieves an upconversion of the laser radiation to up to about 100 eV photon energy. This conveniently covers the region of the transition metal M-edges. In Section 5.2, we have demonstrated that an XMLD reflection experiment at the M-edges of Fe and Ni may yield very high magnetic contrast. Combining these two aspects leads to the first element-selective investigations of spin dynamics in the laboratory [149].

The pump-probe data reproduced in Fig. 38 were obtained with a particular grating sample, consisting of interdigital lines of Fe and Ni. This grating not only serves as a sample, but disperses the spectrum of the HHG radiation onto the detector. Upon excitation with the infrared laser pulse, one clearly observes

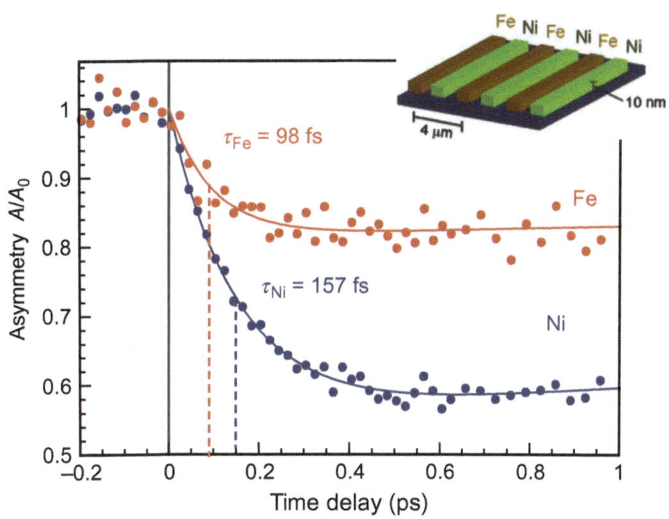

FIGURE 38

Ultrafast spin dynamics in a grating of alternating Fe and Ni lines in a TMOKE experiment with p-polarized light. The element selectivity comes about by employing high-harmonic generation (HHG) radiation covering the Fe and Ni M-edges. The angle of incidence is about $\theta_i \simeq 45°$. The demagnetization traces for Fe (red) and Ni (blue) reveal different quenching and demagnetization times. *Taken from Ref. [150].*

the demagnetization on both the Fe and Ni M-edge. We note that although the excitation is the same for both Fe and Ni lines, the quenching of the magnetization is stronger for Ni than for Fe. This may be related to the fact that the Curie temperature of Fe is almost twice that of Ni. Also, the demagnetization times τ_{Fe} and τ_{Ni}, which are determined from fitting the decay of the magnetization by an exponential function, differ by about 50%. In this case Fe demagnetizes faster than Ni.

The demagnetization behavior is usually interpreted on the basis of the three-temperature model [147]. The decay of the magnetic order is suggested to be due to spin-flip scattering processes (Elliot–Yafet scattering) of the excited electrons. Further studies, however, revealed the importance of superdiffusive spin-polarized currents which leave the excited volume [150,151].

The light emitted from FELs has some characteristics which are quite different from those of slicing or HHG radiation. The repetition rate is at least three orders of magnitude lower (1–100 Hz), but the brilliance of a single light pulse is six to seven orders of magnitude higher than that of regular synchrotron radiation. This property makes FEL radiation uniquely suited for single-shot experiments. On the downside, the high intensity may actually lead to significant damage of the sample [152]. Preliminary experiments have demonstrated that scattering experiments are particularly suitable for the single-shot approach [153].

One of the first magnetic scattering experiments with FEL radiation is depicted in Fig. 39. The FEL pulse is synchronized with an IR laser pulse pumping the sample, which is a CoPt thin-film system with perpendicular magnetization.

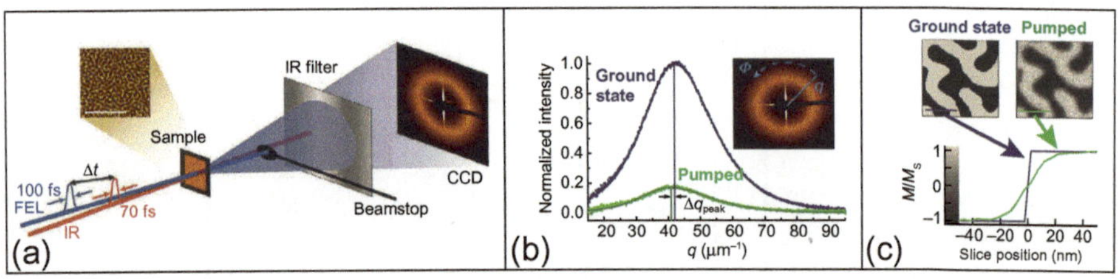

FIGURE 39

Pump-probe small-angle X-ray scattering. (a) Experimental setup showing the sample being pumped by an infrared laser pulse and probed by a delayed FEL X-ray pulse. The magnetic scattering pattern is related to the characteristic length scales in the labyrinth-domain state (magnetic force micrograph with 2 μm scale bar). (b) Scattering profiles are obtained by azimuthally integrating the scattering intensity. Compared to the unpumped state (blue), the pumped one (green) at 1.3 ps time delay exhibits a reduction of the intensity and a shift of the peak position by Δq_{peak}. (c) Simulated action of the superdiffusive spin current on a domain pattern leading to a smearing of the domain walls (top) and a broadening of the wall magnetization profile (bottom). *Taken from Ref. [15].*

At room temperature the thin film assumes a labyrinth-like magnetic domain structure. The domain walls have a characteristic separation which serves as a relevant length scale for a resonant scattering process. In the equilibrium state the intensity is scattered into a ring, because the periodicity of the domains is the same in arbitrary directions within the sample plane. The scattering profile is obtained by azimuthally integrating the scattered intensity. The maximum of the scattering profile is found at a scattering vector $q_{\text{peak}} \simeq 42\,\mu\text{m}^{-1}$ (Fig. 39b) and indicates the average periodicity in the domain pattern. Upon excitation with the IR laser pulse, the maximum of the scattering profile is strongly reduced, reflecting the quenching of the magnetic order in the system. The situation in Fig. 39b reflects the domain pattern about 1.3 ps after the excitation. A closer inspection of the data also reveals a small shift of the scattering maximum by Δq_{peak} to smaller scattering vectors. This shift cannot be related to a simple change of the domain periodicity, as the velocity needed to displace domain walls on this short timescale is unphysically high ($\sim 10^5$m/s). Typical values for domain wall velocities lie below 10^3 m/s [5]. Instead, the optically excited superdiffusive spin currents [154] which pass through the domain walls rather cause a broadening of the domain wall and the respective magnetization profile. This broadening translates into a change of the scattering form factor, causing a shift of the scattering maximum to smaller q-values. This is another example of the importance of superdiffusive spin transport in ultrafast magnetization processes.

8 SUMMARY AND CONCLUSIONS

The combination of element selectivity and magnetic sensitivity has become an indispensable requirement in the study of complex magnetic structures. A comprehensive and detailed understanding of the static and dynamic properties of a magnetic device in data storage or spintronics is mandatory for a critical evaluation and optimization of its performance. In this contribution, we have tried to show that photons offer a unique versatility as far as element-selective studies of magnetic phenomena and properties are concerned. Modern light sources such as synchrotron radiation sources, FELs, and even sophisticated table-top laser systems can provide light with adjustable polarization character and photon energies ranging from the ultraviolet up to hard X-rays. Based on the characteristics of this radiation a large variety of magnetically sensitive absorption, reflection, as well as photon and photoelectron emission techniques have been developed. This suite of techniques can be used to study magnetism on all relevant length scales reaching down to atomic dimensions. As an added benefit, most of these light sources produce well-defined light pulses rather than a continuous wave. Therefore, often the same techniques can be applied to questions addressing quasistatic and dynamic aspects in magnetism even down to the femtosecond regime.

The selection of experimental approaches discussed in this contribution must be necessarily limited. We have therefore focused only on the most prominent approaches, while still trying to touch upon all relevant areas. In terms of element-selective approaches, RIXS [155], X-ray fluorescence [156], and resonant nuclear scattering, which is a synchrotron radiation-based Mössbauer spectroscopy [157], for example, have been omitted. We have also omitted electron beam-induced techniques. In transmission electron microscopy, an electron magnetic circular dichroism has been shown to exist [158,159]. It appears in the electron energy-loss channel and provides quantitative analyses via sum rules [160]. Given the high lateral resolution in contemporary TEMs, investigations on a sub-10 nm level seem to be feasible in the near future [161]. A widespread element-selective analysis technique in a scanning electron microscope is Auger electron spectroscopy. In dedicated Auger microprobe instruments, a lateral resolution of less than 10 nm can be achieved [162]. It has been known for more than 20 years that Auger electrons emitted from ferromagnetic materials are spin-polarized [163]. Given the current developments and achievements in highly efficient spin polarimeters (see Section 3.1.1), a spin-polarized Auger microprobe technique is within reach.

Finally, it should be pointed out that this contribution can only be a snapshot of the actual situation. The ongoing rapid development in photon- and electron-based spectroscopies and microscopies will bring about new approaches for the element-selective investigation of magnetic properties and processes in the future.

Acknowledgments

The author is indebted to R. Adam, M. Aeschlimann, L. Baumgarten, D. Bürgler, S. Döring, M. Eschbach, I. P. Krug, C. S. Fadley, A. X. Gray, P. Grychtol, H. C. Kapteyn, C. La-O-Vorakiat, S. Mathias, H.-C. Mertins, M. M. Murnane, M. Müller, H. Nembach, F. Nickel, P. M. Oppeneer, M. Patt, L. Plucinski, D. Rudolf, Ch. Schmitz, J. Shaw, T. J. Silva, E. Turgut, Ch. Weier, and C. Wiemann for their collaboration. Financial support through the Deutsche Forschungsgemeinschaft (DFG) and the Bundesministerium für Bildung und Forschung (BMBF) is acknowledged.

References

[1] Zangwill A. Physics at surfaces. Cambridge: Cambridge University Press; 1989.

[2] Siegbahn K. Electron spectroscopy for atoms, molecules, and condensed matter. Rev Mod Phys 1982;54:709.

[3] Koningsberger DC, Prins R. X-ray absorption: principles, applications, techniques of EXAFS, SEXAFS, and XANES. New York: John Wiley; 1988.

[4] Stöhr J, Siegmann HC. Magnetism: from fundamentals to nanoscale dynamics. In: Cardona M, Fulde P, von Klitzing K, Merlin R, Queisser H-J, Störmer H, editors. Volume 152 of Springer series in solid-state sciences. Berlin: Springer; 2006.

[5] Kronmueller H, Parkin SSP, editors. Handbook of magnetism and advanced magnetic materials. New York: John Wiley; 2007.

[6] Coey J. Magnetism and magnetic materials. Cambridge: Cambridge University Press; 2009.

[7] Busch G, Campagna M, Cotti P, Siegmann H. Observation of electron polarization in photoemission. Phys Rev Lett 1969;22:597.

[8] Duò L, Finazzi M, Ciccacci F, editors. Magnetic properties of antiferromagnetic oxide materials. Berlin: Wiley-VCH; 2010.

[9] Erskine J, Stern E. Calculation of the M_{23} magneto-optical absorption spectrum of ferromagnetic nickel. Phys Rev B 1975;12:5016.

[10] van der Laan G, Thole B, Sawatzky G, Goedkoop J, Fuggle J, Esteva J, et al. Experimental proof of magnetic x-ray dichroism. Phys Rev B 1986;34:6529.

[11] Schütz G, Wagner W, Wilhelm W, Kienle P, Zeller R, Frahm R, et al. Absorption of circularly polarized x rays in iron. Phys Rev Lett 1987;58:737.

[12] Wiedemann H. Synchrotron radiation, advanced texts in physics. Berlin: Springer; 2003.

[13] O'Shea PG, Freund HP. Free-electron lasers: status and applications. Science 2001;292:1853.

[14] Ayvazyan V, Baboi N, Bähr J, Balandin V, Beutner B, Brandt A, et al. First operation of a free-electron laser generating GW power radiation at 32 nm wavelength. Eur Phys J D 2006;37:297.

[15] Pfau B, Schaffert S, Müller L, Gutt C, Al-Shemmary A, Büttner F, et al. Ultrafast optical demagnetization manipulates nanoscale spin structure in domain walls. Nat Commun 2012;3:1100.

[16] Graves CE, Reid AH, Wang T, Wu B, de Jong S, Vahaplar K, et al. Nanoscale spin reversal by non-local angular momentum transfer following ultrafast laser excitation in ferrimagnetic GdFeCo. Nat Mater 2013;12:293.

[17] Mattis D. The theory of magnetism. In: Cardona M, Fulde P, von Klitzing K, Merlin R, Queisser H-J, Störmer H, editors. Volume 1 of Springer series in solid-state sciences. 2nd ed. Berlin: Springer; 1988.

[18] Mohn P. Magnetism in the solid state. In: Cardona M, Fulde P, von Klitzing K, Merlin R, Queisser H-J, Störmer H, editors. Volume 134 of Springer series in solid-state sciences. Berlin: Springer; 2006.

[19] Hüfner S. Photoelectron spectroscopy. In: Volume 82 of advanced texts in physics. Berlin: Springer; 2003.

[20] de Groot F, Kotani A. Core level spectroscopy of solids. Boca Raton: CRC Press; 2008.

[21] Ebert H, Baumgarten L, Schneider CM, Kirschner J. Polarization dependence of the 2p-core-level photoemission spectra of Fe. Phys Rev B 1991;44:4406.

[22] Schneider C. Photoelectron emission spectroscopy. In: Blügel S, Brückel Th, Schneider CM, Waser R, editors. 41st IFF Spring School "Electronic oxides". Lecture notes. Jülich: Research Center Jülich; 2010.

[23] Schattke W, Van Hove MA. Solid-state photoemission and related methods. Weinheim: Wiley-VCH; 2003.

[24] Kobayashi K. Hard x-ray photoemission spectroscopy. Nucl Instrum Methods Phys Res A 2009;601:32.

[25] Fadley C. X-ray photoelectron spectroscopy: progress and perspectives. J Electron Spectrosc Relat Phenomena 2010;178:2.

[26] Special issue on hard x-ray photoemission, J Electron Spectrosc Rel Phenomena 190, Part B 2013. http://dx.doi.org/10.1016/S0368-2048(13)00212-0.

[27] Sedaghat A, Cracknell A. On the group-theoretical labelling of electronic energy bands in crystalline solids. Physica 1974;71:615.

[28] Gray AX, Minar J, Ueda S, Stone PR, Yamashita Y, Fujii J, et al. Bulk electronic structure of the dilute magnetic semiconductor $Ga_{1-x}Mn_xAs$ through hard x-ray angle-resolved photoemission. Nat Mater 2012;11:957.

[29] Seah M, Dench W. Quantitative electron spectroscopy of surfaces: a standard data base for electron inelastic mean free paths in solids. Surf Interface Anal 1979;1:1.

[30] Woodruff D, Delchar T. Modern techniques of surface science. Cambridge: Cambridge University Press; 1986.

[31] Sekiyama A, Fujiwara H, Imada S, Suga S, Eisaki H, Uchida SI, et al. Mutual experimental and theoretical validation of bulk photoemission spectra of $Sr_{1-x}Ca_xVO_3$. Phys Rev Lett 2004;93:156402.

[32] Tanuma S, Powell CJ, Penn DR. Calculations of electron inelastic mean free paths. IX. Data for 41 elemental solids over the 50 eV to 30 keV range. Surf Interface Anal 2011;43:689.

[33] Thiess S, Kunz C, Cowie BCC, Lee TL, Renier M, Zegenhagen J. Hard x-ray photoelectron spectroscopy from 5–14.5 keV. Solid State Commun 2004;132:589.

[34] Fadley CS. X-ray photoelectron spectroscopy: from origins to future directions. Nucl Instrum Methods Phys Res A 2009;601(1–2):8.

[35] Plucinski L, Oelsner A, Matthes F, Schneider CM. A hemispherical photoelectron spectrometer with 2-dimensional delay-line detector and integrated spin-polarization analysis. J Electron Spectrosc Relat Phenomena 2010;181(2–3):215.

[36] Kolbe M, Lushchyk P, Petereit B, Elmers HJ, Schönhense G, Oelsner A, et al. Highly efficient multichannel spin-polarization detection. Phys Rev Lett 2011;107:207601.

[37] Moulder J, Stickle W, Sobol P, Bomben K. Handbook of X-ray photoelectron spectroscopy. Eden Prairie, MN: Perkin-Elmer; 1992.

[38] Grant JT, Briggs D. Surface analysis by Auger and X-ray photoelectron spectroscopy. Chichester: IM Publications; 2003.

[39] Guenther S, Kaulich B, Gregoratti L, Kiskinova M. Photoelectron microscopy and applications in surface and materials science. Prog Surf Sci 2002;70:187.

[40] Locatelli A, Bauer E. Recent advances in chemical and magnetic imaging of surfaces and interfaces by XPEEM. J Phys Condens Matter 2008;20:093002.

[41] Wiemann C, Patt M, Krug IP, Weber NB, Escher M, Merkel M, et al. A new nanospectroscopy tool with synchrotron radiation: NanoESCA@Elettra. e-J Surf Sci Nanotechnol 2011;9:395.

[42] Tromp RM, Hannon JB, Wan W, Berghaus A, Schaff O. A new aberration-corrected, energy-filtered LEEM/PEEM instrument. II. Operation and results. Ultramicroscopy 2013;127:25.

[43] Wiemann C, Patt M, Cramm S, Escher M, Merkel M, Gloskovskii A, et al. Probing buried layers by photoelectron spectromicroscopy with hard x-ray excitation. Appl Phys Lett 2012;100:223106.

[44] Stöhr J. NEXAFS spectroscopy. Berlin: Springer; 1992.

[45] Margui E, van Grieken R. X-ray fluorescence spectrometry and related techniques. New York: Momentum Press; 2013.

[46] http://xraysweb.lbl.gov/peem2/webpage/Project/TutorialContrast.shtml.

[47] Chen CL, Dong CL, Chen JL, Guo JH, Yang WL, Hsu CC, et al. X-ray spectra and electronic correlations of $FeSe_{1-x}Te_x$. Phys Chem Chem Phys 2011;13:15666.

[48] Bondino F, Magnano E, Malvestuto M, Parmigiani F, McGuire MA, Sefat AS, et al. Evidence for strong itinerant spin fluctuations in the normal state of $CeFeAsO_{0.89}F_{0.11}$ iron-oxypnictide superconductors. Phys Rev Lett 2008;101:267001.

[49] Pietsch U, Holý V, Baumbach T. High-resolution X-ray scattering: from thin films to lateral nanostructures, advanced texts in physics. Berlin: Springer; 2004.

[50] Filatova EO, Jonnard P, André JM. X-ray reflection spectroscopy of the $HfO_2/SiO_2/Si$ system in the region of the O-K absorption edge. Thin Solid Films 2006;500:219.

[51] Filatova EO, Sokolov AA, Kozhevnikov IV, Taracheva EY, Grunsky OS, Schäfers F, et al. Investigation of the structure of thin HfO_2 films by soft x-ray reflectometry techniques. J Phys Condens Matter 2009;21:185012.

[52] Kessler J. Polarized electrons. Ecker G, Lambropoulos P, Walther H, editors. 2nd ed. Springer series on atoms and plasmas. Berlin: Springer; 1985.

[53] Kirschner J. Sources and detectors for spin polarized electrons. In: Feder R, editor. Polarized electrons in surface physics. Singapore: World Scientific; 1985.

[54] Escher M, Weber NB, Merkel M, Plucinski L, Schneider CM. Ferrum: a new highly efficient spin detector for electron spectroscopy. e-J Surf Sci Nanotechnol 2011;9:340.

[55] Meier F, Dil H, Lobo-Checa J, Patthey L, Osterwalder J. Quantitative vectorial spin analysis in angle-resolved photoemission: Bi/Ag(111) and Pb/Ag(111). Phys Rev B 2008;77:165431.

[56] Hillebrecht FU, Jungblut R, Kisker E. Spin polarization of the metallic Fe 3s photoemission spectrum. Phys Rev Lett 1990;65:2450.

[57] Carbone C, Kachel T, Rochow R, Gudat W. Spin polarized photoemission from the Fe 3s core levels. Z Phys B Condens Matter 1990;79:325.

[58] Kirschner J. Inelastic electron scattering by ferromagnets. In: Feder R, editor. Polarized electrons in surface physics. Singapore: World Scientific; 1985.

[59] Klebanoff L, Van Campen D, Pouliot R. Spin-resolved and high-energy-resolution XPS studies of cobalt metal and a cobalt magnetic glass. Phys Rev B 1994;49:2047.

[60] Kakehashi Y. Theory of photoelectron spectra from inner-core states. II. Temperature effect and role of d-electron correlations. Phys Rev B 1985;32:1607.

[61] See AK, Klebanoff LE. Spin-resolved x-ray photoemission from ferromagnetic nickel. Phys Rev B 1995;51:11002.

[62] Clauberg R, Gudat W, Kisker E, Kuhlmann E, Rothberg GM. Nature of the resonant 6-eV satellite in Ni: photoelectron spin-polarization analysis. Phys Rev Lett 1981;47:1314.

[63] See AK, Klebanoff LE. Nature of extra-atomic core-hole screening in ferromagnetic nickel. Phys Rev Lett 1995;74:1454.

[64] Menchero J. Spin polarization and magnetic circular dichroism in photoemission from the 2p core level of ferromagnetic Ni. Phys Rev Lett 1996;76:3208.

[65] Baumgarten L, Schneider C, Schäfers F, Petersen H, Kirschner J. Magnetic x-ray dichroism in core-level photoemission from ferromagnets. Phys Rev Lett 1990;65(4):492.

[66] Fano U. Spin orientation of photoelectrons ejected by circularly polarized light. Phys Rev 1969;178:131.

[67] Fano U. Spin orientation of photoelectrons: erratum and addendum. Phys Rev 1969;184:250.

[68] Meier F, Zakharchenya BP. Optical orientation. Volume 8 of modern problems in condensed matter sciences. Amsterdam: North-Holland; 1984.

[69] Rose HB, Fanelsa A, Kinoshita T, Roth C, Hillebrecht FU, Kisker E. Spin-orbit-induced spin polarization in W 4f photoemission. Phys Rev B 1996;53:1630.

[70] Fadley CS. Recent developments in photoelectron diffraction. In: Kanamori J, Kotani A, editors. Core-level spectroscopy in condensed systems. Volume 81 of Springer series in solid-state sciences. Berlin: Springer; 1988. p. 236.

[71] Schneider C, Venus D, Kirschner J. Strong x-ray magnetic circular dichroism in a "forbidden geometry" observed via photoemission. Phys Rev B 1992;45:5041.

[72] Venus D, Baumgarten L, Schneider C, Boeglin C, Kirschner J. Crystalline symmetry effects in x-ray magnetic circular dichroism in angle-resolved core-level photoemission. J Phys Condens Matter 1993;5:1239.

[73] Hillebrecht FU, Rose HB, Kinoshita T, Idzerda YU, van der Laan G, Denecke R, et al. Photoelectron diffraction in magnetic linear dichroism. Phys Rev Lett 1995;75:2883.

[74] Venus D. Interrelation of magnetic-dichroism effects seen in the angular distribution of photoelectrons from surfaces. Phys Rev B 1994;49:8821.

[75] Venus D. Magnetic circular dichroism in angular distributions of core-level photoelectrons. Phys Rev B 1993;43:6144.

[76] Schellenberg R, Kisker E, Fanelsa A, Hillebrecht F, Menchero J, Kaduwela A, et al. Photoelectron diffraction: a source for magnetic dichroism in angle-resolved photoemission from ferromagnets. Phys Rev B 1998;57:14310.

[77] Roth C, Hillebrecht F, Rose H, Kisker E. Linear magnetic dichroism in angular resolved Fe 3p core level photoemission. Phys Rev Lett 1993;70:3479.

[78] Schneider CM, Pracht U, Kuch W, Chasse A, Kirschner J. Magnetic dichroism in photoemission as a spin-resolving probe for electronic correlations. Phys Rev B 1996;54:15618.

[79] Sirotti F, Panaccione G, Rossi G. Atom-specific surface magnetometry. Phys Rev B 1995;52: R17063.

[80] Starke K, Navas E, Baumgarten L, Kaindl G. Strong magnetic circular dichroism in 4f photoemission from gadolinium metal. Phys Rev B 1993;48:1329.

[81] Weller D, Alvarado S, Gudat W, Schröder K, Campagna M. Observation of surface-enhanced magnetic order and magnetic surface reconstruction on Gd(0001). Phys Rev Lett 1985;54:1555.

[82] Berger M, Hubbell J, Seltzer S, Chang J, Coursey J, Sukumar R, et al. NIST XCOM: photon cross sections database; 2011.

[83] Kozina X, Fecher GH, Stryganyuk G, Ouardi S, Balke B, Felser C, et al. Magnetic dichroism in angle-resolved hard x-ray photoemission from buried layers. Phys Rev B 2011;84:054449.

[84] Wende H. Recent advances in x-ray absorption spectroscopy. Rep Prog Phys 2004;67:2105.

[85] Okamoto K, Nagamatsu S-i, Voronina EV, Miyanaga T, Fujikawa T. Fe K-edge x-ray magnetic circular dichroism analyses of Fe/M (M = Si, Al, Sn) alloys. J Electron Spectrosc Relat Phenomena 2004;141:5.

[86] Chen C, Sette F, Ma Y, Modesti S. Soft-x-ray magnetic circular dichroism at the $L_{2,3}$ edges of nickel. Phys Rev B 1990;42:7262.

[87] Boeglin C, Beaurepaire E, Halte V, Lopez-Flores V, Stamm C, Pontius N, et al. Distinguishing the ultrafast dynamics of spin and orbital moments in solids. Nature 2010;465:458.

[88] Thole B, Carra P, Sette F, van der Laan G. X-ray circular dichroism as a probe of orbital magnetization. Phys Rev Lett 1992;68:1943.

[89] Carra P, Thole BT, Altarelli M, Wang X. X-ray circular dichroism and local magnetic fields. Phys Rev Lett 1993;70:694.

[90] Thole B, van der Laan G. Sum rules for magnetic dichroism in rare earth 4f photoemission. Phys Rev Lett 1993;70:2499.

[91] Wu R, Freeman A. Limitation of the magnetic-circular-dichroism spin sum rule for transition metals and importance of the magnetic dipole term. Phys Rev Lett 1994;73:1994.

[92] Chen CT, Idzerda YU, Lin H-J, Smith NV, Meigs G, Chaban E, et al. Experimental confirmation of the x-ray magnetic circular dichroism sum rules for iron and cobalt. Phys Rev Lett 1995;75:152.

[93] Huang DJ, Chang CF, Chen J, Lin HJ, Chung SC, Jeng HT, et al. Orbital moments of CrO_2 and Fe_3O_4 studied by MCD in soft x-ray absorption. J Electron Spectrosc Relat Phenomena 2004;137–140:633.

[94] Krug IP. Magnetic proximity effects in highly-ordered transition metal oxide heterosystems studied by soft x-ray photoemission electron microscopy. Ph.D. thesis, Essen: University Duisburg-Essen; 2008.

[95] Gambardella P, Dallmeyer A, Maiti K, Malagoli MC, Eberhardt W, Kern K, et al. Ferromagnetism in one-dimensional monatomic metal chains. Nature (London) 2002;416:301.

[96] Wende H, Bernien M, Luo J, Sorg C, Ponpandian N, Kurde J, et al. Substrate-induced magnetic ordering and switching of iron porphyrin molecules. Nat Mater 2007;6:516.

[97] Gambardella P, Dhesi S, Gardonio S, Grazioli C, Ohresser P, Carbone C. Localized magnetic states of Fe, Co, and Ni impurities on alkali metal films. Phys Rev Lett 2002;88:047202.

[98] Bernien M, Miguel J, Weis C, Ali ME, Kurde J, Krumme B, et al. Tailoring the nature of magnetic coupling of Fe-porphyrin molecules to ferromagnetic substrates. Phys Rev Lett 2009;102:047202.

[99] Kunes J, Oppeneer P, Valencia S, Abramsohn D, Mertins H-C, Gudat W, et al. Understanding the XMLD and its magnetocrystalline anisotropy at the $L_{2,3}$-edges of 3d transition metals. J Magn Magn Mater 2004;272–276:2146.

[100] Alders D, Vogel J, Levelut C, Peacor S, Hibma T, Sacchi M, et al. Magnetic x-ray dichroism study of the nearest-neighbor spin-spin correlation function and long-range magnetic order parameter in antiferromagnetic NiO. Europhys Lett 1995;32:259.

[101] Alders D, Tjeng LH, Voogt FC, Hibma T, Sawatzky G, Chen C, et al. Temperature and thickness dependence of magnetic moments in NiO epitaxial films. Phys Rev B 1998;57:11623.

[102] Nogues J, Schuller IK. Exchange bias. J Magn Magn Mater 1999;192:203.

[103] Berkowitz A, Takano K. Exchange anisotropy—a review. J Magn Magn Mater 1999;200:552.

[104] Krug IP, Hillebrecht FU, Gomonaj H, Haverkort MW, Tanaka A, Tjeng LH, et al. Magnetic coupling in highly ordered $NiO/Fe_3O_4(110)$: ultrasharp magnetic interfaces vs. long-range magnetoelastic interactions. Europhys Lett 2008;81:17005.

[105] Krug IP, Hillebrecht FU, Haverkort MW, Tanaka A, Tjeng LH, Gomonay H, et al. Impact of interface orientation on magnetic coupling in highly ordered systems: a case study of the low-indexed Fe_3O_4/NiO interfaces. Phys Rev B 2008;78:064427.

[106] Fink J, Schierle E, Weschke E, Geck J. Resonant elastic soft x-ray scattering. Rep Prog Phys 2013;76:056502.

[107] Hannon JP, Trammell GT, Blume M, Gibbs D. X-ray resonance exchange scattering. Phys Rev Lett 1988;61:1245.

[108] Tonnerre JM, Sève L, Raoux D, Soullié G, Rodmacq B, Wolfers P. Soft x-ray resonant magnetic scattering from a magnetically coupled Ag/Ni multilayer. Phys Rev Lett 1995;75:740.

[109] Hecker M, Valencia S, Oppeneer P, Mertins H-C, Schneider C. Polarized soft-x-ray reflection spectroscopy of giant magnetoresistive Co/Cu multilayers. Phys Rev B 2005;72:054437.

[110] Bürgler DE, Grünberg P, Demokritov S, Johnson M. Interlayer exchange coupling in layered magnetic structures. In: Buschow K, editor. Handbook of magnetic materials, vol. 13. Amsterdam: Elsevier Science; 2001. p. 3.

[111] Tonnerre JM, De Santis M, Grenier S, Tolentino HCN, Langlais V, Bontempi E, et al. Depth magnetization profile of a perpendicular exchange coupled system by soft-x-ray resonant magnetic reflectivity. Phys Rev Lett 2008;100:157202.

[112] Kao C-C, Chen C, Johnson E, Hastings J, Lin H, Ho G, et al. Dichroic interference effects in circularly polarized soft-x-ray resonant magnetic scattering. Phys Rev B 1994;50:9599.

[113] Mertins H-C, Oppeneer PM, Kunes J, Gaupp A, Abramsohn D, Schäfers F. Observation of the x-ray magneto-optical Voigt effect. Phys Rev Lett 2001;87:047401.

[114] Valencia S, Kleibert A, Gaupp A, Rusz J, Legut D, Bansmann J, et al. Quadratic x-ray magneto-optical effect upon reflection in a near-normal-incidence configuration at the M edges of 3d-transition metals. Phys Rev Lett 2010;104:187401.

[115] Mathias S, La-O-Vorakiat C, Grychtol P, Granitzka P, Turgut E, Shaw JM, et al. Probing the timescale of the exchange interaction in a ferromagnetic alloy. Proc Natl Acad Sci USA 2012;109:4792.

[116] Oppeneer PM, Mertins H-C, Abramsohn D, Gaupp A, Gudat W, Kunes J, et al. Buried antiferromagnetic films investigated by x-ray magneto-optical reflection spectroscopy. Phys Rev B 2003;67:052401.

[117] Mertins H-C, Valencia S, Gaupp A, Gudat W, Oppeneer P, Schneider C. Magnetooptical polarization spectroscopy with soft x-rays. Appl Phys A Mater Sci Process 2005;80:1011.

[118] Nolting F, Legut D, Rusz J, Oppeneer PM, Woltersdorf G, Back CH. Anisotropy of the $L_{2,3}$ x-ray magnetic linear dichroism of Fe films on GaAs: experiment and ab initio theory. Phys Rev B 2010;82:184415.

[119] Keil UD, Gerritsen HJ, Haverkort JEM, Wolter JH. Generation of ultrashort electrical pulses with variable pulse widths. Appl Phys Lett 1995;66:1629.

[120] Stoll H, Puzic A, van Waeyenberge B, Fischer P, Raabe J, Buess M, et al. High-resolution imaging of fast magnetization dynamics in magnetic nanostructures. Appl Phys Lett 2004;84:3328.

[121] Fischer P, Denbeaux G, Ono T, Okuno T, Eimüller T, Goll D, et al. Study of magnetic domains by magnetic soft x-ray transmission microscopy. J Phys D 2002;35:2391.

[122] Fischer P. Studying nanoscale magnetism and its dynamics with soft x-ray microscopy. IEEE Trans Magn 2008;44:1900.

[123] van Waeyenberge B, Puzic A, Stoll H, Chou K, Tyliszczak T, Hertel R, et al. Magnetic vortex core reversal by excitation with short bursts of an alternating field. Nature 2006;444:461.

[124] Chou KW, Puzic A, Stoll H, Dolgos D, Schütz G, Van Waeyenberge B, et al. Direct observation of the vortex core magnetization and its dynamics. Appl Phys Lett 2007;90:202505.

[125] Kammerer M, Weigand M, Curcic M, Noske M, Sproll M, Vansteenkiste A, et al. Magnetic vortex core reversal by excitation of spin waves. Nat Commun 2011;2:279.

[126] Krasyuk A, Oelsner A, Nepijko S, Kuksov A, Schneider C, Schönhense G. Time-resolved photoemission electron microscopy of magnetic field and magnetization changes. Appl Phys A Mater Sci Process 2003;76:836.

[127] Schneider C, Kuksov A, Krasyuk A, Oelsner A, Neeb D, Nepijko S, et al. Incoherent magnetization rotation observed in sub-nanosecond time-resolving XPEEM studies. Appl Phys Lett 2004;85:2562.

[128] Hubert A, Schäfer R. Magnetic domains. Berlin: Springer; 1998.

[129] Stiles MD, Miltat J. Spin-transfer torque and dynamics. In: Hillebrands B, Thiaville A, editors. Spin dynamics in confined magnetic structures. Berlin: Springer Verlag; 2006. p. 225.

[130] Farle M. Ferromagnetic resonance of ultrathin metallic layers. Rep Prog Phys 1998;61:755.

[131] Belmeguenai M, Martin T, Woltersdorf G, Maier M, Bayreuther G. Frequency- and time-domain investigation of the dynamic properties of interlayer-exchange-coupled $Ni_{81}Fe_{19}/Ru/Ni_{81}Fe_{19}$ thin films. Phys Rev B 2007;76:104414.

[132] Beguhn S, Zhou Z, Rand S, Yang X, Lou J, Sun NX. A new highly sensitive broadband ferromagnetic resonance measurement system with lock-in detection. J Appl Phys 2012;111:07A503.

[133] Sun NX, Wang SX, Silva TJ, Kos AB. High-frequency behavior and damping of Fe-Co-N-based high-saturation soft magnetic films. IEEE Trans Magn 2002;38(1):146.

[134] Bailey WE, Cheng L, Keavney DJ, Kao CC, Vescovo E, Arena DA. Precessional dynamics of elemental moments in a ferromagnetic alloy. Phys Rev B 2004;70:172403.

[135] Boero G, Mouaziz S, Rusponi S, Bencok P, Nolting F, Stepanow S, et al. Element-resolved x-ray ferrimagnetic and ferromagnetic resonance spectroscopy. New J Phys 2008;10:013011.

[136] Beaurepaire E, Merle JC, Daunois A, Bigot JY. Ultrafast spin dynamics in ferromagnetic nickel. Phys Rev Lett 1996;76:4250.

[137] Pfeifer T, Spielmann C, Gerber G. Femtosecond x-ray science. Rep Prog Phys 2006;69:443.

[138] Brabec T, Krausz F. Intense few-cycle laser fields: Frontiers of nonlinear optics. Rev Mod Phys 2000;72:545.

[139] McKinnie I, Kapteyn H. High-harmonic generation: ultrafast lasers yield x-rays. Nat Photonics 2010;4:149.

[140] Ganeev RA. Strong field-induced frequency conversion of laser radiation in plasma plumes: recent achievements. Sci World J 2013;2013:18.

[141] Schoenlein RW, Chattopadhyay S, Chong HHW, Glover TE, Heimann PA, Shank CV, et al. Generation of femtosecond pulses of synchrotron radiation. Science 2000;287:2237.

[142] Labat M, Hosaka M, Mochihashi A, Shimada M, Katoh M, Lambert G, et al. Coherent harmonic generation on UVSOR-II storage ring. Eur Phys J D 2007;44:187.

[143] Stupakov G. Using the beam-echo effect for generation of short-wavelength radiation. Phys Rev Lett 2009;102:074801.

[144] Düsterer S, Radcliffe P, Bostedt C, Bozek J, Cavalieri AL, Coffee R, et al. Femtosecond x-ray pulse length characterization at the linac coherent light source free-electron laser. New J Phys 2011;13:093024.

[145] Stamm C, Kachel T, Pontius N, Mitzner R, Quast T, Holldack K, et al. Femtosecond modification of electron localization and transfer of angular momentum in nickel. Nat Mater 2007;6:740.

[146] Carva K, Legut D, Oppeneer PM. Influence of laser-excited electron distributions on the x-ray magnetic circular dichroism spectra: implications for femtosecond demagnetization in Ni. EPL 2009;86:57002.

[147] Kirilyuk A, Kimel AV, Rasing T. Ultrafast optical manipulation of magnetic order. Rev Mod Phys 2010;82:2731.

[148] Stöhr J. Exploring the microscopic origin of magnetic anisotropies with x-ray magnetic circular dichroism (XMCD) spectroscopy. J Magn Magn Mater 1999;200:470.

[149] La-O-Vorakiat C, Siemens M, Murnane MM, Kapteyn HC, Mathias S, Aeschlimann M, et al. Ultrafast demagnetization dynamics at the M edges of magnetic elements observed using a tabletop high-harmonic soft x-ray source. Phys Rev Lett 2009;103:257402.

[150] Turgut E, La-O-Vorakiat C, Shaw JM, Grychtol P, Nembach HT, Rudolf D, et al. Controlling the competition between optically induced ultrafast spin-flip scattering and spin transport in magnetic multilayers. Phys Rev Lett 2013;110:197201.

[151] Rudolf D, La-O-Vorakiat C, Battiato M, Adam R, Shaw JM, Turgut E, et al. Ultrafast magnetization enhancement in metallic multilayers driven by superdiffusive spin current. Nat Commun 2012;3:1037.

[152] Chapman HN, Barty A, Bogan MJ, Boutet S, Frank M, Hau-Riege SP, et al. Femtosecond diffractive imaging with a soft-x-ray free-electron laser. Nat Phys 2006;2:839.

[153] Gutt C, Stadler LM, Streit-Nierobisch S, Mancuso AP, Schropp A, Pfau B, et al. Resonant magnetic scattering with soft x-ray pulses from a free-electron laser operating at 1.59 nm. Phys Rev B 2009;79:212406.

[154] Battiato M, Carva K, Oppeneer PM. Superdiffusive spin transport as a mechanism of ultrafast demagnetization. Phys Rev Lett 2010;105:027203.

[155] Magnuson M, Duda LC, Butorin SM, Kuiper P, Nordgren J. Large magnetic circular dichroism in resonant inelastic x-ray scattering at the Mn L-edge of Mn-Zn ferrite. Phys Rev B 2006;74:172409.

[156] Yablonskikh MV, Yarmoshenko YM, Grebennikov VI, Kurmaev EZ, Butorin SM, Duda LC, et al. Origin of magnetic circular dichroism in soft x-ray fluorescence of Heusler alloys at threshold excitation. Phys Rev B 2001;63:235117.

[157] Schlage K, Röhlsberger R. Nuclear resonant scattering of synchrotron radiation: applications in magnetism of layered structures. J Electron Spectrosc Relat Phenomena 2013;189:187.

[158] Schattschneider P, Rubino S, Hebert C, Rusz J, Kuneš J, Novák P, et al. Detection of magnetic circular dichroism using a transmission electron microscope. Nature 2006;441:486.

[159] Muto S, Rusz J, Tatsumi K, Adam R, Arai S, Kocevski V, et al. Quantitative characterization of nanoscale polycrystalline magnets with electron magnetic circular dichroism. Nat Commun 2014;5:3138.

[160] Rusz J, Eriksson O, Novák P, Oppeneer PM. Sum rules for electron energy loss near edge spectra. Phys Rev B 2007;76:060408.

[161] Schattschneider P, Stöger-Pollach M, Rubino S, Sperl M, Hurm C, Zweck J, et al. Detection of magnetic circular dichroism on the two-nanometer scale. Phys Rev B 2008;78:104413.

[162] See, for example, http://www.ulvac-phi.com.

[163] Landolt M, Allenspach R, Taborelli M. Spin polarized Auger electrons from magnetic surfaces. Surf Sci 1986;178:311.

CHAPTER 3

Magnetization Dynamics

Eric Montoya*, Thomas Sebastian†, Helmut Schultheiss†, Bret Heinrich*, Robert E. Camley‡, Zbigniew Celinski‡

**Surface Science Lab, Department of Physics, Simon Fraser University, Burnaby, BC, Canada*

†Helmholtz-Zentrum Dresden—Rossendorf, Institute of Ion Beam Physics and Materials Research, Dresden, Germany

‡Center for Magnetism and Magnetic Nanostructures, University of Colorado at Colorado Springs, Colorado Springs, CO, USA

1 INTRODUCTION

Magnetization dynamics, where one investigates the time-dependent motion of the magnetization rather than its equilibrium behavior, are fundamental to the properties of magnetic materials. It is worthwhile remembering that the concept of spin waves was introduced by Bloch [1] in 1930 in order to calculate the magnetic contribution to the specific heat in macroscopic samples. Much more recently, thermal measurements in multilayers and nanoparticles have been connected to the spin waves that are allowed in confined structures.

In addition, magnetization dynamics also determine the interaction of magnetic materials with electromagnetic radiation. In simple cases, the magnetization dynamics (basically the precessional motion of the magnetization) can be accounted for by a frequency-dependent permeability tensor. An incident electromagnetic wave then interacts with the magnetic system through the permeability, and can have its wavelength, phase, transmission, reflection, and absorption all influenced by the properties of the magnetic material or by an external magnetic field that acts on the material.

Finally, magnetization dynamics are also fundamental for understanding and measuring phase transitions. In particular, the behavior of low-frequency spin waves can be used to indicate phase transitions because they are often associated with a soft mode where the frequency is driven to zero as a function of applied field or other parameter. A detailed understanding of magnetization

Magnetism of Surfaces, Interfaces, and Nanoscale Materials. http://dx.doi.org/10.1016/B978-0-444-62634-9.00003-5

reversal, vital to the magnetic recording industry, also depends on the magnetization dynamics.

Dynamical measurements provide a wealth of information about magnetic materials. They are capable of determining the magnetic parameters (magnetization, bulk and surface anisotropies, exchange constants, dipole fields) of individual films and can also determine the strength of the exchange coupling (antiferromagnetic and ferromagnetic!) between coupled films. As pointed out above, spin-wave frequencies can also be sensitive tools to measure the existence of phase transitions.

Dynamical measurements provide one additional parameter, the width of the absorption line, that characterizes the damping in the system and measures how it can be influenced by composition, currents, spatial variations, etc. For example, ultrathin magnetic structures or magnetic nanoparticles are not ideal systems and have many imperfections, such as edges, atomic steps, and dislocations. As a consequence, different local environments result in different effective fields acting on the magnetic moments from one atomic position to another. This means that the material is not magnetically homogeneous. In a ferromagnetic resonance (FMR) experiment, this would ultimately result in a change in the linewidth of the microwave absorption signal, and only FMR is capable of detecting such changes. Narrower linewidths, for example, can indicate an increased magnetic quality for an individual ferromagnetic film. Similarly, linewidth is an important parameter for measuring the changes in damping introduced by spin (polarized) currents that also carry angular momentum.

As pointed out earlier, the magnetization dynamics play an important role in the electromagnetic properties of magnetic materials. This has a fundamental application in signal processing devices such as phase shifters, band-pass filters, band-stop filters, isolators, and circulators. The operational frequency range depends on the internal and external fields in the magnetic material. Typical frequencies range from a few megahertz to a few terahertz, depending on the material or structure.

In recent decades, a new field of electronics that utilizes the electron's spin degree of freedom, known as spintronics, has received much attention in both fundamental physics research and industrial applications. Spintronics requires the development of magnetic thin-film structures having a wide range of magnetic properties. The bulk, interface, and exchange biased magnetic anisotropies [2], the exchange stiffness [3,4], and interlayer exchange coupling are important for high-density perpendicular magnetic recording [5,6], magnetic tunneling junction (MTJ) sensors [7,8], and spin-transfer torque (STT) devices [9–13]. In particular, ultrathin film structures allow one to create a controllable perpendicular anisotropy having the magnetic moment oriented perpendicular to the film surface with a variable magnetic energy. Equally important are dynamic properties: the Landé factor g and magnetic damping. STT devices utilize spin (polarized) currents to switch the magnetization of a ferromagnetic layer in a spin-valve structure. The critical current at which the free layer in an STT device will switch direction is dependent upon the Gilbert damping parameter [13]. Also of importance for future spintronics devices is determining the spin-diffusion length in a variety of materials. This can be done by examining additional damping in ferromagnet/normal metal and ferromagnet/normal metal/ferromagnet structures [14–17].

FMR is a well-understood experimental technique [18] that has proven to be an invaluable tool to probe the static and dynamic magnetic properties of ultrathin films and multilayer nanostructures. In order to determine the magnetic parameters, one needs to carry out FMR measurements at a wide range of frequencies. For this reason, it is of great advantage to have a broadband system where measurements can be made at many frequencies without the need to remount samples in different waveguides or cavities.

In this chapter, we concentrate on experimental techniques and results for the measurement of the dynamical motion of the magnetization. We discuss a variety of instruments and techniques that involve measurements in the frequency domain or directly in the time domain. The chapter is organized as follows. In Sections 2 and 3, we give some review material which is helpful to understand the broader picture of magnetization dynamics, primarily in ferromagnets. Section 2 collects some analytic formulas that are commonly used in analyzing experimental data and provides a general background on several types of excitations that may be seen. Section 3 discusses the kinds of initial measurements that one might use to characterize an unknown magnetic sample. Section 4 examines FMR in more detail, introducing magnetic damping and discussing both field-swept and frequency-swept FMR. Finally, Section 5 provides a review of a second common technique for measuring magnetization dynamics—Brillouin light scattering (BLS). Although the main emphasis is on experimental methods, some analytic developments are also included. Details on the analytic techniques [19] are given in other references and in this handbook.

2 ANALYTIC RESULTS

In this section, we collect some standard analytic results that are helpful in understanding the connection between a measured frequency and the magnetic parameters. Among these formulas, probably the most important is the so-called Kittel equation, which gives the frequency of a uniform mode excitation in a ferromagnetic ellipsoid. This is given by

$$f = \gamma' \sqrt{(H + (N_y - N_z)4\pi M_s)(H + (N_x - N_z)4\pi M_s)}. \qquad (1)$$

Here M_s is the saturation magnetization (assumed to be in the z direction), γ' is the reduced gyromagnetic ratio ($\gamma' = \gamma/2\pi$), and H is the magnitude of the applied field, assumed to be in the z direction. N_x, N_y, and N_z are demagnetizing factors that are dependent on the shape of the ellipsoid. One reason why this equation is so important is that the uniform mode, a mode that is spatially uniform, often is the magnetic mode that interacts most strongly with a spatially uniform driving microwave field. Thus, it frequently results in the strongest signal and can be easily measured and identified.

We now give some additional results for specific cases, again for ferromagnets.

2.1 Infinitely Extended Flat Plate: In-Plane Field

We consider a system with fourfold in-plane anisotropy and where the magnetization and external field are along one of the anisotropy axes. In this case, the Kittel equation is modified and the frequency is given by the following formula:

$$f = \gamma' \sqrt{(H + H_a)(H + H_a + 4\pi M_s)}. \qquad (2)$$

where H_a is an anisotropy field.

2.2 Spin Waves Propagating Perpendicular to the Surface, Conducting Films

In a thin film one may expect that the wavelengths for spin waves propagating perpendicularly to the thickness of the film are quantized. This leads to standing-wave modes where the effects of the exchange field can be dramatic and gives a method to sensitively measure the exchange constant [20]. The frequency of the standing-wave modes in a thin film with magnetization parallel to the surface is given by the formula:

$$f = \gamma' \sqrt{(H + Dk_\perp^2)(H + Dk_\perp^2 + 4\pi M_s)}. \qquad (3)$$

Here D is the exchange stiffness constant and $k_\perp$ is the wave vector perpendicular to the surfaces of the film. If one uses the idea that there are standing waves with free-end boundary conditions for the magnetization, then a reasonable approximation is given by

$$k_{\perp,N} = \frac{N\pi}{d}, \qquad (4)$$

where $N = 0, 1, 2, \ldots$ and d is the thickness of the film.

The exact boundary conditions and quantization conditions (Eq. 4) depend on the energy at the boundary between the film and the surrounding medium (ambient, nonferromagnetic film). Often the primary contribution comes from a surface uniaxial anisotropy perpendicular to the film surface:

$$E_u^{int} = -\frac{1}{2} K_s \cos^2(\theta), \tag{5}$$

where K_s is the interface energy in erg/cm^2 and θ is the angle between the magnetic moment and the film normal. The effective field of the interface anisotropy and exchange field then results in boundary conditions for spin waves for the in-plane FMR (the magnetization in the plane) on the front side of the film:

$$D\frac{\partial m_\perp}{\partial n} + \frac{K_s}{M_s} m_\perp = 0 \tag{6}$$

and

$$\frac{\partial m_\parallel}{\partial n} = 0, \tag{7}$$

where $\perp$ and $\|$ correspond to the perpendicular and parallel components of the rf magnetization with respect to n, the direction of inner normal of the film surface. The boundary condition for the back side of the film is given by the same equations but with a reversed sign for the anisotropy term. The case of zero interface anisotropy leads to free surface spins, resulting in the spin-wave spectrum shown in Eq. (4).

In the presence of conducting films, the energy of spin waves changes and requires a full simultaneous solution of the LLG equations and Maxwell's equations satisfying the electromagnetic boundary conditions (continuity of the rf electric field, E, and magnetic fields H), and the magnetic boundary conditions given by Eqs. (6) and (7) on both sides of the film.

A solution of Maxwell's and the LLG equations for spin waves with wave vector $k_\perp$ is sought in the form $e^{i\omega t - k_\perp z}$, where z is the direction perpendicular to the film surface. This leads to a secular equation for $k_\perp^2$. The equation is cubic in $k_\perp^2$, meaning three waves propagate forward (along $+z$) and three waves propagate backwards (along $-z$) between the film boundaries. The simultaneous solution of equations for the rf electric, magnetic, and magnetization components requires the determination of six amplitudes for the propagating waves inside the film plus two amplitudes of the rf electric field at the thin film surfaces. That represents eight unknown parameters: one has four electromagnetic boundary conditions (two on each side of the film) and four magnetic boundary conditions, given by Eqs (6) and (7), from both sides of the film. Solution of these equations is straightforward but cumbersome; see Frait and Fraitova in the chapter on low-energy spin-wave excitation in highly conducting thin films and surfaces [21].

When there is a homogeneous rf field inside the film and for zero interface energy $K_s = 0$, one can excite only the homogeneous spin wave mode, $k_\perp = 0$. However, in metallic films the rf driving field is inhomogeneous due to eddy currents which damp the field as it penetrates into the material. In this case one can also excite spin waves with $N \geq 1$. An inhomogeneous rf field can also be generated by proper insertion of the sample in a microwave system. The film substrates are usually semiconductors with low conductivity. If such a sample is inserted outside of the microwave waveguide without any metallic foil backing, then the distribution of the rf field inside the film is triangular even when its thickness is much less than the skin depth (see Fig. 17); in this case one obtains again spin waves with $N \geq 1$ in FMR. Spin waves are usually more easily observed in perpendicular FMR. Further reading on measurements of spin waves using FMR can be found in Ref. [22].

It is interesting to note that interface or surface anisotropy can create a surface FMR mode. In this mode, the rf magnetization amplitude decreases exponentially going away from the surface. From the above equations, one can derive both the surface-mode wave number k_{sur} and the shift in FMR field. The exponential decay length for the surface mode in in-plane FMR requires positive K_s.

The decay wave vector k_{sur} is given from Eq. (6) by $k_{sur} = K_s/DM_s$. This leads to an effective positive field $H_{sur} = Dk_{sur}^2 = (1/D)(K_s^2/M_s^2)$ that is independent of the film conductivity and microwave frequency. In this case H_{sur} has a negative sign in the FMR resonance condition, leading to a higher resonance field than that corresponding to the first bulk excitation mode. The surface mode can be very intense and easily confused with the main FMR mode (see Figure 5 in Ref. [21]). In fact, for films thicker than $1/k_{sur}$ the intensity of surface modes is independent of the film thickness.

2.3 Surface Modes for Spin Waves Propagating Parallel to the Film Surface

In addition to the spin waves discussed above, there is another class of magnetic excitations which may occur. These are surface modes, modes which are localized near one surface of the material. We consider a semi-infinite material, with propagation parallel to the surface and at an angle θ with respect to the applied field [23]. As illustrated schematically in Fig. 1, the propagation directions for the surface modes are restricted to a range of angles where the critical angle is given by

$$\theta_c = \sin^{-1}\left(\frac{H}{H + 4\pi M_s}\right). \tag{8}$$

The frequency of the surface mode depends on the angle of propagation and is given by

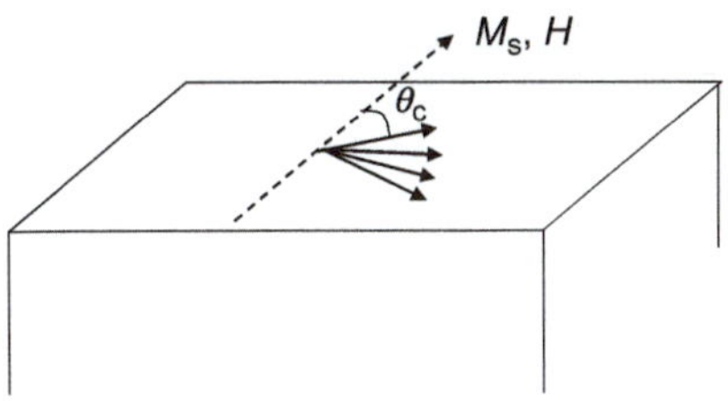

FIGURE 1
Illustration of the allowed directions of propagation for a surface spin wave on a semi-infinite ferromagnet.

$$f = \frac{1}{2}\gamma'[H/\sin\theta + (H + 4\pi M_s)\sin\theta]. \tag{9}$$

In the most typical case, one examines propagation perpendicular to the applied field. In this case, the previous equation reduces to

$$f = \gamma'(H + 2\pi M_s). \tag{10}$$

We note that these magnetic surface modes are generally nonreciprocal [24] in that $\omega(+k_{\|}) \neq \omega(-k_{\|})$, i.e., reversing the direction of the wave vector parallel to the surface, $k_{\|}$, generally results in propagation at a different frequency.

For a film of finite thickness, one can obtain the frequency of a surface mode for propagation perpendicular to the applied field:

$$f = \gamma'\sqrt{(H + 2\pi M_s)^2 - (2\pi M_s)^2 e^{-2|k_{\|}|d}}. \tag{11}$$

The equation describes the frequency of the Damon–Eshbach (DE) mode. Note that in the limit of a thick film, $\boldsymbol{k}_{\|}\boldsymbol{d} \gg 1$, the frequency corresponds to the semi-infinite result of Eq. (10). It is interesting to note that it is also the frequency for a mode in a cylinder with the magnetization parallel to the cylinder axis.

Notice that the DE surface mode has nothing to do with the FMR surface mode described earlier, where $k_{\|} = 0$; in FMR, the surface mode is driven by the interface anisotropy with positive K_s in Eq. (6). The surface mode in FMR is a consequence of the boundary condition with $k_{\|} = 0$, while the DE mode can have a large in-plane wave vector $k_{\|}$.

For ultrathin films where $k_{\|}d \ll 1$, the DE surface mode essentially becomes a bulk mode. One can rewrite Eq. (11) as

$$\left(\frac{f}{\gamma'}\right)^2 = H_1 H_2, \tag{12}$$

where

$$H_1 = H + Dk^2 + \left(\frac{2\pi M_s k_y^2 d}{k}\right) \tag{13}$$

and

$$H_2 = H + 4\pi M_s + Dk^2 - 2\pi M_s dk. \tag{14}$$

Here k_x and k_y are the components of the in-plane wave vector, $k_\| = \left(k_x^2 + k_y^2\right)^{1/2}$, parallel and perpendicular to the saturation magnetization, respectively. Notice that in this case one gets a bulk-like FMR mode which is shifted by a dipolar fields in the equation of motion for an in-plane spin wave with the k wave vector at an angle θ with respect to the in-plane saturation magnetization, $\sin\theta = k_y/k$ (see further details in Ref. [25]).

2.4 Infinitely Extended Flat Plate—Perpendicular-to-Plane Magnetization

If one sets the magnetization, uniaxial anisotropy, and applied magnetic field all perpendicular to the plane, one finds the following frequency:

$$f = \gamma'(H + H_a - 4\pi M_s). \tag{15}$$

It is interesting to note that as H is reduced from a large value, the frequency can be driven to zero, indicating a phase transition, where the magnetization will no longer lie perpendicular to the surface, but will begin to cant away from the surface normal. We will see an example of how this can be useful in Section 3.

2.5 Nanostructures

We have already pointed out that the Kittel equation can be used for ferromagnetic ellipsoids, so it is immediately suited for use with nanoparticles. Here, we want to give some special case results and discuss its application and validity for nonellipsoid particles. It is worthwhile comparing the frequencies for three standard geometries—the sphere, a long rod, and a magnetic plate.

$$\text{Sphere } f = \gamma' H \tag{16}$$

$$\text{Long rod with magnetization along the length } f = \gamma'(H + 2\pi M_s) \tag{17}$$

$$\text{Large magnetic plate with in-plane magnetization } f = \gamma' \sqrt{H(H + 4\pi M_s)}. \tag{18}$$

Typical results for the frequency as a function of applied field using the three different geometries are given in Fig. 2. The parameters for this calculation are for Fe, $M_s = 1.7\,\text{kG}$ and $\gamma' = 2.9\,\text{GHz/kOe}$.

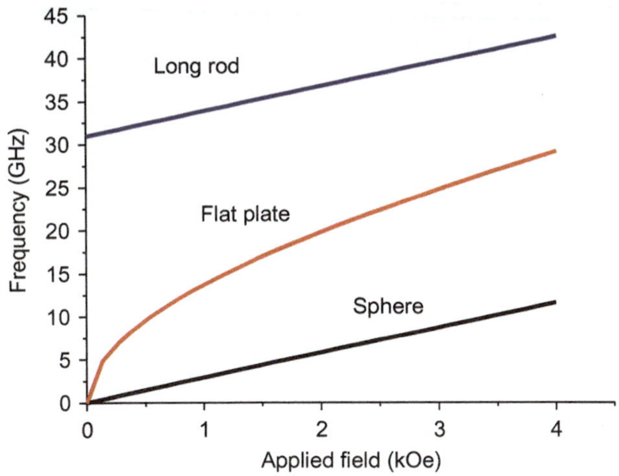

FIGURE 2
Frequency as a function of applied field for a ferromagnet with different shapes. The curves are calculated for an Fe film.

We notice that the frequency for both the sphere and the flat plate goes to zero as the applied field is reduced to zero. In contrast, the frequency for the long rod remains quite high, even in the absence of an applied field. This can be useful, for example, in designing signal processing devices to work at high frequencies [26], without an applied field.

In principle, the Kittel equation is valid only for ellipsoids where the demagnetizing fields produced by the surface poles are spatially constant. In rectangular structures, however, this is not the case. Despite this, one can obtain reasonable estimates [27] for the frequency of the uniform mode based on the Kittel equation, using approximate demagnetizing factors that can be calculated based on a method developed by Aharoni [28].

Clearly, the above only provides a very simple overview of typical frequencies and typical geometries. For example, in addition to the frequency of the uniform mode, one can calculate the frequencies of all the standing modes and edge modes in a nanosized object. More complex analytic results are available [29], and in many cases one also needs to use dynamic micromagnetic calculations [30].

A particularly interesting class of systems are the small disks and squares for which the equilibrium magnetic structure is a vortex state [31]. The magnetic dynamics of this structure are particularly interesting. In addition to standing wave-like modes (radial and azimuthal quantization), there is also a low-frequency motion of the vortex core [32,33]. Vortex motion has been studied both in cylindrical dots and in a square dot [34]. The dynamic behavior of small nanosized objects, including vortex structures, has been reviewed recently in Ref. [35].

2.6 Antiferromagnets

We note that the results presented above are appropriate for ferromagnets or ferrimagnets. However, there are many different types of magnetic materials (antiferromagnets, helical magnets, multilayers) and they each will have their own set of characteristic frequencies. We briefly present some results for the antiferromagnets [36–39]. For example, if the external field is applied parallel to the uniaxial axis of a semi-infinite antiferromagnet and the surface mode propagates perpendicular to this direction, the frequency of the surface mode is given by

$$\omega_s = \gamma' \sqrt{H_a(2H_{ex} + H_a + 4\pi M_s)} \pm \gamma' H. \tag{19}$$

Here H_a is the anisotropic field, H_{ex} is the exchange field, $4\pi Ms$ is the saturation magnetization of one sublattice, and the sign in the $\pm\gamma' H$ term depends on which way the surface is traveling, i.e., from right to left across the magnetic field, or from left to right. In addition, there are also bulk spin waves in antiferromagnets. In the absence of a field, and for propagation perpendicular to the uniaxial direction, there are two bulk frequencies, which lie above and below the surface wave frequency.

$$\omega_{B1} = \gamma' \sqrt{H_a(2H_{ex} + H_a)} \tag{20}$$

$$\omega_{B2} = \gamma' \sqrt{H_a(2H_{ex} + H_a + 8\pi M_s)}. \tag{21}$$

The dynamical modes in antiferromagnets have been confirmed experimentally through reflection measurements [40] and attenuated total reflection measurements [41,42] in the infrared frequency range.

3 EXAMPLES OF EXPERIMENTAL CHARACTERIZATION OF FERROMAGNETIC SAMPLES

We would first like to provide the reader with a set of examples for FMR measurements on a sample with unknown properties. We will assume that the reader has access to a broadband FMR system described in the section 4.2. Such a system can allow one to carry out FMR measurements with a magnetic field oriented either in-plane or out-of-plane of the sample at different values of the applied magnetic field.

In principle, there are two different ways to do the FMR experiment. One method is to keep the external magnetic field constant and scan the frequency to search for a peak in the absorption. The second method uses a constant frequency and sweeps the magnetic field, looking for the field value which gives the maximum absorption. Each of these methods has advantages. Frequency scans in a constant applied field are very fast and a convenient way of collecting data. Such measurements allow one to determine the basic magnetic parameters of the sample such as $4\pi M_{eff}$, the gyromagnetic ratio γ' or the effective magnetic anisotropy. In very rare events there can be a situation where the FMR resonance will be outside of the accessible frequency/magnetic field range, and of course in such a situation the FMR characterization of the sample will not be possible. One advantage of a field scan is that the linewidth measurement is simpler in the sense that it is directly related to the damping, as we will see later.

We will start with a description of the measurements carried out with an applied magnetic field directed normal to the film surface. There are two distinct possibilities. Figure 3 depicts the behavior of a sample that has magnetization in-plane in zero applied field. In such a case, one observes a nonmonotonic dependence of the resonance frequency as a function of applied field. Initially, the FMR resonance frequency decreases with increasing applied magnetic field, up to a critical value of the field (demagnetizing field). For higher fields the resonance frequency increases again. If one does a field scan (sweeping the magnetic field at a constant frequency), one can observe resonance at two different values of the applied field for the given frequency. One represents the case when magnetization is fully aligned with the applied field (higher value of the field), while the second represents the case when magnetization and the applied field are not collinear. Figure 3 was generated using magnetic parameters of Permalloy ($4\pi M_s = 10$ kG, $\gamma' = 2.9$ GHz/kOe).

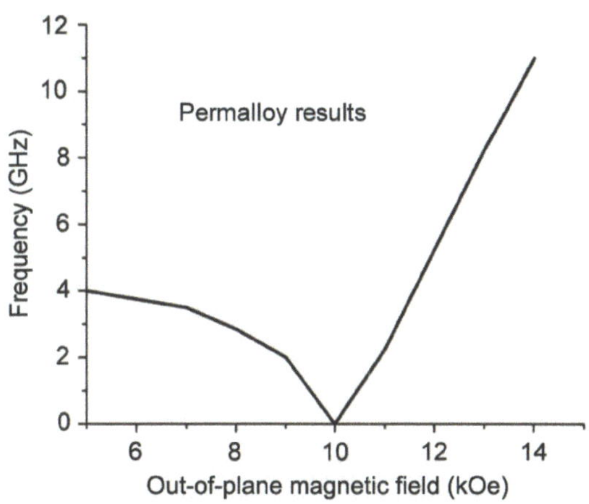

FIGURE 3

FMR frequency as a function of applied field *H* for a thin Permalloy film.

The situation is very different in the case when the sample has a magnetization that is directed out-of-plane at zero applied field. Figure 4 shows the expected frequency as a function of out-of-plane magnetic field in this case, using, as an example, FMR measurements carried out on a hexagonal ferrite film. The linear dependence of the FMR frequency is a clear indication that magnetization is perpendicular to the film surface. By fitting experimental data to Eq. (15), one can determine two parameters—the gyromagnetic ratio γ' (using slope) and the effective anisotropic field ($H_{\rm eff} = H_{\rm a} - 4\pi M_{\rm s}$).

In order to determine the value of the anisotropic field, H_a, one needs to know the value of $4\pi M_s$. There are two ways of achieving this: one can carry out FMR measurements with the magnetic field in the plane of the films or one can use the results of magnetometry measurements (SQUID, VSM, or AGM). We will discuss the FMR measurements with field applied in the plane of the film below.

Figure 5 shows the results of an FMR measurements with magnetic field applied in the plane of the film in the simplest case, for a sample with the magnetization in the plane of the film and without any anisotropy. We illustrate this example with FMR measurements of Permalloy film. One can clearly see the experimental data points can be fitted using Kittel's formula (Eq. 18).

We note that the curve in Fig. 5 passes through the origin. If there were some anisotropy in the system, this would not generally be the case. For example, in-plane fourfold anisotropy or in-plane uniaxial anisotropy would generally result in a finite frequency at $H = 0$. To verify the absence of an anisotropic field,

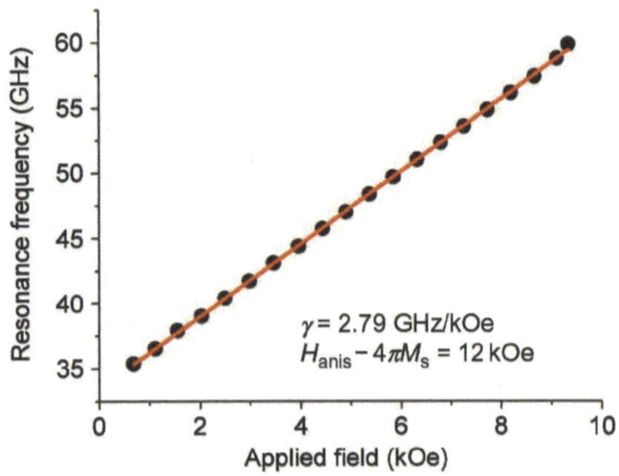

FIGURE 4

Frequency as a function of an out-of-plane magnetic field for a sample with out-of-plane magnetization. The results here are for a hexagonal ferrite film.

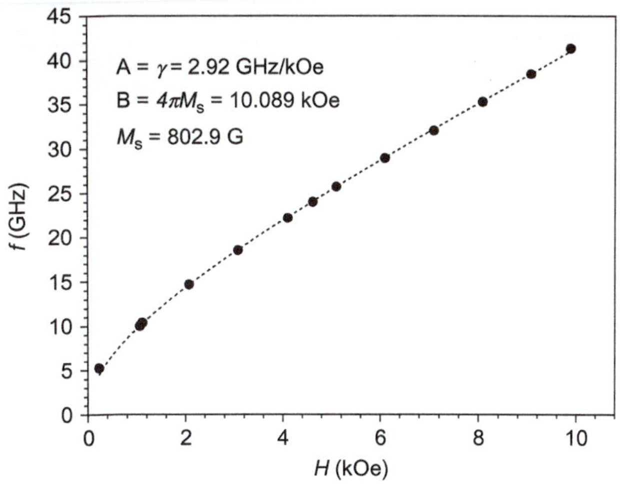

FIGURE 5
FMR frequency as a function of applied field *H* for a Permalloy film.

one needs to rotate the sample in the plane with respect to the direction of the applied DC magnetic field.

In fact, even if the frequency is zero at $H = 0$, this does not prove the absence of anisotropic fields in the sample. We demonstrate this using simulated data for a single-crystal film of Fe which has fourfold anisotropy in Fig. 6. As parameters we used $\gamma = 2.9$ GHz/kOe, $4\pi M_s = 21.5$ kG, and fourfold anisotropic field $H_a = 0.6$ kOe. For an Fe (001) film, the FMR resonance condition with magnetization along the direction of the applied field is given by

$$f = \gamma' \sqrt{\left(H + 4\pi M_s + \frac{H_a}{4}(3 + \cos 4\Theta)\right)(H + H_a \cos 4\Theta)}, \tag{22}$$

where Θ represents the angle between the Fe [100] crystallographic direction and the direction of the applied field H. Figure 6 shows the behavior of the FMR frequency as a function of applied field for three Θ angles: 0° (easy axis), 22.5°, and 45° (hard axis). We note that the special case for $\Theta = 22.5°$ ($\cos 4\Theta = 0$) behaves similarly to the sample without any magnetocrystalline anisotropies in that the frequency goes to zero at zero field. This clearly demonstrates the necessity of carrying out angular-dependent measurements.

In Fig. 6, one can see that for most values of the applied field the highest FMR resonance frequency is associated with the magnetic field along the easy axis while the lowest frequency has the magnetic field along the hard axis. In the example of Fig. 6 the magnetization is not always along the field direction for values of H below about 0.6 kOe and therefore Eq. (22) is no longer valid.

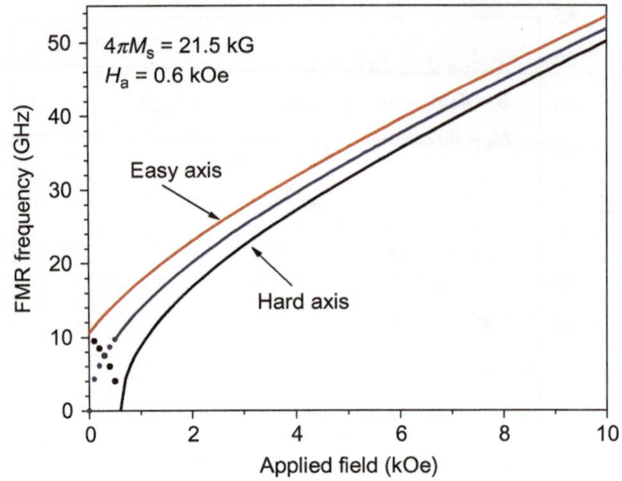

FIGURE 6

Frequency as a function of applied field for an Fe film with fourfold in-plane anisotropy. We show results for three different cases—the magnetic field is applied along an easy axis, along a hard axis, and along an axis that is 22.5° away from the easy axis. The dotted portions of the curve indicate the regions where Eq. (22) is not appropriate.

Because of this, we have drawn the expected behavior for the frequencies when the magnetization and field are not collinear as dashed lines. At lower fields the FMR frequency approaches zero for a magnetic field applied along the hard axis (so-called mode softening) and for a specific low frequency there are two values of field at which FMR resonance occurs. First, at higher field, at which magnetization is along the direction of the applied field, and second at lower field, when magnetization is not aligned with the applied magnetic field.

In order to determine the nature (symmetry) of the anisotropies present in the sample one needs to carry out angular-dependent FMR measurements. It is better to use higher values of applied field to eliminate the so-called dragging effect, which we will discuss later. Figure 7 depicts the behavior of the FMR frequency for an Fe (100) film with only a fourfold anisotropy present. The fourfold symmetry is clearly visible. In order to determine the all-magnetic parameters (γ', $4\pi M_s$, H_a), angular FMR measurements at two different frequencies need to be carried out.

It is very rare that there is only one anisotropic field in a single-crystal film (in addition to the demagnetizing field). Let's include a second anisotropic field, uniaxial anisotropy, to modify the example of the Fe (100) film discussed above. For an Fe (100) film a uniaxial in-plane anisotropy could be induced, for example, by a series of steps present in the substrate or by the growth of

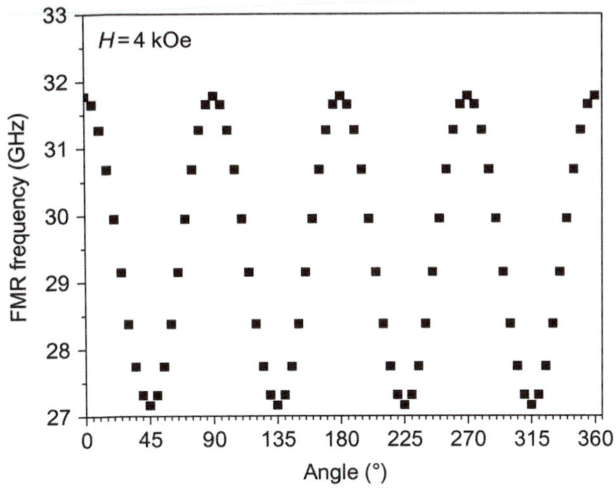

FIGURE 7
FMR frequency as a function of the angle between the magnetic field and an in-plane easy axis for an Fe film.

Fe directly on the surface of the GaAs (100) substrate. The resonance condition in this case is given by [43]:

$$\left(\frac{f}{\gamma'}\right)^2 = \left(H + 4\pi M_s + \frac{H_a}{4}(3 + \cos 4\Theta) + \frac{H_u}{2}(1 + \cos(2(\Theta - \varphi)))\right) \times (H + H_a \cos 4\Theta + H_u \cos(2(\Theta - \varphi))), \quad (23)$$

where φ is the angle between the preferred axis of the uniaxial anisotropy and the [100] crystallographic direction. The angle Θ measures the angle between the magnetic field and the [100] direction. This formula also assumes that the magnetization is saturated along the direction of the applied field. Figure 8 depicts the behavior of the FMR frequency as a function of sample orientation for two values of $\varphi = 0°$ and 45°. We have assumed the same magnetic parameters as those used in Figs. 6 and 7 with the addition of a uniaxial anisotropic field $H_u = 0.3$ kOe. Such behavior clearly indicates that we have superposition of patterns, with periodicity of 90° and 180°. Note that for $\varphi = 0°$ the easy axes are not equivalent (some are "easier" than others), while for $\varphi = 45°$ hard axes are not identical (some are "harder" than others).

In general, every time we have some form of magnetic energy, for example, the interaction energy between two magnetic films, there is an effective field associated with it which will modify the resonance conditions. With sufficient measurements, this allows one to obtain values for all the relevant fields.

We have mentioned above the dragging effect associated with the fact that magnetization in the low fields may not be collinear with the applied field.

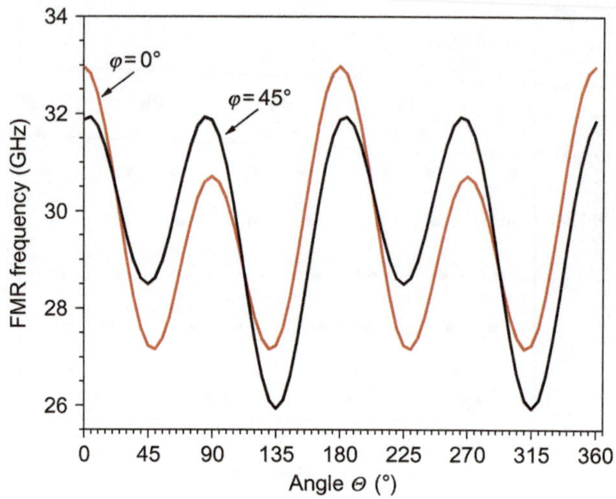

FIGURE 8

FMR frequency as a function of the angle for an applied field $H=4$ kOe for an Fe film with uniaxial anisotropy.

This affects the value of the FMR frequency; however, for the frequency-swept measurements this effect is relatively small. The result is different for the field-swept FMR measurements, as shown below. We will demonstrate this effect using as an example a single-crystal Fe (001) film with fourfold in-plane anisotropy, but without any additional uniaxial in-plane anisotropy. The requirement that $M \times H_{\mathrm{eff}} = 0$ in equilibrium, where H_{eff} is the sum of all effective fields in the plane, determines the angle φ between magnetization M_s and the direction of the applied field H. This condition can be expressed as

$$H_a \sin 4\Theta - 4H \sin\varphi = 0. \tag{24}$$

Thus,

$$\sin\varphi = \frac{H_a \sin 4\Theta}{4H}. \tag{25}$$

As the applied field is increased, the angle φ between M_s and H decreases. This dragging effect therefore is more pronounced in low applied fields. The projection of H in the direction of the magnetization ($H\cos\varphi$) effectively decreases the value of the applied field and as a result modifies the resonance condition (see Eq. 22), which can now be expressed as

$$f_{\mathrm{FMR}} = \gamma' \sqrt{\left(H\cos\varphi + 4\pi M_s + \frac{H_a}{4}(3 + \cos 4\Theta)\right)(H\cos\varphi + H_a \cos 4\Theta)}. \tag{26}$$

We illustrate the dragging effect using the parameters above for the Fe (001) film and an applied field of $H = 0.75$ kOe. At this value of the applied field, the magnetization is aligned with the applied field when the applied field is along a hard axis ($\Theta = 45°$). Figure 9 shows the FMR frequencies calculated using Eqs. (22) and (26). It is almost impossible to distinguish between these two curves. The inset shows the difference between these two results ($\Delta f = f - f_{FMR}$). At most, the difference is slightly above 0.1 GHz and it will decrease with increasing values of the applied field H.

Field-swept FMR measurements at constant frequency can show a more pronounced dragging effect, especially at low frequencies and when the applied magnetic field is much smaller than the anisotropic field. To illustrate this, we will consider a Permalloy film with a magnetic field applied at an angle with respect to the film plane. For the Permalloy film, we have only one effective anisotropic field, which is the demagnetizing field. We will consider examples for FMR in this case for both 10 and 36 GHz frequencies. Solving the Landau–Lifshitz equation for geometry shown in the inset of Fig. 10a leads to two equations. The first equation shows an equilibrium condition for angle ϕ, the angle between the direction of the applied field H, and the direction of magnetization M:

$$\sin\phi = \frac{2\pi M_S \sin 2\Theta}{H}. \quad (27)$$

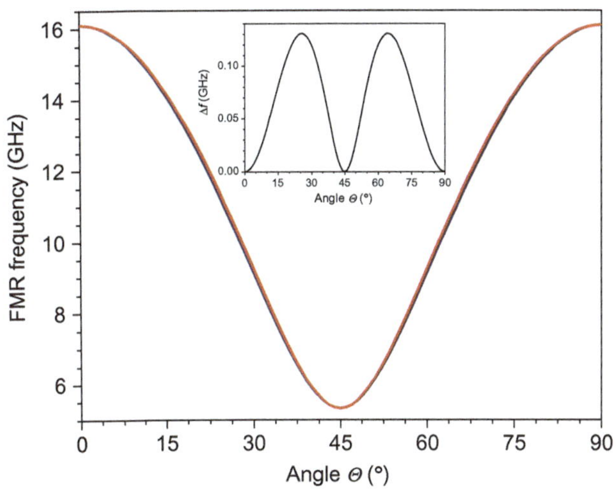

FIGURE 9
Comparison between the FMR frequencies determined using Eqs. (25) and (26). The inset shows the difference between these frequencies.

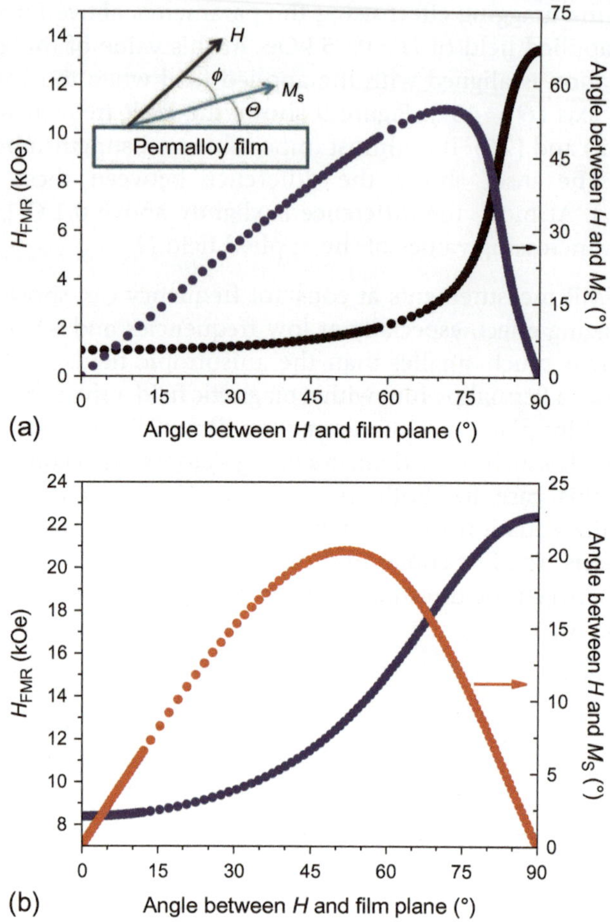

FIGURE 10

The simulated FMR behavior of dragging effects. The resonance frequency and angle between applied magnetic field and magnetization as a function of the angle the applied magnetic field makes with the film plane: (a) for 10 GHz and (b) for 36 GHz.

The angle Θ is the angle between the plane of the film and the direction of magnetization M. The second equation determines the resonance conditions:

$$\left(\frac{f}{\gamma'}\right)^2 = (H\cos\phi + 4\pi M_s\cos 2\Theta)(H\cos\phi - 4\pi M_s\sin^2\Theta). \tag{28}$$

In order to observe FMR, both equations must be satisfied simultaneously. Note that in the limiting cases, for $\Theta = 0°$ and $90°$, the resonance condition reduces to Eqs. (18) and (15) (with $H_a = 0$), respectively. In these two cases, the magnetization and applied field are collinear. The frequency used in the

FMR experiment significantly changes the observed dragging effects as seen in Fig. 10. For the experiment carried out at 10 GHz (Fig. 10a), the magnetization in Permalloy is basically in the plane of the film until the applied magnetic field makes an angle of 40–45° with the film plane (see Fig. 10a). To be specific, when the applied field makes an angle of 45° ($\phi + \Theta$) then the magnetization has an angle Θ smaller than 5° with respect to the film plane. As a result, initially the resonance field increases very slowly with increasing angle of the applied field. The largest angle between the applied magnetic field and the magnetization (~55°) occurs when the magnetic field is applied approximately 70° out of the film plane. With a further increase of the angle between the applied magnetic field and the film plane, the resonance field rapidly increases while the angle ϕ decreases. For 36 GHz, the observed dragging effects are qualitatively similar, but much smaller because the magnetic field is larger in magnitude and the magnetization lies closer to the applied field. These two examples clearly show that dragging effect can be significant at lower frequencies. It is therefore good practice to carry out FMR experiments at a frequency which is as high as possible to minimize dragging effects.

4 FMR TECHNIQUES

Up to this point, we have concentrated on topics related primarily to the resonance frequency or the resonance field of various excitations. With these simple considerations, we have already seen how the magnetization direction and magnitude can be determined, as well as the values of the exchange and various anisotropy constants. In this section, we extend this initial overview by developing a discussion of FMR including damping. We use this to discuss and compare two experimental techniques in detail—broadband frequency-swept FMR and field-swept FMR. In particular, we show that a simple field-swept technique using a coplanar waveguide can produce results equivalent to that of a more expensive vector network analysis technique. At the end of this section, we discuss the effects of a spatially nonuniform microwave field on the FMR absorption efficiency.

4.1 Magnetic Damping

In an FMR experiment, the key measurements are the location of the absorption peak and its linewidth. The linewidth is a measure of the magnetic damping in the system; however, it is not always obvious how the damping is related to the linewidth, particularly for FMR measurements using the frequency scan method discussed later in this section. In this section, we give an overview of damping and its connection to the linewidth.

The magnetization dynamics are described by the phenomenological Landau–Lifshitz–Gilbert (LLG) equation of motion:

$$\frac{\partial \boldsymbol{M}}{\partial t} = -\gamma(\boldsymbol{M} \times \boldsymbol{H}_{\text{eff}}) + \frac{\alpha}{M_s}\left(\boldsymbol{M} \times \frac{\partial \boldsymbol{M}}{\partial t}\right), \tag{29}$$

where $\boldsymbol{M}$ is the instantaneous magnetization vector, $\boldsymbol{H}_{\text{eff}}$ is the total effective magnetic field, α is the Gilbert damping parameter, M_s is the saturation magnetization, and γ is the absolute value of the gyromagnetic ratio,

$$\gamma = \frac{g\mu_B}{\hbar}, \tag{30}$$

where μ_B is the Bohr magneton and $\hbar$ is the reduced Planck constant. The first term on the RHS of Eq. (29) describes the precession of the magnetization about $\boldsymbol{H}_{\text{eff}}$ and is mainly responsible for the position of the resonance (resonance field or resonance frequency). The second term describes a viscous-like damping and is mainly responsible for the linewidth of the resonance. It is also worth noting that one needs to be careful with definitions of γ in the literature, as one may either see γ or γ' referred to as γ. Furthermore, sometimes γ is explicitly negative and sometimes it is taken as positive. In this chapter, we will take γ as a positive number. A simple analysis of the resonance condition given should be enough to resolve any ambiguity.

The total effective field acting on the magnetic moments within a film is given by

$$H^i_{\text{eff}} = -\frac{\partial}{\partial M_i}\mathcal{F}, \tag{31}$$

where $i = x, y, z$ and $\mathcal{F}$ is the free energy. $\boldsymbol{H}_{\text{eff}}$ is the sum of external and internal fields. The external fields involved in the FMR experiment are the applied DC magnetic field $\boldsymbol{H}$ and the rf microwave magnetic field $\boldsymbol{h}_{\text{rf}}$. The internal fields are a consequence of the anisotropy of $\mathcal{F}$ satisfying the film's crystalline structure (symmetry) and its shape. The individual parts of $\mathcal{F}$ are called magnetic anisotropies. These anisotropies arise from either the spin–orbit interaction, which leads to magnetocrystalline and magnetoelastic contributions, or from the magnetic dipole–dipole interaction, which is usually treated by magnetostatic approximation. A detailed overview of common crystalline symmetries and related anisotropies is given by Lindner and Farle [44] and Heinrich [45]; the magnetoelastic anisotropies by Sander [46]; and the dipolar interaction by Hurben and Patton [47].

In the FMR experiment, the microwave response signal ε_χ is proportional to the rf magnetic susceptibility $\chi = \chi' + i\chi''$ in the direction of an applied rf field $\boldsymbol{h}_{\text{rf}}$. For the following we assume $\boldsymbol{h}_{\text{rf}} \propto e^{i\omega t}$, where $\omega = 2\pi f$ is the microwave angular frequency. In the linear response regime of Eq. (29), it can be shown that the rf signal for the in-phase component (dispersion) is given by

$$\varepsilon_{\chi} = \mathcal{A}\left(\frac{1}{B_{\mathrm{FMR}}} + \frac{H - H_{\mathrm{FMR}}}{\Delta H^2 + (H - H_{\mathrm{FMR}})^2}\right), \tag{32}$$

and for the out-of-phase component (absorption) by

$$\varepsilon_{\chi''} = \mathcal{A}\frac{\Delta H}{\Delta H^2 + (H - H_{\mathrm{FMR}})^2}. \tag{33}$$

Equation (33) describes a simple Lorentzian absorption peak, where H is the external applied DC field, H_{FMR} is the resonance field, ΔH is the half-width-at-half-maximum (HWHM) of the Lorentzian absorption peak for χ'', and $\mathcal{A}$ is an amplitude.

Equations (32) and (33) are obtained from a simple expansion of χ around the resonant field. Their validity for a wide range of fields can be tested by fitting a full equation for the in-plane rf susceptibility for a simple configuration where one does FMR on a thin magnetic film and the external field is parallel to the surface of the film. The calculation for the susceptibility includes only the external field and a perpendicular demagnetizing field, appropriate for a thin film, and the susceptibility is given by

$$\chi = \frac{M_s\left(H + 4\pi M_s - \mathrm{i}\alpha\frac{\omega}{\gamma}\right)}{\left(H + 4\pi M_s - \mathrm{i}\alpha\frac{\omega}{\gamma}\right)\left(H - \mathrm{i}\alpha\frac{\omega}{\gamma}\right) - \left(\frac{\omega}{\gamma}\right)^2}. \tag{34}$$

Figure 11 shows the quality of fit using Eq. (33) for simulated data generated by taking the imaginary part of Eq. (34). The error in fitting does not reach 1% even far away from the FMR peak, justifying using the simple ansatz Eqs. (32) and (33) for fitting experimental data.

In our analysis of data, we are concerned mainly with line position and shape. The line position, or resonance field H_{FMR}, is used to determine the anisotropic fields H_{eff} and γ. The lineshape ΔH is used to determine the Gilbert damping parameter α as it is a linear function of microwave frequency:

$$\Delta H(\omega) = \alpha\frac{\omega}{\gamma} + \Delta H(0), \tag{35}$$

where $\Delta H(0)$ is the zero frequency offset due to inhomogeneity-induced line broadening [48]. It is important to realize that a single measurement of linewidth will not generally allow one to determine α because $\Delta H(0)$ is unknown.

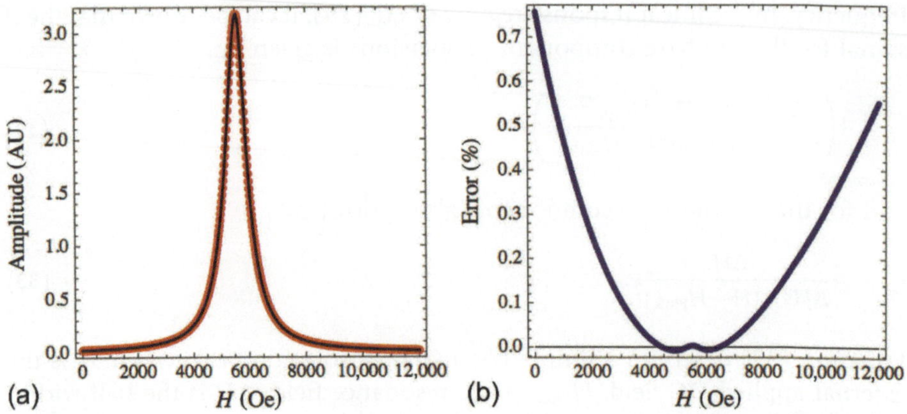

FIGURE 11

A comparison of the absorption described by Eqs. (33) and (34). (a) The simulated data χ'' given by the imaginary part of Eq. (34) are shown by red points and the black line is the fit using Eq. (33). The input parameters were similar to those of an Fe ultrathin film, except that the damping was made artificially large. This emphasizes the possible differences between the two sets of results. They were $M_s = 1740$ G, $\alpha = 0.037$, and $\omega/\gamma = 12{,}300$ Oe. The corresponding resonance field $H_{FMR} = 5530$ with exact linewidth $\Delta H = 455$ Oe. The resulting fit parameters were $H_{FMR} = 5530$ and $\Delta H = 455$ Oe. (b) Percent error of the fit with respect to the simulated data. The percent error is calculated by taking 100 times the difference between the fit (black line) and the χ'' rf susceptibility simulated data (red points) and dividing by the simulated data. In this manner positive percent error represents overestimate and negative percent error represents underestimate of the actual rf susceptibility.

In general, the determination of the intrinsic damping constant, α, can be more complex than described by Eq. (35). Often magnetic inhomogeneity will contribute to FMR line broadening (extrinsic contribution) and this is treated by a two-magnon scattering process. Two-magnon scattering involves scattering of the resonance FMR mode into degenerate spin waves due to the presence of magnetic inhomogeneities. It is a dephasing effect which is narrowed down by the dipolar and exchange interaction, as discussed in Ref. [49]. McMichael *et al.* [50] addressed the validity of this approach by using a magnetic film consisting of size D grains and an anisotropic field H_a, which randomly changes from grain to grain. It was shown that two-magnon scattering is correct for describing an additional extrinsic FMR linewidth for small inhomogeneity, $H_aD \ll \pi M_s d$ where d is the film thickness. For large inhomogeneity, $H_aD \gg \pi M_s d$, the FMR precession becomes localized and then the two-magnon model does not apply and the FMR line is given by superposition of the FMR lines in the individual sample grains.

Two-magnon scattering in thin films has a characteristic angular feature. It is absent in the perpendicular FMR configuration (magnetic moment perpendicular to the film plane) because there are no degenerate modes in that geometry.

For other angles, the contribution of two-magnon scattering in thin films usually saturates at high microwave frequencies and can be described by $\arctan(\omega/\omega_0)$ with a characteristic microwave frequency ω_0 (see Arias and Mills [51,52], Lindner *et al.* [53], and Woltersdorf and Heinrich [54]). For low frequencies, this leads to a gradual downward trend in the FMR linewidth, approaching zero in a linear dependence.

The long-range inhomogeneity distributed around the mean value, with a Lorentzian distribution, results in a Lorentzian FMR lineshape with a linewidth contribution from $\Delta H(0)$ and a linear dependence on the microwave frequency. If one measures linewidth only at higher frequencies, a large $\Delta H(0)$ (compared to the intrinsic contribution) could result in an apparently smaller slope for the linear dependence on frequency. This would suggest a decreased effective damping parameter α_{eff}. Instead, for large $\Delta H(0)$ one has to find, in a self-consistent manner, the true contribution to the Gilbert damping parameter α and that caused by magnetic inhomogeneity. Therefore, using Eq. (35) requires careful inspection of the measured FMR linewidth as a function of microwave frequency. In the presence of a downward trend in $\Delta H(\omega)$ from a linear dependence one can fit the measured FMR linewidth $\Delta H(\omega)$ as

$$\Delta H(\omega) = \Delta H(0) + a \times \arctan\left(\frac{\omega}{\omega_0}\right) + \alpha_{\text{eff}}\frac{\omega}{\gamma}. \tag{36}$$

Usually in ultrathin films, the FMR linewidth is a linear function of the microwave frequency above 20 GHz (see Ref. [47]). Further reading can be found in Ref. [55].

It is less obvious how to connect α to a linewidth for the frequency-swept FMR measurement [56,57]. One can relate the frequency-swept linewidth, $\Delta\omega$, to the field-swept linewidth, ΔH, by the general relation:

$$\Delta\omega = \frac{\partial\omega}{\partial H}\Delta H. \tag{37}$$

So, one might expect that as ΔH increases the frequency-swept linewidth will increase also, and this is often the case. However, there are important exceptions. There are two important factors for the frequency-swept linewidth: the slope $\partial\omega/\partial H$ and the precise form of ΔH. From Figs. 2, 3, and 6, for example, one can see that the slope of the ω versus H curve changes substantially depending on the field and the physics of the problem. For example, for an infinitely extended film (Fig. 6) the slope is large near $H = 0$, and one would thus expect a large frequency-swept linewidth. In this case, the frequency-swept linewidth can be large at low field (or frequencies) and actually get smaller as the field (or frequency) is increased. This is contrary to what one might naively expect by just assuming that the frequency-swept linewidth would be directly proportional to the field-swept linewidth at a given frequency.

4.2 Broadband FMR—Frequency Swept

The FMR technique has been used for many years in the characterization of magnetic materials. Initially, FMR measurements were carried out in cavity-based systems with standard waveguide components [58]. The development of modern high-frequency electronics (vector network analyzers (VNAs), microwave generators, etc.) and probe stations for on-wafer measurements has stimulated recent advances in the construction of modern broadband FMR systems. Specifically, it is the incredible sensitivity of the VNAs, to a level below −110 dB over a wide frequency range, that has allowed construction of the new FMR systems. There are many experimental broadband FMR systems reported in the literature which use VNAs [59–61]. We will first mention two cryogenic broadband FMR systems based on VNAs and then describe in detail an inexpensive broadband FMR design based on a simple microwave generator.

The system developed by Denysenkov and Grishin [62] is based on a helium cryostat that can operate in the temperature range between 4 and 420 K. It allows one to carry out angular-dependent FMR measurements at frequencies up to 40 GHz and in fields up to 2.4 T. Their instrument combines the use of phase-sensitive detection (magnetic field modulation) with the use of a VNA. As a result, their instrument has very high sensitivity, with the ability to detect and measure as many as 10^{11} spins.

A different approach to cooling of the sample was developed by Harward *et al.* [63]. They used a closed-cycle He refrigeration system which allowed FMR measurements in the 28- to 360-K temperature range and employed a VNA. They measured *S* parameters to determine FMR signals in the frequency range 1–70 GHz in both frequency- and field-swept modes. This system also allowed measurement of two samples simultaneously due to construction of a cold finger which accommodates two coplanar waveguides, one designed to carry out angular in-plane field measurements, while the second allows measurements with the magnetic field in the out-of-plane geometry.

Some drawbacks of VNA–FMR systems include the need for careful calibration. For example, systems where one uses a flip-chip scheme to place the sample on a coplanar waveguide may show signals that arise from geometric factors which include multiple reflections from the edges of the sample or the coplanar waveguide itself. This must be excluded from the results, and for magnetic systems can often be done by doing a measurement at lower or higher external magnetic fields. Signals that are independent of the magnetic field need to be subtracted from the results. An additional drawback is that a VNA can be quite expensive, about $50,000 for a system that goes up to 40 GHz. Furthermore, the analysis of data can be somewhat complicated, involving the calculation of the susceptibility from scattering parameters.

4.3 Broadband FMR—Field Swept

In this section, we aim to demonstrate how conventional microwave sources, e.g., gun diodes, klystrons, and signal generators, together with a coplanar waveguide and broadband microwave detector, can be used to measure FMR signals in a broadband system and how the analysis of data can be carried out in a quite simple and straightforward manner.

This section is organized as follows. First, we provide a detailed description of the system. We then carefully analyze the FMR signals obtained on a high-quality crystalline 16 atomic layer (AL)-thick Fe (001) ultrathin film deposited on a GaAs (001) template by means of molecular beam epitaxy (MBE). This test sample was also measured in a standard multimode resonance cavity to compare the results from different microwave systems. This is indeed important because the coupling of ferromagnetic films to the rf signal is dramatically different for a coplanar microwave transmission line and a microwave cavity. We then show the sensitivity of the system for measuring structures similar to those found in industrial applications by measuring sputter-deposited Co/Ru superlattices, with atomically thick Co layers. We discuss the obtained data in terms of magnetic anisotropies, FMR linewidths, and measurement time required to obtain similar signal-to-noise ratios (SNRs) and comment on the advantages and disadvantages of each technique.

We are going to demonstrate that one is able to create a broadband FMR system without a VNA, and this system has the sensitivity capable of detecting FMR signals in ultrathin films, even for structures with large FMR linewidths. This system is based on broadband, low-noise components: an Anritsu MG3696B microwave generator (0.1–67 GHz) and an Agilent 8474E planar-doped barrier diode microwave detector (0.01–50 GHz) (Fig. 12). FMR measurements were carried out using two microwave spectrometers:

(1) A typical microwave spectrometer based on rectangular microwave waveguides with samples inserted inside a microwave cavity.
(2) A coaxial microwave track with samples mounted on a coplanar waveguide. Despite the different coupling between the film and microwave cavity compared to that of a coplanar waveguide, the FMR results can be analyzed in a simple manner. The magnetic properties measured with these two techniques are in very good agreement, as shown in Figs. 13 and 14 and Table 1.

In a typical FMR setup, measurements are made at a fixed frequency while the H field is swept over a range of fields, which includes the resonance field. The FMR signals from ultrathin films are usually weak; for this reason, it is common practice to use lock-in detection to enhance the SNR. It is most common to use field modulation (HM) to this end. In our setup, HM was achieved by using a Kepco

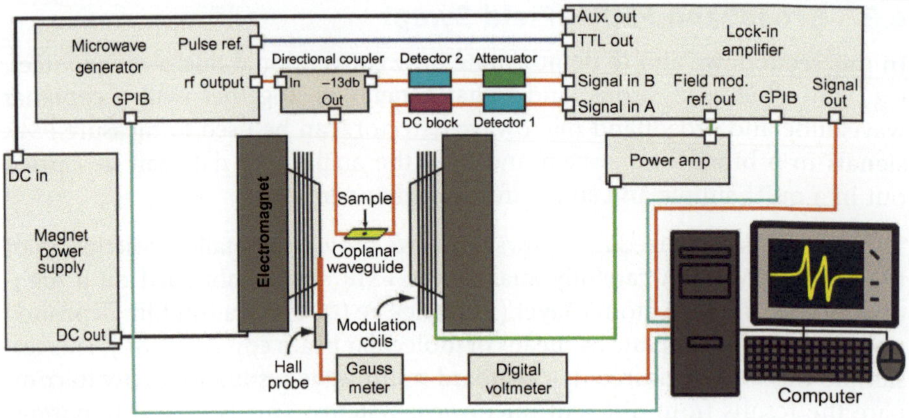

FIGURE 12

Schematic of the FMR spectrometer for a coplanar waveguide (CPW) with field modulation (HM) or microwave power amplitude pulse modulation (PM). The coplanar waveguide is mounted in the center of an electromagnet that provides the DC field and two Helmholtz coils (modulation coils) that can provide HM. The rf microwave source is a broadband microwave generator. The rf FMR signal is converted to a low-frequency output signal, with the frequency determined by HM or PM, by means of a planar-doped barrier diode detector (0.01–50 GHz) (Detector 1). The lock-in amplifier is used to provide both the modulation reference signal and to control the electromagnet to step the DC H field. CPW-HM: The 105-Hz modulation signal is provided by the reference out (field mod. ref. out.) of the lock-in amplifier and is amplified by a power amplifier that supplies current to the modulation coils. A DC block is used to filter out signals that can be induced by modulation. CPW-PM: The 10-kHz pulse modulation is supplied by the lock-in amplifier TTL out port to the microwave generator pulse reference port. The result is an on/off chopping of the microwave signal. To enhance sensitivity, a directional coupler is used to take a −13 dB portion of the microwave signal that bypasses the coplanar waveguide. This microwave signal is turned into a low-frequency signal by a planar-doped barrier diode detector (0.01–50 GHz) (Detector 2). The signals from Detector 1 and Detector 2 are detected by the lock-in amplifier in A-B mode.

bipolar operational power supply/amplifier (model: BOP72-5M) to drive a set of Helmholtz modulation coils that slightly alter the swept external field. We used a modulation frequency of 105 Hz, as this system is limited by the inductive reactance of the coils. Lock-in detection is performed with a Stanford Research Systems SR830 DSP Lock-In Amplifier.

4.3.1 Multimode Cavity Configuration

In the multimode cavity with field modulation configuration (Cavity-HM), the experimental setup is the same as described by Celinski *et al.* [58] except with the Anritsu MG3696B microwave generator replacing the Klystron, Klystron Power Supply, and Feedback Voltage Supply. High sensitivity in this case can be achieved using a microwave cavity. The cavity used is a 26-mm resonant section of WR-28 waveguide and is coupled to the waveguide track through a small aperture. The sample is mounted in the center of a terminating plate on the other side of the rectangular cavity. The cavity modes used were TE_{00n}, where $n=3$, 4, 5, and 6. The cavity therefore has four resonances at $f \approx 27.23$, 31.22, 35.68, and 40.45 GHz.

4.3.2 Coplanar Waveguide Configuration

The experimental setup for the coplanar waveguide configuration (CPW) is shown in Fig. 12. The system is designed to be compatible with an He flow cryostat to allow FMR measurements between room and liquid He temperatures (see Ref. [64]). The rf signal is brought down to the center of the electromagnet by using semi-rigid coax cable. The rf transmission output is converted to a DC voltage by means of a planar-doped barrier diode detector.

The sample filling factor for the CPW is much higher than that in the microwave cavity, allowing one to achieve high sensitivity. FMR measurements were carried out using two modes of operation: (i) field modulation mode, HM, and (ii) microwave power amplitude pulse modulation (PM).

(i) The coplanar waveguide with field modulation mode (CPW-HM) is essentially the same as the Cavity-HM mode with the WR-28 waveguides and cavity being replaced by broadband coax cables and the coplanar waveguide.
(ii) The coplanar waveguide-based system can also operate in power amplitude pulse modulation mode (CPW-PM). For this case, the low-frequency square wave signal generator TTL output port of the lock-in is connected to the Pulse Trig port of the microwave generator. The microwave output is generally pulsed at 10 kHz. This has advantages over the HM mode of operation as it moves away from $1/f$ noise, allowing a shorter time constant for the lock-in amplifier than that for HM. In ultrathin films, the FMR signal is usually very small compared to the direct transmission signal. In order to enhance the sensitivity in the PM mode, a microwave directional coupler is used to take a -13 dB portion of the direct signal before the coplanar waveguide and is detected by a second planar-doped barrier detector diode. The lock-in will compare two signals, one from the coplanar waveguide with data about the sample, and the signal from the microwave generator. Because the microwave signal is much larger, an attenuator is used to make the sizes to the signals nearly equal. The data collection is performed in A–B mode through the lock-in amplifier. The A–B mode allows one to use the lock-in at higher sensitivity in order to avoid being digital noise limited. Another advantage of measuring the transmission of modulated microwave power, as opposed to the derivative of the transmission, is that one measures

the total transmission signal. Because the FMR signal is only a very small fraction of the A–B signal, it allows one to add both phases of the lock-in amplifier in quadrature. In this case one does not have to optimize the phase of the reference signal. This fact alone can greatly speed up the setup of the measurement, especially for very thin samples with wide FMR lines.

4.3.3 *Results*

4.3.3.1 Single-Crystal Samples

To compare the different FMR techniques, we use a single-crystal sample with small damping and in-plane magnetic anisotropy. The GaAs (001)/16(2.3 nm) Fe/20(4 nm)Au sample was prepared by means of MBE, where the integers refer to the layer thickness in ALs. Details of sample preparation can be found in Ref. [15].

In order to assess the performance of the described system in the three different modes, we briefly discuss the origin of the anisotropy in Fe films on GaAs substrates. The free energy $\mathcal{F}$ (magnetic anisotropies) needs to follow the symmetry of the ultrathin Fe grown on the GaAs (001) reconstructed surface. The magnetic anisotropies are as follows:

(i) The fourfold anisotropic energy due to the cubic symmetry of Fe, $K_1\left(\alpha_x^2\alpha_y^2+\alpha_x^2\alpha_z^2+\alpha_y^2\alpha_z^2\right)$, where $\alpha_{x,y,z}$ are the directional cosines of $\vec{M}$ with respect the Fe cube axes. A positive K_1 results with the easy axes parallel to the cube axes{100}.

(ii) The reconstructed GaAs (001) surface results in an in-plane uniaxial anisotropic energy $-K_\mathrm{u}^{\|}\left(\vec{M}\cdot\vec{n}_\mathrm{u}^{\|}\right)^2/M_\mathrm{s}^2$, where $\vec{n}_\mathrm{u}^{\|}=1/\sqrt{2}(1,-1,0)$ is parallel to the As dimers at the Fe/GaAs interface. By this definition, a negative $K_\mathrm{u}^{\|}$ corresponds to a hard uniaxial axis.

(iii) Magnetoelastic energy due to the lattice mismatch between the Fe (001) and GaAs (001) and the lack of cubic symmetry at the interfaces results in the perpendicular uniaxial anisotropic energy $-K_\mathrm{u}^{\perp}\alpha_z^2$ [44].

(iv) For ultrathin films, the shape leads to a demagnetization energy $2\pi M_\mathrm{s}^2\alpha_z^2$.

For more details, see Kardasz *et al.* [15] and references therein.

In the in-plane configuration, the transverse magnetic susceptibility is given by [44]

$$\chi=\frac{M_\mathrm{s}\left(B_\mathrm{eff}-\mathrm{i}\alpha\frac{\omega}{\gamma}\right)}{\left(B_\mathrm{eff}-\mathrm{i}\alpha\frac{\omega}{\gamma}\right)\left(H_\mathrm{eff}-\mathrm{i}\alpha\frac{\omega}{\gamma}\right)-\left(\frac{\omega}{\gamma}\right)^2}. \tag{38}$$

Resonance occurs when the denominator is a minimum. Neglecting the small contributions from α, the resonance condition is

$$\left(\frac{\omega}{\gamma}\right)^2 = B_{\text{eff}} H_{\text{eff}}\big|_{H_{\text{FMR}}}, \tag{39}$$

where

$$B_{\text{eff}} = H\cos(\phi_{\text{M}} - \phi_{\text{H}}) + \frac{K_1}{2M_{\text{s}}}\cos(3 + 4\phi_{\text{M}}) + \frac{K_{\text{u}}^{\parallel}}{M_{\text{s}}}\cos(1 + 2(\phi_{\text{M}} - \phi_{\text{u}})) + 4\pi M_{\text{eff}}, \tag{40}$$

$$H_{\text{eff}} = H\cos(\phi_{\text{M}} - \phi_{\text{H}}) + \frac{2K_1}{M_{\text{s}}}\cos(4\phi_{\text{M}}) + \frac{2K_{\text{u}}^{\parallel}}{M_{\text{s}}}\cos(2(\phi_{\text{M}} - \phi_{\text{u}})), \tag{41}$$

and

$$4\pi M_{\text{eff}} = 4\pi M_{\text{s}} - \frac{2K_{\text{u}}^{\perp}}{M_{\text{s}}}, \tag{42}$$

where d is the film thickness, ϕ_{M} is the angle of the DC magnetic moment, ϕ_{H} is the angle of the applied external field H, ϕ_{u} is the angle of the in-plane uniaxial hard magnetic axis, with the angles being measured from the [100] crystallographic direction.

Note that in FMR experiments in thin films, one measures the effective anisotropic fields H_K and not the anisotropy parameters K directly. Thus, in this sample one measures $H_{K_1} = 2K_1/M_{\text{s}}$ and $H_{K_{\text{u}}^{\parallel}} = 2K_{\text{u}}^{\parallel}/M_{\text{s}}$. Note that since (iii) and (iv) are both perpendicular uniaxial anisotropies (dependent on α_z^2), they therefore enter into the rf susceptibility on an equal footing; thus, in the FMR experiment one measures an effective demagnetization field $4\pi M_{\text{eff}}$. Other magnetometry techniques such as superconducting quantum interference devices (SQUID) and vibrating sample magnetometry (VSM) can be used to determine M_{s}. Then from the FMR measured effective fields, one can determine the energy of crystalline anisotropies.

In order to determine the in-plane anisotropy of the sample, the resonance field H_{FMR} was measured as a function of angle ϕ_{H} at 35.6802 GHz. Note that we have to fit the FMR spectra as a mixed signal,

$$\varepsilon_{\chi''}^{\text{mix}} = -\varepsilon_{\chi'}\sin\varphi + \varepsilon_{\chi''}\cos\varphi, \tag{43}$$

where the angle φ represents the deviation from perfect adjustment of the FMR system. For $\varphi = 0$, the measured signal would be proportional to only the imaginary part of the rf susceptibility; however, such a situation is uncommon during FMR measurements. Thus, measuring only one signal is equivalent to having an admixture of the in-phase and out-of-phase components of the rf susceptibility signal. In systems that can simultaneously measure both projections,

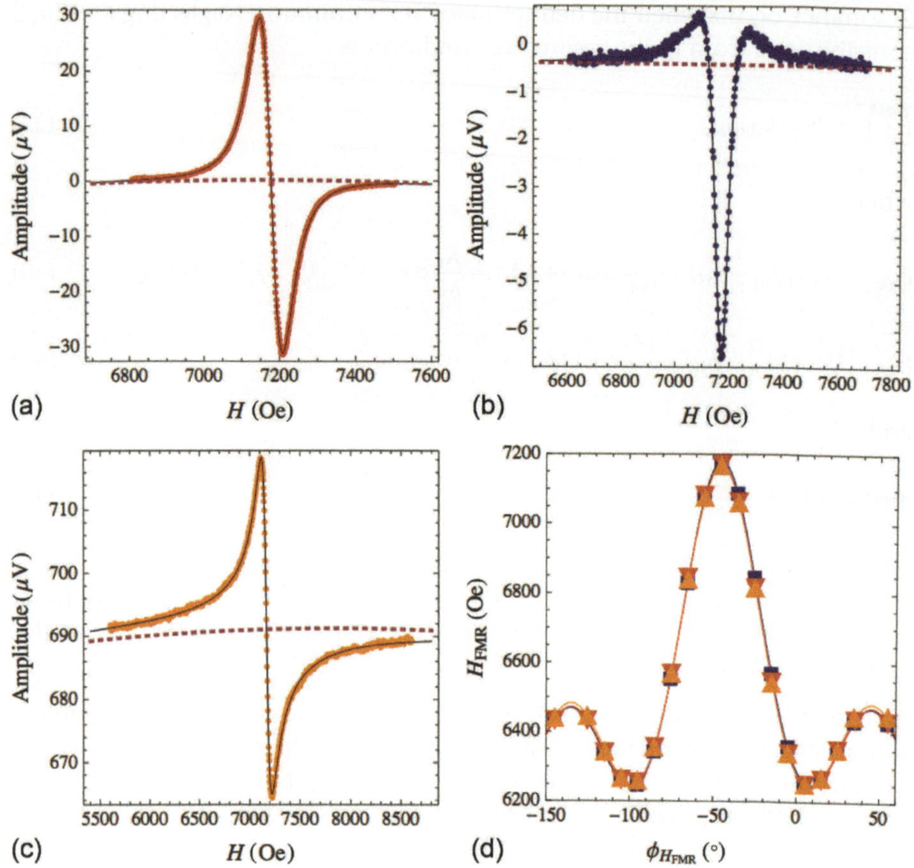

FIGURE 13

FMR measurements of GaAs/[16]Fe/[20]Au at $f = 35.6802$ GHz along the hard uniaxial axis [1$\bar{1}$0] direction are shown for (a) the multimode cavity with field modulation (Cavity-HM) configuration with relevant fit parameters $H_{FMR} = 7179$ Oe, $\Delta H = 51.6$ Oe, $\varphi = 3.3°$, with collection time $t = 240$ s; (b) the coplanar waveguide with field modulation configuration (CPW-HM) with $H_{FMR} = 7170$ Oe, $\Delta H = 52.1$ Oe, $\varphi = 83.7°$, $t = 704$ s; and (c) the coplanar waveguide with microwave power amplitude pulse modulation (CPW-PM) configuration with HFMR = 7166 Oe, $\Delta H = 52.9$ Oe, $\varphi = 88.9°$, $t = 335$ s. The colored dots are the collected data and the corresponding solid lines are the theoretical fit as given by Eq. (43) and a second-order polynomial background (dotted black lines). From an experimental standpoint, it is interesting not only to examine what parameters can be extracted from a technique but also how quickly measurements can be performed (see the parameter t). (d) Angular dependence of H_{FMR} for Cavity-HM (blue squares), CPW-HM (red down triangles), and CPW-PM (orange up triangles). Angles are measured from the [100] direction. The corresponding solid lines are the theoretical fit using Eq. (39). Fitting anisotropy parameters are shown in Table 1. As can be seen, all three methods are in excellent agreement.

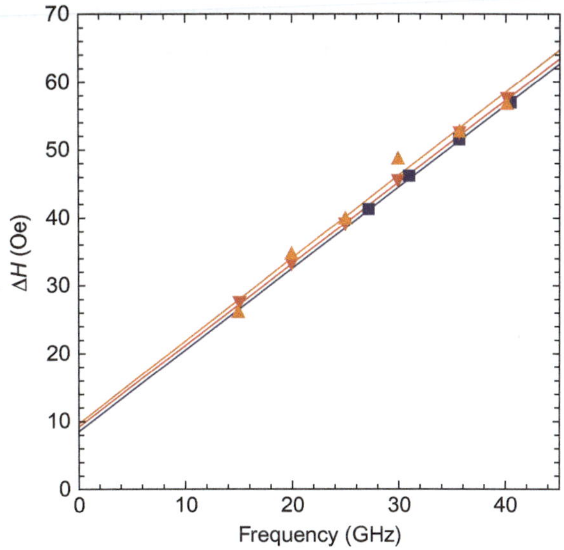

FIGURE 14

FMR linewidth measurements of GaAs/^{16}Fe/^{20}Au along the hard uniaxial axis [1$\bar{1}$0] direction are shown for Cavity-HM (blue squares), CPW-HM (red down triangles), and CPW-PM (orange up triangles). The corresponding lines are the theoretical fit using Eq. (35). For both field modulation techniques, all points lie on a straight line and the fitted α values are nearly identical. The CPW-PM measurements have a wider spread due to drift-induced error in each measurement; however, measurements at a number of microwave frequencies average out the power fluctuations and the fitted Gilbert damping parameter α agrees with the field modulation techniques within 1%.

Table 1 Magnetic Properties of GaAs/^{16}Fe/^{20}Au

Measurement configuration	$4\pi M_{eff}$(kOe)	$H_{K_1^{\parallel}}$(Oe)	$H_{K_u^{\parallel}}$(Oe)	ϕ_u(°)	$\alpha(10^{-3})$	$\Delta H(0)$(Oe)
Cavity-HM	16.07	313	−399	−44.6	3.52	8.6
CPW-HM	16.08	313	−393	−45.1	3.52	9.2
CPW-PM	16.07	315	−384	−45.3	3.56	9.8

such as field-swept VNA-FMR, one is able to measure both components of the rf susceptibility, but care must be taken to fit the correct functions to the in-phase and out-of-phase components to maintain the correct handedness under rotation. For more details we direct readers to Refs. [58,64].

An example of the FMR data for each technique, together with fits using Eq. (43) and an instrumental background (second-order polynomial), are shown in Fig. 13a–c, the angular dependences of H_{FMR} are shown in Fig. 13d, and the resulting fitted anisotropic fields are listed in Table 1. Clearly, the results of all three techniques agree very well and are described well by Eqs. (40) and (41). Notice that the FMR measurements using a CPW can result in signals which are far away from measuring the microwave absorption χ'' (common to measurements in the microwave cavity); in fact, in Fig. 13b and c, the measured FMR signal is very close to χ' and yet the fitting parameters used for the FMR signals obtained in the microwave cavity (Fig. 13a) and using CPW (Fig. 13b and c) are identical, testifying to the fact that fitting using Eq. (43) is a very reliable procedure. This is particularly important with large samples flipped over the transmission line. This propagation across the transmission line affects both the damping and the phase. One needs to also note that the above descriptions in this section are applicable when the change in transmitted signal is only weakly effected by the presence of the sample at FMR.

Additionally, the frequency dependence of the linewidth of the FMR signal was measured along the $[1\bar{1}0]$ direction. The data are shown in Fig. 14. The three methods are obviously in excellent agreement.

4.3.3.2 Polycrystalline Samples

In order to show the usefulness of this technique for samples similar to those which might be found in industrial applications, Si/3 nm Ta/8 nm Ru/10 × [*d* Co/8 nm Ru] samples were prepared by means of sputter deposition, where $d=0.6$, 0.8, 1.0, 1.2, 1.4, 1.6 is the thickness in nm for the Co layer and the 10 × [] indicates that the structure in the square brackets was repeated 10 times as a superlattice. The layer thicknesses were controlled using deposition rate calibrations and verified by means of high-resolution X-ray diffraction (XRD).

Note that 8-nm Ru spacer layers magnetically decouple each Co layer; therefore, FMR is carried out on an assembly of 10 independent ultrathin Co layers. In this thickness range, even the thinnest Co layers, $d=0.6$ nm (3 AL), were already ferromagnetic. However, their room temperature saturation magnetic moment $M_s \simeq 550$G was appreciably smaller than that of bulk Co with $M_s = 1400$G and their Currie temperature of $T_C \simeq 650$ K was lowered from the bulk value of $T_C = 1388$K [65]. This sample represents an ultimate test for the sensitivity of the described broadband FMR system for measuring samples that would be useful in material studies, spintronics, and high-density memory applications (Fig. 15).

Typical FMR signals for sputtered samples are shown in Fig. 15 comparing the thinnest Co layers, $d=0.6$ nm (3AL), to the thickest $d=1.6$ nm (8 AL). In all of these samples, the FMR lines are very broad, especially compared to the single-crystal Fe samples discussed above. The FMR linewidths as a function of frequency displayed a significant contribution due to inhomogeneity with the average $\Delta H(0) \simeq 250$ Oe—see Eq. (35)—and the damping parameter α on average was an order of magnitude larger than that in the Fe sample. Additionally, the benefit of CPW-PM over CPW-HM for these broad lines is evident from the

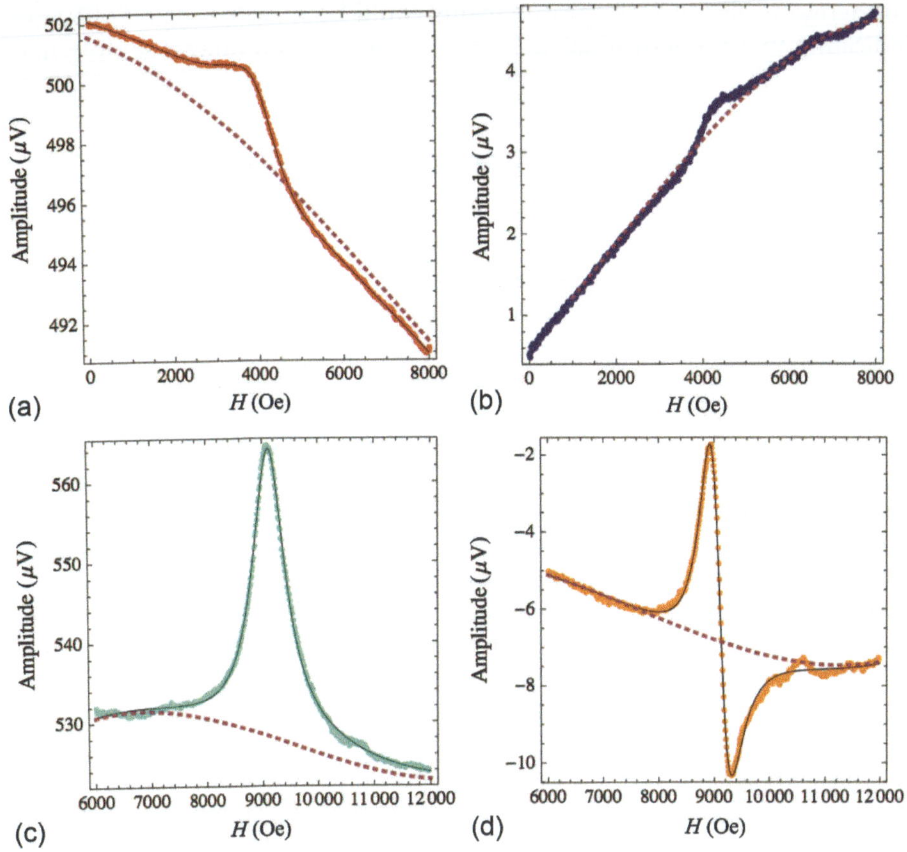

FIGURE 15
Coplanar waveguide configuration (CPW) measurements for sputtered samples at $f = 15$ GHz in the perpendicular-to-plane orientation: (Top row) Si/3 nm Ta/8 nm Ru/10 × [0.6 nm Co/8 nm Ru] for (a) PM with relevant fit parameters $H_{FMR} = 4103$ Oe, $\Delta H = 782$ Oe, $\varphi = 52°$, and collection time $t = 360$ s, and (b) HM with $H_{FMR} = 4119$ Oe, $\Delta H = 791$ Oe, $\varphi = 45°$, and collection time $t = 2315$ s. (Bottom row) Si/3 nm Ta/8 nm Ru/10 × [1.6 nm Co/8 nm Ru] for (c) PM with $H_{FMR} = 9094$ Oe, $\Delta H = 357$ Oe, $\varphi = -6°$, and collection time $t = 132$ s, and (d) HM with $H_{FMR} = 9069$ Oe, $\Delta H = 336$ Oe, $\varphi = -20°$, and collection time $t = 224$ s. Notice that the measurements using PM require a significantly shorter time for collection of data than those using HM. The colored dots are the collected data and the corresponding solid lines are the theoretical fit as given by a third-order polynomial background (dotted black lines).

collection times required to obtain similar SNRs. It is worth noting that the maximum modulation amplitude in our FMR spectrometer $\simeq 25$ Oe, which is not the optimum magnitude for such wide lines. However, at large DC H fields this much modulation can already lead to vibrations in the system.

Notice that even for these FMR lines with large inhomogeneity broadening, the simple ansatz Eqs. (32) and (33) can be used to fit the data very well. This is indicative that the distribution of inhomogeneous internal fields is Lorentzian.

Furthermore, all the FMR measurements could be fit with single FMR lines, indicating that the individual Co films within each multilayer sample have the same basic magnetic properties.

4.4 Inhomogeneous rf Driving in FMR

In multilayer films, where there are a minimum of two ferromagnetic layers separated by a normal metal spacer(s) (NM) FM1/NM/FM2, one can encounter problems with the efficiency of the rf magnetic field driving. In a flip-chip configuration, it is common that the lateral size of the sample is comparable to the lateral size of the coplanar waveguide. If an NM spacer is much thicker than the FM2 layer (located away from the transmission line), significantly decreased strength of the FMR signal in FM2 as compared to FM1 is observed. This is caused by the distribution of the rf field across the sample thickness. One can determine this distribution by taking into account the characteristic impedances in metallic layers and that of the substrate [66] and solving the propagation of the rf fields using Maxwell's equations with appropriate boundary conditions. The characteristic impedance of the semi-insulating wafer is large, approximately 100 Ω, while the characteristic impedance in the metallic layers is significantly smaller, in of the order of milliohms. For samples much thinner than the skin depth (the usual case in ultrathin films) the spatial distribution of h_{rf} follows a triangular shape with a maximum at the FM1 film facing the transmission line and a minimum at FM2 film. If the NM spacer is much thicker than the thickness of FM2, then the drop of the rf field occurs mostly across the NM. This results in a very small h_{rf} in FM2 compared to that in FM1 and can even lead to the absence of the FMR absorption in FM2. Yet the same film structure inserted in a microwave cavity will have FMR signal observed in both FM layers; in this case the rf field is driven symmetrically, resulting in a homogeneous distribution of the rf field across the FM1/NM/FM2 structure. The coplanar waveguide does not create this problem if the sample is inserted between the central and ground leads (see, for example, Woltersdorf *et al.* [14], where the rf driving is symmetric).

A significant decrease of the FMR signal in a normal cavity in the FM2 layer can also be observed in samples grown on heavily doped semiconductor substrates. Due to the high conductivity of the semiconductor wafer, a diminishing rf field at the FM2 layer can occur, resulting in asymmetric rf driving (see Fig. 16); this leads to the same problem as that in transmission lines.

Asymmetric driving can also occur when a sample is inserted outside of the cavity. It is well known that samples grown on Si wafers create problems by decreasing the cavity Q factor. Only samples less than $1 \times 1\ mm^2$ across are suitable for use inside the cavity. For larger lateral size samples,

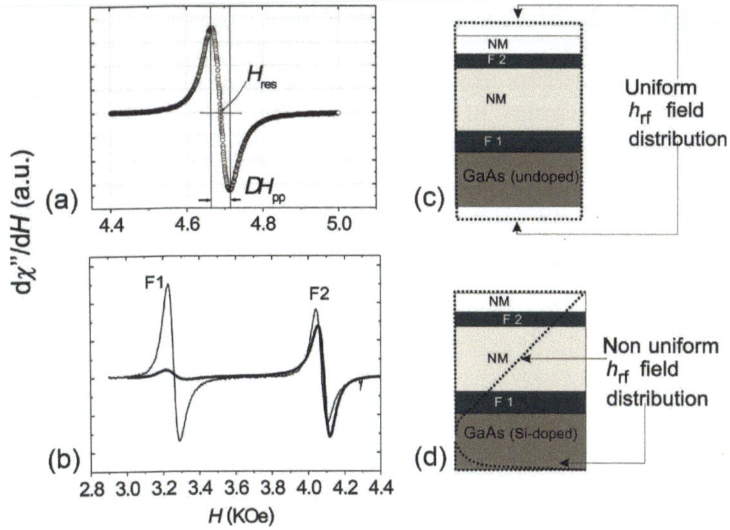

FIGURE 16

(a) A typical FMR spectrum of ^{20}Au/^{16}Fe/GaAs sample measured at 24 GHz. (b) The FMR spectra for a ^{20}Au/^{12}Fe/^{900}Ag/^{16}Fe/GaAs (thin solid line) and ^{20}Au/^{12}Fe/^{450}Ag/^{16}Fe/GaAs (thick solid line) samples at 24 GHz. The thin solid line shows the FMR spectrum acquired for a sample deposited on a semi-insulating GaAs (undoped) wafer. The resonance fields for the individual layers are well separated due to the strong in-plane uniaxial anisotropy of the ^{16}Fe film grown on GaAs. The FMR signal at the lower field corresponds to the ^{16}Fe bottom layer and the high-field signal to the ^{12}Fe top layer. The sample was symmetrically driven on both sides of the sample by a uniform rf field (h_{rf}) inside the cavity. The thick solid line shows the FMR signal for a similar specimen but grown on semi-conducting GaAs (Si-doped). Clearly, the signal from the ^{16}Fe layer grown on GaAs (Si-doped) is much smaller than that obtained from the ^{16}Fe layer grown on the undoped GaAs. This can be explained by the skin effect in the GaAs (Si-doped) substrate. (c) A schematic sketch of the uniform h_{rf} distribution inside the specimen grown on insulating GaAs and shown by the dotted line rectangle. (d) A schematic representation of the inhomogeneous rf driving field in the sample grown on GaAs (Si-doped). The skin depth of $\sigma \sim 16$ μm leads to the attenuation of h_{rf} inside the 0.375-mm-thick GaAs. Also, the impedance mismatch between the metallic layers and the GaAs substrate leads to a triangular distribution of the rf field inside the sample: The field is strongest at the top of the ^{12}Fe film and decays linearly to a small value at the ^{16}Fe/semiconductor interface.

one has to put them outside of the cavity and couple them through an aperture in the wall of the cavity. If the sample is not backed by a conducting foil then one again gets asymmetric rf driving with a triangular distribution across the sample thickness; this can result in different FMR signals even in a single film, where one can observe spin waves. In fact, spin waves with higher n (larger q vector) have larger FMR signals than those with low n because the rf field has a triangular distribution across the FM film thickness.

5 BRILLOUIN LIGHT SCATTERING

In this section, we describe BLS, an alternative experimental technique for the detection of spin dynamics. BLS involves the inelastic scattering of both photons and magnons. BLS and the electrical detection based on microwave absorption as described in the previous section 4 are complementary experimental approaches. While the spectral resolution provided by electrical detection techniques is superior, BLS spectroscopy enables wave vector as well as spatially resolved observation of spin waves. In addition, the remarkable sensitivity of BLS spectroscopy even allows for the detection of thermally activated spin-wave modes without external excitation.

The realization of frequency- and wave-vector-resolved BLS measurements for the investigation of the spin-wave dispersion as well as the determination of material parameters will be discussed in Section 5.1. In Section 5.2, the implementation of a microscopic spatial resolution is described followed by a discussion of recent studies on spin-wave propagation on the microscale. Section 5.3 is an introduction to the additional functionalities given by phase and time-resolved BLS on the basis of recent experiments.

5.1 BLS Spectroscopy

The BLS process is schematically depicted in Fig. 17a. In the Stokes process, an incident photon of frequency ν_{in} and momentum $\mathbf{k}_{\mathrm{i}}$ creates a magnon of frequency ν_{m} and momentum $\mathbf{k}_{\mathrm{m}}$. Due to energy and momentum

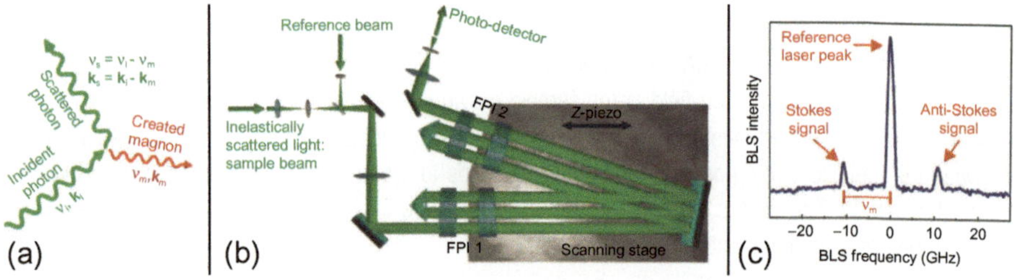

FIGURE 17

(a) Sketch of the BLS Stokes process where the incident photon creates a magnon. Due to the energy and momentum conservation described in Eqs. (44) and (45), the scattered photon carries the frequency and momentum information of the observed spin-wave mode. (b) Light path in a Tandem Fabry–Pérot interferometer (TFPI). The wavelength and, thus, the frequency of incoming photons are analyzed by scanning the mirror distance of the two etalons FPI1 and FPI2. In the actual (3 + 3) arrangement, each etalon is passed three times to ensure high contrast of the instrument. (c) Typical BLS spectrum. The reference beam—not subject to any frequency shift—is used for the stabilization of the TFPI and therefore marks the zero position of the spectrum. In addition, Stokes and anti-Stokes signals created in the BLS process can be observed for negative and positive frequency shifts, respectively.

conservation, the observation of the inelastically scattered photon yields the frequency and momentum information of the magnon according to the following equations:

$$\nu_S = \nu_i \pm \nu_m, \tag{44}$$

$$\mathbf{k}_S = \mathbf{k}_i \pm \mathbf{k}_m, \tag{45}$$

where the minus and plus signs denote the Stokes and anti-Stokes process, respectively.

Before we turn to the experimental realization of a BLS spectroscope based on these fundamental equations (44) and (45), we would like to draw attention to the extensive studies on light scattering from spin-wave excitations in thin ferromagnetic films [67–70]. These studies provide an excellent overview of the theoretical background of BLS that is beyond the scope of this chapter. In particular, an excellent and very transparent quantitative description of BLS in an arbitrary ferromagnetic film was carried out by Cochran and Dutcher [71].

The energy analysis of the inelastically scattered photons—and, thus, the frequency analysis of the observed spin-wave modes—is the basis for BLS spectroscopy. The typical spin-wave frequencies of a few gigahertz as well as the small scattering cross-section of BLS make this frequency analysis challenging. Thus, the major requirements for a BLS setup are a laser with a narrow spectral linewidth and an interferometer combining a high resolution better than $0.05\ \text{cm}^{-1}$ with a very high contrast. In the last decades, the Tandem Fabry–Pérot interferometer (TFPI) developed by Sanderdock and illustrated in Fig. 17b became the standard instrument for the frequency analysis in BLS spectroscopy [72].[1] The high contrast of the TFPI (better than 10^{10}) is achieved by the utilization of two etalons FPI1 and FPI2 consisting of pairs of highly reflective, parallel mirrors. These etalons are mounted on the same scanning stage to enable frequency filtering. Transmission through each of the etalons is only allowed for a mirror spacing that is a multiple of half of the wavelength of the probing beam. Thus, wavelength and, therefore, frequency analysis is realized by moving the scanning stage to control the mirror distance. To increase the contrast of the instrument, each of the etalons is passed three times by the probing beam before it is finally detected by a photo-detector. The accessible frequency range of the TFPI is up to several hundred gigahertz, which is hard to achieve by any competing methods. The frequency resolution of the instrument depends on this maximum frequency range that can be adjusted to the needs of the actual experiments. Values in the range of typical spin-wave frequencies are $\Delta f = 273\ \text{MHz}$ for $f_{max} = 30\ \text{GHz}$ and $\Delta f = 45\text{MHz}$ for $f_{max} = 5\ \text{GHz}$.

---

[1] www.tablestable.com.

An exemplary BLS spectrum is illustrated in Fig. 17c. The zero position of this spectrum is defined by the signal of the reference beam that is not subject to any inelastic scattering process and, therefore, has a well-defined frequency. This beam is used to stabilize the relative alignment and position of the mirrors in the TFPI during the scanning process. Stokes and anti-Stokes signals of the inelastically scattered light can be found for negative and positive frequencies, respectively. Due to energy conservation, these frequencies directly correspond to the frequencies of the investigated spin-wave modes. The intensity of the inelastically scattered light is proportional to the intensity of the investigated spin-wave mode.

In addition to the frequency resolution and in contrast to typical FMR experiments, BLS spectroscopy offers wave-vector resolution and, thus, direct access to the spin-wave dispersion. Key to the realization of wave-vector-resolved BLS measurements, based on Eq. (45), is the control over the momentum transfer between photon and magnon in the scattering process. This momentum transfer can be quantified by the following geometric considerations.

The beam path of a BLS experiment in backscattering geometry—where the scattered light is collected by the same objective that already focused the beam on the sample—is shown in Fig. 18a. The large working distance of the chosen lens results in a well-defined wave-vector direction of the incident beam that is crucial for the following determination of the spin-wave wave vector.

According to Noether's theorem, momentum conservation is only valid in the case of translational symmetry. This symmetry is broken by the sample surface. Thus, only the in-plane wave-vector components of magnon and photon are conserved in the scattering process and will be considered in the following.

As illustrated in Fig. 18a, the in-plane components of the incident and the scattered photons are completely defined by the scattering geometry and, in particular, the angle of incidence φ. Therefore, the momentum transfer to the investigated magnon mode can be determined as a function of φ. For the geometry discussed here, the wave vector of the investigated spin-wave mode is given by:

$$k_{\mathrm{SW},\parallel} = 2\frac{2\pi}{\lambda_{\mathrm{L}}}\sin\varphi, \tag{46}$$

where $\lambda_{\mathrm{L}} = 532\,\mathrm{nm}$ is the wavelength of typically used laser systems. By changing the angle of incidence φ, wave-vector sensitivity for wave vectors up to $k^{\max}_{\mathrm{SW},\parallel} = 23.6\,\mathrm{rad}/\mu\mathrm{m}$ (or a minimum wavelength $\lambda^{\min}_{\mathrm{SW},\parallel} = 266\,\mathrm{nm}$) can be realized.

In contrast to electrical detection methods, wave-vector-resolved BLS spectroscopy enables the direct observation of the spin-wave dispersion relation. The left panel of Fig. 18b illustrates the dispersion in a thin film of the Heusler

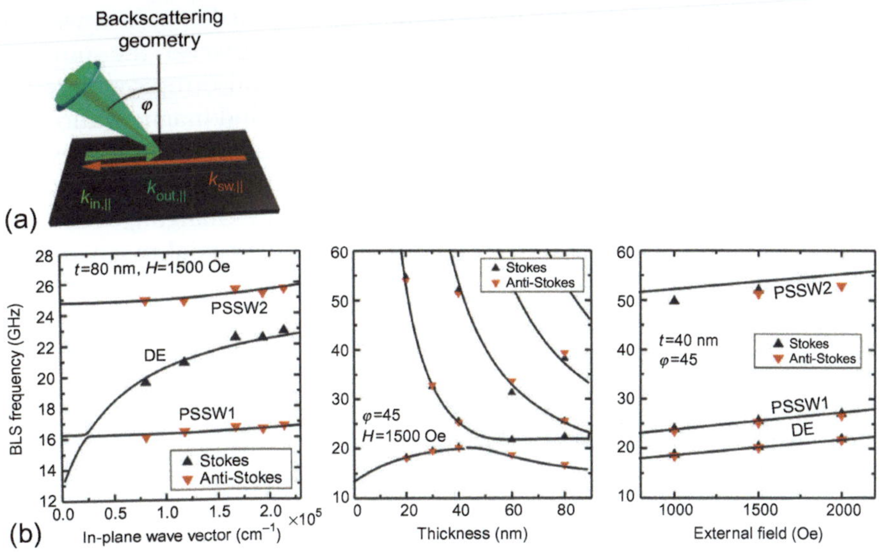

FIGURE 18

(a) Sketch of a BLS experiment in the backscattering geometry, where the inelastically scattered light takes the same path as the incident beam. The in-plane wave-vector components of incident and scattered light ($k_{in,||}$ and $k_{out,||}$) as well as of the investigated spin-wave mode $k_{SW,||}$ are defined by the angle of incidence φ. (b) BLS study of the Heusler compound Co_2MnSi. From left to right: Spin-wave dispersion relation as well as thickness and field dependence of the Damon–Eshbach (DE) mode and two perpendicular standing spin waves (PSSW1 and PSSW2). The symbols represent the experimental data, whereas the solid lines correspond to numerical calculations of the spin-wave dispersion relation according to Ref. [73]. *Graphs are taken from Ref. [68].*

compound Co_2MnSi measured by BLS (symbols) [74]. It should be noted that all data were obtained by the observation of thermally activated spin waves without external excitation, which indicates the high sensitivity of BLS spectroscopy. The solid lines in the graph are based on numerical modeling of the dispersion to extract the material parameters of the investigated Heusler compound [67]. The BLS spectra exhibit not only the DE mode but also two perpendicular standing spin-wave (PSSW) modes. While the existence of these modes can also be observed by FMR, the latter does not allow for the detection of nonzero wave vectors and, thus, for the analysis of the clearly distinct dispersion characteristics. To allow for a thorough investigation of the spin-wave dispersion and a reliable determination of materials parameters, the additional dependencies of the spin-wave modes on the film thickness and the external field can be used (middle and right panels of Fig. 18b, respectively).

Finally, it should be noted that the study discussed above takes into account only a part of the full spin-wave dispersion. In all the measurements presented above, the magnetization was oriented perpendicular to the plane of incidence of the probing beam. Thus, the investigation was restricted to spin-wave wave vectors perpendicular to the magnetization direction. However, the spin-wave dispersion of in-plane magnetized films is intrinsically anisotropic with respect to the relative orientation $\theta = \angle(\mathbf{k}_{\parallel}, \mathbf{M})$ of wave vector and magnetization. For the full investigation of this complex dispersion surface, changing just one wave-vector component via the angle of incidence φ is not sufficient. The investigation of the entire dispersion surface by individually changing both wave-vector components parallel and perpendicular to the external magnetic field is described in Ref. [75].

5.2 BLS Microscopy

As an optical technique, BLS enables space-resolved measurements and, in particular, submicrometer spatial resolution. To realize this microscopic resolution, the experimental setup needs to be extended by a microscope lens with high numerical aperture, an illumination and camera system to monitor the sample position, and a positioning stage. In the following, we will briefly introduce the major components of a BLS microscope followed by a discussion of recent BLS studies on spin dynamics on the microscale.

The light path of the incident probing beam in a BLS microscope is shown in Fig. 19a. To achieve maximum spatial resolution, the beam diameter is expanded with a telescope before it is guided to the microscope lens via the polarizing beam splitter cube **BS3**. The incident beam is focused onto the

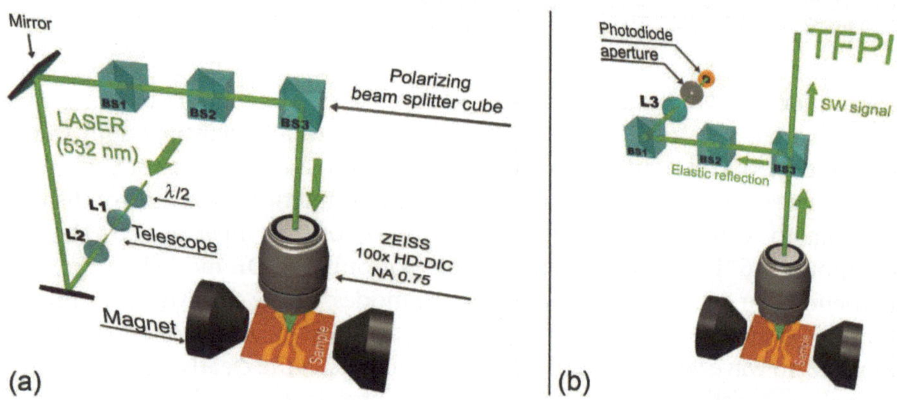

FIGURE 19
Beam path in a BLS microscope. (a) Path of the incident probing beam. (b) Light path of the elastically as well as inelastically scattered photons. The inelastically scattered photons that carry the spin-wave information are transmitted by the polarizing beam splitter cube **BS3**, whereas the elastically scattered photons are deflected toward a photodiode for the stabilization of the focal position (see text). This separation via polarization-dependent optics is possible due to a polarization rotation of 90° in the BLS process.

sample and gathered and collimated again by the same objective. Thus, the microscope works in the backscattering geometry. As can be seen in Fig. 19b, the scattered light reaches **BS3** a second time, where the elastically and inelastically scattered photons are separated. This separation is enabled by the polarization dependence of BLS. In the scattering process with spin waves, the polarization direction of photons is rotated by 90°. Therefore, only the inelastically scattered photons pass **BS3** toward the TFPI for frequency analysis whereas the elastically scattered light with perpendicular polarization is deflected. This separation drastically increases the contrast of the measurement and reduces the influence of additional signals arising from higher longitudinal laser modes or from the inelastic scattering of photons and phonons (which does not result in a change of the polarization direction).

The elastically scattered and, therefore, deflected beam can be used to determine the exact probing position of the laser and to find and stabilize the focal position of the sample via an intensity measurement at an additional photodiode behind the beam splitter cube **BS1**. Finally, **BS2** is used to integrate the illumination into the existing beam path of the laser and to monitor the sample position via a CCD camera.

In BLS microscopy, spatial resolution is achieved by scanning the sample position below the fixed probing beam and acquiring a BLS spectrum at each point. The spatial resolution is therefore defined by the spot size of the probing beam. In the studies presented in the following, the spatial resolution was 250 nm. For an automated scanning process, the sample is mounted on an *xyz*-stage with nanometer accuracy to control the lateral as well as the focal position of the sample.

The scanning process for the 2D BLS intensity distributions can be a very time-consuming task, in particular regarding the observation of thermally activated spin-wave modes with weak spin-wave signals. BLS microscopy—like any other scanning microscopy technique—therefore requires a good stability against, e.g., thermal drifts for long periods of time. Automated stabilization of the sample position is implemented in the software package TFPDAS-Micro[2] that also allows for the automated control of the TFPI, and up to three peripheral devices like, e.g., a microwave signal generator for the external excitation of spin waves or a power supply for setting the external magnetic bias field. This enables complex measurement sequences in a wide parameter space and the corresponding data acquisition.

For the stabilization process, a picture recognition algorithm is used to compare a user-defined reference image with the real-time image of the sample that is acquired via the CCD camera. Possible drifts are detected and

[2]www.tfpdas.de.

translated into a corresponding movement of the sample stage. Therefore, this software-based procedure ensures stabilization of the sample position for arbitrary periods of time. In addition, it enables the definition of scan arrays relative to the reference image for a space-resolved mapping of the spin-wave intensity. The applicability of this method and the great sensitivity of BLS microscopy was demonstrated by the observation of the 2D intensity distribution different thermally activated spin-wave modes in nanoscale ring structures [76].

To illustrate both the performance of a BLS microscope as well as recent developments in the research field of *magnonics* [77–81] we will present recent studies based on BLS microscopy in the following. The basic idea behind magnonics is the utilization of spin waves as information carriers to establish an alternative spin-wave-based logic to extend or even substitute charge-based information technologies. In recent years, several concepts and studies in the emerging field of magnonics have aimed to construct the basic building blocks for information transport and processing via spin waves [82–87]. One of the main challenges regarding the realization of complex magnon-based devices is spin-wave transport in two-dimensional structures. To understand the 2D propagation characteristics thoroughly, a space-resolved detection technique like BLS microscopy is necessary.

The control of spin-wave transport in 2D is mainly complicated by the anisotropic dispersion relation in in-plane magnetized samples. The spin-wave dispersion of an in-plane magnetized thin film based on an approximative, analytical model [88] is illustrated in Fig. 20a. Spin-wave frequencies on this dispersion surface in k-space depend on the angle between the wave vector and the magnetization direction $\theta = \angle(\mathbf{k}_{\|}, \mathbf{M})$. This anisotropy results in a preferred propagation direction $\theta = 90°$ with the spin-wave wave vector oriented perpendicular to the magnetization. This direction is indicated by the blue line on the dispersion surface. In this geometry, the group velocity, given by the gradient of the dispersion, is maximal and—as a result of this—the spatial decay of propagating spin-wave modes is minimal. In addition, the external excitation of spin dynamics via dynamic Oersted fields is most efficient in this geometry and the experimental realization of alternative geometries turns out to be difficult [90]. Therefore, almost all studies in spin-wave propagation on the microscale are based on $\mathbf{k}_{\|} \perp \mathbf{M}$.

Vogt and coworkers addressed the issue of two-dimensional spin-wave transport in a curved spin-wave waveguide as illustrated in Fig. 20b [91]. Spin waves are launched at the top of the structure by an antenna and propagate toward the bend at the bottom. In the case of a global external magnetic bias field as shown on the left side of the figure, the spin-wave propagation suddenly stops at the beginning of the bend region. The reason for this rapid spatial decay is the magnetization configuration in the spin-wave waveguide. The entire structure is

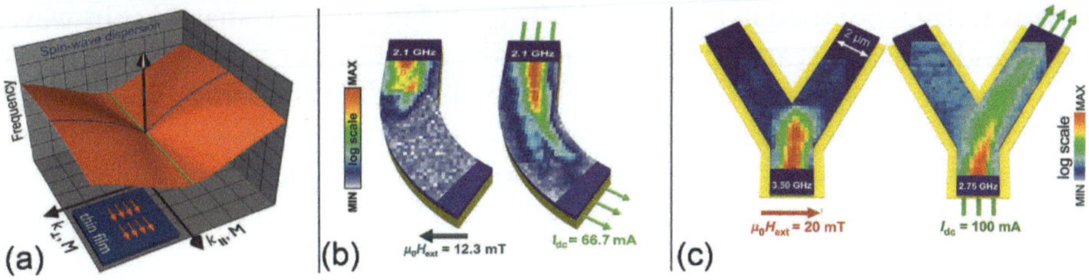

FIGURE 20

Control of 2D spin-wave transport. (a) Anisotropic spin-wave dispersion surface of a magnetic thin film. Spin-wave frequencies not only depend on the modulus of the corresponding wave vector but also on its relative orientation with respect to the magnetization $\theta = \angle(\mathbf{k}_{\parallel}, \mathbf{M})$. Highest (lowest) group velocity and thus lowest (highest) spatial decay is observed for $\theta = 90°$ ($\theta = 0°$) as indicated by the blue (green) line [77]. (b) Turning spin waves around a corner. Due to the anisotropic dispersion relation spin-wave propagation along a bended waveguide is strongly suppressed in the case of a global magnetic bias field (left). In contrast, 2D spin-wave transport is enabled by locally rotating the magnetization direction via locally generated magnetic fields to ensure $\mathbf{k}_{\parallel} \perp \mathbf{M}$ along the entire propagation path [79]. (c) Realization of a spin-wave multiplexer. The concept of local magnetic fields not only enables spin-wave transport in 2D but also switching of the propagation path in a Y-shaped spin-wave waveguide [89].

magnetized in the same direction along this bias field. Thus, spin waves that follow the geometric confinement at the bend leave the region with $\mathbf{k}_{\parallel} \perp \mathbf{M}$ and can no longer propagate in the preferred geometry. As a result a rapid decay without a noticeable change of direction of the investigated spin-wave mode is observed. Geometric confinement alone is therefore not sufficient to steer spin-wave propagation.

Instead, the anisotropic dispersion has to be taken into account. A possible scheme to address the anisotropy is to substitute the global magnetic field by local magnetic fields to enable propagation with $\mathbf{k}_{\parallel} \perp \mathbf{M}$ along the entire preferred path. Vogt and coworkers realized this scheme via local Oersted fields created by a direct current flowing through a gold conductor just below the spin-wave waveguide. The resulting Oersted field aligns the magnetization perpendicular to the propagation direction along the entire waveguide structure. Indeed, and as illustrated on the right side of Fig. 20b, spin waves clearly follow the curved waveguide structure in this case. Thus, using the concept of local magnetic fields enables one to turn spin waves around a corner.

In fact, the same scheme can also be used to go one step further. The underlying physical mechanism that hinders—at first sight—the propagation of spin waves in two dimensions, namely the anisotropy of the dispersion, also enables an interesting mechanism to not only turn spin waves around a corner but also to switch their propagation path. In an additional, subsequent study that builds upon the concept of local magnetic fields, Vogt and coworkers illustrated this possibility by the realization of a spin-wave multiplexer [80].

The sample layout used in this study is shown in Fig. 20c. A Y-shaped spin-wave waveguide was fabricated on top of a gold conduit of the same shape. In the case of a global external magnetic field (see left side of Fig. 20c), spin waves propagate from the antenna at the bottom of the structure toward the intersection in their preferred geometry $\mathbf{k}_{\|} \perp \mathbf{M}$. However, in the arms of the structure, this condition is no longer fulfilled and the spin waves are stopped at the intersection, similar to the previous case of the bend.

In the actual scenario with the Y-shaped sample layout it is possible to restrict the generation of the local magnetic fields to either the left or the right arm of the structure by adjusting the current path in the Au conduit. This extension of the previous experiment leads to new functionality: spin-wave propagation can be switched in either the left or right arms whereas it is strongly suppressed in the other direction (right side of Fig. 20c). The local magnetic field generated by the DC results in perpendicular orientation of the magnetization and spin-wave wave vector along the entire current path, allowing for propagation with high group velocity and thus low spatial decay. In contrast, in the arm without current flow the magnetization is aligned along the spin-wave waveguide by the demagnetizing field and no spin-wave intensity is observed. The presented scheme to control spin-wave propagation in 2D therefore enabled the realization of a spin-wave multiplexer as a fundamental building block of magnon-based logic.

An alternative way to affect the propagation characteristics of spin waves is the utilization of nonlinear phenomena [92–95]. The spin-wave system is an intrinsically nonlinear system and various nonlinear effects can be observed if there is sufficient excitation power. Since the thresholds of most nonlinearities can be easily reached with typical microwave signal generators, the spin-wave system can be regarded as a model system for general nonlinear phenomena that also occur in other fields of physics—like (room temperature) Bose–Einstein condensation [96] or the formation of solitons, which directly affects the propagation characteristics of the investigated spin-wave mode [97].

An important pre-condition for the observation of most nonlinear phenomena is a broadband spectral detection that is independent of the external excitation. Since the frequency analysis via the TFPI covers a frequency range up to several hundred gigahertz and is completely decoupled from external microwave generators, BLS is a very useful and flexible tool for the observation of nonlinearities in the spin system. The following study combines this simultaneous observation of different spin-wave frequencies with the spatial resolution offered by a BLS microscope.

This study opened the perspective for the frequency-controlled propagation of spin waves in 2D by the observation of nonlinearly generated higher harmonic spin-wave beams in a low-damping $Co_2Mn_{0.6}Fe_{0.4}Si$ waveguide [81].

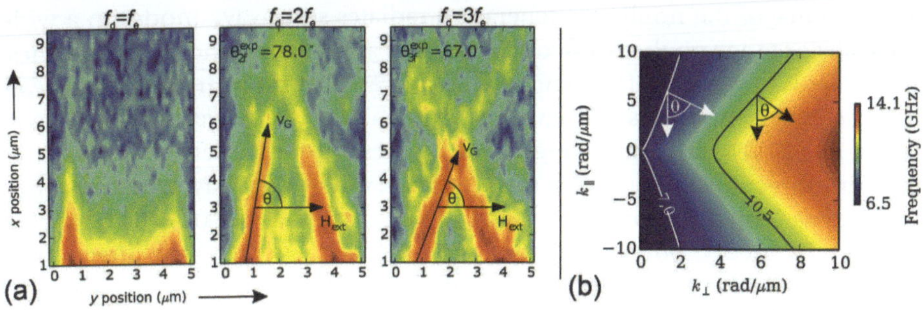

FIGURE 21
(a) Nonlinear emission of spin-wave caustics with frequency-dependent propagation characteristics. The left panel shows the directly excited, localized edge mode which is the source of the higher harmonic generation. The higher harmonics at twice and three times the excitation frequency form spin-wave caustic beams with a well-defined propagation direction [98]. Color-coded spin-wave dispersion in *k*-space with the iso-frequency curves at the higher harmonic frequencies is shown. The formation of spin-wave beams is promoted by the existence of several spin-wave modes with collinear group velocity.

Figure 21a shows the space-resolved BLS intensity detected in the Heusler waveguide at an excitation frequency $f_e = 3.5$ GHz for different detection frequencies f_d. The left panel illustrates the intensity profile of the directly excited mode ($f_d = f_e$), whereas the other panels show the nonlinearly generated higher harmonics at twice and three times the excitation frequency ($f_d = 2f_e$ and $f_d = 3f_e$). Each mode exhibits its own distinct propagation characteristics. The directly excited mode is strongly localized at the edges and strongly damped for increasing distance from the antenna at the bottom of the structure. The reason for this localization is the excitation frequency below the cut-off of the dispersion for modes propagating along the waveguide. At the edges of the waveguide, demagnetizing fields result in a strong reduction of the effective field with respect to the center area. As a result, low-frequency modes can exist there localized in spin-wave wells formed by the gradient of the effective field [99,100].

This situation changes in the case of higher harmonics. At the higher harmonic frequencies, the dispersion allows for spin-wave propagation in the center of the waveguide. As can be seen in Fig. 21a, the higher harmonics indeed exhibit propagation in the center area of the waveguide. They are emitted from the localized edge mode in the form of spin-wave beams with frequency-dependent propagation direction. Their propagation directions can be described via the angle $\theta = \angle(\mathbf{v}_G, \mathbf{H}_{ext})$ between the corresponding group velocity and external bias field.

This strongly anisotropic emission from an approximately point-like source—with actually isotropic radiation characteristics in *k*-space—is a result of the anisotropic spin-wave dispersion. The strongly driven, localized edge mode as the source of the nonlinear generation radiates spin-wave modes in a wide wave-vector range on the iso-frequency curves at higher harmonic frequencies. The analysis of these iso-frequency curves that are illustrated in Fig. 21b reveals the existence of various spin-wave modes with different wave vectors but collinear group velocity. The energy flow of these modes, therefore, points in the same direction and results in the formation of the spin-wave caustic beams. The propagation direction $\theta = \angle(\mathbf{v}_G, \mathbf{H}_{ext})$ of caustics can not only be predicted precisely by theory, but its dependence on the frequency and the direction of the external field even enables the control of their propagation characteristics in 2D [101–103].

5.3 Phase- and Time-Resolved BLS

The examples described above illustrate the basic operation of a BLS microscope, namely the frequency-resolved mapping of the spin-wave intensity in microscopic structures. However, the ongoing developments regarding BLS microscopy enable additional functionalities of the instrument: the phase- as well as the time-resolved detection of spin waves. In the following, we will briefly introduce the experimental realization of these detection methods.

In BLS spectroscopy, the wave vector of the investigated spin-wave modes is directly accessible via the scattering geometry. The pre-condition for the geometric considerations summarized in Eq. (46) is a well-defined direction of incident and detected beams. This pre-condition is fulfilled in the case of objectives with a large working distance. However, in BLS microscopy the objective is chosen with respect to maximum spatial resolution, which results in a high numerical aperture and a short working distance. For the chosen objectives, photons are focused onto the sample within a cone with an opening angle of $\varphi = 49°$ to the film normal. Consequently, the wave-vector selectivity is lost in BLS microscopy. This general phenomenon, shared by any microscopic technique, is a result of the Heisenberg uncertainty principle, which hinders the simultaneous determination of space and momentum information with arbitrary precession.

A possible way to address this issue is the phase sensitivity of the BLS process. The inelastically scattered photons carry not only the frequency and momentum information of the corresponding spin wave, but also the phase information. To extract this information in phase-resolved BLS microscopy, two-beam interference is used—similar to the approach taken in a Mach–Zehnder interferometer [104,74]. For this purpose, a reference beam is needed that is temporally coherent to the sample beam. The inelastically scattered photons of the sample beam are subject to a frequency shift. Thus, to achieve temporal coherence, the reference beam needs to be shifted in frequency as well. This is achieved via an electro-optical modulator that is driven by the same microwave signal that is also used for the excitation of the investigated spin-wave mode. The interference of reference and sample beams is now observed at the photodetector in the TFPI.

In this arrangement, any change of their relative phase and, thus, of the interference signal can be attributed to the changing phase of the sample beam. Since the phase of the sample beam is defined by the investigated spin-wave mode, this means that a space-resolved interference pattern of reference and sample beams directly reflects the wavelength of the spin wave. Such a space-resolved interference pattern dependent on the distance from the antenna is shown in the left panel of Fig. 22a [87]. The interference signal clearly exhibits minima and maxima corresponding to the destructive and constructive interference of the two beams. The distance between neighboring minima or maxima directly reveals the wavelength of the investigated spin-wave mode. Thus, phase-resolved BLS enables investigation of the spin-wave dispersion relation even with microscopic spatial resolution. An example is shown in the right panel of Fig. 22a.

After this introduction to phase-resolved BLS microscopy, the following paragraphs are devoted to the discussion of time-resolved BLS. BLS spectroscopy works directly in the frequency domain via the detection of the frequency shift in the inelastic scattering of photons and magnons. Therefore, the observation of spin precession in the time domain is no longer needed. However, in many cases it is useful to investigate the temporal evolution of the spin-wave intensity. In time-resolved BLS spectroscopy and microscopy, this is realized in a

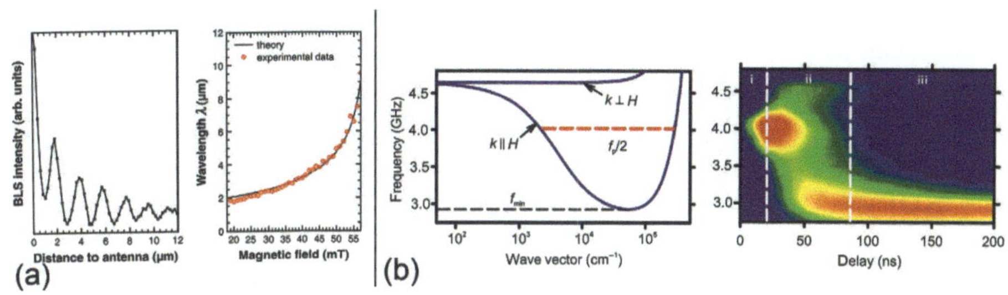

FIGURE 22

(a) Left panel: Interference signal of sample and reference beam as a function of the distance from the antenna. The distance of minima and maxima is given by the wavelength of the investigated spin-wave mode. Right panel: Experimentally determined wavelength compared to analytical calculation [105]. (b) Left panel: Spin-wave dispersion in a 5-μm-thick YIG film. The graph shows the two cases of $k_{\perp} H$ and $k \parallel H$ (solid lines). These two cases of perpendicular and parallel orientations of wave vector and effective field form the border of the manifold of allowed spin-wave modes for intermediate angles. Magnons are injected at the frequency $f_p/2$ via parametric excitation (dashed line). Right panel: Temporal evolution of the spectral occupancy of magnon modes after the parametric generation by a 30-ns microwave pulse [106].

time-of-flight measurement. In this detection scheme an external trigger defines the starting time t_0 of the measurement. In addition to the usual frequency analysis of the inelastically scattered photons, their relative arrival time t_i is then detected via a time-of-flight data acquisition card connected to the photodetector of the TFPI.

In the following example of a time-resolved BLS study from Demokritov and coworkers, the redistribution of energy pumped into the spin-wave system via parametric excitation is investigated [88]. This thermalization finally leads to the formation of a Bose–Einstein condensate at the bottom of the spin-wave spectrum. In the parametric process using a microwave pulse of 30 ns, microwave photons generate magnons at half of the pumping frequency $f_p = 8.1\,\text{GHz}$ [107]. This frequency $f_p/2$ of injected magnons is indicated by the dashed line in the calculated spin-wave dispersion of the investigated yttrium–iron–garnet (YIG) film illustrated in the left panel of Fig. 22b.

The time evolution of the spectral occupancy of spin-wave modes is shown in the right panel of Fig. 22b. In the first stage of the overall observation (i), the pulsed parametric excitation results in the injection of initial magnons at 4.05 GHz and, thus, a non-equilibrium magnon density. After the pumping pulse is switched off, magnon–magnon interactions cause the redistribution of the injected energy within a broad spectral range (ii). Finally, this thermalization results in a predominant occupancy of the lowest-energy state of the spin-wave dispersion (iii). In fact, the quantitative analysis of this final magnon distribution presented by Demidov and coworkers reveals the formation of a Bose–Einstein condensate at the bottom of the spectrum. In addition to this time-resolved observation of the spectral occupancy of spin waves, the time resolution can also be combined with space resolution offered by BLS to analyze the temporal characteristics of the spin-wave propagation [108].

6 SUMMARY AND FUTURE OUTLOOK

It is clear that magnetization dynamics play an important role in understanding and characterizing magnetic materials. The examples here have concentrated primarily on the dynamics in thin ferromagnetic films. But there has also been extensive work on the dynamics in magnetic nanostructures, including dot [109] and antidot [110,111] arrays and in magnetic nanowires [112]. In nanostructures one also has edge modes and corner modes in addition to surface modes [30,113]. Furthermore, nanostructures can concentrate incident power into very small regions, leading to large microwave fields and resulting in large precession angles and nonlinear behavior [114]. This has opened up studies of magnetic nonlinear dynamics in metallic samples. The study of new materials will certainly continue to use magnetic dynamics as a probe. As an example, one can consider the Dzyaloshinskii–Moriya interaction (DMI), where it has recently been predicted that an interfacial DMI could lead to nonreciprocal propagation of spin waves in a variety of situations [115,116]. Based on this, it seems very likely that the measurement of magnetic dynamics will continue to be significant in developing and understanding the properties of magnetic materials.

Acknowledgments

This chapter is based in part on the paper by Montoya *et al.* [64]. The authors would like to thank our colleagues at the Surface Science Lab at Simon Fraser University, namely Tommy McKinnon, Atieh Zamani, and Erol Girt, for their collaboration and stimulating discussions during this work. We would also like to extend our gratitude to the Natural Sciences and Engineering Research Council of Canada (NSERC) for providing the majority of the funding that makes our work possible. Portions of the work by R.E.C. and Z.C. were funded by the ARO (Grant No. W911NF-10-1-0225) and by the NSF (Grant No. DMR 0907063).

References

[1] Bloch F. Zur Theorie des Ferromagnetismus. Z Phys 1930;61:206.

[2] Nogués J, Schuller IK. Exchange bias. J Magn Magn Mater 1999;192:203–32. http://dx.doi.org/10.1016/S0304-8853(98)00266-2.

[3] Girt E, Huttema W, Mryasov ON, Montoya E, Kardasz B, Eyrich C, et al. A method for measuring exchange stiffness in ferromagnetic films. J Appl Phys 2011;109:07B765. http://dx.doi.org/10.1063/1.3565203.

[4] Eyrich C, Huttema W, Arora M, Montoya E, Rashidi F, Burrowes C, et al. Exchange stiffness in thin film Co alloys. J Appl Phys 2012;111:07C919. http://dx.doi.org/10.1063/1.3679433.

[5] Girt E, Wu S, Lu B, Ju G, Nolan T, Harkness S, et al. Influence of oxide on the structural and magnetic properties of CoPt alloy. J Appl Phys 2006;99:08E715. http://dx.doi.org/10.1063/1.2177395.

[6] Weller D, Mosendz O, Parker G, Pisana S, Santos TS. L1 0 FePtX-Y media for heat-assisted magnetic recording. Phys Status Solidi A 2013;210:1245–60. http://dx.doi.org/10.1002/pssa.201329106.

[7] Parkin SSP, Kaiser C, Panchula A, Rice PM, Hughes B, Samant M, et al. Giant tunnelling magnetoresistance at room temperature with MgO (100) tunnel barriers. Nat Mater 2004;3:862–7. http://dx.doi.org/10.1038/nmat1256.

[8] Yuasa S, Nagahama T, Fukushima A, Suzuki Y, Ando K. Giant room-temperature magnetoresistance in single-crystal Fe/MgO/Fe magnetic tunnel junctions. Nat Mater 2004;3:868–71. http://dx.doi.org/10.1038/nmat1257.

[9] Slonczewski J. Current-driven excitation of magnetic multilayers. J Magn Magn Mater 1996;159:L1–L7. http://dx.doi.org/10.1016/0304-8853(96)00062-5.

[10] Myers EB, Ralph DC, Katine JA, Louie RN, Buhrman RA. Current-induced switching of domains in magnetic multilayer devices. Science 1999;285:867–70. http://dx.doi.org/10.1126/science.285.5429.867.

[11] Katine J, Albert F, Buhrman R, Myers E, Ralph D. Current-driven magnetization reversal and spin-wave excitations in Co/Cu/Co pillars. Phys Rev Lett 2000;84:3149–52. http://dx.doi.org/10.1103/PhysRevLett.84.3149.

[12] Mangin S, Ravelosona D, Katine JA, Carey MJ, Terris BD, Fullerton EE. Current-induced magnetization reversal in nanopillars with perpendicular anisotropy. Nat Mater 2006;5:210–5. http://dx.doi.org/10.1038/nmat1595.

[13] Katine J, Fullerton EE. Device implications of spin-transfer torques. J Magn Magn Mater 2008;320:1217–26. http://dx.doi.org/10.1016/j.jmmm.2007.12.013.

[14] Woltersdorf G, Mosendz O, Heinrich B, Back C. Magnetization dynamics due to pure spin currents in magnetic double layers. Phys Rev Lett 2007;99:246603. http://dx.doi.org/10.1103/PhysRevLett.99.246603.

[15] Kardasz B, Heinrich B. Ferromagnetic resonance studies of accumulation and diffusion of spin momentum density in Fe/Ag/Fe/GaAs(001) and Ag/Fe/GaAs(001) structures. Phys Rev B 2010;81:1–7. http://dx.doi.org/10.1103/PhysRevB.81.094409.

[16] Montoya E, Kardasz B, Burrowes C, Huttema W, Girt E, Heinrich B. Spin transport in Au films: an investigation by spin pumping. J Appl Phys 2012;111:07C512. http://dx.doi.org/10.1063/1.3676026.

[17] Montoya E, Heinrich B, Girt E. Quantum well state induced oscillation of pure spin currents in Fe/Au/Pd(001) systems. Phys Rev Lett 2014;113:136601. http://dx.doi.org/10.1103/PhysRevLett.113.136601.

[18] Heinrich B. Radio frequency techniques. In: Heinrich B, Bland J, editors. Ultrathin magnetic structures II: measurement techniques and novel magnetic properties. Springer; 2005. p. 195–296, ISBN: 354027166X. http://dx.doi.org/10.1007/3-540-27166-X-3.

[19] Camley RE. Magnetization dynamics in thin films and multilayers. J Magn Magn Mater 1999;200:583. http://dx.doi.org/10.1016/S0304-8853(99)00345-5.

[20] Wolfram T, DeWames RE. Surface dynamics of magnetic materials. Prog Surf Sci 1972;2:233. http://dx.doi.org/10.1016/0079-6816(72)90001-9.

[21] Wigen P, Baryachtar V, Lesnik N, editors. Frontiers in magnetism of reduced dimension systems. In: NATO ASI Series 3, vol. 49. Dordrecht: Kluwer Academic; 1998. p. 121–52. http://dx.doi.org/10.1007/978-94-011-5004-0_5.

[22] Schreiber F, Frait Z. Spin wave resonance in high conductivity films: the FeCo alloy systems. Phys Rev B 1996;54:6473. http://dx.doi.org/10.1103/PhysRevB.54.6473.

[23] Damon RW, Eshbach JR. Magnetostatic modes of a ferromagnet slab. J Phys Chem Solids 1960;308. http://dx.doi.org/10.1016/0022-3697(61)90041-5.

[24] Camley RE. Nonreciprocal surface waves. Surf Sci Rep 1987;7:103. http://dx.doi.org/10.1016/0167-5729(87)90006-9.

[25] Cochran JFC. In: Heinrich B, Bland JAC, editors. Ultrathin magnetic structures II. Springer; 1984, 1995. http://dx.doi.org/10.1007/978-3-540-27166-6 [section 3.2.3.2].

[26] Kuanr BK, Camley RE, Celinski Z. Effect of shape anisotropy on stop-band response of Fe and permalloy based tunable microstrip filters. IEEE Trans Magn 2004;40:2841–3. http://dx.doi.org/10.1109/TMAG.2004.834197.

[27] Camley RE, McGrath BV, Khivintsev Y, Celinski Z, Adam R, Schneider CM, et al. Effect of cell size in calculating frequencies of magnetic modes using micromagnetics: special role of the uniform mode. Phys Rev B 2008;78:024425. http://dx.doi.org/10.1103/PhysRevB.78.024425.

[28] Aharoni A. Demagnetizing factors for rectangular ferromagnetic prisms. J Appl Phys 1998;83:3432. http://dx.doi.org/10.1063/1.367113.

[29] Guslienko KY, Chantrel RW, Slavin AN. Dipolar localization of quantized spin-wave modes in thin rectangular magnetic elements. Phys Rev B 2003;68:024422. http://dx.doi.org/10.1103/PhysRevB.68.024422.

[30] Grimsditch M, Leaf GK, Kaper HG, Karpeev DA, Camley RE. Normal modes of spin excitations in magnetic nanoparticles. Phys Rev B 2004;69(174428). http://dx.doi.org/10.1103/PhysRevB.69.174428.

[31] Shinjo T, Okuno T, Hassdorf R, Shigeto K, Ono T. Magnetic vortex core observation in circular dots of Permalloy. Science 2000;289:930. http://dx.doi.org/10.1126/science.289.5481.930.

[32] Guslienko KY, Ivanov BA, Novosad V, Otani Y, Shima H, Fukamichi K. Eigenfrequencies of vortex state excitations in magnetic submicron-size disks. J Appl Phys 2002;91:8037. http://dx.doi.org/10.1063/1.1450816.

[33] Choe S-B, Acremann Y, Scholl A, Bauer A, Doran A, Stohr J, et al. Vortex core-driven magnetization dynamics. Science 2004;304:420. http://dx.doi.org/10.1126/science.1095068.

[34] Yan M, Leaf G, Kaper H, Camley R, Grimsditch M. Spin-wave modes in a cobalt square vortex: micromagnetic simulations. Phys Rev B 2006;73:014425. http://dx.doi.org/10.1103/PhysRevB.73.014425.

[35] Vukadinovic N. High-frequency response of nanostructured magnetic materials. J Magn Magn Mater 2009;321:2074–81. http://dx.doi.org/10.1016/j.jmmm.2009.01.049.

[36] Camley RE. Long-wavelength surface spin waves on antiferromagnets. Phys Rev Lett 1979;45:283. http://dx.doi.org/10.1103/PhysRevLett.45.283.

[37] Stamps RL, Camley RE. Bulk and surface spin waves in thin-film antiferromagnets. J Appl Phys 1984;56:3497. http://dx.doi.org/10.1063/1.333915.

[38] Stamps RL, Camley RE. Dipole-exchange spin-wave modes in very-thin-film antiferromagnets. Phys Rev B 1987;35:1919. http://dx.doi.org/10.1103/PhysRevB.35.1919.

[39] Camley RE, Mills DL. Surface polaritons on uniaxial antiferromagnets. Phys Rev B 1982;26:1280. http://dx.doi.org/10.1103/PhysRevB.26.1280.

[40] Remer L, Luthi B, Sauer H, Geick R, Camley RE. Nonreciprocal optical reflection of the uniaxial antiferromagnet MnF_2. Phys Rev Lett 1986;56:2752. http://dx.doi.org/10.1103/PhysRevLett.56.2752.

[41] Jensen MRF, Parker TJ, Abraha K, Tilley DR. Experimental observation of magnetic surface polaritons in FeF_2 by attenuated total reflection. Phys Rev Lett 1995;75:3756. http://dx.doi.org/10.1103/PhysRevLett.75.3756.

[42] Jensen MRF, Feiven SA, Parker TJ, Camley RE. Experimental observation and interpretation of magnetic polariton modes in FeF_2. J Phys Condens Matter 1997;9:7233. http://dx.doi.org/10.1088/0953-8984/9/34/013.

[43] Heinrich B, Arrott AS, Cochran JF, Urquhart KB, Myrtle K, Celinski Z, et al. In situ techniques for studying epitaxially grown layers and determining their magnetic properties. Materials research society symposium proceedings—fall meeting, San Diego 1989, vol. 151. p. 177. http://dx.doi.org/10.1557/PROC-151-177.

[44] Linder J, Farle M. Magnetic anisotropy of heterostructures. In: Zabel H, Bader SD, editors. Magnetic Heterostructures, Springer Tracts in Modern Physics, vol. 227. Berlin, Heidelberg: Springer; 2008.

[45] Heinrich B. Basics of nano-thin film magnetism. In: Tsymbal E, Zutic I, editors. Handbook of spin transport and magnetism. Chapman-Hall/CRC; 2011. 21–46, ISBN: 978-1-4398-0377-6.

[46] Sander D. The magnetic anisotropy and spin reorientation of nanostructures and nanoscale films. J Phys Condens Matter 2004;16:R603–R636. http://dx.doi.org/10.1088/0953-8984/16/20/R01.

[47] Hurben M, Patton C. Theory of magnetostatic waves for in-plane magnetized isotropic films. J Magn Magn Mater 1995;139:263–91. http://dx.doi.org/10.1016/0304-8853(95)90006-3.

[48] Heinrich B, Cochran JF. Ultrathin metallic magnetic films: magnetic anisotropies and exchange interactions. Adv Phys 1993;42:523–639. http://dx.doi.org/10.1080/00018739300101524.

[49] Bland JAC, Heinrich B, editors. Ultrathin magnetic structures III. Berlin, Heidelberg: Springer-Verlag; 2005. p. 202. http://dx.doi.org/10.1007/b138703 [Section 5.7].

[50] McMichael RD, Twisselmann DJ, Kunz A. Localized ferromagnetic resonance in inhomogeneous thin films. Phys Rev Lett 2003;90:227601. http://dx.doi.org/10.1103/PhysRevLett.90.227601.

[51] Arias R, Mills D. Extrinsic contributions to the ferromagnetic resonance response of ultrathin films. Phys Rev B 1999;60:7395–409. http://dx.doi.org/10.1103/PhysRevB.60.7395.

[52] Arias R, Mills DL. Extrinsic contributions to the ferromagnetic resonance response of ultrathin films. J Appl Phys 2000;87:5455. http://dx.doi.org/10.1063/1.373370.

[53] Lindner J, Lenz K, Kosubek E, Baberschke K, Spoddig D, Meckenstock R, et al. Non-Gilbert-type damping of the magnetic relaxation in ultrathin ferromagnets: importance of magnon-magnon scattering. Physics (College Park Md) 2003;68:060102. http://dx.doi.org/10.1103/PhysRevB.68.060102.

[54] Woltersdorf G, Heinrich B. Two-magnon scattering in a self-assembled nanoscale network of misfit dislocations. Phys Rev B 2004;69:184417. http://dx.doi.org/10.1103/PhysRevB.69.184417.

[55] McMichael RD. A mean-field model of extrinsic line broadening in ferromagnetic resonance. J Appl Phys 2008;103:07B114. http://dx.doi.org/10.1063/1.2837887.

[56] Patton CE. Linewidth and relaxation processes for the main resonance in the spin-wave spectra of Ni-Fe alloy films. J Appl Phys 1968;39:3060–8. http://dx.doi.org/10.1063/1.1656733.

[57] Kuanr B, Camley RE, Celinski Z. Narrowing of the frequency-linewidth in structured magnetic strips: experiment and theory. Appl Phys Lett 2005;87:012502. http://dx.doi.org/10.1063/1.1968433.

[58] Celinski Z, Urquhart KB, Heinrich B. Using ferromagnetic resonance to measure magnetic moments of ultrathin films. J Magn Magn Mater 1997;166:6. http://dx.doi.org/10.1016/S0304-8853(96)00428-3.

[59] Kalarickal SS, Krivosik P, Wu M, Patton CE, Schneider ML, Kabos P, et al. Ferromagnetic resonance linewidth in metallic thin films: comparison of measurement methods. J Appl Phys 2006;99:093909. http://dx.doi.org/10.1063/1.2197087.

[60] Bilzer C, Devolder T, Crozat P, Chappert C, Cardoso S, Freitas PP. Vector network analyzer ferromagnetic resonance of thin films on coplanar waveguides: comparison of different evaluation methods. J Appl Phys 2007;101:074505. http://dx.doi.org/10.1063/1.2716995.

[61] Bilzer C, Devolder T, Crozat P, Chappert C. Open-circuit one-port network analyzer ferromagnetic resonance. IEEE Trans Magn 2008;44:3265–8. http://dx.doi.org/10.1109/TMAG.2008.2004379.

[62] Denysenkov VP, Grishin AM. Broadband ferromagnetic resonance spectrometer. Rev Sci Instrum 2003;74:3400. http://dx.doi.org/10.1063/1.1581395.

[63] Harward I, O'Keevan T, Hutchison A, Zagorodnii V, Celinski Z. A broadband ferromagnetic resonance spectrometer to measure thin films up to 70 GHz. Rev Sci Instrum 2011;82:095115. http://dx.doi.org/10.1063/1.3641319.

[64] Montoya E, McKinnon T, Zamani A, Girt E, Heinrich B. Broadband ferromagnetic resonance system and methods for ultrathin magnetic films. J Magn Magn Mater 2014;356:12–20. http://dx.doi.org/10.1016/j.jmmm.2013.12.032.

[65] Kittel C. Introduction to solid state physics. 8th ed. Hoboken, NJ: Wiley; 2004. ISBN: 9780471415268238.

[66] Cheng D. Field and wave electromagnetics. 2nd ed. Reading, MI: Addison-Wesley; 1992. ISBN: 0201128195.

[67] Cochran J. Heinrich B, Bland JAC, editors. Ultrathin magnetic structures II: light scattering from ultrathin magnetic layers and bilayers. Berlin: Springer; 1994. http://dx.doi.org/10.1007/978-3-540-27166-6.

[68] Camley RE, Mills DL. Surface response of exchange- and dipolar-coupled ferromagnets: application to light scattering from magnetic surfaces. Phys Rev B 1978;18:4821–41. http://dx.doi.org/10.1103/PhysRevB.18.4821.

[69] Camley RE, Rahman TS, Mills DL. Theory of light scattering by the spin-wave excitations of thin ferromagnetic films. Phys Rev B 1981;23:1226–44. http://dx.doi.org/10.1103/PhysRevB.23.1226.

[70] Camley RE, Grimsditch M. Brillouin scattering from magnons in ferromagnetic thin films of polycrystalline iron. Phys Rev B 1980;22:5420–5. http://dx.doi.org/10.1103/PhysRevB.22.5420.

[71] Cochran JF, Dutcher JR. Light scattering from thermal magnons in thin metallic ferromagnetic films. J Appl Phys 1988;63:3814–6. http://dx.doi.org/10.1063/1.340623.

[72] Sandercock J. Brillouin scattering study of SbSI using a double-passed, stabilised scanning interferometer. Opt Commun 1970;2:73–6. http://dx.doi.org/10.1016/0030-4018(70)90047-7.

[73] Hillebrands B. Spin-wave calculations for multilayered structures. Phys Rev B 1990;41:530–40. http://dx.doi.org/10.1103/PhysRevB.41.530.

[74] Fohr F, Serga AA, Schneider T, Hamrle J, Hillebrands B. Phase sensitive Brillouin scattering measurements with a novel magneto-optic modulator. Rev Sci Instrum 2009;80(043903). http://dx.doi.org/10.1063/1.3115210.

[75] Sandweg CW, Jungfleisch MB, Vasyuchka VI, Serga AA, Clausen P, Schultheiss H, et al. Wide-range wavevector selectivity of magnon gases in Brillouin light scattering spectroscopy. Rev Sci Instrum 2010;81:073902. http://dx.doi.org/10.1063/1.3454918.

[76] Schultheiss H, Sandweg CW, Obry B, Hermsdörfer S, Schäfer S, Leven B, et al. Dissipation characteristics of quantized spin waves in nano-scaled magnetic ring structures. J Phys D Appl Phys 2008;41:164017. http://dx.doi.org/10.1088/0022-3727/41/16/164017.

[77] Demokritov SO, Slavin AN. Magnonics: from fundamentals to applications. Berlin: Springer; 2013. http://dx.doi.org/10.1007/978-3-642-30247-3.

[78] Kruglyak VV, Demokritov SO, Grundler D. Magnonics. J Phys D Appl Phys 2010;43:260301. http://dx.doi.org/10.1088/0022-3727/43/26/260301.

[79] Serga AA, Chumak AV, Hillebrands B. YIG magnonics. J Phys D Appl Phys 2010;43:264002. http://dx.doi.org/10.1088/0022-3727/43/26/264002.

[80] Lenk B, Ulrichs H, Garbs F, Münzenberg M. The building blocks of magnonics. Phys Rep 2011;507:107–36. http://dx.doi.org/10.1016/j.physrep.2011.06.003.

[81] Hoffmann A, Schultheiß H. Mesoscale magnetism. Curr Opin Solid State Mater Sci 2014; http://dx.doi.org/10.1016/j.cossms.2014.11.004.

[82] Hertel R, Wulfhekel W, Kirschner J. Domain-wall induced phase shifts in spin waves. Phys Rev Lett 2004;93:257202. http://dx.doi.org/10.1103/PhysRevLett.93.257202.

[83] Hansen U-H, Demidov VE, Demokritov SO. Dual-function phase shifter for spin-wave logic applications. Appl Phys Lett 2009;94:252502. http://dx.doi.org/10.1063/1.3159628.

[84] Schneider T, Serga AA, Leven B, Hillebrands B, Stamps RL, Kostylev MP. Realization of spin-wave logic gates. Appl Phys Lett 2008;92:022505. http://dx.doi.org/10.1063/1.2834714.

[85] Kostylev MP, Serga AA, Schneider T, Leven B, Hillebrands B. Spin-wave logical gates, Appl Phys Lett 2005;87:153501. http://dx.doi.org/10.1063/1.2089147.

[86] Khitun A, Bao M, Wang K. Spin wave magnetic nanofabric: a new approach to spin-based logic circuitry. IEEE Trans Magn 2008;44:2141–52. http://dx.doi.org/10.1109/TMAG.2008.2000812.

[87] Chumak AV, Serga AA, Hillebrands B. Magnon transistor for all-magnon data processing. Nat Commun 2014;5:4700. http://dx.doi.org/10.1038/ncomms5700.

[88] Kalinikos BA, Slavin AN. Theory of dipole-exchange spin wave spectrum for ferromagnetic films with mixed exchange boundary conditions. J Phys C Solid State Phys 1986;19:7013. http://dx.doi.org/10.1088/0022-3719/19/35/014.

[89] Vogt K, Fradin F, Pearson J, Sebastian T, Bader S, Hillebrands B, et al. Realization of a spin-wave multiplexer. Nat Commun 2014;5:3727. http://dx.doi.org/10.1038/ncomms4727.

[90] Brächer T, Pirro P, Westermann J, Sebastian T, Lägel B, Van de Wiele B, et al. Generation of propagating backward volume spin waves by phase-sensitive mode conversion in two-dimensional microstructures. Appl Phys Lett 2013;102:132411. http://dx.doi.org/10.1063/1.4800005.

[91] Vogt K, Schultheiss H, Jain S, Pearson JE, Hoffmann A, Bader SD, et al. Spin waves turning a corner. Appl Phys Lett 2012;101:042410. http://dx.doi.org/10.1063/1.4738887.

[92] Schultheiss H, Vogt K, Hillebrands B. Direct observation of nonlinear four-magnon scattering in spin-wave microconduits. Phys Rev B 2012;86:054414. http://dx.doi.org/10.1103/PhysRevB.86.054414.

[93] Pirro P, Sebastian T, Brächer T, Serga A, Kubota T, Naganuma H, et al. Non-Gilbert-damping mechanism in a ferromagnetic Heusler compound probed by nonlinear spin dynamics. Phys Rev Lett 2014;113:227601. http://dx.doi.org/10.1103/PhysRevLett.113.227601.

[94] Demidov VE, Kostylev MP, Rott K, Krzysteczko P, Reiss G, Demokritov SO. Generation of the second harmonic by spin waves propagating in microscopic stripes. Phys Rev B 2011;83:054408. http://dx.doi.org/10.1103/PhysRevB.83.054408.

[95] Demidov VE, Ulrichs H, Demokritov SO, Urazhdin S. Nonlinear scattering in nanoscale magnetic elements: overpopulation of the lowest-frequency magnon state. Phys Rev B 2011;83:020404. http://dx.doi.org/10.1103/PhysRevB.83.020404.

[96] Demokritov SO, Demidov VE, Dzyapko O, Melkov GA, Serga AA, Hillebrands B, et al. Bose-Einstein condensation of quasi-equilibrium magnons at room temperature under pumping. Nature 2006;443:430–3. http://dx.doi.org/10.1038/nature05117.

[97] Demokritov SO, Serga AA, Demidov VE, Hillebrands B, Kostylev M, Kalinikos B. Experimental observation of symmetry-breaking nonlinear modes in an active ring. Nature 2003;426:159–62. http://dx.doi.org/10.1038/nature02042.

[98] Sebastian T, Brächer T, Pirro P, Serga AA, Hillebrands B, Kubota T, et al. Nonlinear emission of spin-wave caustics from an edge mode of a microstructured $Co_2Mn_{0.6}Fe_{0.4}Si$ waveguide. Phys Rev Lett 2013;110:067201. http://dx.doi.org/10.1103/PhysRevLett.110.067201.

[99] Bayer C, Park J, Wang H, Yan M, Campbell C, Crowell P. Spin waves in an inhomogeneously magnetized stripe. Phys Rev B 2004;69:134401. http://dx.doi.org/10.1103/PhysRevB.69.134401.

[100] Gubbiotti G, Conti M, Carlotti G, Candeloro P, Fabrizio ED, Guslienko KY, et al. Magnetic field dependence of quantized and localized spin wave modes in thin rectangular magnetic dots. J Phys Condens Matter 2004;16:7709–21. http://dx.doi.org/10.1088/0953-8984/16/43/011.

[101] Demidov VE, Demokritov SO, Birt D, O'Gorman B, Tsoi M, Li X. Radiation of spin waves from the open end of a microscopic magnetic-film waveguide. Phys Rev B 2009;80:014429. http://dx.doi.org/10.1103/PhysRevB.80.014429.

[102] Veerakumar V, Camley RE. Magnon focusing in thin ferromagnetic films. Phys Rev B 2006;74:214401. http://dx.doi.org/10.1103/PhysRevB.74.214401.

[103] Schneider T, Serga AA, Chumak AV, Sandweg CW, Trudel S, Wolff S, et al. Nondiffractive subwavelength wave beams in a medium with externally controlled anisotropy. Phys Rev Lett 2010;104:197203. http://dx.doi.org/10.1103/PhysRevLett.104.197203.

[104] Serga AA, Schneider T, Hillebrands B, Demokritov SO, Kostylev MP. Phase-sensitive Brillouin light scattering spectroscopy from spin-wave packets. Appl Phys Lett 2006;89:063506. http://dx.doi.org/10.1063/1.2335627.

[105] Vogt K, Schultheiss H, Hermsdoerfer S, Pirro P, Serga AA, Hillebrands B. All-optical detection of phase fronts of propagating spin waves in a $Ni_{81}Fe_{19}$ microstripe. Appl Phys Lett 2009;95:182508. http://dx.doi.org/10.1063/1.3262348.

[106] Demidov V, Dzyapko O, Demokritov S, Melkov G, Slavin A. Observation of spontaneous coherence in Bose-Einstein condensate of magnons. Phys Rev Lett 2008;100:047205. http://dx.doi.org/10.1103/PhysRevLett.100.047205.

[107] L'Vov V. Wave turbulence under parametric excitation. Berlin: Springer; 1994. http://dx.doi.org/10.1007/978-3-642-75295-7.

[108] Serga AA, Demokritov SO, Hillebrands B, Slavin AN. Self-generation of two-dimensional spin-wave bullets. Phys Rev Lett 2004;92:117203. http://dx.doi.org/10.1103/PhysRevLett.92.117203.

[109] Kakazei GN, Wigen PE, Guslienko KY, Novosad V, Slavin AN, Golub VO, et al. Spin-wave spectra of perpendicularly magnetized circular submicron dot arrays, Appl Phys Lett 2004;85:443. http://dx.doi.org/10.1063/1.1772868.

[110] Yu C, Pechan MJ, Mankey GJ. Dipolar induced, spatially localized resonance in magnetic antidot arrays, Appl Phys Lett 2003;83:3948. http://dx.doi.org/10.1063/1.1625104.

[111] Martyanov ON, Yudanov VF, Lee RN, Nepijko SA, Elmers HJ, Hertel R, et al. Ferromagnetic resonance study of thin film antidot arrays: experiment and micromagnetic simulations, Phys Rev B 2007;75:174429. http://dx.doi.org/10.1103/PhysRevB.75.174429.

[112] Ding J, Demidov VE, Cottam MG, Demokritov SO, Adeyeye AO. Dynamic behavior of $Ni_{80}Fe_{20}$ nanowires with controlled defects, Appl Phys Lett 2014;104:143105. http://dx.doi.org/10.1063/1.4870837.

[113] Guo F, Belova LM, McMichael RD. Spectroscopy and imaging of edge modes in permalloy nanodisks, Phys Rev Lett 2013;110:017601. http://dx.doi.org/10.1103/PhysRevB.74.024424.

[114] Khivintsev Y, Kuanr B, Fal TJ, Haftel M, Camley RE, Celinski Z, et al. Nonlinear ferromagnetic resonance in permalloy films: a nonmonotonic power-dependent frequency shift, Phys Rev B 2010;81:054436. http://dx.doi.org/10.1103/PhysRevB.81.054436.

[115] Moon J-H, Seo S-M, Lee K-J, Kim K-W, Ryu J, Lee H-W, et al. Spin-wave propagation in the presence of interfacial Dzyaloshinskii-Moriya interaction, Phys Rev B 2013;88:184404. http://dx.doi.org/10.1103/PhysRevB.88.184404.

[116] Garcia-Sanchez F, Borys P, Vansteenkiste A, Kim J-V, Stamps RL. Nonreciprocal spin-wave channeling along textures driven by the Dzyaloshinskii-Moriya interaction. Phys Rev B 2014;89:224408. http://dx.doi.org/10.1103/PhysRevB.89.224408.

CHAPTER 4

Nonlinear Behavior in Metallic Thin Films and Nanostructures

Karen Livesey
Center for Magnetism and Magnetic Materials, Department of Physics, University of Colorado at Colorado Springs, Colorado Springs, CO, USA

1 INTRODUCTION

Nonlinear behavior in magnetic materials has been studied since the 1950s when Bloembergen and Damon first measured strange absorption properties in magnetic materials that were driven by magnetic fields at very high power levels [1, 2] and Suhl explained them in terms of first- and second-order nonlinear instabilities [3]. Magnetization dynamics are intrinsically nonlinear. However, to see these nonlinear dynamics it is often necessary to drive a system above some threshold power. This is so there is sufficient energy going in as to counteract the natural losses or damping from the magnetic system. If driven far enough into the nonlinear regime, magnetic systems may display chaotic behavior. That is, very small changes in the initial magnetization conditions lead to vastly different dynamics.

Why should we be interested in the nonlinear and chaotic behavior of magnetic elements? Firstly, from a fundamental physics standpoint, magnetic materials represent a model system in which to study chaos. Unlike so many other chaotic dynamical systems governed by differential equations, such as atmospheres [4], stock markets [5] and neuron pathways [6] to name just a few, magnetic systems can have their parameters very accurately tuned and therefore provide a great test-bed for studying how various factors affect the onset and the type of chaos that results.

Secondly, there are many technological applications of magnetic elements for which operation may be affected by nonlinear effects. For example, as switching speeds in magnetic memories get larger, nonlinear damping of the magnetization vector's motion becomes more important to consider in order to understand the resulting dynamics and to optimize a device's performance. Spin torque nano-oscillators and other novel spintronic devices are also affected

Magnetism of Surfaces, Interfaces, and Nanoscale Materials. http://dx.doi.org/10.1016/B978-0-444-62634-9.00004-7

by nonlinear dynamics in certain situations so it is advantageous for physicists and engineers to understand these processes. In some cases, the nonlinear effects are in fact desirable and allow for the design of signal processing devices such as frequency modulators or filters that operate in the microwave regime.

Many reviews and book chapters have been devoted to the topic of nonlinear behavior in magnetic materials [7–10] and so the task of providing a new one is a little intimidating. In this work I will focus on recent progress in nanometer-thin metallic films and nanostructures as this emphasis was perhaps lacking in the previous reviews and is of interest with the advent of smaller and smaller devices based on such physics. Much of the early work on nonlinear processes in magnets focused on insulators but that has changed over the past 10 years. Typically, magnetic insulators are larger in size and are more difficult to integrate into devices than metals. A discussion of the latest progress in magnetic insulators and their nonlinear behaviors will be left almost entirely out of this review but can be found in some recent articles [11, 12]. The technological issues aside, yttrium iron garnet (YIG) is still the go-to material for studying fundamental nonlinear phenomena such as soliton propagation and regeneration due to very low losses compared to metals. In these studies their large size is actually favorable as it allows for easier detection of nonlinear excitations. Probing excitations and nonlinear processes on the nanoscale can be far more challenging.

It turns out that the theoretical and experimental study of nonlinear dynamics in a magnetic object becomes increasingly more complicated when the object is nanometer-sized in one, two and three dimensions. Although the end goal is to explain some of these more complicated systems, first in Section 2 a general introduction to nonlinear behavior in magnets will be provided by looking at the case of a single macrospin. Then in Section 3 thin films will be discussed and finally in Section 4 the nonlinear processes in nanostructures that are confined in two or three dimensions will be described.

2 THEORY OF ONE MACROSPIN

In this section the equations of motion governing the magnetization, the ways to drive a system into the nonlinear regime, and also the ways to characterize and study nonlinear and chaotic dynamics will be outlined using the special case of a macrospin. The macrospin model is valid when the magnetization vector has the same magnitude and direction throughout an object. This simple case allows an overview of many of the important descriptors of nonlinear behavior and chaos before detailing the more complicated physics of films and nanostructures where the magnetization varies throughout the material.

2.1 Equations of Motion

Let us consider an ellipsoid or a rectangular sample with arbitrary demagnetizing factors N_x, N_y and N_z, and treat the magnetization $\vec{M}$ as a macrospin. In other words, all the magnetic moments move in-phase with one another. The magnetic energy has contributions from the demagnetizing field and from a static applied field H applied along the z direction. There is also a driving time-dependent r.f. field $h(t) = h\cos(\omega_d t)$ applied parallel to the x direction, where ω_d is the angular driving frequency. A possible geometry for a rectangular sample is shown in Fig. 1.

In CGS units the total energy density of the magnet is

$$E = 2\pi\left(N_x M_x^2 + N_y M_y^2 + N_z M_z^2\right) - M_z H - M_x h\cos(\omega_d t). \tag{1}$$

The demagnetizing factors add to 1; i.e., $N_x + N_y + N_z = 1$. For our rectangular geometry, this is an approximation as the demagnetizing field actually varies within the sample, unlike the case of an ellipsoid where the demagnetizing fields are uniform throughout. For a small enough rectangular sample, the approximate demagnetizing factors are calculated by either averaging the fields through the sample dimensions or by taking the values at the center of the sample [13, 14].

The torque equation which describes the dynamics of the magnetization vector if there were no damping is given by [15]

$$\frac{d\vec{M}}{dt} = -|\gamma|\,\vec{M} \times \left(-\frac{\delta E}{\delta \vec{M}}\right), \tag{2}$$

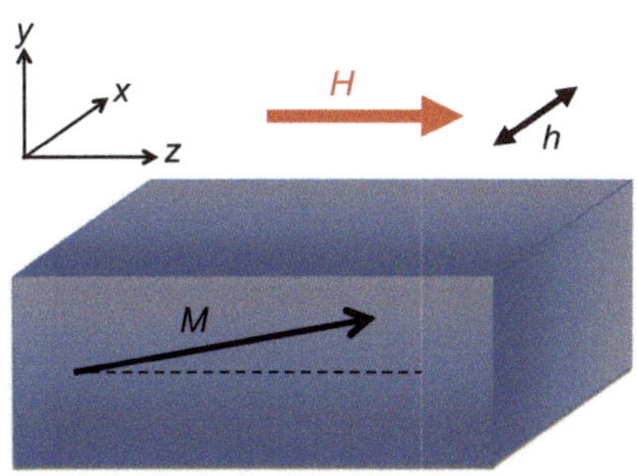

FIGURE 1
A macrospin geometry corresponding to the equations of motion presented in this section. One magnetization vector is drawn to represent the whole rectangular sample. There is a static field *H* and a transverse, oscillating driving field *h*.

where γ is the gyromagnetic ratio and the functional derivative $-\delta E/\delta \vec{M}$ gives the total effective field felt by the magnetization vector. Really the effective field is only a functional derivative when we allow $\vec{M}$ to vary throughout space, but this notation is most general. Substituting the energy Eq. (1) into Eq. (2) results in a set of three differential equations for the three components of the magnetization, of which only two are independent:

$$\frac{dM_x(t)}{dt} = -|\gamma| M_y(t)\left(H + 4\pi(N_y - N_z)M_z(t)\right) \tag{3}$$

$$\frac{dM_y(t)}{dt} = |\gamma| M_x(t)(H + 4\pi(N_x - N_z)M_z(t)) - |\gamma| h M_z(t)\cos(\omega_d t) \tag{4}$$

$$\frac{dM_z(t)}{dt} = -|\gamma| 4\pi(N_x - N_y)M_x(t)M_y(t) + |\gamma| h M_y(t)\cos(\omega_d t). \tag{5}$$

It can be seen quite clearly that each of the three coupled equations has a nonlinear term that is quadratic in components of the magnetization. These arise due to the demagnetizing energy for the object. There are additional nonlinear terms involving products of the driving field h with the magnetization components.

For small deviations of the magnetization from its equilibrium direction (along the z axis), it can be assumed that $M_x \ll M_z$ and $M_y \ll M_z$, with $M_z \sim M$, the constant saturation magnetization. This allows Eqs. (3)–(5) to be linearized. Assuming precessional motion about the z axis such that $M_{x/y}(t) \sim M_{x/y}e^{i\omega_d t}$ the equations of motion become

$$\begin{aligned} \frac{i\omega_d}{|\gamma|} M_x &= -M_y\left(H + 4\pi(N_y - N_z)M\right) \\ \frac{i\omega_d}{|\gamma|} M_y &= M_x(H + 4\pi(N_x - N_z)M) - hM. \end{aligned} \tag{6}$$

The result for the magnitude of the precession for the x and y components of the magnetization is

$$\begin{aligned} M_x &= \frac{-|\gamma|^2 hM\left(H + 4\pi(N_y - N_z)M\right)}{\omega_d^2 - \omega^2} \\ M_y &= \frac{i\omega_d |\gamma| hM}{\omega_d^2 - \omega^2}, \end{aligned} \tag{7}$$

where ω is the natural angular frequency of the magnetic object, also known as the ferromagnetic resonance (FMR) frequency, and is given by

$$\omega = |\gamma| \sqrt{[H + 4\pi M(N_x - N_z)]\left[H + 4\pi M(N_y - N_z)\right]}. \tag{8}$$

When the driving field's frequency matches the natural frequency of the material, the response of the magnetization vector will be strongest. In fact, by not including damping, Eq. (7) suggests infinite response since the denominator goes to zero, which is not physical. Also, if the element has the same demagnetizing factors in the x and y directions, then the precessional motion is circular under these conditions. In other words, the amplitude in the x and y directions will be equal. If the driving field's frequency is not on resonance or if the demagnetizing factors are not equal, then the precession will be elliptical with a larger maximum amplitude along one axis compared to the other.

When this assumption (M_x and $M_y \ll M_z$) is not valid then the nonlinear terms in Eqs. (3)–(5) are important to consider and precession of the magnetization is no longer a valid solution. It is interesting to note that for a sphere or a cube ($N_x = N_y = N_z = \frac{1}{3}$), the nonlinear terms of the form M_iM_j all disappear and the only nonlinear terms are those that couple the r.f. field to the magnetization, such as hM_x. In experiments, however, a perfect sphere or cube is very difficult to produce and so the demagnetization nonlinear terms remain.

Damping has not been included in the equations above but is vital in order to have a physical response at resonance and a nonlinear threshold. The nonlinear terms are always present in the equations of motion for the magnetization components. But if insufficient energy is being pumped into the system to counteract the natural losses (via phonons, electronic excitation or some other source) then no nonlinear dynamics may be observed. It is in general only at the point where sufficient energy is fed in to negate losses that the system may go nonlinear. A way to phenomenologically include damping is to add a term to the Torque Eq. (2) in what is known as the Landau–Lifshitz–Gilbert (LLG) equation [16]:

$$\frac{d\vec{M}}{dt} = -|\gamma|\vec{M} \times \left(-\frac{\delta E}{\delta \vec{M}}\right) + \frac{\alpha}{M}\left(\vec{M} \times \frac{d\vec{M}}{dt}\right), \qquad (9)$$

where α is the dimensionless Gilbert damping parameter. For NiFe (Permalloy), α is typically around 0.01 but may be more or less depending on the quality of the sample. Inhomogeneities in the material or surface roughness will usually contribute to a higher damping parameter as two-magnon scattering is more likely to occur. The additional Gilbert term in Eq. (9) acts to bring the magnetization dynamics towards an equilibrium. Without it, even very small amplitude motions would persist forever. Since it is a phenomenological parameter, it encompasses the effects of energy loss via a number of pathways including phonon losses and two-magnon scattering (due to magnetic defects). It does not include nonlinear magnetic damping, which will be discussed later.

2.2 Ways of Entering the Chaotic Regime

As the strength or amplitude of the oscillating driving field h increases, the components of the magnetization that are perpendicular to the equilibrium direction (M_x and M_y) start to increase and become comparable in size to M_z. This corresponds to large-angle precession and causes the quadratic nonlinear terms to become increasingly dominant in the damped equation of motion, Eq. (9). Above a certain threshold, the magnetization's trajectory may become chaotic. That is, small differences in the initial conditions of the vector lead to very different trajectories.

An example of the three-dimensional magnetization trajectory for a chaotic system is shown in Fig. 2. The position of the magnetization vector tip $(m_x, m_y, m_z) = (M_x, M_y, M_z)/M$ is plotted in this 3D space as a function of time. The trajectory is calculated using a fourth-order Runge–Kutta time integration of the LLG Eq. (9). The magnetization components are normalized so take on values between -1 and 1. The trajectory is that of the magnetization in an NiFe film and the same thin film parameters taken from [17] will be used to elucidate a number of points in this section and the next. The film lies in the x–z plane and therefore $N_y = 1$ and $N_x = N_z = 0$. It can be noticed in Fig. 2 that there is very little variation of the magnetization in the y direction as the demagnetizing

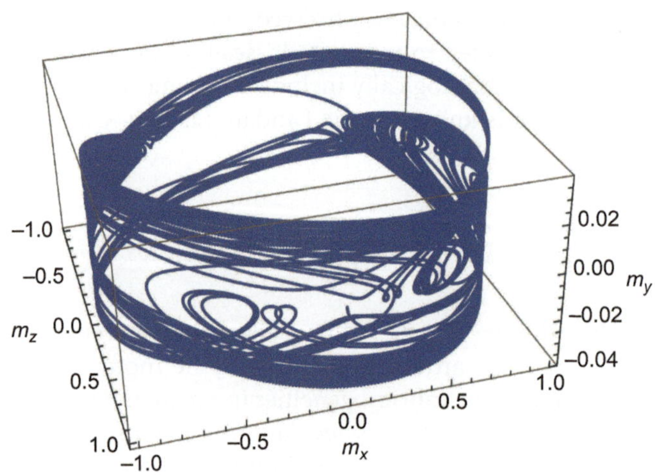

FIGURE 2

An example of a macrospin's nonlinear or chaotic trajectory. Each of the three axes corresponds to a component of the magnetization vector. Each component is normalized so may take on values between -1 and 1. The line traces out the path of the magnetization vector's tip. A NiFe film with finite width in the y direction is used to produce this simulation data. Hence there is a very small amplitude of the magnetization in the y direction.

energy is prohibitively large when the magnetization points along that axis. A static magnetic bias field $H = 1.72$ Oe is applied along the z axis, the magnetization is $M = 860$ G, the damping constant $\alpha = 0.01$, and the gyromagnetic ratio is taken to be $|\gamma| = 2\pi \cdot 2.83$ rad·GHz/kOe. In Fig. 2, the driving field has amplitude $h = 7.74$ Oe and is applied with a period of 2.59 ns.

In Fig. 2 it can be seen that each time the magnetization passes through a region, its trajectory is slightly changed and occasionally these changes are dramatic, opening up new loops. In contrast, when the macrospin is in the linear regime, its trajectory is a single closed loop: the motion traces out the same loop each period. Such chaotic motion can be induced by exciting the magnetic system in three general ways that will be discussed in detail in this section. Those methods involve microwave magnetic fields, electric currents, and large-angle magnetization switching respectively.

The usual geometry to drive the magnetization into such a chaotic motion is to apply a microwave driving field perpendicular to the static field that saturates the magnetization, as described in the equations above and illustrated in Fig. 1. For the most efficient transfer of energy from the driving field to the magnetic system, the frequency used for driving should be near the FMR frequency, as was seen in Section 2.1. The FMR frequency depends on the magnetization, the element shapes and the static applied magnetic field strength. It is typically in the low gigahertz regime for most ferromagnets. It should be noted that only ferromagnets are discussed in this chapter and the nonlinear behavior of ferrimagnets and antiferromagnets plus other more exotic spin orderings is left out. Such materials may have much larger resonant frequencies in the terahertz range.

It is also possible to drive the magnetization dynamics into the nonlinear regime by applying the driving field along the same direction as the saturation magnetization and this is known as the "parallel pump" geometry, as first used by Schlömann, Green and Milano in 1960 [18]. Since for precessional motion the z component of the magnetization oscillates at double the frequency of the other two transverse magnetization components, the driving frequency should now be double that of the system's FMR frequency in order to most efficiently excite nonlinear behavior. Above the nonlinear threshold, the FMR frequency shifts.

Rather than using a microwave field to excite such large-angle precession and induce the onset of chaos, one may instead drive the magnetization using current-induced spin-transfer torque. The main mechanism involves two magnetic layers of material: a polarizing or hard layer and a magnetically soft layer. An electric current impinging on the hard layer becomes spin-polarized as it passes through the layer. This spin-polarized current then impinges on the soft layer and transfers spin torque to the magnetization leading to precession in certain geometries. The spin torque may act with or against damping [19–21].

Over the past few years, the study of nonlinear dynamics in nanoelements has increasingly been concerned with such spin-torque systems [22–24]. Spin valves or magnetic tunnel junctions have potential applications in data storage, low-power microwave signal processing and sensing [25]. Spin valve multilayer structures are typically nanometer-sized (under one micron) in all directions and contain a magnetic soft layer that is just a few nanometers thick. Driving of the magnetization in this soft layer has two contributions: firstly, from the spin torque transferred from the magnetic hard layer as described above, and secondly due to the circular Oersted magnetic field created by the applied current that provides the spin torque. A microwave magnetic field may also be applied. Whether the current is applied through the whole element or is applied using a point-contact influences the resulting dynamics as a uniform or nonuniform magnetic mode will be excited [26]. For the rest of this section concerning macrospin dynamics, spin torque-induced dynamics will be mostly ignored although the discussion is kept general in order to apply to any method of nonlinear driving. However, a particular example will be outlined in Section 4.3.1 when the focus moves to cases where macrospin models break down.

A third option of how to drive a magnetic system nonlinear has been studied by Livesey *et al.* and others [27, 28]. In order for the quadratic terms in Eqs. (3)–(5) to be large, as mentioned before, we need to have a large-angle precession. Such large angles exist for one or two precession periods when the magnetization is switched suddenly by application of a large, static applied field. The motion is damped and the magnetization settles into its new equilibrium direction quickly. An example is shown in Fig. 3. The magnetization vector's trajectory is traced for a large-angle switch of over 90 degrees in a thin NiFe film with finite thickness in the y direction. The macrospin initially points to the bottom-left of Fig. 3a but then a static field is switched to be along the z direction. It takes the macrospin roughly 2 ns to complete most of its reorientation, then there is small-angle precession and damping towards the new equilibrium direction (see Fig. 3b and c). The total time for the trajectory shown is 6 ns.

Since the large-angle precession exists for a few nanoseconds, the nonlinear terms must be very large in order to see any chaotic behavior in the system on such small timescales [27]. Under certain conditions, this has been shown to be possible in metallic films [28–31] with apparent threshold angles that are far larger than in continuously-driven ferromagnetic resonance (FMR) experiments on the same materials [32]. In fact, it was proposed that the nonlinear processes are the dominant damping mechanism for the magnetization in large-angle switching [33–35].

No matter which of these three methods are used to drive the magnetic system, the resulting physics is the same. As the driving field/current/angle is increased,

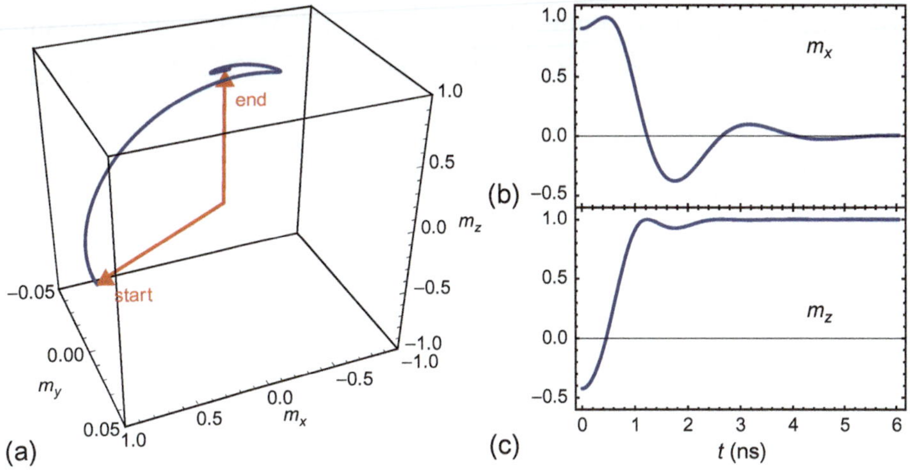

FIGURE 3

An example of large-angle switching in a thin film of NiFe. Initially the magnetization points primarily along the x direction, as shown in panel (a). A static field H is then applied along the z direction and reorients the magnetization. The initial and final vectors are shown along with a line tracing out the vector tip's trajectory over a period of 6 ns. In panels (b) and (c) the explicit time dependence of the x and z components of the magnetization respectively are shown. The magnetization components are normalized to the saturation magnetization so take on values only between − 1 and 1.

the behavior of the macrospin becomes increasingly nonlinear, and for some systems may eventually become chaotic.

2.3 Ways to Characterize the Chaos

Now that we know how power may be fed into a magnetic system to drive it nonlinear, the next task is to be equipped with tools that can characterize the route to chaos and also tools that can quantify the extent to which the system is behaving chaotically. In general, chaotic systems may be characterized using a variety of techniques including Lyapunov exponents, bifurcation diagrams, and Fourier analysis. These techniques will be illustrated below using the macrospin model. Some examples will utilize the simulations for an NiFe film that have already been used to illustrate points in this chapter, while other examples will be taken from published articles.

2.3.1 Lyapunov Exponents

The first method, calculating a Lyapunov exponent, corresponds to mathematically stating how far apart two initial trajectories that are very close together may end up being from one another. In other words, if at time $t = 0$ we start with two close initial configurations $\vec{M}_1(0)$ and $\vec{M}_2(0)$ with separation $|\vec{M}_1(0)-\vec{M}_2(0)|=\sqrt{(M_{1x}-M_{2x})^2+(M_{1y}-M_{2y})^2+(M_{1z}-M_{2z})^2}$, we can find how far apart they end up being a time t later, $|\vec{M}_1(t)-\vec{M}_2(t)|$. The Lyapunov exponent λ is given by making the correspondence

$$|\vec{M}_1(t)-\vec{M}_2(t)|\sim e^{\lambda t}|\vec{M}_1(0)-\vec{M}_2(0)|. \tag{10}$$

A different λ will be found depending on where the trajectory is started from and what time t is chosen. Therefore many calculations need to be done and typically it is the largest value of λ that is used to represent the system, the so-called maximal Lyapunov exponent.

A recent example of such a study on a magnetic macrospin is provided by Laroze *et al.* [36]. An anisotropic magnetic nanoparticle is considered with a static field applied along its easy anisotropy axis and a periodic magnetic field applied perpendicular to that. The maximal Lyapunov exponent as a function of the amplitude and frequency of the applied oscillating field is plotted to examine where chaotic regions of phase-space lie. In Fig. 4 some of their results are reproduced. The Lyapunov exponent is indicated by the color with darker shades indicating larger exponents. The horizontal axis indicates increasing r.f. field strength (in dimensionless units) and the vertical axis represents the

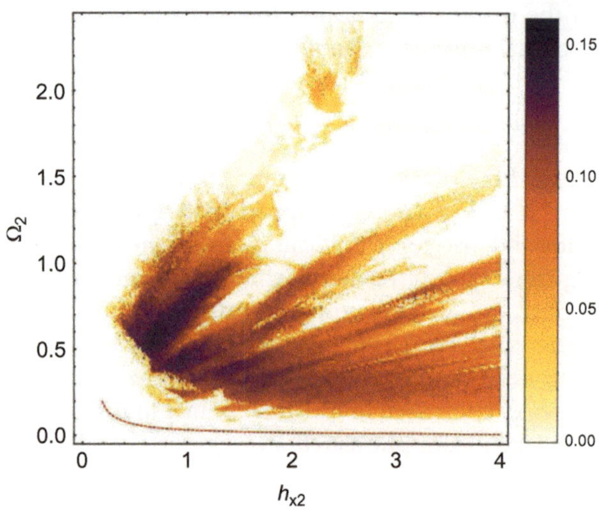

FIGURE 4

Maximal Lyapunov exponent as a function of r.f. field strength h_{x2} and r.f. field angular frequency Ω_2 for a driven anisotropic magnetic nanoparticle. The bar on the right shows that larger Lyapunov exponents correspond to darker colors. Taken from [36].

oscillation frequency (again in dimensionless units). It can be seen that there are distinct bands of color that indicate regions where the behavior is most chaotic. It is not simply the case that larger driving fields result in more chaotic systems. The dynamics are more complicated than that, even for a simple macrospin.

Another recent example by Yang *et al.* [37] studied the free-layer of a spin-valve oscillator as a macrospin and calculated the Lyaponov exponent of the dynamics under different fields and currents. Larger exponents correspond to a "more chaotic" system. In this article some of the other methods to be discussed below were also used to characterize chaos in the spin valve.

2.3.2 Bifurcation Diagrams

Bifurcation diagrams serve to show how chaos develops as the so-called bifurcation parameter is varied. For systems with an r.f. driving field, the driving field strength serves as an excellent bifurcation parameter. If Eq. (9) is integrated forward in time, then a trajectory like that seen in Fig. 2 is calculated. If this trajectory is probed once every period of the driving field ($t = 1/f$), then whether the trajectory is the same each cycle or starts to develop nonlinear and chaotic dynamics can be seen clearly.

In Fig. 5 such an analysis is done on simulated data generated for a thin NiFe film following the calculation of [17]. The parameters used for the simulation were given earlier at the start of Section 2.2. The linear FMR frequency of the film is

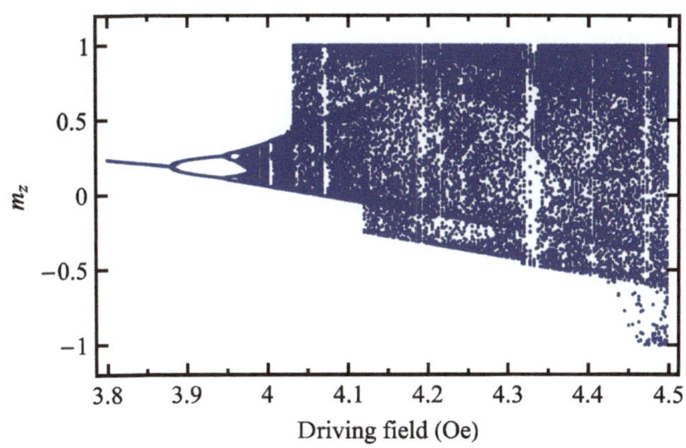

FIGURE 5

A bifurcation diagram for a magnetic macrospin with parameters the same as those used in [17]. The z component of the magnetization is stroboscopically recorded every time interval t equal to the driving field's period to build up an idea of the amount of phase space that the magnetization vector can explore. As the driving field amplitude increases, the system becomes more chaotic and more values of m_z are recorded.

$$f = \frac{|\gamma|}{2\pi}\sqrt{H(H + 4\pi M)} = 0.386 \text{ GHz}. \tag{11}$$

An r.f. driving field h in the transverse x direction is applied at precisely this frequency but with varying field strength. As the field strength in Oersteds is increased, the system becomes more nonlinear. Note that the static field is very small ($H = 1.72$ Oe) and in reality may not be sufficient to saturate such a film. However, it is no problem for the simulation. The simulation must be run for sufficient time such that transient dynamics are not included in the data; it may take some time for the system to settle into a periodic motion.

The z component of the magnetization is recorded once each period for many hundreds of periods and for many different values of the driving field strength. The results are shown in Fig. 5. The z component of $\vec{M}$ is divided by the saturation magnetization so a dimensionless number between 1 and − 1 is plotted. As the amplitude of the driving field increases, and therefore the energy being pumped into the macrospin increases, the motion becomes increasingly nonlinear and then chaotic as can be seen because more values of m_z are visited by the macrospin for a given value of h. To the left of the map, a bifurcation (two values of m_z) is seen around $h = 3.9$ Oe, indicating that the period has doubled and the system has entered further into the nonlinear regime.[1] Above $h = 4.0$ Oe, the system is mainly chaotic, visiting a continuous range of m_z values. A so-called "periodic window" around $h = 4.33$ Oe is seen as intermittent amongst the chaos. That is, for a very short range of driving field amplitude, chaos is replaced by a periodic orbit where only a finite number of m_z values are visited. A few such regions appear in Fig. 5. Such regions are undesirable for applications where the chaotic behavior is required to be robust over a broad range of parameters [38].

The aforementioned work by Yang *et al.* showed that for a spin-valve's free layer, the approach to chaos as a function of the driving current strength involved a series of periodic doublings [37]. The strength of the current at which the system first goes nonlinear is predicted within the theory.

These observations of bifurcation, period doubling, chaos, and periodic windows are typical for all kinds of systems. Similar looking diagrams exist, for example, for oscillation amplitude in electrical circuits as a function of the input current strength or, as another example, for chemical reaction products as a function of some input reactant amount. Many dynamical systems governed by simple differential equations have similar bifurcating behavior [39].

---

[1]Note that the system may already be nonlinear at values of the driving field amplitude below that where the bifurcation takes place. For example, the frequency of oscillation may depend on the amplitude of the driving field before the bifurcation occurs, which is a classic sign of nonlinearity in a magnetic system.

2.3.3 Fourier Transforms

One may use Fourier transforms to visualize this same data. When the macrospin is still in the linear regime, its motion is periodic and a Fourier transform in time will have one peak or possibly two (one at the driving frequency and another at the natural frequency of the system). When first behaving nonlinearly, the motion remains periodic but the frequency may *shift*, which can be seen in the Fourier transform of the time-dependent data. For many magnetic systems, being driven further into the nonlinear regime results in the appearance of other peaks in the Fourier spectrum. The frequencies at which peaks appear reveal which nonlinear terms in the equations of motion are most important for driving the system nonlinear and therefore give insight into the system [40]. Far into the nonlinear regime, it is common for the Fourier spectrum to display a broad range of frequencies that have large amplitude because the macrospin no longer behaves in a periodic way.

To demonstrate how Fourier Transforms, or more precisely Fast Fourier Transforms (FFT), may be useful, some representative data is given in Fig. 6. Once again we use material parameters taken from [17] and perform simulations. Time series data can then be Fast Fourier Transformed and the peak frequencies examined. In Fig. 6, five different values of the driving field are illustrated, with the smallest value at the top and the largest at the bottom. The panels on the left show a trace of the normalized x component of the macrospin as a function of time. The panels on the right show the corresponding FFT amplitudes as a function of frequency.

Panels (a) are for $h = 0.086$ Oe. The magnetization precession is small-amplitude (left panel) and so the nonlinear terms are likely unimportant. This is reflected in the FFT (right panel), where there is one large peak at precisely the driving frequency. Panels (b) are for a larger driving field $h = 3.44$ Oe. The time trace (left panel) shows that m_x now takes on all values from -1 to 1 and so the dynamics most definitely represent large-angle precession. In the FFT data (right panel), a peak appears at precisely three times the driving frequency at $f = 1.16$ GHz. This tells us that the dominant nonlinear process is a so-called "four-wave process"—a process whereby three input modes with the driving frequency produce one output mode with three times that frequency in order that energy is conserved in the process. Such "three-" and "four-wave processes" will be discussed in further detail in Section 3 on thin films.

Next, panels (c) ($h = 3.87$ Oe) show some new, higher-frequency structure in the time trace, near $m_x = -1$. The change in the FFT is even more dramatic, with many new frequency peaks appearing. In panels (d) the driving field is $h = 4.3$ Oe, which is truly in the chaotic regime, as also seen on the bifurcation diagram Fig. 5. The time trace (left) is completely random looking and the FFT

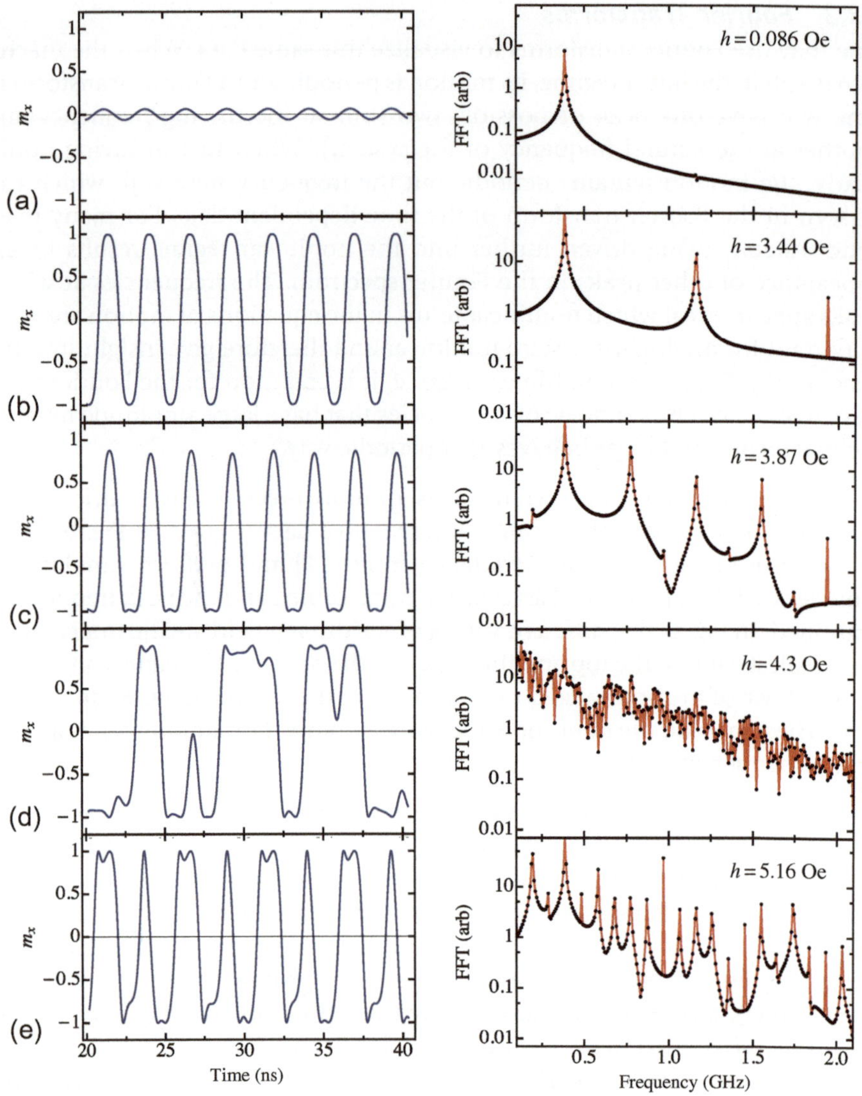

FIGURE 6

Progression into the nonlinear regime. For five different driving field strengths, the magnetization dynamics as a function of time (left) and the Fast Fourier Transform (right) are shown for an NiFe thin film's magnetization treated as a macrospin. The driving field strength increases from top to bottom. The FFT data is shown by dots, which are connected by a line as a guide for the eyes.

data (right) is simply a broad mess with no clear peaks. One more driving field value $h = 5.16$ Oe is shown in panels (e), as an example of a periodic window. Although the system is being driven harder than in the panels above, both the time (left) and FFT data (right) show that there is some periodicity and order in the magnetization dynamics.

2.3.4 Other Methods

Many more sophisticated mathematical tools exist to quantify and analyze chaos. These include Ponincaré maps, Poincaré sections and fractal dimensions of strange attractors. However, only one more method will be discussed here and the reader is referred to other texts to learn more about the mathematics of the others [41, 42]. The final method to be discussed is a phase-space portrait.

A phase-space portrait is created by collecting time series data for one of the magnetization components (equivalently, one of the two angles that define the magnetization direction could be chosen). Taking, for example, m_x, we record $m_x(t + n\tau)$, where n is a positive integer and τ is the period of the time series. Some time on the order or less than the driving period works best for τ. Also, the driving period should not be an integer multiple of τ for best results. Then ordered points $(m_x(t+n\tau), m_x(t+[n+1]\tau))$ are formed and plotted. We use the same simulation data as has been used throughout this section for an NiFe film to demonstrate such a diagram; however the same method may be used on experimental data also.

A time-series period equal to 0.2469 ns (9.531% of the driving period) is chosen and results are shown in Fig. 7 for different driving field strengths that correspond to those used in Fig. 6 (c–e). For panel (a) of Fig. 7, $h = 3.87$ Oe and the phase diagram is a closed loop showing that the system is periodic. This matches what we found analyzing the time and FFT data in Fig. 6c. In Fig. 7b, the system is seen to be chaotic with no closed loops formed in the phase diagram but rather points lying all over the place. In panel (c), $h = 5.16$ Oe and the phase diagram shows the quasi-periodic dynamics that were also seen in the FFT data.

2.4 Breakdown of the Macrospin Model

The macrospin model allows us to make simple predictions about the non-linear processes in nanoelements. For example, as we have seen, we can predict the onset field and the route to chaos via bifurcations. As another example, one may study how varying the shape of the element (varying N_x, N_y, and N_z) changes the field strength at which the onset of chaos may occur.

But, unsurprisingly, the macrospin model eventually breaks down. To give an example of this, the resonant frequency of a system as a function of r.f. field h is

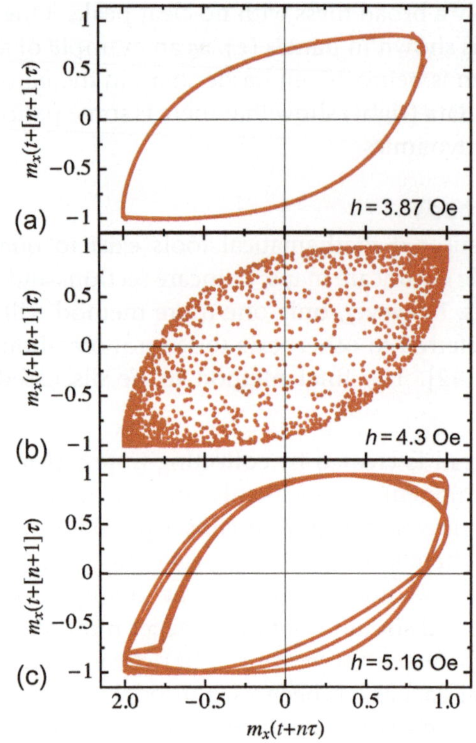

FIGURE 7

Phase-space diagrams for the nonlinear dynamics of an NiFe thin film. Three different values of the driving field strength are shown displaying very different dynamical behaviors: (a) $h = 3.87$ Oe periodic, (b) $h = 4.3$ Oe chaotic, and (c) $h = 5.16$ Oe quasi-periodic. The parameters used are the same as throughout this section. The time series period τ is 9.531% of the driving period 2.59 ns.

discussed. The equations of motion (3)–(5) may be linearized as discussed in Section 2.1 for small-amplitude motion. Then the angular frequency for an element is:

$$\omega = |\gamma| \sqrt{\left[H + 4\pi M_z (N_x - N_z)\right]\left[H + 4\pi M_z (N_y - N_z)\right]}. \tag{12}$$

This expression differs from Eq. (8) only in that the saturation magnetization M is replaced by the vector component M_z. Typically, as the r.f. field strength is increased, the precession has a larger angle and the component M_z along the equilibrium direction is decreased. Therefore, using this expression for the frequency that is valid only for the geometry shown in Fig. 1, Eq. (12), it can be seen that the trend is for the frequency to decrease as the r.f. field strength increases, as is seen in some experiments [43].

Of course, Eq. (12) is only an approximation as it is only valid in the linear regime. But macrospin numerical calculations also support this trend [44]. The problem is that in experiments this trend is not always seen [44]! The macrospin model (whether analytic or numeric) clearly breaks down and a more sophisticated model is needed. One such semi-analytic calculation is presented in [45] for a spin-torque nanopillar and predicts both positive and negative frequency shifts depending on the driving current and voltage.

Numerically, that more sophisticated method is to perform a micromagnetic simulation. The magnetic element is split up into pieces within which the magnetization is assumed constant. Sufficient pieces are needed so that the magnetization dynamics are clearly resolved in spatial dimensions. A study by Berkov and Gorn shows that while smaller structures may in some limits be treated as macrospins, slightly larger nanostructures must be treated using micromagnetics to capture properly their route to chaos [46]. The results of such calculations will be discussed in Section 4, which will address particularly structures that are nanometer-scale in two or three dimensions. However, in the next section, thin films will be discussed and for such structures we can make some predictions about the nonlinear behavior using analytic tools.

3 NONLINEAR PROCESSES IN THIN FILMS

Thin films are of interest because they are used to create many magnetic devices such as giant magnetoresistance (GMR) read/write heads and sensors, and tunable on-wafer devices such as filters and isolators. They provide a good model for any magnetic element that is confined to the nanometer-scale in the y direction, with thickness L, but is relatively large in the other two directions, as shown in Fig. 8a. Because of their large lateral dimensions, the films may be assumed to have translational invariance in the x and z directions. This vastly simplifies the analytic study of the possible magnetization oscillations (or spin waves) that are allowed in a structure and therefore which of these oscillations interact in the initial stages of nonlinear behavior. We will first discuss these linear spin waves before discussing the analytic theory that can be used to gain insight into the onset of chaos. At the end of the section, the implications to magnetic damping will be discussed.

3.1 Linear Spin Waves in a Thin Film

If the magnetization at both the top and bottom surfaces of a thin film is assumed to oscillate freely, then the standing spin waves shown in Fig. 8b are allowed in the confined y direction. Here the dynamic magnetization's deviation from zero is shown schematically through the thickness of the

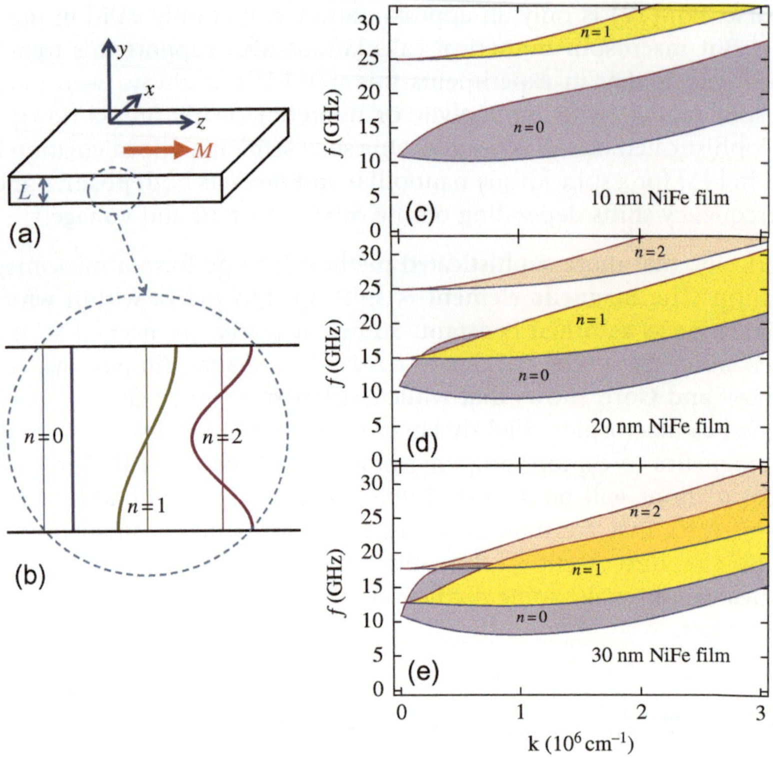

FIGURE 8

Panel (a) shows the thin film geometry with magnetization along the z axis. Panel (b) shows the first three standing spin waves through the film thickness, indexed by n. Panel (c) shows the spin wave bands for a 10 nm $Ni_{80}Fe_{20}$ film; allowed frequencies are shaded as a function of the in-plane wave vector k. Only the $n = 1$ and $n = 2$ spin wave bands are visible for frequencies below 30 GHz. In panels (d) and (e) the spin wave bands are shown for a 20 nm and a 30 nm film respectively. The spin wave bands corresponding to different thickness modes start to overlap more as the film gets thicker. Only the first three bands are shown. The parameters used are: magnetization $4\pi M = 10300$ Oe, static field $H_z = 1300$ Oe, exchange energy $A = 10^{-6}$ erg/cm, and gyromagnetic ratio $\gamma = 2.8$ GHz/kOe.

film with antinodes at the top and bottom surfaces. Discrete wavenumbers k_y are allowed and these are indexed by integer n. Only the first three standing waves ($n = 0,1,2$) are shown in Fig. 8b. Mathematically, they are given by:

$$\Phi_n(y) = \left(\frac{2}{1+\delta_{0n}}\right)^{1/2} \cos\left(\frac{\pi n}{L}y\right), \quad n = 0, 1, \ldots, \tag{13}$$

where $y = 0$ is the bottom of the film and $y = L$ is the top, and $\delta_{nn'}$ is the discrete Kronecker delta function equal to 1 when $n = n'$ and 0 otherwise.

In the x and z directions, on the other hand, any wavenumber is allowed due to the effectively infinite size. Kalinikos and Slavin presented a method to find all the possible dipole-exchange spin waves allowed for arbitrary magnetization pinning at the surfaces [47]. Calculated dispersion relations are shown in Fig. 8c–e for an $Ni_{80}Fe_{20}$ film with thickness varying between (c) 10 nm, (d) 20 nm and (e) 30 nm. Each standing spin wave gives rise to a band of frequencies, which are shaded. The bottom of this band corresponds to in-plane propagation along the z direction ($k_x = 0$), parallel to the equilibrium magnetization direction. Spin waves propagating in this direction are known as Backward Volume Magnetostatic Spin Waves (BVMSWs) since their phase velocity is negative for small wave numbers k_z. The top of each spin wave band corresponds to waves propagating perpendicular to the equilibrium magnetization direction ($k_z = 0$). These are known as Magnetostatic Surface Spin Waves (MSSWs). The shaded region between these two extremes show the frequencies allowed for propagation in all other directions. Note that in this model surface modes are ignored and only standing spin waves through the film thickness are calculated.

For the thinnest film shown in Fig. 8c, only the spin wave bands corresponding to $n = 0$ and $n = 1$ standing waves are shown. This is because the $n = 2$ band has a *much* larger frequency around 60 GHz. As the film thickness is made smaller, the exchange energy contribution becomes larger for the thickness modes with $n > 0$. This will be important later when we discuss the onset of chaos in films of varying thickness.

In Fig. 8d and e, three standing spin wave bands are shown for the thicker films. The shading for each of the three bands overlaps, especially in panel (e), the thickest film. As the film is made thicker, all the bands start to overlap increasingly. The frequency that corresponds to a particular choice of k_y, k_x, and k_z can be written $f(n,k_x,k_z)$.

When driving such a magnetic film with an r.f. field, it makes sense to choose a frequency for which the system is on-resonance. This will ensure that the nonlinear threshold or onset is at the lowest field amplitude possible. Moreover, a uniform r.f. field couples most strongly to the coherent precession mode of the film with frequency $f(0,0,0)$ (i.e., $n = k_x = k_z = 0$). This is the Ferromagnetic Resonce (FMR) mode of the system. Therefore a frequency close to this should be chosen in order to induce chaotic dynamics. In the case where the chaos is onset by some short-lived large-angle precession, as discussed earlier, the precession will also have frequency given by $f(0,0,0)$. Therefore, our task is to find how this mode of the thin film couples to traveling spin waves in order to examine the most likely route to chaos.

Note that these calculated frequencies are all for the *linear* system and break down when the system goes nonlinear. As discussed before, typically the

FMR frequency shifts downwards for in plane magnetization with the static field in the plane of the film/element as well.

3.2 Perturbative Expansion

There are few analytic tools available to study systems that are truly behaving chaotically. An exception is S-theory, developed in the 1970s [48, 49]. However, *at* threshold (when the nonlinear terms start growing in the equations of motion) then perturbation expansions may be used to discern certain characteristics of the nonlinearity. The first- and second-order corrections to the linearized equations of motion may be considered, corresponding to Suhl's so-called first- and second-order instabilities. Far beyond threshold, the higher order terms dominate and it becomes nearly impossible to make any analytic predictions for the system.

The aim is to transform the magnetization equation of motion, Eq. (2) to an equation that explicitly shows the interactions between the thin film's spin waves. Through a series of transformations detailed later in this subsection, the equation of motion becomes [7]

$$\frac{\partial b_{nk}}{\partial t} = -i\frac{\delta \mathcal{H}}{\delta b_{nk}^*}, \tag{14}$$

where b_{nk} is the elliptical spin wave amplitude with thickness index n and in-plane wave vector $\boldsymbol{k} = (k_x, k_z)$, and $\mathcal{H}$ is the Hamiltonian for the thin film. The asterisk denotes the complex conjugate and the right-hand side of Eq. (14) involves a functional derivative. The spin wave amplitudes are the semi-classical analog of spin wave quantum creation and annihilation operators.

The Hamiltonian is written in terms of a perturbative expansion of spin wave terms [7,50]:

$$\begin{aligned}
\mathcal{H} = & \sum_{nk} \omega_{nk} b_{nk}^* b_{nk} + \frac{1}{3}\sum_{1,2,3} U_{123}^{(3)} \ \delta(\boldsymbol{k}_1 + \boldsymbol{k}_2 + \boldsymbol{k}_3)[b_1 b_2 b_3 + \text{c.c.}] \\
& + \sum_{1,2,3} U_{123}^{(1)} \ \delta(\boldsymbol{k}_1 + \boldsymbol{k}_2 - \boldsymbol{k}_3)[b_1^* b_2^* b_3 + \text{c.c.}] \\
& + \sum_{1,2,3,4} V_{1234}^{(1)} \ \delta(\boldsymbol{k}_1 - \boldsymbol{k}_2 - \boldsymbol{k}_3 - \boldsymbol{k}_4)[b_1^* b_2 b_3 b_4 + \text{c.c.}] \\
& + \frac{1}{2}\sum_{1,2,3,4} V_{1234}^{(2)} \ \delta(\boldsymbol{k}_1 + \boldsymbol{k}_2 - \boldsymbol{k}_3 - \boldsymbol{k}_4) b_1^* b_2^* b_3 b_4 \\
& + \frac{1}{4}\sum_{1,2,3,4} V_{1234}^{(4)} \ \delta(\boldsymbol{k}_1 + \boldsymbol{k}_2 + \boldsymbol{k}_3 + \boldsymbol{k}_4)[b_1 b_2 b_3 b_4 + \text{c.c.}] + \cdots
\end{aligned} \tag{15}$$

The coefficient of the linear, two-wave term in the Hamiltonian ω_{nk} is the angular frequency corresponding to a linear spin wave with thickness index n and in-plane wave vector $\boldsymbol{k}$. The next term is a three-wave term with coefficient $U^{(3)}(n_1,\boldsymbol{k}_1,n_2,\boldsymbol{k}_2,n_3,\boldsymbol{k}_3)$ that we write in shorthand notation as $U^{(3)}_{123}$ for the sake of brevity. The c.c. denotes that the complex conjugate of the same term is added. The summation is also written in shorthand notation and represents that the variables $(n_1,\boldsymbol{k}_1,n_2,\boldsymbol{k}_2,n_3,\boldsymbol{k}_3)$ are all being summed over. This three-wave term in Eq. (15) involves a Dirac delta function defined such that it acts on a function $F(\boldsymbol{k})$ as

$$\sum_{k}\delta(\boldsymbol{k})F(\boldsymbol{k})=F(0). \tag{16}$$

This Dirac delta function ensures that momentum is conserved in all of the possible spin wave interactions that are being summed.

The subsequent terms in Eq. (15) are different three- and four-wave interactions with the scattering coefficients written with similar shorthand notation. The expansion has higher order terms, but at threshold only the three- and/or four-wave interactions need to be considered, depending on the geometry of the problem. In Fig. 9 a schematic of the terms in Eq. (15) is shown to illustrate what spin wave interactions the different terms correspond to. In each process, energy and momentum must be conserved.

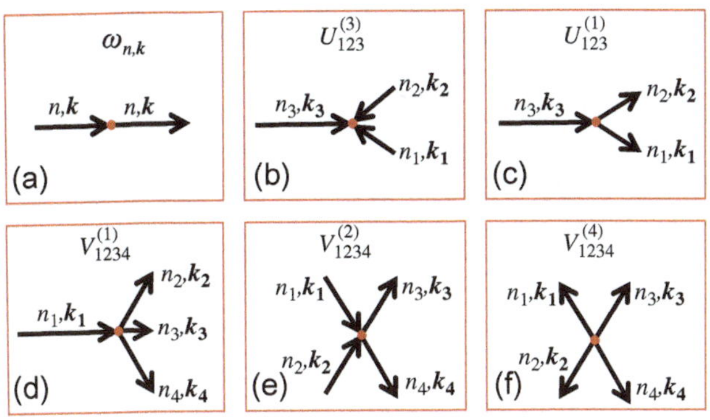

FIGURE 9
Pictures of the six different types of spin wave interactions corresponding in order to the terms in the Hamiltonian, Eq. (15). Energy and momentum must be conserved in each process. Panel (a) shows the linear spin wave or self-interaction term. Panel (b) shows a three-wave process where three spin waves annihilate (the complex conjugate term is three waves spontaneously forming). Panel (c) shows one wave annihilating and two forming. Panels (d), (e), and (f) are the different allowed four-wave interactions. Higher order interactions are not discussed here and are not illustrated.

Each of the coefficients to the three- and four-wave processes can be found elsewhere [7, 50]. They depend on the magnetization, film thickness, exchange constant, static applied field strength, and any magnetocrystalline anisotropy energy density. The expressions are complicated but are found directly by transforming the energy, such as Eq. (1), into the Hamiltonian form of Eq. (15). The details of how to transform the magnetization vector $\vec{M}$ into elliptical spin wave amplitudes $b_{n\,k}$ (and therefore the energy into the perturbative expansion) have been described in many works [7, 27] so here we will just provide the outline of the steps.

- Step 1. Use a semi-classical version of the Holstein–Primakoff transformation [51] to change the magnetization into spin wave variables $a(\boldsymbol{r})$ and $a(\boldsymbol{r})^*$, where $\boldsymbol{r}$ is the position in the thin film.
- Step 2. Use a Fourier-like expansion for the spin wave variables to get amplitudes for spin waves with thickness index n and in-plane wave vector $\boldsymbol{k}$. The expansion is a true Fourier transformation in the plane of the thin film plane and is an expansion in terms of orthogonal standing waves (pictured in Fig. 8b) in the out-of-plane direction [47]. More precisely, the expansion is:

$$a(\boldsymbol{r}) = \sum_{n,\boldsymbol{k}} a_{n\boldsymbol{k}} \Phi_n(y) e^{i\boldsymbol{k}\cdot\boldsymbol{R}}, \tag{17}$$

  where the thickness modes were given above in Eq. (13) and $\boldsymbol{R}$ denotes a two-dimensional in-plane position vector.
- Step 3. Use a Bogoliubov transformation to change the spin wave variables into elliptical spin wave variables $b_{n\boldsymbol{k}}$ and $b^*_{n\boldsymbol{k}}$ [7]. This step diagonalizes the two-wave part of the Hamiltonian. It is necessary because the x and y transverse magnetization components are not identical due to the thin film geometry and magnetization precession is highly elliptical.

Depending on the geometry of the problem, one further transformation may be necessary. When the thin film is sufficiently thin, or when large enough static field is applied, then three-wave nonlinearity is not possible [27] and an additional transformation is required. Why this is the case will be briefly explained.

Below threshold, only the uniform spin wave mode ($n = 0$ and $\boldsymbol{k} = 0$) typically has a large amplitude since it is the one driven by a uniform r.f. field or by a uniform spin torque or by large-angle motion of the macrospin (if the magnetization is moving coherently). Therefore, the initial nonlinear process must involve this mode. The largest coefficient is typically $U^{(1)}_{\boldsymbol{k},-\boldsymbol{k},0}$, corresponding to decay of the uniform precession into two equal but opposite spin wave excitations with half the frequency. However, sometimes there are no spin waves at half the frequency available to the system. For example, in Fig. 8c the dispersion

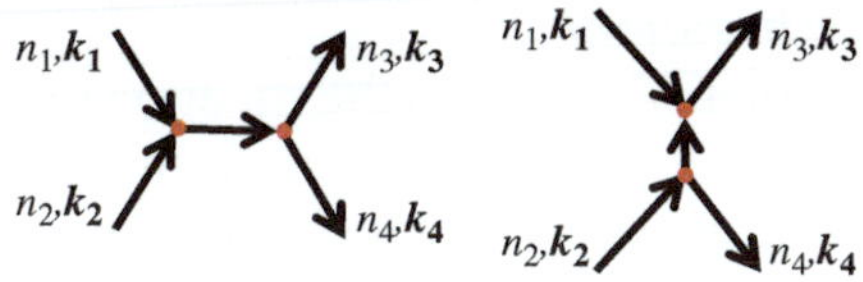

FIGURE 10
Two examples of "cascaded processes," whereby a four-wave interaction is made up of two three-wave interactions. These processes can be incorporated into a renormalized four-wave coefficient using a quasi-linear transformation of the elliptical spin wave variables.

curves for a 10 nm NiFe film are shown. The uniform mode has frequency of roughly 11 GHz but the minimum frequency possible for the film is just under 10 GHz. Therefore, this three-wave process is nonresonant in the system and it will instead be a four-wave process that is the first to take the system into the nonlinear regime.

However, two three-wave processes can together create a four-wave process in what is known as a "cascaded process." A diagram is shown in Fig. 10. Therefore, a quasi-linear transformation first used by Krasitskii [7, 50, 52] is performed on the Hamiltonian Eq. (15) in order to eliminate the direct three-wave terms and incorporate them in the four-wave terms when this is the case. The new spin wave amplitudes are labeled c_{nk}. This becomes particularly important to do when the geometry and applied fields are just right and one of the higher-frequency spin wave bands is exactly twice the frequency as the lowest band and therefore a cascaded process may be on-resonance [27]. At other times, the cascaded processes are negligible and the quasi-linear transformation does not significantly alter the threshold results.

3.3 Nonlinear Thresholds in Thin Films

Once the correct Hamiltonian expansion has been obtained, the equation of motion, Eq. (14), can be examined to find the point at which the system goes nonlinear.

Depending on the method of driving the system, the threshold may be best given as a threshold microwave radiation power or a threshold current density. However, perhaps the most universal parameter to use is the average precession angle of the magnetization as this can be used for all types of experiments. For small-angle motion about the equilibrium direction the system behaves linearly but above some critical angle it enters the nonlinear regime.

First the equation of motion will be examined for three-wave and four-wave processes respectively. Then the thresholds will be written in terms of the average magnetization precession angle θ around one period of oscillation.

3.3.1 Three-Wave Threshold

The Hamiltonian (15) is substituted into the equation of motion (14) and then only the largest term in the expansion is kept. As mentioned before, this term corresponds to the uniform mode producing two spin waves with half frequency traveling in opposite directions with wave vectors $\boldsymbol{k}$ and $-\boldsymbol{k}$ so that energy and momentum are conserved. The out-of-plane index is $n = 0$ (lowest spin wave branch) for each of the three waves in this interaction. The equation we get for the elliptical spin wave amplitude of the produced wave is

$$\frac{\partial b_{0\boldsymbol{k}}}{\partial t} + i\omega_{0\boldsymbol{k}} = -2i \;\; U^{(1)}_{0,\boldsymbol{k},-\boldsymbol{k}} \;\; b_{00} b^{*}_{0-\boldsymbol{k}'} \tag{18}$$

when there is no r.f. driving field. This equation applies therefore to cases where there is some large-amplitude motion of the magnetization due to cases such as switching. The case where there is a driving field involves the addition of a term linear in $h(t)$ on the right-hand side of the equation and is discussed, for example, in Chapter 10 of [10].

In Eq. (18) there is currently no damping term. To include the effect of losses phenomenologically (otherwise the threshold will be at a precession angle of zero degrees), the linear frequency $\omega_{0\boldsymbol{k}}$ is taken to have an imaginary part:

$$\omega_{0\boldsymbol{k}} \rightarrow \omega_{0\boldsymbol{k}} - i\eta_{\boldsymbol{k}}. \tag{19}$$

The damping parameter can be related to the Gilbert damping parameter that was mentioned earlier according to $\eta_{0\boldsymbol{k}} = 2\alpha\omega_{0\boldsymbol{k}}$.

Writing such an equation of motion (18) for the elliptical spin wave amplitude is only valid *right* on threshold. Once traveling spin waves are present at large amplitudes in the system, then it no longer makes sense to truncate the perturbative expansion to one nonlinear term and many more terms must be included on the right-hand side of Eq. (18). The equation is then unable to be solved analytically!

But precisely on threshold we can make progress by assuming that the elliptical spin wave amplitude takes the form:

$$b_{0\boldsymbol{k}} = B_{0\boldsymbol{k}} e^{\Gamma_{\boldsymbol{k}} t} e^{-i\omega_{0\boldsymbol{k}} t}. \tag{20}$$

The oscillating part of this solution is the solution to the linear problem where the magnetization precesses in an elliptical, repeatable orbit. There will be an exponentially increasing solution as long as the so-called "incremental rate of parametric excitation" $\Gamma_{\boldsymbol{k}}$ is greater than zero. Putting Eq. (20) into (18) and requiring that this be the case, it is found that the system may go nonlinear if [53]

$$|B_{00}|^2 > \frac{\eta_k^2}{4|U^{(1)}_{0,k,-k}|^2}. \tag{21}$$

In words, what this tells us is that the threshold amplitude of the elliptical precession is proportional to the damping constant squared and is inversely proportional to the strength of the three-wave interaction strength. Physically, this makes a great deal of sense as we would expect the nonlinear threshold to go up for a film with stronger losses or one with weaker three-wave interactions.

Now this strange variable B_{00} must be converted to a physical angle of precession. This can be done using the details of the various transformations and also using trigonometry. The details are given in [27]. For thin films the precession is highly elliptical, meaning that the magnetization forms a very small angle with the z axis when it is out-of-plane but a very large angle when it is in-plane. By averaging around one precession cycle, it is found that the three-wave nonlinear threshold is given by

$$\theta_{3wave} > 2\arcsin\left(\sqrt{\frac{\gamma(u_0^2+v_0^2)}{2M}}\frac{\eta_k}{2U^{(1)}_{0,k,-k}}\right), \tag{22}$$

where γ is the gyromagnetic ratio and u_0 and v_0 are coefficients of the third (Bogolyubov) transformation [27] that represent the ellipticity.

There are many traveling spin waves that may take part in a three-wave interaction with the uniform precession. However, one will have the highest three-wave interaction coefficient $U^{(1)}$ and will therefore determine the lowest threshold. Typically this is for magnetostatic spin waves (MSSWs), those that propagate perpendicular to the static field's direction.

3.3.2 Four-Wave Threshold

Assume that the three-wave interactions have been eliminated via the quasi-linear transformation discussed above and the new spin wave amplitudes are c_{nk}. The new perturbative Hamiltonian is substituted into the equation of motion and only the largest nonlinear term is kept, as before. The resulting equation with phenomenological damping included is:

$$\frac{\partial c_{0k}}{\partial t} + (i\omega_{0k} + \eta_k)c_{0k} = -i\widetilde{V}^{(2)}_{0,0,k,-k}c_{00}^2 c^*_{0-k'} \tag{23}$$

where $\widetilde{V}^{(2)}_{0,0,k,-k}$ represents the four-wave coefficient where two spin waves are destroyed and two are created (see Fig. 9e), which is renormalized due to the quasi-linear transformation and hence has the tilde on top. Once again, the interaction with a driving field or current is ignored here.

Assuming the same kind of solution for $c_{0k}(t)$ as was assumed above for $b_{0k}(t)$ in Eq. (20), namely

$$c_{0k} = C_{0k} e^{\Gamma_k t} e^{-i\omega_{0k} t}, \tag{24}$$

and requiring that the spin wave amplitude increase over time, the four-wave threshold is found to be:

$$|C_{00}|^2 > \frac{\eta_k}{|\widetilde{V}^{(2)}_{0,0,k,-k}|}. \tag{25}$$

As with the three-wave threshold, we find that the uniform precession amplitude must be larger than some value which is directly proportional to damping and inversely proportional to the interaction strength. In terms of the average precession angle of the magnetization about an elliptical orbit, the four-wave nonlinear threshold can be written as [27]

$$\theta_{4wave} > 2\arcsin\sqrt{\frac{\gamma(u_0^2 + v_0^2)\eta_k}{2M|\widetilde{V}^{(2)}_{0,0,k,-k}|}}. \tag{26}$$

Once again, a range of spin waves can be produced that satisfy conservation of energy and momentum. But the spin wave with the largest interaction with the uniform mode has the lowest nonlinear threshold and is the first to be "parametrically excited." For four-wave processes this is typically a backward volume spin wave (BVSW) propagating parallel or antiparallel to the total internal field of the magnet. Typical threshold angles are only a few degrees for NiFe films [27,32].

3.4 Nonlinear Damping

Although here the focus has been on using the perturbative Hamiltonian equation (15) to calculate nonlinear thresholds, the three- and four-wave interaction terms can also be used to study such things as nonlinear shifts in the resonant frequencies of spin waves (discussed earlier for the FMR or uniform mode) and nonlinear damping [7, 10, 54].

The phenomenological parameters α and η describe energy loss to external sources but an individual spin wave may also be damped if energy moves from it and into other spin wave modes. Let us define another phenomenological damping parameter $\tilde{\alpha}$ that encompasses damping to phonon or electron baths, as well as to other spin wave modes. Therefore, $\tilde{\alpha}$ is greater than α. This is most pronounced for the FMR mode, where it has been shown that in very thin films four-wave production of traveling spin waves can dominate all other forms of damping [28, 33–35]. This results in a faster approach to equilibrium for large-angle switching experiments than would otherwise be expected. This is perhaps counter-intuitive in that the further you move the magnetization from its equilibrium, the quicker it may reach equilibrium. It also has important implications for any devices that rely on magnetization switching, including magnetic data storage and magnetic hyperthermia (heating magnetic particles by placing them in a strong r.f. field as raising the temperature may aid certain cancer therapies).

The damping parameter $\tilde{\alpha}$ for the FMR mode may be determined from experiments as it is related to the linewidth of the microwave absorption at this resonant frequency. The broader the linewidth, the larger the effective damping or losses from the uniform precessional motion. For a frequency-swept linewidth Δf, the effective damping parameter for a thin film is given by [55, 56]

$$\tilde{\alpha} = \frac{\Delta f}{\gamma(H + 2\pi M_s)}, \tag{27}$$

in CGS units, where the gyromagnetic ratio has units of Hz/Oe rather than rad·Hz/Oe. In the linear regime, $\tilde{\alpha} \rightarrow \alpha$, the Gilbert damping parameter. But in the nonlinear regime it will be larger. This will be discussed further in Section 4.2.1, where the damping parameter is seen to increase by roughly 20% when a particular nonlinear process is resonant in a rectangular magnetic strip [57].

It may be difficult to resolve the linear two-magnon damping due to inhomogeneities from the intrinsic nonlinear three- or four-magnon scattering as all these processes produce similar spin waves with nonzero wave vector. For two-magnon scattering, inhomogeneities, surface roughness [58] or other defects in a material [59] allow coupling between the FMR mode and a traveling spin wave. This typically leads to a broadening of the absorption linewidth in ferromagnetic resonance experiments, which can be related back to the two-magnon scattering amplitude [60, 61]. The strength of two-magnon scattering depends on geometry and the angle of the applied field. For example, out-of-plane static fields applied to a thin film may lead to no two-magnon scattering events being on resonance in the system [62].

Such traveling spin waves can also be detected indirectly using the magneto-optical Kerr effect (MOKE) as they produce an *apparent* reduction in the magnetization vector length that is reconstructed from the MOKE signal [28]. This is because the optical signal is found using a finite-sized laser spot with width that is larger than the wavelength of the excited waves. This is demonstrated schematically in Fig. 11. In panel (a) the case of uniform precession is shown within the elliptical laser spot that may be a few tens of microns wide. The magnetization at each location in the film, shown by the arrows, are all in-phase about their elliptical orbit. Therefore MOKE reconstructs a magnetization vector that has the proper length. In panel (b) the case of a traveling spin wave propagating

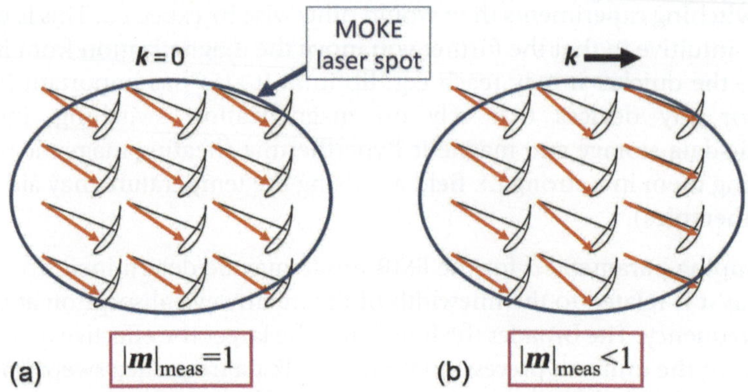

FIGURE 11

Schematic showing how MOKE experiments can detect short wavelength spin waves that have been excited via two, three, or four-magnon processes. Panel (a) shows spatially uniform magnetization dynamics within the laser spot and a magnetization $|\vec{m}| = |\vec{M}|/M_s = 1$ is measured. Panel (b) shows a spin wave propagating to the right with wavelength of the order of the laser spot width. This results in a spatially averaged MOKE measurement that perceives that the magnetization vector is shortened.

to the right is shown. The magnetization is at different points in its elliptical orbit at different locations within the laser spot, as demonstrated by the vectors. This means that the spatially averaged signal measures an apparent magnetization vector length that is less than the saturation magnetization.

Brillouin light scattering (BLS) experiments can be used to directly probe whether waves with larger wavenumbers are present in the system and so may in this case show advantages over MOKE. This method measures in reciprocal rather than real space. A neat example is given in [63] where the authors use BLS to probe three-wave processes involving BVMSWs. Since the YIG strip used is quite thick (6.3 μm), there are six spin wave branches that may can take part in three-wave confluence or splitting processes. This is in stark contrast to our discussion above that focused on very thin films and only allowed on-resonance interactions with the lowest order thickness branch in general.

One way to distinguish which type of process (two or three magnon) is dominant in the damping is to examine the small-amplitude or linear dynamics of the system. For linear dynamics, only two-magnon scattering processes are possible. The rate of two-magnon processes will not change when the magnetization dynamics are forced nonlinear but the three- and four-magnon processes will turn on. Hence, depending on the amplitude of magnetization precession, one type of damping may dominate over the other. For example, in large-angle magnetization switching experiments on thin NiFe films it was found that the four-wave damping was dominant over the two-wave scattering [28]. The mechanism is that the system was so far above the nonlinear threshold—with large precession angles around 20°—the characteristic decay time for four-wave scattering was of the order of nanoseconds but the decay time for two-magnon scattering was over 100 times longer.

4 NONLINEAR PROCESSES IN NANOSTRUCTURES

When it comes to small structures, we often lose our intuition about which modes of excitation couple together. Firstly, the ground state magnetization configuration of single-domain elements may be nontrivial, for example, flower and S states in small rectangular elements[64, 65] and vortex states in cylindrical or spherical elements. The aspect ratio and the magnetic material used to make the element may determine which ground states are present in the nano-object since there is a competition between the exchange energy wanting the magnetization to be aligned and shape anisotropy which wants the stray field of the magnet to be minimized. Some of the states possible in a rectangular element are illustrated schematically in Fig. 12, where the local magnetization

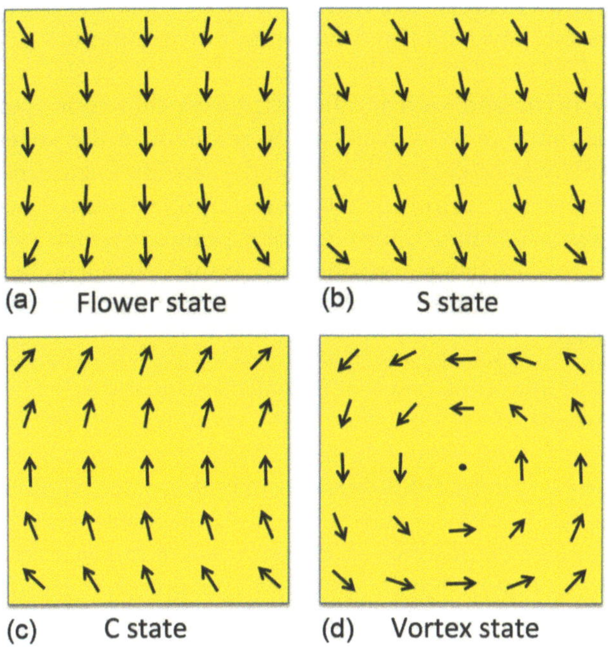

FIGURE 12
Schematic of four possible magnetization states in a rectangular nanoelement. The arrows show the direction of the local magnetization. These diagrams are not the result of simulations, simply a guide to the types of states that exist. In the vortex state, the magnetization points into or out of the page at its center, known as the vortex core.

direction is shown by small arrows. If a sufficiently large magnetic field is applied, then the magnetization may be saturated and these states are no longer seen. However, that typically also means that the system requires more energy to drive it nonlinear.

Furthermore, the linear excitations in such a complicated ground state are not always obvious. Some progress may be made using analytic functions adapted to the geometry of the element to calculate the linear spin waves (see, for example, [66] that deals with cylindrical elements) but otherwise, one must resort to numerical calculations in order to gain insight into the resonant modes and their interactions that give rise to chaos. Some of the techniques to characterize chaos described in Section 2 can also be used to shed light on the situation. For example, Fourier transforms of micromagnetic data in the time domain show whether three- or four-wave interactions may be the dominant nonlinear terms depending on whether there are frequency peaks at double or triple the fundamental frequency. Fourier transforms in the spatial domain can indicate what the excited modes look like. For example, if they have maximum amplitude near the edges or in the middle of an element.

Another complication for small structures is that there is not a continuous band of frequencies available for the driven mode to couple to, as was the case for thin films, illustrated in the dispersion relations of Fig. 8. Only discrete frequencies may occur and so a nonlinear process that conserves energy and momentum is much more unlikely to be resonant in the system compared to if our sample was bulk in at least one dimension. A great demonstration of this trend is illustrated in the first figure of [67], reprinted here in Fig. 13. The panels show schematically the effect of making a magnetic cylinder smaller and smaller in size. Exchange contributions to the frequencies become large as the dimensions of a structure shrink and so the resonant frequencies may also be widely spaced.

As an example of this point, consider a small element and a three-wave decay of the uniform precession at the onset of the nonlinear regime. As mentioned, three-wave decay occurs resonantly and with the lowest threshold so as to generate spin waves at half the FMR frequency in bulk and thin film samples. However, in a small element there may be no modes available to the system at this half-frequency! (See Fig. 13d for an extreme case of this in a very small structure.) This does not mean that the system will not go nonlinear. The three-wave process that occurs may be a decay into two spin waves with frequencies either side of the half frequency (one above and one below) so that energy is still conserved. Such resonant processes were recently demonstrated in a micromagnetics calculation for small NiFe elements [68]. Another option is that the system is driven very strongly and nonresonant processes are able to take place which have much higher thresholds than resonant processes [69].

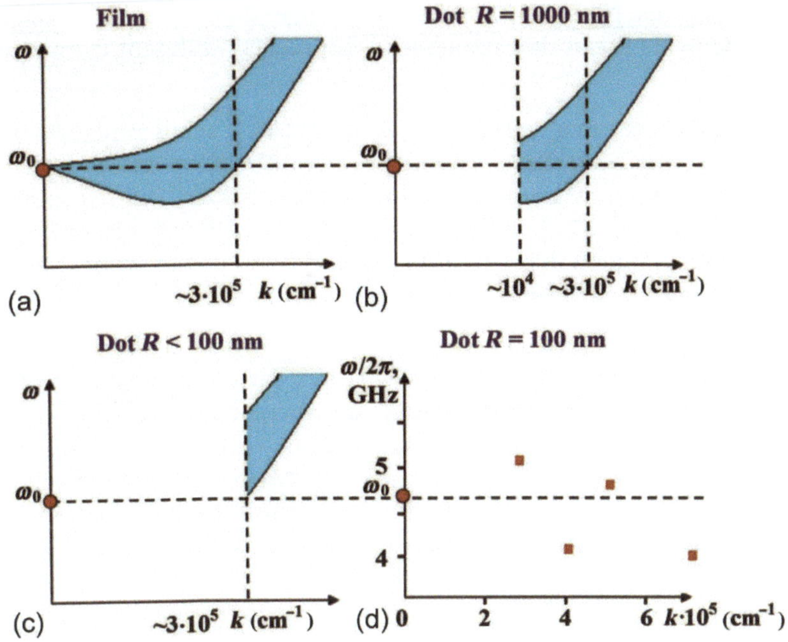

FIGURE 13

Schematic of the dispersion relations for different magnetic geometries, taken from [67]. Panel (a) shows that of a thin film, as in Figure 8. Panels (b) and (c) show that for a disk with radius R much larger than the thickness L, with $R = 1000$ and 100 nm respectively. Panel (d) has the discrete modes for a disk with $R = 100$ nm and $L = 10$ nm, calculated using the analytic methods of [66].

The issues discussed above are all related to how one may understand the data from small elements driven into the nonlinear regime. However, there are also problems in experimentally measuring such systems! Ways to measure the nonlinear dynamics in small structures will be discussed in the following subsection, then some results will be presented for nanostrips and then nanoelements. In particular, we discuss situations where the magnetization cannot be treated as a macrospin as this was already discussed in Section 2.

4.1 Experimental Techniques

The study of small samples that have submicron length in two or in all three dimensions has only been achieved in the last decade. Although such elements can be fabricated using lithographic methods, techniques to study their nonlinear behavior were not readily available. There is not sufficient magnetic material in one element of this size to obtain a signal using traditional ferromagnetic resonance experiments involving cavities. Creating arrays of elements will alleviate this problem but the long-range dipole–dipole interactions between elements mean that the physics can be very different from that of a single element.

Methods to study the magnetization dynamics in such small structures include micro-focused Brillouin light scattering (micro-BLS) [70–73] and electrical-based methods such as giant magnetoresistance (GMR) [74, 75] and photo voltage techniques [76, 77]. In fact, if it is a current which induces the nonlinear behavior in the element, then a resistance-based technique to probe the dynamics is often the natural choice.

In general, the micro-BLS technique is the best for obtaining information of the position-dependent amplitudes of magnetic excitations. Spatial resolution down to around 200–250 nm is possible and frequency resolution is around 50 MHz [78]. The modes in small structures can be complicated and it is often useful to know what mode is being excited in order to elucidate some information on the nonlinear processes that are dominant. Although the spatial resolution is diffraction limited, it is possible to discern edge modes from breathing modes and other typical modes in nano-objects.

Another option is to use a microstrip waveguide with a sample attached under the current-carrying strip, as illustrated in Fig. 14. This is exactly what is done in the lab downstairs from where I currently sit writing this chapter. Such a geometry can be prepared using sputtering or molecular beam epitaxial deposition, followed by lithography to create the correct shape and attach the necessary electrodes. The microstrip geometry is advantageous for two reasons: Firstly, the Oersted field generated around the microstrip may be huge and hence inducing nonlinear dynamics is no problem at all. An estimate shows that this *dynamic* microwave field may have an amplitude of up to 300 Oe [79].

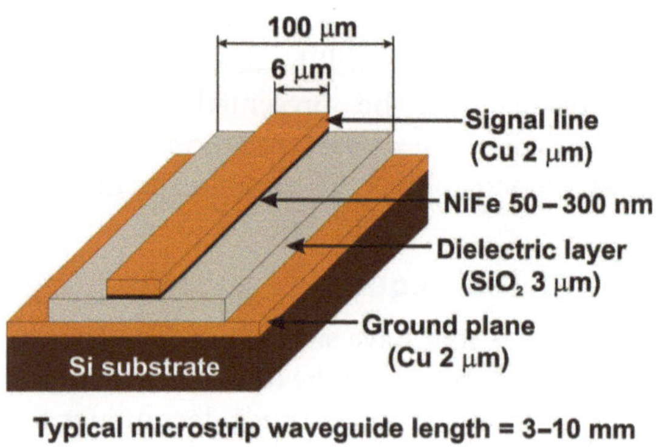

FIGURE 14

A typical microstrip geometry. The magnetic sample lies directly below the metallic signal line. Illustration courtesy of R. Camley.

Secondly, the sample is of a comparable size to the exciting wire and so any change in the sample's behavior will be noticed in the absorption spectrum of radiation passing through the waveguide [79].

Finally, X-rays can be used to image the magnetization in a small magnetic sample. X-ray magnetic circular dichroism (XMCD) means that areas of the sample having magnetization pointing in different directions will deflect left- and right-circularly-polarized X-ray light differently. By subtracting the two signals, an image is produced that shows regions with different magnetization directions. Spatial resolution of 30 nm is possible and time resolution is below 1 ns. For example, [80] reports X-ray studies of a magnetic vortex in an NiFe square that is 2 μm on the sides and 20 nm thick. The vortex can be seen to gyrate around the center of the square with a period around 16 ns. Although X-ray imaging has fantastic size- and temporal-resolution, the disadvantage is that time must be purchased or awarded to run experiments at a large facility, such as the Advanced Light Source in Berkeley or BESSY in Berlin, and often "in-house" equipment is preferable for testing samples.

4.2 Nanostrips

Nanostrips, which are nanometer-scaled in two dimensions only, will be discussed first as we can extend some of the discussions of thin films. Strips with rectangular cross-sections will be considered. Two interesting phenomena will be described that illustrate how nonlinear processes may occur in such ribbons or strips, producing interesting effects. The first involves seeing an increase in the effective damping at particular values of applied field or frequency. The second involves generating a "comb" of frequencies when just two high-power signals are fed into a sample.

4.2.1 Increase in Damping

The fact that the transfer of energy from the FMR mode and into traveling spin wave modes can be viewed as a form of nonlinear damping was discussed in Section 3.4. It has also been mentioned previously that nanometer dimensions tend to create discrete excitation spectra. These two ingredients can be added together and we can expect to see a peak or increase in the phenomenological damping parameter $\tilde{\alpha}$ at very precise values of the magnetic field and driving frequency (or driving current) corresponding to when a certain spin wave interaction is on resonance. This is precisely what was seen in [57] for NiFe strips.

If the nanostrip is thin enough, then magnetization may be considered uniform through the thickness and there is only variation across the strip's width and along the strip's length. The width can be treated using discrete thickness modes, as was done for thin films earlier, with pinning at the boundaries that depends on the sample. Any wavenumber is allowed along the strip's length. In

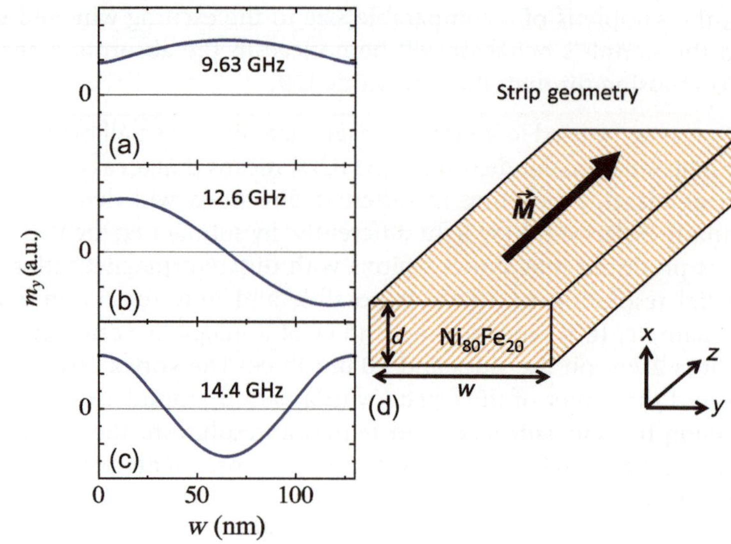

FIGURE 15

Panels (a)–(c) show the lowest frequency standing spin wave modes in an NiFe strip that is 130 nm wide and 25 nm thick with zero applied field. Panel (d) shows the strip geometry with width *w* in the *y* direction and thickness *d* in the *x* direction. The static magnetization is along the *z* axis. The standing waves of dynamic magnetization are across the width.

fact, the allowed spin waves in this geometry can be treated similarly to those in the thin film, apart for needing an appropriate change in the demagnetizing energy to account for the finite size in two dimensions. In Fig. 15 the first three standing modes in an NiFe nanostrip are shown, as calculated using the methods presented in [81]. The corresponding frequencies are given in each panel. There is strong dynamic dipolar pinning at the edges for the first two standing waves [82], but for the third mode (and higher frequency modes) the standing waves are dominated by the exchange energy contribution and are assumed to have free boundary conditions.

Changing the applied field strength along the z (longest) direction or changing the thickness d and width w of the strip changes the resonant frequencies. When one of the higher order mode frequencies ($n = 2$ or 3) is exactly twice the first mode's frequency, then a three-wave interaction is on resonance. The process of one wave in and two out is shown in Fig. 9c and the process described here is exactly the inverse process with two waves in and one out, such that energy is conserved. The effective damping $\tilde{\alpha}$ of the quasi-uniform (lowest frequency) mode is maximal at precisely this geometry and applied field value. This can be seen in the experimental data of [57], reprinted in Fig. 16. For a wider strip the peak occurs at a lower value of the magnetic field; that is, the peak is around

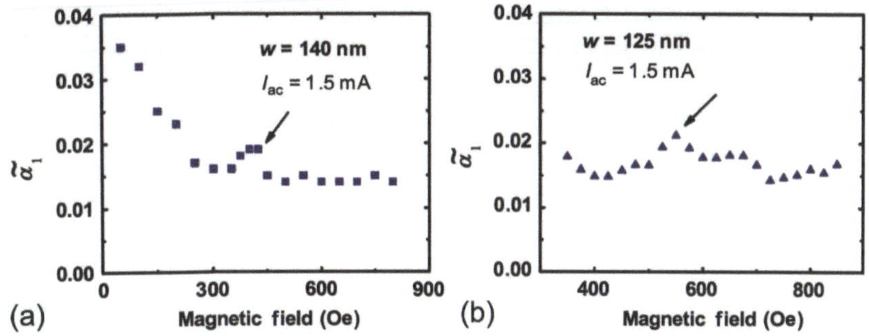

FIGURE 16
The phenomenological damping constant $\tilde{\alpha}$ plotted as a function of the applied static magnetic field, for two different strips with width (a) 140 nm and (b) 125 nm. Both are 6 nm thick and a driving current of 1.5 mA is applied to excite magnetization dynamics. Figure is taken from [57].

400 Oe for a 140-nm-wide strip (panel a) and is around 550 Oe for the thinner 125-nm-wide strip (panel b). This is because the exchange contribution to the spin wave frequencies is weaker for a wider strip and thus all of the modes have lower frequencies. The conservation law for the three-wave process to be resonant is therefore fulfilled at lower applied field values.

As mentioned in Section 3.4, the phenomenological damping parameter $\tilde{\alpha}$ is inferred from experiments by fitting a Gaussian function to the ferromagnetic resonance absorption peak and finding the width of the peak. Line broadening indicates an increase in the damping parameter due to a variety of losses, including nonlinear spin wave interactions. An upper bound on the damping is given by the thin film expression in Eq. (27). However, this expression should be changed slightly to take into account the different demagnetizing fields of the strip geometry.

4.2.2 Two Mode Mixing and Frequency Combs

Very recently, papers by various authors have emerged that study two microwave frequencies being input into a waveguide containing a nanostrip of magnetic metal. It has been shown that an input signal may be amplified [83] and a "comb" of frequencies made up of linear combinations of the input frequencies may be produced [79]. NiFe was used as the magnetic material in these studies.

These kinds of processes are hardly new. In the 1980s and 1990s there was much work on ferrites like Yttrium Iron Garnet (YIG), including studies that showed spin wave amplification in a sample with micronthickness and millimeter width [84]. In fact, most of the work reported in the book on nonlinear phenomena by Wigen [8] concerned YIG. In the past decade, studies on metallic films have predominated. Although metallic magnets have greater losses than ferrites and must be driven harder to see nonlinear effects, their resonant frequencies are generally higher since their saturation magnetization is higher. For applications in the 10–30 GHz range (typical for radar and other communications), metals are therefore better suited than ferrites. They are also better able to be incorporated in on-wafer devices and may be made much smaller (nanometer rather than micron thickness). This also allows any Oersted field from a neighboring microstripline to penetrate through the whole thickness of the magnetic material. This is not to say that the study of insulating magnetic materials is dead! In fact, there have been exciting developments in both the fabrication and the integration of these materials over the past few years [85], making their applications both smaller (although not nanometer-sized) and more feasible.

A frequency comb produced using two waves input into a microstrip waveguide containing a 300-nm-thick NiFe strip is shown in Fig. 17 and is taken from [79]. The plot is for the output power exiting the microstrip waveguide. In panels (a–c), the input power of wave 1 ($f_1 = 10.0$ GHz) is increased while that of wave 2 ($f_2 = 9.99$ GHz) is kept constant. In panel (d), both waves have large input power. The first thing to notice is that although the input power of tone 2

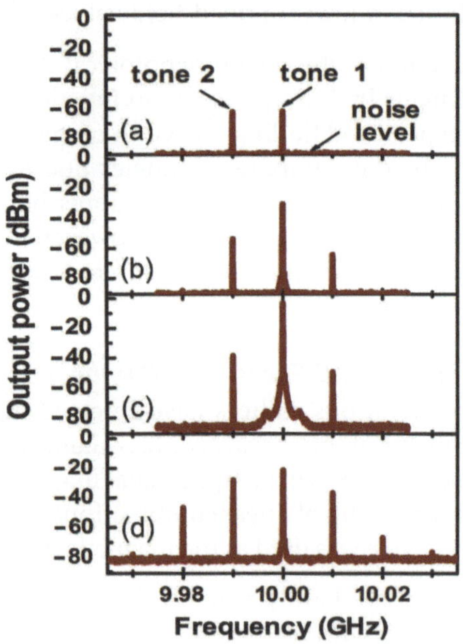

FIGURE 17

An example of a nonlinear frequency comb that is generated from just two input frequencies or tones. The power of each input frequency is varied through the four panels. They are (a) − 29 dBm for both, (b) − 2 dBm for tone 1 and − 29 dBm for 2, (c) +10 dBm for tone 1 and − 29 dBm for 2, and (d) both tones 1 and 2 have large input power. Taken from [79].

is kept constant in panels (a)–(c), its output power increases. In other words, this wave is parametrically amplified by tone 1. The second thing to notice in panel (d) is that new waves are excited that are all equally spaced along the frequency axis ($\Delta f = 0.01$ GHz). Increasing the power of the input waves sees this comb grow in both height and in width. The new frequencies are all linear combinations of the input frequencies, for example $(2f_2 - f_1)$ and $(2f_1 - f_2)$, suggesting a four-wave process is dominant in this situation. A perturbative theory to describe the results is presented in [86].

A macrospin model proves to be adequate in explaining the frequencies and amplitudes of the newly created modes in these experiments [86]. (I'm sorry that I promised not to discuss macrospins any more in this chapter.) This is argued to be because the wavelengths of the input waves are of the order of millimeters and are therefore much larger than the sample size. Also, for a 300-nm-thick NiFe film the spin wave bands corresponding to different thickness modes overlap so there is not a clear process that can be seen to be on-resonance, as was the case in the previous section.

4.3 Nanoelements

In the introduction to this section, the problems with studying nonlinear spin waves in nanoelements were discussed briefly. In particular, the magnetization ground state may be nonuniform and it is difficult for different modes to couple together in resonant processes due to a discrete frequency spectrum[87] and so nonlinear studies are much more rare for systems confined in all three dimensions. This may be an advantage in certain situations. For example, if one wishes to transfer a high-power microwave signal down a magnetic waveguide, power is ofter lost to spin wave modes. However, if the waveguide consists of nano-sized magnetic elements, then nonlinear processes are suppressed and the power that passes through the waveguide may be much larger [67]. It can be said that a nonlinear metameterial has been designed.

Suppressing nonlinear processes may sometimes be advantageous, but we would also like to see some unique nonlinear processes in very small elements. In this section, just two examples will be described. Firstly, the nonlinear dynamics of a magnetic vortex core in cylindrical nanoelements will be detailed. Secondly, if an element is driven hard enough then nonresonant nonlinear processes may occur and the fact that a limited number of processes are resonant in the system does not present a hurdle to nonlinear dynamics occurring.

4.3.1 *Nonlinear Vortex Dynamics*

Magnetic vortices may be the magnetic ground state for thin, small cylindrical elements in zero field. A phase diagram demonstrating when a vortex forms—

rather than the magnetization being aligned in or out of the disk plane—as a function of the radius-to-height ratio is provided in [88, 89]. Typically vortices form in NiFe disks on the order of 100–200 nm in diameter and around 10 nm thick. The vortex state is of interest as the direction of the core's magnetization (into or out of the disk) may be used to encode information. The information may be written using a magnetic field or a spin-transfer electric current to switch the core's magnetization direction [90]. For the case of spin-transfer devices, the thin NiFe disk that contains the magnetic vortex is the free magnetic layer in a larger spintronic nanopillar or stack.

The elementary magnetic excitation on this magnetic vortex ground state is circular vortex core gyration. If driven by spin torque or by oscillating magnetic fields, the vortex core will gyrate or move in a circle about its equilibrium position in the center of the disk [80, 91, 92]. The gyration can be seen in X-ray experiments and is illustrated schematically in Fig. 18. The resonant frequency of this translational motion is typically in the tens to hundreds of megahertz for NiFe disks corresponding to radio frequencies, just below the microwave range. No matter the material, such translational modes have frequencies below the spin wave bands (roughly 10 GHz in NiFe). The line width of the gyration is small [92], as has been noted for other spin torque nano-oscillators where the magnetization does not form a vortex.

With larger driving fields or currents, the vortex core moves to larger distances from the center of the cylindrical element. Micromagnetic simulations and experiments show that this results in a nonlinear shift in the resonant frequency to higher values and nonlinear contributions to the damping too [93]. The driving may come from two inter-related sources: the circular Oersted field of the

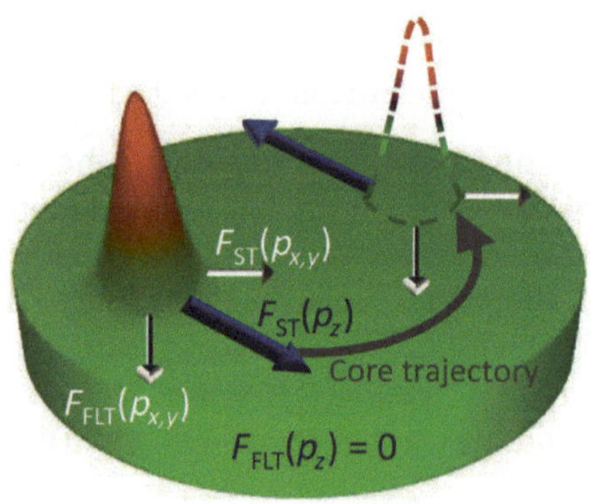

FIGURE 18

A schematic of the vortex gyration. The lump represents the location of out-of-plane magnetization within the cylindrical magnetic dot. The various force vectors can be ignored for the purposes of the discussion here. Taken from [93].

applied electric current or the spin-transfer torque from the current. These nonlinear shifts occur before the onset of any kind of chaotic motion.

Parametric excitation of the magnetization dynamics can occur most easily when the current and therefore the Oersted field are applied with exactly twice the frequency of the resonant vortex gyration frequency [94]. This is important as it opens up all the traditional applications of parametric excitation in magnetic systems (filters, amplifiers and frequency selectors for microwave signal processing) to miniature spin torque devices. Moreover, compared to spin torque stacks where the magnetization ground state is uniform, the frequency of operation is lower and therefore may find application in low-microwave and radio devices.

It should be noted that only a small part of the larger body of work concerning nonlinear spin torque oscillators has been touched on here by focusing on vortices. This was deliberate as the wish was to provide another obvious example where macrospin models are inadequate to describe the nonlinear processes occurring in a nanostructure. For the case of uniformly magnetized free layers, the macrospin model is often useful [46] but can break down in some instances, as described in [26].

4.3.2 Nonresonant Nonlinear Processes

If elements are sufficiently small, then no mode may have a frequency below that of the FMR mode. Let the angular frequency of the FMR mode be denoted by ω_0 and the next lowest frequency mode be $\omega_1 = \omega_0 + \Omega$, where Ω is the positive frequency gap. Let the damping parameter in the system be η. A nonresonant process may occur such that two FMR modes produce two of the higher-frequency modes with some energy needed to be provided into the magnetic system. The threshold for this nonlinear process will be larger than the threshold for four-wave processes in a thin film of the same material, roughly by a factor of $(\Omega/\eta)^{1/2}$ [69].

In other words, the smaller the element, the larger the exchange contributions to the resonant frequencies, the larger the frequency gap Ω and therefore the larger the threshold for nonlinearity. In [69], methods for studying resonant spin wave instabilities [7] are adapted to these nonresonant processes and fold-over effects are seen. The amplitude of the FMR mode increases as the magnitude of an external driving field increases but above the nonresonant threshold for the driving field, the precession increases at a decreased rate due to energy losses to other modes in the system. It may be important to know how the uniform precession or FMR mode behaves for applications such as high-power microwave transmission, as mentioned previously.

5 CONCLUSION

In this chapter, nonlinear and chaotic processes and phenomena have been discussed in thin films and nano-scaled objects. Starting with a macrospin, the techniques used to analyze nonlinear and chaotic systems were discussed. Next, a theory of thin films near the nonlinear threshold was presented and its implications to various experiments discussed. Finally, some challenges to study nanostructures and some interesting results were discussed. The process through the chapter was to add finite width in one, then two, then finally all three dimensions of a magnetic object in order to guide the reader through progressively more difficult systems to study both experimentally and theoretically.

In the future, it is certain that a variety of magnetic devices for communication, information storage and sensing purposes will be further miniaturized. Therefore, the dynamics—linear and nonlinear—of such devices will be increasingly of interest. Although nonlinear processes are less likely to be resonant in small structures, the structures will require operation at large power levels and there is sure to be nonlinear effects at play. Nonlinear processes may be desirable (tunable microwave frequency combs, a faster approach to equilibrium, efficient frequency conversion) or may be undesirable (loss of the main signal's power, destruction of information) depending on the device application.

Acknowledgments

I would like to thank Bob Camley who discussed some of the material in this chapter. Bob Stamps and Mikhail Kostylev were involved in much of the work detailed in the section on thin films, some years ago. Matthew Phelps contributed many useful discussions on nonlinear macrospins while this chapter was being prepared. Miriam DeJong was helpful in discussing demagnetizing effects. Any omission of relevant articles is solely the fault of the author.

References

[1] Bloembergen N, Damon RW. Relaxation effects in ferromagnetic resonance, Phys Rev 1952;85:699. URL: http://journals.aps.org/pr/abstract/10.1103/PhysRev.85.699.

[2] Bloembergen N, Wang S. Relaxation effects in para- and ferromagnetic resonance, Phys Rev 1954;93:72–83. URL: http://journals.aps.org/pr/abstract/10.1103/PhysRev.93.72.

[3] Suhl H. The theory of ferromagnetic resonance at high signal powers, J Phys Chem Solids 1957;1:209–27. URL: http://www.sciencedirect.com/science/article/pii/0022369757900100.

[4] Lorenz EN. Deterministic nonperiodic flow, J Atmos Sci 1963;20:130–48. URL: http://journals.ametsoc.org/doi/abs/10.1175/1520-0469%281963%29020%3C0130:DNF%3E2.0.CO;2.

[5] Barkoulas J, Travlos N. Chaos in an emerging capital market? The case of the Athens stock exchange, Appl Finan Econ 1998;8:231–43. URL: http://www.tandfonline.com/doi/abs/10.1080/096031098332998.

[6] Canavier C, Clark J, Byrne J. Routes to chaos in a model of a bursting neuron. Biophys J 1990;57:1245–51. URL: http://www.sciencedirect.com/science/article/pii/S0006349590826436.

[7] L'vov VS. Wave turbulence under parametric excitation. Berlin: Springer-Verlag; 1994.

[8] Wigen PE. Nonlinear phenomena and chaos in magnetic materials. Singapore: Springer-Verlag, World Scientific; 1994.

[9] Bertotti G, Mayergoyz ID, Serpico C. Nonlinear magnetization dynamics in nanosystems. Amsterdam: Elsevier; 2009.

[10] Gurevich AG, Melkov GA. Magnetization oscillations and waves. Boca Raton: CRC Press; 1996.

[11] Wu M. Nonlinear spin waves in magnetic film feedback rings. In: Camley RE, Stamps RL, editors. Solid state physics, 62. Amsterdam: Elsevier; 2011. p. 163–224.

[12] Ustinov AB, Drozdovskii AV, Kalinikos BA. Multifunctional nonlinear magnonic devices for microwave signal processing. Appl Phys Lett 2010;96:142513. URL: http://scitation.aip.org/content/aip/journal/apl/96/14/10.1063/1.3386540.

[13] Brown Jr WF. Magnetostatic principles in ferromagnetism. Amsterdam: North-Holland Publishing Company; 1962.

[14] Aharoni A. Demagnetizing factors for rectangular ferromagnetic prisms, J Appl Phys 1998;83:3432–4. URL: http://scitation.aip.org/content/aip/journal/jap/83/6/10.1063/1.367113.

[15] Aharoni A. Introduction to the theory of ferromagnetism. 2. Oxford: Clarendon Press; 1996.

[16] Gilbert T. A phenomenological theory of damping in ferromagnetic materials, IEEE Trans Magn 2004;40:3443–9. URL: http://ieeexplore.ieee.org/xpls/abs_all.jsp?arnumber=1353448&tag=1.

[17] Smith RK, Grabowski M, Camley RE. Period doubling toward chaos in a driven magnetic macrospin, J Magn Magn Mater 2010;322:2127–34. URL: http://www.sciencedirect.com/science/article/pii/S0304885310000636.

[18] Schlömann E, Green JJ, Milano U. Recent developments in ferromagnetic resonance at high power levels, J Appl Phys 1960;31:386. URL: http://scitation.aip.org/content/aip/journal/jap/31/5/10.1063/1.1984759.

[19] Slonczewski JC. Current-driven excitation of magnetic multilayers, J Magn Magn Mater 1996;159:L1–L7. URL: http://www.sciencedirect.com/science/article/pii/0304885396000625.

[20] Berger L. Emission of spin waves by a magnetic multilayer traversed by a current, Phys Rev B 1996;54:9353. URL: http://journals.aps.org/prb/abstract/10.1103/PhysRevB.54.9353.

[21] Ralph D, Stiles M. Spin transfer torques. J Magn Magn Mater 2008;320:1190. URL: http://www.sciencedirect.com/science/article/pii/S0304885307010116.

[22] Lee KJ, Deac A, Redon O, Nozières JP, Dieny B. Excitations of incoherent spin-waves due to spin-transfer torque, Nat Mater 2004;3:877–81. URL: http://www.nature.com/nmat/journal/v3/n12/abs/nmat1237.html.

[23] Chen W, de Loubens G, Beaujour JML, Sun JZ, Kent AD. Spin torque driven ferromagnetic resonance in a nonlinear regime. Appl Phys Lett 2009;95:172513. URL: http://scitation.aip.org/content/aip/journal/apl/95/17/10.1063/1.3254242.

[24] Cheng X, Katine JA, Rowlands GE, Krivorotov IN. Nonlinear ferromagnetic resonance induced by spin torque in nanoscale magnetic tunnel junctions. Appl Phys Lett 2013;103:082402. URL: http://scitation.aip.org/content/aip/journal/apl/103/8/10.1063/1.4819179.

[25] Locatelli N, Cros V, Grollier J. Spin-torque building blocks, Nat Mater 2014;13:11–20. URL: http://www.nature.com/nmat/journal/v13/n1/full/nmat3823.html.

[26] Berkov D, Miltat J. Spin-torque driven magnetization dynamics: micromagnetic modeling, J Magn Magn Mater 2008;320:1238–59. URL: http://www.sciencedirect.com/science/article/pii/S0304885307010207.

[27] Livesey KL, Kostylev MP, Stamps RL. Parametric spin wave excitation cascaded processes during switching in thin films, Phys Rev B 2007;75:174427. URL: http://journals.aps.org/prb/abstract/10.1103/PhysRevB.75.174427.

[28] Nembach HT, Livesey KL, Kostylev MP, Martin-Pimentel P, Hermsdoerfer SJ, Leven B, et al. Magneto-optical observation of four-wave scattering in a 15-nm $Ni_{81}Fe_{19}$ film during large-angle magnetization precession, Phys Rev B 2011;84:184413. URL: http://journals.aps.org/prb/abstract/10.1103/PhysRevB.84.184413.

[29] Silva TJ, Kabos P, Pufall MR. Detection of coherent and incoherent spin dynamics during the magnetic switching process using vector-resolved nonlinear magneto-optics, Appl Phys Lett 2002;81:2205–7. URL: http://scitation.aip.org/content/aip/journal/apl/81/12/10.1063/1.1508163.

[30] Gerrits T, Silva TJ, Nibarger JP, Rasing T. Large-angle magnetization dynamics investigated by vector-resolved magnetization-induced optical second-harmonic generation, J Appl Phys 2004;96:6023–8. URL: http://scitation.aip.org/content/aip/journal/jap/96/11/10.1063/1.1811783.

[31] Gerrits T, Schneider ML, Kos AB, Silva TJ. Large-angle magnetization dynamics measured by time-resolved ferromagnetic resonance, Phys Rev B 2006;73:094454. URL: http://journals.aps.org/prb/abstract/10.1103/PhysRevB.73.094454.

[32] An SY, Krivosik P, Kraemer MA, Olsen HM, Nazarov AV, Patton CE. High power ferromagnetic resonance and spin wave instability processes in Permalloy thin films, J Appl Phys 2004;96:1572. URL: http://scitation.aip.org/content/aip/journal/jap/96/3/10.1063/1.1763996.

[33] Dobin AY, Victora RH. Intrinsic nonlinear ferromagnetic relaxation in thin metallic films, Phys Rev Lett 2003;90:167203. URL: http://journals.aps.org/prl/abstract/10.1103/PhysRevLett.90.167203.

[34] Dobin AY, Victora RH. Intrinsic nonlinear ferromagnetic relaxation (invited), J Appl Phys 2004;95:7139–44. URL: http://scitation.aip.org/content/aip/journal/jap/95/11/10.1063/1.1651812.

[35] Dobin AY. Ferromagnetic relaxation and interlayer exchange coupling in magnetic films and heterostructures. Ph.D. thesis; 2003. PhD in Physics, University of Minnesota.

[36] Laroze D, Becerra-Alonso D, Gallas J, Pleiner H. Magnetization dynamics under a quasiperiodic magnetic field, IEEE Trans Magn 2012;48:3567–70. URL: http://ieeexplore.ieee.org/xpls/abs_all.jsp?arnumber=6332802.

[37] Yang Z, Zhang S, Li YC. Chaotic dynamics of spin-valve oscillators. Phys Rev Lett 2007;99:134101. URL: http://link.aps.org/doi/10.1103/PhysRevLett.99.134101.

[38] Banerjee S, Yorke J, Grebogi C. Robust chaos, Phys Rev Lett 1998;80:3049–52. URL: http://journals.aps.org/prl/abstract/10.1103/PhysRevLett.80.3049.

[39] Fečkan M. Introduction. Bifurcation and chaos in discontinuous and continuous systems, nonlinear physical science. Heidelberg: Springer; 2011. p. 1–7.

[40] Smith RK, Grabowski M, Camley R. Nonlinear behavior in magnetic transients. J Magn Magn Mater 2009;321:3472–7. URL: http://www.sciencedirect.com/science/article/pii/S0304885309006763.

[41] Alligood KT, Sauer TD, Yorke JA. Chaos: an introduction to dynamical systems. New York: Springer; 1996.

[42] Broer H, Takens F. Dynamical systems and chaos. New York: Springer; 2011.

[43] Olson HM, Krivosik P, Srinivasan K, Patton CE. Ferromagnetic resonance saturation, second order Suhl spin wave instability processes in thin permalloy films. J Appl Phys 2007;102:023904. URL: http://scitation.aip.org/content/aip/journal/jap/102/2/10.1063/1.2756481.

[44] Khivintsev Y, Kuanr B, Fal TJ, Haftel M, Camley RE, Celinski Z, Mills DL. Nonlinear ferromagnetic resonance in permalloy films: a nonmonotonic power-dependent frequency shift,

Phys Rev B 2010;81:054436. URL: http://journals.aps.org/prb/abstract/10.1103/PhysRevB.81.054436.

[45] Rodríguez-Suárez RL, Matos-Abiague A, Azevedo A, Rezende SM. Frequency shift of spin waves in tunnel-junction spin-transfer nano-oscillators, Phys Rev B 2010;82:132410. URL: http://journals.aps.org/prb/abstract/10.1103/PhysRevB.82.132410.

[46] Berkov D, Gorn N. Transition from the macrospin to chaotic behavior by a spin-torque driven magnetization precession of a square nanoelement. Phys Rev B 2005;71:052403. URL: http://link.aps.org/doi/10.1103/PhysRevB.71.052403.

[47] Kalinikos BA, Slavin AN. Theory of dipole-exchange spin wave spectrum for ferromagnetic films with mixed exchange boundary conditions, J Phys C: Solid State Phys 1986;19:7013–33. URL: http://iopscience.iop.org/0022-3719/19/35/014.

[48] Zakharov VE, L'vov VS, Starobinets SS. Stationary nonlinear theory of parametric excitation of waves, Sov Phys JETP 1971;32:656. URL: http://jetp.ac.ru/cgi-bin/dn/e_032_04_0656.pdf.

[49] L'vov VS, Musher SL, Starobinets SS. Theory of magnetization self-oscillations on parametric excitation of spin waves. Sov Phys JETP 1973;37:546.

[50] Slavin AN, Rojdestvenski IV. 'Bright' and 'dark' spin wave envelope solitons in magnetic films, IEEE Trans Magn 1994;30:37–45. URL: http://ieeexplore.ieee.org/xpls/abs_all.jsp?arnumber=272513.

[51] Holstein T, Primakoff H. Field dependence of the intrinsic domain magnetization of a ferromagnet, Phys Rev 1940;58:1098–113. URL: http://journals.aps.org/pr/abstract/10.1103/PhysRev.58.1098.

[52] Krasitskii VP. Canonical transformation in a theory of weakly nonlinear waves with a nondecay dispersion law, Sov Phys JETP 1990;71:921–7. URL: http://www.jetp.ac.ru/cgi-bin/dn/e_071_05_0921.pdf.

[53] Boardman AD, Nikitov SA. Three- and four-magnon decay of nonlinear surface magnetostatic waves in thin ferromagnetic films, Phys Rev B 1988;38:11444–51. URL: http://journals.aps.org/prb/abstract/10.1103/PhysRevB.38.11444.

[54] Filho RNC, Cottam MG, Farias GA. Microscopic theory of dipole-exchange spin waves in ferromagnetic films: linear and nonlinear processes, Phys Rev B 2000;62:6545–60. URL: http://journals.aps.org/prb/abstract/10.1103/PhysRevB.62.6545.

[55] Slavin A, Tiberkevich V. Spin wave mode excited by spin-polarized current in a magnetic nanocontact is a standing self-localized wave bullet. Phys Rev Lett 2005;95:237201. URL: http://link.aps.org/doi/10.1103/PhysRevLett.95.237201.

[56] Nibarger JP, Silva RLTJ. Damping as a function of pulsed field amplitude and bias field in thin film permalloy. Appl Phys Lett 2003;82:2112–4. URL: http://scitation.aip.org/content/aip/journal/apl/82/13/10.1063/1.1564866.

[57] Boone CT, Katine JA, Childress JR, Tiberkevich V, Slavin A, Zhu J, et al. Resonant nonlinear damping of quantized spin waves in ferromagnetic nanowires: a spin torque ferromagnetic resonance study. Phys Rev Lett 2009;103:167601. URL: http://link.aps.org/doi/10.1103/PhysRevLett.103.167601.

[58] Arias R, Mills DL. Extrinsic contributions to the ferromagnetic resonance response of ultrathin films. Phys Rev B 1999;60:7395–409. URL: http://link.aps.org/doi/10.1103/PhysRevB.60.7395.

[59] Sparks M, Loudon R, Kittel C. Ferromagnetic relaxation. I. theory of the relaxation of the uniform precession and the degenerate spectrum in insulators at low temperatures, Phys Rev 1961;122:791–803. URL: http://journals.aps.org/pr/abstract/10.1103/PhysRev.122.791.

[60] Haas CW, Callen HB. Ferromagnetic relaxation and resonance line widths. Magnetism, 1. New York: Academic Press; 1963. p. 450–549.

[61] Krivosik P, Mo N, Kalarickal S, Patton CE. Hamiltonian formalism for two magnon scattering microwave relaxation: theory and applications, J Appl Phys 2007;101:083901. URL: http://scitation.aip.org/content/aip/journal/jap/101/8/10.1063/1.2717084.

[62] Srivastava AK, Hurben MJ, Wittenauer MA, Kabos P, Patton CE, Ramesh R, et al. Angle dependence of the ferromagnetic resonance linewidth and two magnon losses in pulsed laser deposited films of yttrium iron garnet, MnZn ferrite, and NiZn ferrite. J Appl Phys 1999;85:7838–48. URL: http://scitation.aip.org/content/aip/journal/jap/85/11/10.1063/1.370595.

[63] Romero CLO, Kalinikos BA, Krivosik P, Tong W, Kabos P, Patton CE. Three-magnon splitting and confluence processes for spin-wave excitations in yttrium iron garnet films: wave vector selective brillouin light scattering measurements and analysis. Phys Rev B 2009;79:144428. URL: http://link.aps.org/doi/10.1103/PhysRevB.79.144428.

[64] Rave W, Hubert A. Magnetic ground state of a thin-film element, IEEE Trans Magn 2000;36:3886–99. URL: http://ieeexplore.ieee.org/xpls/abs_all.jsp?arnumber=914337.

[65] Cowburn RP, Welland ME. Phase transitions in planar magnetic nanostructures. Appl Phys Lett 1998;72:2041–3. URL: http://scitation.aip.org/content/aip/journal/apl/72/16/10.1063/1.121258.

[66] Zivieri R, Stamps RL. Theory of spin wave modes in tangentially magnetized thin cylindrical dots: a variational approach. Phys Rev B 2006;73:144422. URL: http://link.aps.org/doi/10.1103/PhysRevB.73.144422.

[67] Kobljanskyj Y, Melkov G, Guslienko K, Novosad V, Bader SD, Kostylev M, Slavin A. Nanostructured magnetic metamaterial with enhanced nonlinear properties, Sci Rep 2012;2:478. URL: http://www.nature.com/srep/2012/120628/srep00478/full/srep00478.html.

[68] Camley RE. Three-magnon processes in magnetic nanoelements: quantization and localized mode effects. Phys Rev B 2014;89:214402. URL: http://link.aps.org/doi/10.1103/PhysRevB.89.214402.

[69] Melkov G, Slobodianiuk D, Tiberkevich V, de Loubens G, Klein O, Slavin A. Nonlinear ferromagnetic resonance in nanostructures having discrete spectrum of spin-wave modes, IEEE Magn Lett 2013;4:400–0504. URL: http://ieeexplore.ieee.org/xpls/abs_all.jsp?arnumber=6595654.

[70] Demokritov S, Hillebrands B, Slavin A. Brillouin light scattering studies of confined spin waves: linear and nonlinear confinement. Phys Rep 2001;348:441–89. URL: http://www.sciencedirect.com/science/article/pii/S0370157300001162.

[71] Kostylev M, Demidov VE, Hansen UH, Demokritov SO. Nonlinear mode conversion in monodomain magnetic squares, Phys Rev B 2007;76:224414. URL: http://journals.aps.org/prb/abstract/10.1103/PhysRevB.76.224414.

[72] Demidov VE, Jersch J, Rott K, Krzysteczko P, Reiss G, Demokritov SO. Nonlinear propagation of spin waves in microscopic magnetic stripes. Phys Rev Lett 2009;102:177207. URL: http://link.aps.org/doi/10.1103/PhysRevLett.102.177207.

[73] Madami M, Gubbiotti G, Tacchi GS, Carlotti G. Application of microfocused Brillouin light scattering to the study of spin waves in low-dimensional magnetic systems. Solid State Phys 2012;63:79.

[74] Maruyama T, Shiota Y, Nozaki T, Ohta K, Toda N, Mizuguchi M, et al. Large voltage-induced magnetic anisotropy change in a few atomic layers of iron, Nat Nanotechnol 2009;4:158–61. URL: http://www.nature.com/nnano/journal/v4/n3/full/nnano.2008.406.html.

[75] Sankey JC, Braganca PM, Garcia AGF, Krivorotov IN, Buhrman RA, Ralph DC. Spin-transfer-driven ferromagnetic resonance of individual nanomagnets. Phys Rev Lett 2006;96:227601. URL: http://link.aps.org/doi/10.1103/PhysRevLett.96.227601.

[76] Gui YS, Mecking N, Hu CM. Quantized spin excitations in a ferromagnetic microstrip from microwave photovoltage measurements. Phys Rev Lett 2007;98:217603. URL: http://link.aps.org/doi/10.1103/PhysRevLett.98.217603.

[77] Wismayer MP, Southern B, Fan XL, Gui YS, Hu CM, Camley RE. Nonlinear behavior for the uniform mode and horizontal standing spin-wave modes in metallic ferromagnetic microstrips: experiment and theory. Phys Rev B 2012;85:064411. URL: http://link.aps.org/doi/10.1103/PhysRevB.85.064411.

[78] Demokritov SO, Demidov VE. Micro-Brillouin light scattering spectroscopy of magnetic nanostructures, IEEE Trans Magn 2008;44:6–12. URL: http://ieeexplore.ieee.org/xpls/abs_all.jsp?arnumber=4407581.

[79] Khivintsev Y, Marsh J, Zagorodnii V, Harward I, Lovejoy J, Krivosik P, et al. Nonlinear amplification and mixing of spin waves in a microstrip geometry with metallic ferromagnets. Appl Phys Lett 2011;98:042505. URL: http://scitation.aip.org/content/aip/journal/apl/98/4/10.1063/1.3541787.

[80] Bolte M, Meier G, Krüger B, Drews A, Eiselt R, Bocklage L, et al. Time-resolved X-ray microscopy of spin-torque-induced magnetic vortex gyration. Phys Rev Lett 2008;100:176601. URL: http://link.aps.org/doi/10.1103/PhysRevLett.100.176601.

[81] Livesey KL, Ding J, Anderson NR, Camley RE, Adeyeye AO, Kostylev MP, Samarin S. Resonant frequencies of a binary magnetic nanowire, Phys Rev B 2013;87:064424. URL: http://journals.aps.org/prb/abstract/10.1103/PhysRevB.87.064424.

[82] Guslienko KY, Demokritov SO, Hillebrands B, Slavin AN. Effective dipolar boundary conditions for dynamic magnetization in thin magnetic stripes. Phys Rev B 2002;66:132402. URL: http://link.aps.org/doi/10.1103/PhysRevB.66.132402.

[83] Bao M, Khitun A, Wu Y, Lee JY, Wang KL, Jacob AP. Coplanar waveguide radio frequency ferromagnetic parametric amplifier. Appl Phys Lett 2008;93:072509. URL: http://scitation.aip.org/content/aip/journal/apl/93/7/10.1063/1.2975174.

[84] Kolodin PA, Kabos P, Patton CE, Kalinikos BA, Kovshikov NG, Kostylev MP. Amplification of microwave magnetic envelope solitons in thin yttrium iron garnet films by parallel pumping. Phys Rev Lett 1998;80:1976–9. URL: http://link.aps.org/doi/10.1103/PhysRevLett.80.1976.

[85] Recent advances in magnetic insulators–from spintronics to microwave applications, Wu M, Hoffman A, editors. Solid state physics 64:San Diego: Elsevier; 2013. p. 1–392.

[86] Marsh J, Camley RE. Two-wave mixing in nonlinear magnetization dynamics: a perturbation expansion of the Landau–Lifshitz–Gilbert equation. Phys Rev B 2012;86:224405. URL: http://link.aps.org/doi/10.1103/PhysRevB.86.224405.

[87] Ulrichs H, Demidov VE, Demokritov SO, Urazhdin S. Parametric excitation of eigenmodes in microscopic magnetic dots. Phys Rev B 2011;84:094401. URL: http://link.aps.org/doi/10.1103/PhysRevB.84.094401.

[88] Cowburn RP, Koltsov DK, Adeyeye AO, Welland ME, Tricker DM. Single-domain circular nanomagnets. Phys Rev Lett 1999;83:1042–5. URL: http://link.aps.org/doi/10.1103/PhysRevLett.83.1042.

[89] Chung SH, McMichael RD, Pierce DT, Unguris J. Phase diagram of magnetic nanodisks measured by scanning electron microscopy with polarization analysis. Phys Rev B 2010;81:024-410. URL: http://link.aps.org/doi/10.1103/PhysRevB.81.024410.

[90] Uhlíř V, Urbánek M, Hladík L, Spousta J, Im MY, Fischer P, et al. Dynamic switching of the spin circulation in tapered magnetic nanodisks, Nat Nanotechnol 2013;8:341–6. URL: http://www.nature.com/nnano/journal/v8/n5/abs/nnano.2013.66.html.

[91] Guslienko KY, Han XF, Keavney DJ, Divan R, Bader SD. Magnetic vortex core dynamics in cylindrical ferromagnetic dots. Phys Rev Lett 2006;96:067205. URL: http://link.aps.org/doi/10.1103/PhysRevLett.96.067205.

[92] Pribiag VS, Krivorotov IN, Fuchs GD, Braganca PM, Ozatay O, Sankey JC, et al. Magnetic vortex oscillator driven by D.C. spin-polarized current, Nat Phys 2007;3:498–503. URL: http://www.nature.com/nphys/journal/v3/n7/abs/nphys619.html/.

[93] Dussaux A, Khvalkovskiy AV, Bortolotti P, Grollier J, Cros V, Fert A. Field dependence of spin-transfer-induced vortex dynamics in the nonlinear regime. Phys Rev B 2012;86:014402. URL: http://link.aps.org/doi/10.1103/PhysRevB.86.014402.

[94] Bortolotti P, Grimaldi E, Dussaux A, Grollier J, Cros V, Serpico C, et al. Parametric excitation of magnetic vortex gyrations in spin-torque nano-oscillators. Phys Rev B 2013;88:174417. URL: http://link.aps.org/doi/10.1103/PhysRevB.88.174417.

CHAPTER 5

Linear Magnetization Dynamics in an Array of Dipolarly Coupled Magnetic Nanodots

Roman Verba*, Gennadiy Melkov†, Vasil Tiberkevich‡, Andrei Slavin‡
*Institute of Magnetism, National Academy of Sciences of Ukraine, Kyiv, Ukraine
†Faculty of Radiophysics, Electronics and Computer Systems, Taras Shevchenko National University of Kyiv, Kyiv, Ukraine
‡Department of Physics, Oakland University, Rochester, MI, USA

1 INTRODUCTION

Regular arrays of magnetic dots—typically, submicron-sized ferromagnetic elements—have been studied intensively during the last two decades. At some point, magnetic dots were considered as promising candidates for applications in data storage technology as elements of magnetic random access memory and bit-patterned magnetic storage media [1–3]. Recent advances in the fabrication of patterned magnetic media allow one to fabricate large regular arrays of submicron-sized magnetic dots with interdot distances that are sufficiently small to guarantee significant magneto-dipolar interaction between the dots [4–7]. While the interdot dipolar interaction is an undesirable phenomenon for applications of dot arrays in magnetic data storage, the existence of this interaction opens a possibility to develop a new type of an artificial magnetic material—patterned magnonic crystals (MCs).

In general, MCs are the artificial materials with periodic variation of magnetic parameters [8]. Man-made MCs can be classified into two main types. MCs of the first type are formed from continuous magnetic materials by periodic modulation of the material parameters, for instance by variation of a ferromagnetic film thickness [9,10], formation of an antidot lattice [11,12], or application of a spatially periodic external magnetic field [13]. These periodic variations of the magnetic parameters lead to significant changes in the spin wave (SW) spectrum of a periodically modulated medium and, in particular, to the formation of frequency stop-bands (or nonpropagation bands). It has been shown recently [13] that the position and the depth of these stop-bands in a periodically modulated medium can be controlled *dynamically*.

Magnetism of Surfaces, Interfaces, and Nanoscale Materials. http://dx.doi.org/10.1016/B978-0-444-62634-9.00005-9

MCs of the second type are formed by periodic arrays of distinct magnetic elements (e.g., magnetic nanowires [14] or nanodots [15]), usually coupled by the long-range magneto-dipolar interaction. In contrast to the majority of conventional magnetic materials, these MCs can be used without the application of a permanent bias magnetic field. Owing to small sizes and shape anisotropy of the dots, their static magnetic configuration can be well defined even in the absence of the bias magnetic field [16,17]. The interdot interaction in an array leads to the formation of *collective* SW excitations, the frequency of which typically lies in the gigahertz range. Eigenfrequencies and spatial profiles of collective SW excitations significantly depend on the shape of the dots and on the geometry of the array's lattice. Thus, by variation of the array geometry one can create MC with desirable microwave properties [18,19]. Also note that the characteristic wavelengths of the collective SWs propagating in an array are mainly determined by the period of the array's lattice and can lie in the submicron range. All these properties make arrays of coupled magnetic nanodots promising candidates for various applications in microwave micro- and nanoelectronics.

MCs of the second type also have another prominent feature. They are multistable, in the sense that under the same external conditions these MCs can exist in several quasi-stable static magnetic configurations (approximate ground states), separated by energy barriers [18,20,21]. Different static magnetic configurations correspond to the different SW spectra and, as a consequence, to the different microwave properties of an array. Therefore, by application of a pulsed external magnetic field one can switch from one array's configuration to another and therefore one can dynamically *control* the magnetic and microwave properties of this artificial magnetic material. Several possibilities of this dynamic control of the material properties have been demonstrated theoretically [22,23] and experimentally [24,25].

Thus, the investigation of collective SW excitation in arrays of coupled magnetic dots is important for various practical applications. Note that the theoretical understanding of the SW spectra of a dot array allows one not only to evaluate theoretically the array's microwave properties, but also provides information about the stability of the array stationary configurations. Usually, when a certain array's parameter (e.g., interdot distance or applied field) is approaching a value that is critical for the magnetic order in the array, one of the array's collective SW modes is softening, and its frequency vanishes. Moreover, the symmetry of this "soft" mode is usually related to the symmetry of a new stationary state into which the array is switched after the existing state of the array becomes unstable. This happens when the critical parameter (e.g., the applied field) is external and can be changed gradually.

The collective SW excitations in arrays of magnetic dots were investigated experimentally by means of a ferromagnetic resonance (FMR) microscopy [26],

Brillouin light scattering [4,27], and the magneto-optical Kerr effect [28]. The theoretical investigations of the collective SW excitations in magnetic dot arrays are rather complicated due to the long-range nature of the magneto-dipolar interaction existing between the individual dots. Small dot arrays, consisting of no more than 10 dots, were investigated using micromagnetic modeling [29–32]. The arrays of long magnetic stripes and rectangular dots were analyzed using the method of tensorial Green functions [9,33,34]. In the framework of this method, the problem is reduced to a system of integral equations that are solved numerically. Alternatively, collective SW excitations of an array of interacting magnetic elements can be treated as linear combinations of eigenmodes of an isolated magnetic dot. Using this approach, the SW spectra in arrays of spherical magnetic elements [35–37] and rectangular magnetic dots [38] in a saturated state, and arrays of cylindrical dots in a vortex state [39], have been studied theoretically. The lowest (gyrotropic) collective mode of an array of dots existing in a vortex ground state of static magnetization was analyzed using the well-known Thiele equation in [40]. An approximate analytical theory of the SW spectrum in infinite periodic arrays of magnetic dots in ferromagnetic and chessboard antiferromagnetic ground states was developed in [18,41] using the approximation of individual dots as point-like dipoles. A variant of a similar analytical theory that uses a multipole expansion of the dot variable magnetization was developed in [42].

In this chapter, following Ref. [19], we present a general theory of collective SW excitation of an *arbitrary* array of interacting magnetic dots. To simplify the calculations, we use the macrospin approximation, i.e., we assume a spatially uniform distribution of the static magnetization inside dots but, in contrast to the "point-dipole" approximation used in [18,41], we take into account the real shape-dependent demagnetization fields of individual dots. Of course, the macrospin approximation limits the applicability of our calculations only to the lowest branches of collective SW excitations of the array, i.e., branches formed by quasi-uniform modes of individual dots. However, these branches are expected to be dominant in the microwave response of an array and therefore are the most interesting branches from the point of view of practical applications. We also note that accounting for a possible nonuniform spatial profile of a static and/or dynamic magnetization inside the dot does not lead to any fundamental changes in the presented theory—it only complicates the technical calculations (see the example of a detailed consideration of this case and method of approximate calculations of the SW spectra in [43]). The general theory developed in the first part of the chapter is used to study several practically important particular cases in Section 3. First, following Refs. [19,44], we consider collective SW excitations in an array of cylindrical nanodots arranged into either a square or a hexagonal lattice. Then, we consider the influence of different point defects in the array's lattice on the SW mode structure and microwave response of an array [45]. Finally, we formulate the conditions under which the collective SW modes in a periodic dot array become nonreciprocal [46]. In the Section 4, we summarize the obtained results and discuss the limits of their applicability.

2 THEORY OF COLLECTIVE SW EXCITATION IN AN ARRAY OF MAGNETIC DOTS

2.1 Principal Equations

We consider below a two-dimensional array of magnetic dots lying in the same (x–y) plane. The position of a dot in an array is described by the two-dimensional position vector $\boldsymbol{r}_j$, where j is the index which numerates dots in the array. All the dots are assumed to have identical shape and the same saturation magnetization—accounting for a possible variation in the dot parameters does not lead to anything qualitatively new, but leads to significant technical complications. We also note that most of the effects arising from the difference in individual dot parameters can be qualitatively taken into account by introducing different crystallographic anisotropy for different dots, as is done in Section 3.3.

In the framework of a macrospin approximation used in this chapter, the state of each dot is completely described by one magnetization vector $\boldsymbol{M}_j \equiv \boldsymbol{M}_j(t)$ of a constant magnitude $(|\boldsymbol{M}_j| = M_s)$, where M_s is the saturation magnetization. The magneto-dipolar coupling between the dots is taken into account by introduction of an average magneto-dipolar field created by the dot located at the position $\boldsymbol{r}_k$ and acting on a dot located at the position $\boldsymbol{r}_j$:

$$\boldsymbol{B}_{jk} = -\mu_0 \hat{N}(\boldsymbol{r}_j - \boldsymbol{r}_k) \cdot \boldsymbol{M}_k. \tag{1}$$

Here, μ_0 is the vacuum magnetic permeability and $\hat{N}(\boldsymbol{r})$ is the mutual demagnetization tensor, which depends on the size and shape of the dots and on the interdot separation $\boldsymbol{r}$ [47]. In the case $\boldsymbol{r} = \boldsymbol{0}$, the tensor $\hat{N}(\boldsymbol{0})$ coincides with the usual self-demagnetization tensor of an individual dot. We note that in many particular cases the mutual demagnetization tensor can be calculated rather simply (see, e.g., Section 2.4 below and [47,48]).

Below we consider the case when the dot array is placed in a uniform external magnetic field $\boldsymbol{B}_e$. For simplicity, in this section we will not explicitly consider any crystallographic anisotropy of the dot material. This anisotropy can be taken into account by the renormalization of the self-demagnetization tensor. In particular, the uniaxial anisotropy leads to the following renormalization of the tensor:

$$\hat{N}(0) \to \hat{N}(0) + \hat{N}_{an}, \quad \hat{N}_{an} = -\frac{H_{an}}{M_s}\hat{T} \cdot (e_{z'} \otimes e_{z'}) \cdot \hat{T}^{-1}, \tag{2}$$

where H_{an} is the value of the anisotropy field, z' is the anisotropy axis, $\hat{T}$ is the rotation matrix from coordinate system $(x'y'z')$ to (xyz), and symbol $\otimes$ denotes direct product of Cartesian vectors.

Under the above formulated approximations the dynamics of the dots' magnetization vectors M_j is described by the Landau–Lifshitz equation:

$$\frac{dM_j}{dt} = \gamma(B_{\text{eff},j} \times M_j), \tag{3}$$

where $\gamma = 2\pi g \cdot 14$ GHz/T is the modulus of the gyromagnetic ratio and g is the Lande factor ($g \approx 2$ for most ferromagnetic materials). The effective magnetic field $B_{\text{eff},j}$ consisting of the external bias field and magneto-dipolar field created by other dots is given by:

$$B_{\text{eff},j} = B_e - \mu_0 \sum_k \hat{N}(r_j - r_k) \cdot M_k. \tag{4}$$

The total magnetic energy of the array under our assumptions has the form:

$$W = -V\sum_j B_e \cdot M_j + \frac{\mu_0 V}{2}\sum_{j,k} M_j \cdot \hat{N}(r_j - r_k) \cdot M_k, \tag{5}$$

where V is the dot volume.

2.2 General Formalism for Collective SW Excitations in a Dot Array

For small-amplitude (linear) SW excitations, the magnetization of the jth dot can be represented as:

$$M_j = M_s\left(\mu_j + m_j\right) + O(|m_j|^2), \tag{6}$$

where μ_j is a unit vector in the direction of the dot static magnetization and m_j is the small dimensionless deviation of the dot magnetization from the ground state described by the vector μ_j. The condition of conservation of the length of magnetization vector M_j in each dot requires the orthogonality of vectors μ_j and m_j: $\mu_j \cdot m_j = 0$. Vectors μ_j are determined from the stationary solutions of equations (3) that yield the following system of equations:

$$B_j \mu_j = B_e - \mu_0 M_s \sum_k \hat{N}(r_j - r_k) \cdot \mu_k, \tag{7}$$

where B_j is the modulus of the static internal magnetic field acting on the jth dot. By solving Eq. (7), one can find all the possible stationary magnetic configurations of an array in given conditions, but some of these configurations may be unstable (see stability conditions below).

Substituting Eq. (6) for M_j in Eq. (3), using Eq. (7), and keeping only the terms that are linear in m_j, one obtains the following dynamical equations for m_j:

$$\frac{dm_j}{dt} = \mu_j \times \sum_k \hat{\Omega}_{jk} \cdot m_k. \tag{8}$$

Here, the tensor operator $\hat{\Omega}_{jk}$ is defined by the equation:

$$\hat{\Omega}_{jk} = \gamma B_j \delta_{jk} \hat{I} + \gamma \mu_0 M_s \hat{N}(r_j - r_k), \tag{9}$$

where δ_{jk} is the Kronecker delta and $\hat{I}$ is the identity matrix. Recalling that the mutual demagnetization tensor is real, symmetric $(\hat{N} = \hat{N}^T)$ and invariant with respect to spatial inversion $(\hat{N}(r) = \hat{N}(-r))$, it becomes clear that the tensor operator $\hat{\Omega}_{jk}$ is also real, self-adjoint $\left(\hat{\Omega}_{jk} = \hat{\Omega}^T_{kj}\right)$, symmetric with respect to spatial inversion $(\hat{\Omega}_{jk} = \hat{\Omega}_{kj})$, and has the dimension of frequency. The SW frequencies ω_ν and the profiles of collective SW modes $m_{\nu,j}$ can be found as eigenvalues and eigenvectors of Eq. (8) respectively:

$$-i\omega_\nu m_{\nu,j} = \mu_j \times \sum_k \hat{\Omega}_{jk} \cdot m_{\nu,k}. \tag{10}$$

Here, ν is the index or the set of indices used to enumerate different SW modes.

The main properties of the collective SW excitations of an array of dots follow from the properties of the eigenvalue problem Eq. (10) and from the fact that the tensor $\hat{\Omega}_{jk}$ is real and self-adjoint. In particular, these properties result in the following orthogonality relation for the SW modes:

$$\sum_j m^*_{\nu',j} \cdot \mu_j \times m_{\nu,j} = -iA_\nu \delta_{\nu,\nu'}, \tag{11}$$

where the superscript * denotes complex conjugation and A_ν are real constants dependent on the normalization of the eigenvectors $m_{\nu,j}$.

Using the properties of the tensors $\hat{\Omega}_{jk}$ it can be shown that, if $m_{\nu,j}$ is an eigenvector with an eigenfrequency ω_ν, then $m^*_{\nu,j}$ is also an eigenvector with an eigenfrequency $-\omega_\nu$. Such double degeneracy of the SW eigenmodes arises from the fact that the magnetization vectors m_j are real quantities, and their spectra contain both positive and negative frequency components. Only half of the formal solutions of Eq. (10) describe "physical" modes $m_{\nu,j}$, while the rest of the solutions are the formal "conjugated" modes $m^*_{\nu,j}$ that guarantee that the vectors m_j are real valued. Below we call a mode "physical" if its norm is positive, $A_\nu > 0$. In this case, a mode with a positive (negative) frequency ω_ν

increases (decreases) the system energy—see Eq. (13) below. Naturally, in the expansion of the dot dynamical magnetization into a series of the SW eigenmodes, one should take into account *only* the "physical" modes:

$$\boldsymbol{m}_j(t)=\sum_{\nu}\left(\boldsymbol{m}_{\nu,j}c_{\nu}(t)+\text{c.c.}\right), \tag{12}$$

where $c_\nu(t)$ is the complex amplitude of the νth mode and c.c. denotes complex conjugated terms (which include all the "conjugated" modes).

Substituting Eqs. (6) and (12) for magnetization in the expression for energy Eq. (5) and using Eqs. (10) and (11), one can derive the following expression for the change in the magnetic energy of a dot array due to the excitation of SW modes:

$$\Delta W=\frac{M_s V}{\gamma}\sum_{\nu}\omega_\nu A_\nu |c_\nu|^2. \tag{13}$$

If the equilibrium magnetization configuration $\boldsymbol{\mu}_j$ corresponds to a local minimum of the magnetic energy, then ΔW is positive-definite, and the "physical" SW modes ($A_\nu>0$) have positive frequencies $\omega_\nu>0$. On the other hand, if we formally consider dynamics near a *maximum* of the magnetic energy, then $\Delta W<0$, and the frequencies of the "physical" SW modes will be negative, $\omega_\nu<0$, which corresponds to the precession of the magnetization vector in the opposite direction. Thus, Eq. (13) is a convenient tool for the investigation of stationary state stability—only the states having all ω_ν real and $\omega_\nu A_\nu>0$ are stable.

The above-presented general SW formalism can be used for efficient numerical calculations of the SW spectra in finite arrays of magnetic dots in an arbitrary ground state. It follows from Eq. (10) that the problem of calculation of the SW spectrum of a dot array can be reduced to a standard task of finding eigenvalues and eigenvectors of a Hamiltonian matrix. In the case when the ground state of an array is periodic (e.g., ferromagnetic or antiferromagnetic), the theory can be further simplified; this case is considered in Section 2.4.

2.3 Accounting for the Effects of Weak Perturbations

Magnetic damping and other weak perturbative effects, neglected in Eq. (3), can be effectively considered in the framework of a nonstationary perturbation theory [49]. In a general case, the perturbed equations for the dot magnetization $\mathbf{M}_j(t)$ can be written as:

$$\frac{d\mathbf{M}_j}{dt}=\gamma\left(\mathbf{B}_{\text{eff},j}\times\mathbf{M}_j\right)+\gamma\left(\boldsymbol{b}_j\times\mathbf{M}_j\right), \tag{14}$$

where $\boldsymbol{b}_j$ is the effective field of the perturbation which may depend both on time and the dot magnetization $\boldsymbol{M}_k$. Considering only linear processes, we can use expansion Eq. (12), and by substituting into Eq. (14) we obtain the perturbed dynamical equations for the SW amplitudes $c_\nu(t)$:

$$\frac{dc_\nu}{dt} = -i\omega_\nu c_\nu + i\gamma b_\nu - i\gamma \sum_{\nu'} \left(S_{\nu,\nu'} c_{\nu'} + \widetilde{S}_{\nu,\nu'} c_{\nu'}^* \right). \tag{15}$$

The coefficients describing the influence of the perturbation are equal to

$$b_\nu = \frac{1}{A_\nu} \sum_j \boldsymbol{m}_{\nu,j}^* \cdot \boldsymbol{b}_j, \tag{16a}$$

$$S_{\nu,\nu'} = \frac{1}{A_\nu} \sum_j \left(\boldsymbol{m}_{\nu,j}^* \cdot \boldsymbol{m}_{\nu',j} \right) \left(\boldsymbol{\mu}_j \cdot \boldsymbol{b}_j \right), \tag{16b}$$

$$\widetilde{S}_{\nu,\nu'} = \frac{1}{A_\nu} \sum_j \left(\boldsymbol{m}_{\nu,j}^* \cdot \boldsymbol{m}_{\nu',j}^* \right) \left(\boldsymbol{\mu}_j \cdot \boldsymbol{b}_j \right). \tag{16c}$$

Equation (15) allows one to analyze the influence of any type of small perturbations on the SWs in a dot array. In particular, if $\boldsymbol{b}_j$ is an external microwave field, the term $i\gamma b_j$ in Eq. (15) describes the linear excitation of SW modes, while the other two terms describe the parametrical processes under the action of parallel pumping. Note, that if the field $\boldsymbol{b}_j$ of the perturbation depends on the magnetization vectors, one should retain in Eq. (15) only the terms of the zeroth and first order in c_ν. For the investigation of higher-order processes one should take into account changes caused by the SW excitation in the dynamic magnetization component that is parallel to the static magnetization (i.e., the expansion Eq. (12) is not valid for the description of such processes).

As an example of such a perturbative analysis, we consider below two practically important cases: weak damping of collective SW modes and their direct excitation by an external microwave magnetic field. In the former case, the perturbation field that is created by the Gilbert damping has the form $\boldsymbol{b}_j = -(\alpha_G/\gamma M_s) d\boldsymbol{M}_j/dt$, where α_G is the Gilbert damping parameter. In the latter case, the perturbation field has the form $\boldsymbol{b}_j = \left(\boldsymbol{b}_{e,j} e^{-i\omega t} + \text{c.c.} \right)$. Calculating the coefficients in Eqs. (16a)–(16c) one can derive the following first-order perturbative equation for the SW amplitudes:

$$\frac{dc_\nu}{dt} = -i\omega_\nu c_\nu - \Gamma_\nu c_\nu + i\gamma b_{e,\nu} e^{-i\omega t}, \tag{17}$$

where the mode damping rate is equal to

$$\Gamma_\nu = \alpha_G \omega_\nu \left(\frac{1}{A_\nu} \sum_j \boldsymbol{m}_{\nu,j}^* \cdot \boldsymbol{m}_{\nu,j} \right), \tag{18}$$

and the excitation efficiency is defined by

$$b_{e,v} = \frac{1}{A_v} \sum_j m_{v,j}^* \cdot b_{e,j}. \tag{19}$$

As one can see from Eq. (18), the damping rate Γ_v is proportional to $\alpha_G \omega_v$ and the coefficient that describes the averaged precession ellipticity of the mode v. Only if the precession in all the dots has circular polarization do we get $\Gamma_v = \alpha_G \omega_v$. A similar relation between SW ellipticity and its damping rate is well known for continuous magnetic media [50].

Using the above expressions, one can easily calculate the microwave absorption spectra of an array of interacting magnetic dots. In particular, in a practically important case of a spatially uniform external microwave field, $\boldsymbol{b}_{e,j} = \boldsymbol{b}_e$, and the microwave power absorbed by the dot array is given by:

$$P = \frac{\omega V N_d}{\mu_0} \boldsymbol{b}_e^* \cdot \hat{\chi}''(\omega) \cdot \boldsymbol{b}_e, \tag{20}$$

where N_d is the number of dots in the array and the effective susceptibility tensor $\hat{\chi}(\omega) = \hat{\chi}'(\omega) + i\hat{\chi}''(\omega)$ of the array is given by:

$$\hat{\chi}(\omega) = \gamma \mu_0 M_s \sum_v \frac{\hat{\chi}_v}{(\omega_v - \omega) - i\Gamma_v}, \quad \hat{\chi}_v = \frac{1}{N_d A_v} \sum_{j,k} \boldsymbol{m}_{v,j} \otimes \boldsymbol{m}_{v,k}^*. \tag{21}$$

Thus, all the practically important characteristics of an array of magnetic dots can be easily found provided one knows the frequencies ω_v of the collective SW eigenmodes and the corresponding eigenmode profiles $\boldsymbol{m}_{v,j}$.

2.4 Collective SWs of a Dot Array in a Periodic Ground State

The periodic ground states represent an important class of the stationary magnetic configurations of arrays of magnetic dots arranged into a periodic lattice. In particular, these states include a ferromagnetic (FM) state, in which the directions of magnetization are the same in all the nanodots, as well as various antiferromagnetic (AFM) states, which usually are the true ground states of the dot arrays in the absence of an external magnetic field (i.e., they correspond to a global minimum of the total magnetic energy) [20,51]. If the period of a state is substantially shorter than the array dimensions, it is reasonable to consider such arrays as infinite. In this case, the above-presented formalism can be further simplified.

The periodicity of a ground state can be complex, i.e., an elementary cell of a magnetic lattice $\mathcal{L}$ (which describes the periodicity of a stationary magnetic configuration) includes $P > 1$ magnetic dots (so-called lattice with a basis). Note that $P = 1$ only for an array of dots which are arranged into a simple (oblique) lattice and stay in an FM state. In all other cases, the magnetic lattice is complex ($P>1$), and this complex periodicity can be induced by a complex structure of a nanodot lattice, as well as by a complex (nonferromagnetic) ground state.

In general, the periodicity of an array's ground state can be described by two basis vectors, $\tilde{\boldsymbol{a}}_1$, $\tilde{\boldsymbol{a}}_2$, which form a sublattice $\mathcal{SL}$ of an array's magnetic lattice $\mathcal{L}$:

$$\mathcal{SL}=\{n_1\tilde{\boldsymbol{a}}_1+n_2\tilde{\boldsymbol{a}}_2|n_1,n_2\in\mathbb{Z}\}. \tag{22}$$

Each nanodot at the position j_p belongs to a certain sublattice $p\in[1,P]$ of the array, and the position of the dot's center is defined as $\boldsymbol{r}_{j_p}=\boldsymbol{\delta}_p+n_1\tilde{\boldsymbol{a}}_1+n_2\tilde{\boldsymbol{a}}_2$. The relative position of different sublattices is fully described by a set of shift vectors $\boldsymbol{\delta}_{pq}=\boldsymbol{\delta}_p-\boldsymbol{\delta}_q$. The sum of all the P sublattices gives the positions of all the magnetic dots in the array (see the illustration in Fig. 1).

It is clear that the equilibrium directions of the dot magnetizations $\boldsymbol{\mu}_j$ and internal fields B_j depend only on the sublattice index p: $\boldsymbol{\mu}_{j_p}=\boldsymbol{\mu}_p$, $B_{j_p}=B_p$. Then, the general equilibrium condition Eq. (7) reduces to P vector equations,

$$B_p\boldsymbol{\mu}_p=\boldsymbol{B}_e-\mu_0M_s\sum_q\hat{\mathrm{G}}_0(\boldsymbol{\delta}_{pq})\cdot\boldsymbol{\mu}_q, \tag{23}$$

where

$$\hat{\mathrm{G}}_0(\boldsymbol{\delta})=\sum_{\boldsymbol{r}\in\mathcal{SL}}\hat{\mathrm{N}}(\boldsymbol{r}+\boldsymbol{\delta}). \tag{24}$$

The meaning of the subscript **0** will be explained below.

The linear SW excitations in a sublattice have the form of plane waves,

$$\boldsymbol{m}_{j_p}=\boldsymbol{m}_{\boldsymbol{k},p}e^{i\boldsymbol{k}\cdot\boldsymbol{r}_{jp}}, \tag{25}$$

where the wave vector $\boldsymbol{k}$ belongs to the first Brillouin zone of the reciprocal $\mathcal{S}$sublattice $\mathcal{SL}^*$,

$$\mathcal{SL}^*=\left\{n_1\tilde{\boldsymbol{k}}_1+n_2\tilde{\boldsymbol{k}}_2|n_1,n_2\in\mathbb{Z}\right\}, \tag{26}$$

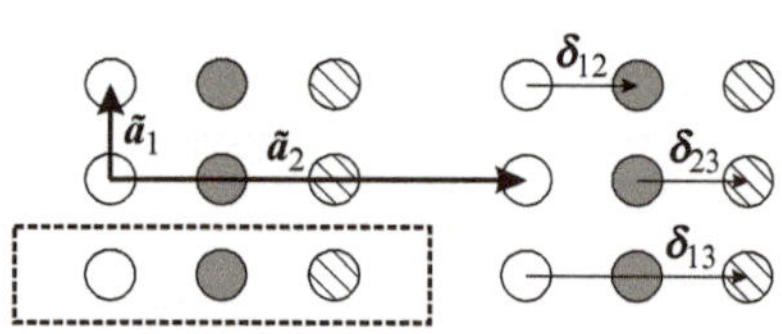

FIGURE 1

An illustration of a complex periodic lattice of a nanodot array with $P=3$ dots in a primitive cell. Circles with different compositions indicate nanodots which belong to different sublattices. Dashed line shows the boundary of a primitive cell of the lattice $\mathcal{L}$.

$$\widetilde{k}_1 = -\left(\frac{2\pi}{S_{SL}}\right) e_z \times \widetilde{a}_2, \quad \widetilde{k}_2 = \left(\frac{2\pi}{S_{SL}}\right) e_z \times \widetilde{a}_1. \tag{27}$$

Here, $S_{SL} = |e_z \cdot (\widetilde{a}_1 \times \widetilde{a}_2)|$ is the area of a unit cell of the sublattice $\mathcal{SL}$. Using the ansatz Eq. (25) in Eq. (10) one can obtain a finite-dimensional eigenvalue problem for ω_k and $m_{k,p}$:

$$-\mathrm{i}\omega_k m_{k,p} = \mu_p \times \sum_q \hat{\Omega}_{k,pq} \cdot m_{k,q}. \tag{28}$$

Here,

$$\hat{\Omega}_{k,pq} = \gamma B_p \delta_{pq} \hat{I} + \gamma \mu_0 M_s \hat{G}_k(\delta_{pq}), \tag{29}$$

where

$$\hat{G}_k(\delta) = \sum_{r \in \mathcal{SL}} \hat{N}(r+\delta) e^{-\mathrm{i}k \cdot (r+\delta)}. \tag{30}$$

As one can see, for $k=0$, the definition Eq. (30) coincides with Eq. (24).

The tensor $\hat{G}_k(\delta)$ is symmetric with respect to transposition of the vector indices $\left(\hat{G}_k^{\mathrm{T}} = \hat{G}_k\right)$, but in general is neither real nor symmetric with respect to inversion. Instead, one can prove the following general symmetry properties of this tensor:

$$\hat{G}_k(\delta) = \hat{G}^*_{-k}(\delta) = \hat{G}^*_k(-\delta) = \hat{G}_{-k}(-\delta). \tag{31}$$

In addition, $\hat{G}_{k+K}(\delta) = \hat{G}_k(\delta) e^{-\mathrm{i}K \cdot \delta}$ for any wave vector of the reciprocal sublattice $K \in \mathcal{SL}^*$.

For a fixed wave vector k, the eigenvalue problem Eq. (28) is a $2P$-dimensional system of linear equations. It has $2P$ solutions—pairs of eigenvectors $m_{k,p}$ and eigenfrequencies ω_k respectively. The real SW mode has the form $m_{j_p} = m_{k,p} \exp\left[\mathrm{i}\left(k \cdot r_{j_p} - \omega_k t\right)\right] + \text{c.c.}$ The vector $m^*_{k,p}$ is a solution of the problem that is conjugate to the problem, Eq. (28). However, as it follows from the properties of the operator $\hat{\Omega}_k$, the vector $m^*_{k,p}$ is also an eigenvector of the problem, but with the opposite direction of the wave vector k, since $\hat{\Omega}_k = \hat{\Omega}^*_{-k}$. As a result, the eigenvalue problem (28) simultaneously describes waves propagating in the direction k (solutions with $\omega_k > 0$) and in the opposite direction (solutions with $\omega_k < 0$). Of course, if we consider waves propagating in a given direction, only half of the solutions should be taken into consideration; i.e., in the general case, the SW spectrum for the array has P branches.

SWs in periodic dot arrays have properties that are similar to the properties of collective SW modes in finite dot arrays. In particular, different SW branches are orthogonal in the sense

$$\sum_p m^*_{k,\lambda,p} \cdot \boldsymbol{\mu}_p \times \boldsymbol{m}_{k,\lambda,p} = -iA_{k,\lambda}\delta_{\lambda,\lambda'}, \tag{32}$$

where the index λ enumerates SW branches. The condition of stability of a stationary state also remains valid: the state is stable only if all the frequencies $\omega_{k,\lambda}$ are real valued, and the inequality $\omega_{k,\lambda}A_{k,\lambda} > 0$ holds true for all the wave vectors in the first Brillouin zone. Also, the SW damping rate could be calculated using Eq. (18), while the absorption spectra can be evaluated using Eqs. (20) and (21), where the summation over the dots is replaced by the summation over the sublattices. Obviously, for the calculation of the absorption spectra one should take into account only the modes with a zero wave vector $\boldsymbol{k}=\boldsymbol{0}$.

Let us make some notes on the numerical calculation of the SW spectra. Using a well-known expression that relates the sums over the direct and reciprocal lattices [52], we can rewrite the tensor $\hat{G}_k(\boldsymbol{\delta})$ in the form:

$$\hat{G}_k(\boldsymbol{\delta}) = \frac{1}{S_{\mathcal{SL}}} \sum_{K \in \mathcal{SL}*} \hat{N}_{k+K} e^{iK\cdot\boldsymbol{\delta}}. \tag{33}$$

Here, $\hat{N}_k$ is a two-dimensional Fourier image of the mutual demagnetization tensor $\hat{N}(\boldsymbol{r})$, i.e., $\hat{N}(\boldsymbol{r}) = \int \hat{N}_k e^{ik\cdot r} d^2\boldsymbol{k}/(2\pi)^2$. For many particular cases of the dots' geometry, the tensor $\hat{N}_k$ can be obtained explicitly [19,47]. For instance, for identical flat dots (dots with constant height along the z axis) lying in the same x–y plane the tensor $\hat{N}_k$ takes the form:

$$\hat{N}_k = \frac{|\sigma_k|^2}{S} \begin{pmatrix} \frac{k_x^2}{k^2} f(kh) & \frac{k_x k_y}{k^2} f(kh) & 0 \\ \frac{k_x k_y}{k^2} f(kh) & \frac{k_y^2}{k^2} f(kh) & 0 \\ 0 & 0 & 1 - f(kh) \end{pmatrix}, \tag{34}$$

where the function $f(kh) = 1 - \left(1 - e^{-kh}\right)/(kh)$, S is the area of the dot, and σ_k is the Fourier image of the dot's shape in the x–y plane, $\sigma_k = \int_S e^{-ik\cdot r} d^2\boldsymbol{r}$, where the integration is done over the area S of the dot. In particular, for a circular dot of radius R we get:

$$S = \pi R^2, \quad \sigma_k = S\frac{2J_1(kR)}{kR}, \tag{35a}$$

where $J_1(x)$ is the Bessel function of the first order. For an elliptical dot having semi-axis a and b (along the x and y axes respectively), we get:

$$S = \pi ab, \quad \sigma_k = S\frac{2J_1\left(\sqrt{k_x^2 a^2 + k_y^2 b^2}\right)}{\sqrt{k_x^2 a^2 + k_y^2 b^2}}, \tag{35b}$$

and for a rectangular dot having sizes l_x and l_y we obtain:

$$S = l_x l_y, \quad \sigma_{\boldsymbol{k}} = S\frac{\sin(k_x l_x/2)}{k_x l_x/2}\frac{\sin(k_y l_y/2)}{k_y l_y/2}, \tag{35c}$$

Also, the tensor $\hat{N}_{\boldsymbol{k}}$ can be generalized to take into account a nonuniform profile of the magnetization inside a dot or different shapes of the interacting dots [43,47]. In conclusion, the expression Eq. (33) gives a numerically efficient way to calculate the SW spectra of a nanodot array in a periodic ground state. The calculation is no more complex than the calculation done in the "point-dipole" model but, in contrast with it, takes into account the real shape of nanodots in an array.

3 APPLICATIONS OF THE GENERAL THEORY

3.1 SW Spectra of a Dot Array with Rectangular and Hexagonal Lattice

As a first example of an application of the above-presented theory, we calculate the SW spectra of a dot array existing in a ferromagnetic (FM) state. The dots are assumed to be cylindrical nanoparticles having height h and radius R and aspect ratio $h/R > 1.81$, which means that the dots have out-of-plane shape anisotropy. We consider two simple lattice structures—square and hexagonal—with the lattice constant a (see Fig. 2a and d); both these lattices are simple, i.e., have one dot per a primitive cell.

For the FM state, the system Eq. (28) is reduced to a single vector equation (effectively two-dimensional), the eigenvalue of which can be easily found analytically:

$$\omega_{\boldsymbol{k}}^2 = \left(\gamma B + \omega_M F_{\boldsymbol{k}}^{(x'x')}\right)\left(\gamma B + \omega_M F_{\boldsymbol{k}}^{(y'y')}\right) - \left(\omega_M F_{\boldsymbol{k}}^{(x'y')}\right)^2, \tag{36}$$

where the tensor $\hat{F}_{\boldsymbol{k}} \equiv \hat{G}_{\boldsymbol{k}}(\boldsymbol{0})$, $\omega_M = \gamma\mu_0 M_s$, and the coordinate system $(x'y'z')$ is chosen so that the z'-axis is aligned to the direction of the dots' static magnetization. For the out-of-plane FM state considered below, the z'-axis is directed perpendicular to the array plane. Noting that the basis vectors of the inverse sublattice $\mathcal{SL}*$ are equal to $\boldsymbol{k}_1 = 2\pi\boldsymbol{e}_x/a$, $\boldsymbol{k}_2 = 2\pi\boldsymbol{e}_y/a$ for a square lattice and $\boldsymbol{k}_1 = -2\pi\boldsymbol{e}_x/a + 2\pi\boldsymbol{e}_y/\sqrt{3}a$, $\boldsymbol{k}_2 = 4\pi\boldsymbol{e}_y/\sqrt{3}a$ for a hexagonal lattice (see Eq. (27) and Fig. 2b and e), one can easily calculate the SW spectra using Eqs. (30) and (36).

The calculated SW spectra are shown in Fig. 2. As one can see, for large interdot distances the SW spectra for both geometries are monotonic in the half of the first Brillouin zone with a minimum in the Γ point ($\boldsymbol{k}=0$; see solid lines).

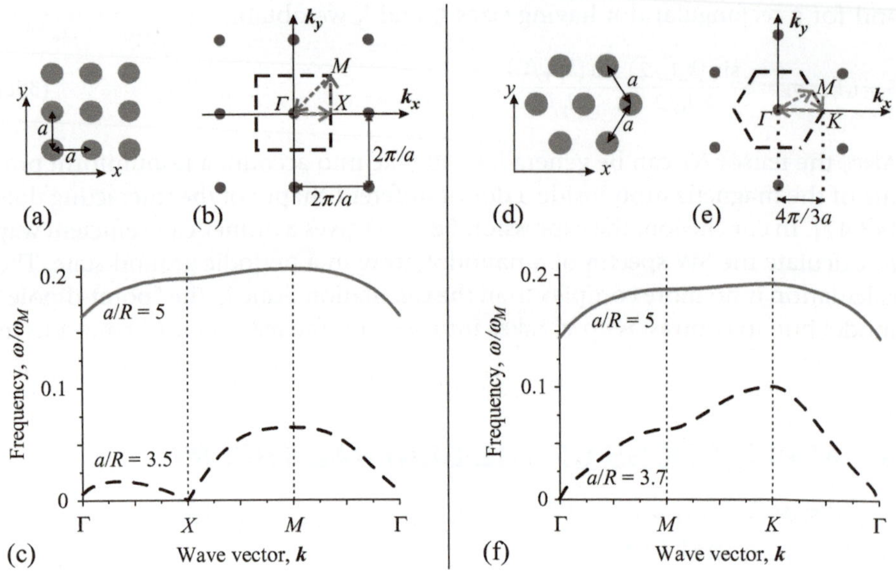

FIGURE 2
(a and d) Sketches of the square (a) and hexagonal (d) dot lattices. (b and e) Inverse sublattice (dots) and the boundary of the first Brillouin zone (dashed line) for the square and hexagonal lattice respectively. (c and f) Corresponding SW spectra of arrays in the out-of-plane FM state in zero external field. Definitions of the symmetric points of the first Brillouin zone and the plot path (dashed arrows) are shown in (b) and (e). Dots aspect ratio $h/R=5$. After [19,44].

However, with the decrease of the interdot separation the spectra start to differ qualitatively. For a hexagonal lattice the spectrum remains monotonic, and at some point ($a \approx 3.5R$) the SW with $\boldsymbol{k}=0$ becomes unstable. In contrast, for a square array the SW spectrum becomes nonmonotonic, and the minimum appears at the X-point ($\boldsymbol{k}=\pi \boldsymbol{e}_x/a$) (see dashed lines). The different location of the spectral minimum means, in particular, that SWs in two lattice types have different types of instability while approaching the boundary of the state stability (for example, by variation of perpendicular bias field). The instability at $\boldsymbol{k}=0$ corresponds to the division of an array into domains with a local FM state, while the instability at the X-point corresponds to the transition to a chessboard antiferromagnetic state. Note also that the FM state in a hexagonal lattice becomes unstable at a larger interdot separation, which is explained by a greater packing density in this type of lattice.

Therefore, the calculation of the SW spectrum gives information not only about eigenfrequencies of the SW, but also about the boundaries of stability of the array's ground state and the type of instability at this boundary.

3.2 Localized SW Modes Caused by a Point Defect in an Array of Magnetic Dots

In this section, we consider the influence of different kinds of defects on the properties of a dot array. The investigation of defect-related effects is of great practical importance, since real-life dot arrays are not perfect. In general, defects lead to such undesirable effects as broadening of the FMR line, random scattering of SWs, etc. [50]. However, defects can also cause the formation of defect bands in the prohibited SW zones [53] and changes in the SW mode profiles [54], which may possibly have their own applications, analogous to various applications of the defects in photonic crystals [55].

We should note that the defects in a nanodot array can be of variable nature. First, there may be common (material) defects caused by these technological imperfections, such as spread of geometrical and/or material parameters of the dots. Then, the multistability of the array's stationary state leads to a second kind of defects—defects of the magnetization order—that break the ideal magnetic periodicity in an array. Such defects, for example, can be formed by a single dot or a small group of dots having inverted direction of the static magnetization (with respect to the majority of dots in the array). Here, we consider the influence of a single "point" defect (a single defect dot in an array) of both kinds on the microwave properties of an array existing in an FM stationary state. Below we restrict our attention to the case of the FMR, since in the majority of standard microwave experiments and in practical applications the characteristic scale of the spatial nonuniformity of external microwave fields is usually much larger than the size of the dot array, and therefore the external microwave fields can be considered spatially uniform.

We consider an array of axially magnetized cylindrical nanodots arranged in a square lattice (see Fig. 3a). All the numerical calculations presented below were made for the following array geometry: dot aspect ratio $h/R=5$, lattice constant $a=4R$ (an array of interacting dots with different parameters will show qualitatively the same features). To avoid the influence of edge effects, we used periodic boundary conditions at the arrays edges, so formally a periodic state of an array with periodic defects was considered. However, if the separation between the defects is sufficiently large, the interaction between them can be considered negligible.

First, let us consider a point defect of the magnetization order. For an FM ground state, this is a single dot with inverted direction of magnetization (see Fig. 3a). The states with such defects are stable in a wide range of bias magnetic fields and dot geometric parameters. In particular, in a zero bias field such "defect" states are more stable than the ideal FM state, since they correspond to a smaller value of the array's total magnetic moment.

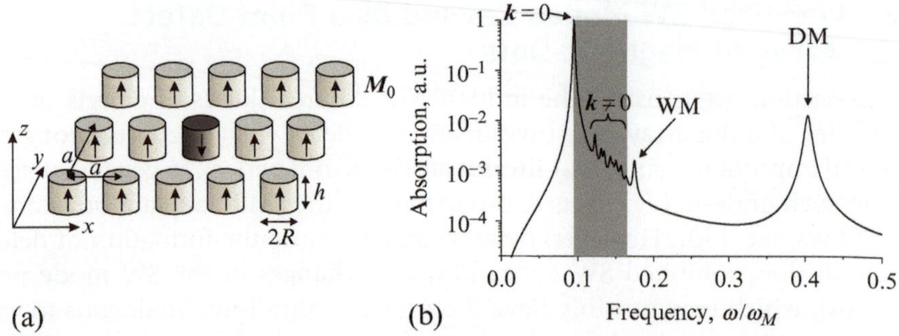

FIGURE 3

(a) Sketch of a dot array containing one point defect of the magnetic order—a dot with inverted magnetization. (b) FMR absorption spectrum in a zero external field of a dot array having one defect per 10×10 dots. The gray area indicates the spectral range of the "volume" SW modes in the array. The polarization of the microwave field is linear. After [45].

It is clear from Fig. 2c that the FMR absorption spectrum of an array in the ideal FM state contains only one absorption peak (since there is only one SW branch). As one can see from Fig. 3b, the presence of a defect leads to a substantial complication of the FMR spectrum. The spectrum of the "volume" SW modes (VMs) in the array is only weakly affected: the mode's frequencies simply slightly increase due to the increase of a static demagnetization field of a single dot forming a point defect. However, the presence of a defect creates a certain spatial nonuniformity in the profiles of the VMs (see Fig. 4a). Due to this nonuniformity, VMs having nonzero wave vectors are excited by a uniform microwave field, resulting in the appearance of many small peaks in the VM range.

Another significant absorption peak in the FMR spectrum of the array, which is seen at a higher frequency, is related to the excitation of a "defect eigenmode" (DM). This mode is mainly localized on the defect dot (see the profile in

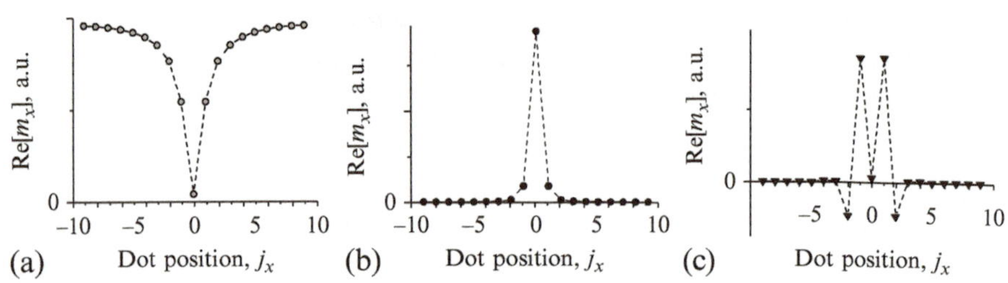

FIGURE 4

Profiles $m_x(j)$ of the SW modes of a dot array containing a single point defect of the magnetic order: (a) the most intensive volume mode, scattered on the defect; (b) defect eigenmode (DM); (c) the most intensive "well mode" (WM). The defect dot is located at the position $j_x=0$. After [19,45].

Fig. 4b), and has a polarization determined by the static magnetization and shape of the defect (left circular polarized for our geometry, while the VMs are mainly right circular polarized). Due to the different polarizations of the DM and the VMs, one can neglect dynamical interactions between them, and calculate the DM frequency as a frequency of the eigenmode of an isolated defect dot placed in a given static field, created by all the other dots. Under this approximation the DM frequency is expressed as

$$\omega_{\mathrm{DM}} = -\gamma B_e + \omega_{\mathrm{M}}\left(F_0^{zz} + N_s^{xx} - 2N_s^{zz}\right), \tag{37}$$

where $\hat{N}_s$ is the self-demagnetization tensor of the defect dot. Obviously, in a zero external field, the DM has a larger frequency due to the larger static internal field at the defect. At the fields $B_e > B_c = \mu_0 M_s\left(F_0^{zz} + N_s^{xx} - 2N_s^{zz}\right)$, the frequency of the DM becomes negative, corresponding to the instability of the stationary state with a point defect. For such bias fields only the ideal FM stationary state remains stable.

However, as one can see from Fig. 3b, there are also other absorption peaks outside the volume SW spectrum of the array. The SW modes corresponding to these peaks are also localized in the vicinity of the defect but, in stark contrast with the DM, these modes have vanishingly small amplitudes at the defect dot itself (see, e.g., Fig. 4c). The physical reason for the appearance of these additional localized modes is the local inhomogeneity of the dipolar field (in our case—the local maximum of this field) caused by the inverted magnetization of the defect dot. Such modes localized in a potential well are usually called "well modes" (WMs) [56]. The small amplitudes of the WMs at the location of the defect dot are caused by the different polarizations of the magnetization oscillations in the defect dot and in all the other dots. The structure of the WMs is very sensitive to the strength of the magneto-dipolar interaction between the dots. Besides the WM shown in Fig. 4c, there are also antisymmetric WMs with similar profiles and also higher WMs with more complicated spatial profiles. The total number of WMs increases with the increase of the depth of the defect-related potential well (which increases with the increase of the zz component of the mutual demagnetization tensor of dots $\hat{N}(\boldsymbol{r})$), but decreases when the dynamic dipolar interaction (xx, xy, and yy components of $\hat{N}(\boldsymbol{r})$) becomes stronger. Therefore, by studying the WM spectra one can gain valuable information about the dipolar coupling between the dots in a dot array. Note also that the spectral position of WMs relative to the VMs remains the same at any external bias field: the WMs are always situated in the vicinity of the upper edge of the volume SW spectrum. This is natural, since both WMs and VMs have vanishingly small amplitudes at the location of a defect dot, in which the internal magnetic field depends differently on the externally applied bias magnetic field B_e, compared to all the other dots. Therefore, in a conventional FMR experiment performed on dot arrays with rare point defects of the magnetic order, all the spectral range of the VMs becomes visible: the lower edge is identified by the uniform VM, while the upper edge is marked by the spectral position of the WMs.

Now, we shall consider a point defect caused by the variation of the dots' material parameters. As a simple example of such a "material" defect, we shall assume that one of the dots in a dot array existing in the FM ground state has a value of the perpendicular uniaxial anisotropy (in the *z*-direction) that is different from the value of the same parameter in all the other dots. According to Eq. (2), this additional anisotropy of a single defect dot leads to the renormalization of the *zz* component of the self-demagnetization tensor of this dot, $N_s^{zz} \rightarrow N_s^{zz} - H_{an}/M_s$, which results in the increase of the static internal field in the defect dot by the value of the additional anisotropy field H_{an}. All the other characteristics of the FM dot array are not affected.

The FMR absorption spectrum of an array with an "anisotropic" point defect is shown in Fig. 5a. As one can see, there are only two absorption peaks, associated with the uniform VM and the DM respectively. The DM is strongly localized; its profile is similar to the one shown in Fig. 4b. Since the "anisotropic" defect does not produce an inhomogeneity of the static internal magnetic field, the WMs are absent. The absence of the field "well" together with the identical polarizations of the oscillations in the "defect" dot and in all the other dots leads to a very weak scattering of the VMs on the defect. Therefore, the nonuniform VMs have a small excitation efficiency, and become practically invisible in the FMR spectrum.

It is interesting to consider how the magnitude of the "material" defect affects the FMR line of an array. For a small value of the defect anisotropy, the DM is situated close to the volume SW spectrum, and hybridizes with the closest VM,

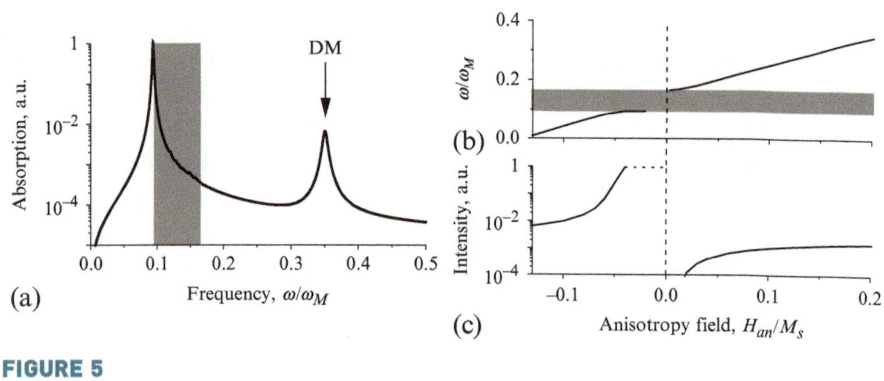

FIGURE 5

(a) FMR absorption spectrum of an array having one "material" (anisotropy) defect per 10×10 dots ($H_{an}/M_s = 0.2$). The gray area indicates the range of the VMs. (b and c) Frequency and excitation efficiency of the DM respectively as functions of the magnitude of the defect anisotropy.

which is not orthogonal to the DM. This hybridization leads to the delocalization of the DM, and to the disappearance of the DM in the volume SW spectrum. Also, the intensity of the DM tends to become similar to the intensity of the VM, with which the DM hybridizes. For $H_{an} > 0$, the DM intensity tends to zero (hybridization with a nonuniform VM), while its frequency differs significantly from ω_{FMR}. For $H_{an} < 0$ (hybridization with the uniform VM), both the intensity and frequency of the DM tend to become similar to the intensity and frequency of the main absorption peak (see Fig. 5b and c). Therefore, low-contrast "material" defects lead only to a relatively weak broadening of the FMR line, in stark contrast to the defects of the magnetization order, which may lead to a significant broadening of the FMR line due to the existence of the WMs.

3.3 Conditions of the Frequency Nonreciprocity of SWs in a Magnetic Dot Array

In this section, we consider the possibility of the nonreciprocity of the SW dispersion relation in a magnetic dot array. The wave nonreciprocity is a widely used phenomenon in signal transmission and information processing techniques. However, the majority of nonreciprocal devices are practically incompatible with conventional microelectronics due to their large sizes. The MCs based on arrays of interacting magnetic nanodots have the potential to become promising artificial magnetic materials for the solution of this problem.

All the properties of the collective SWs in a nanodot array are determined by the tensor $\hat{\boldsymbol{\Omega}}_{\boldsymbol{k}}$ and the general rules of magnetization dynamics, resulting in the structure of the eigenvalue problem in the form of Eq. (28). Thus, analyzing Eqs. (28) and (29), one can find the conditions for the nonreciprocity of the collective SWs in an array without actually solving the eigenvalue problem itself. As was pointed out earlier, the eigenvalue problem Eq. (28) describes SWs propagating in both "positive" and "negative" (respective to $\boldsymbol{k}$) directions (solutions $\omega_{\boldsymbol{k}} > 0$ and $\omega_{\boldsymbol{k}} < 0$ respectively). Therefore, there are two ways to demonstrate the nonreciprocity of the collective SWs described by the eigenvalue problem Eq. (28): (i) either to show that the positive eigenvalues of the problem Eq. (28) and of its complex conjugate are not equivalent, or (ii) to show that the absolute values of the positive and negative eigenvalues of the problem Eq. (28) are different.

Let us start a test for the nonreciprocity of SWs in a dipolarly coupled nanodot array from the analysis of the general symmetric properties of the characteristic matrix $\left(\boldsymbol{\mu}_p \times \hat{\boldsymbol{\Omega}}_{\boldsymbol{k},pq}\right)$ of the eigenvalue problem Eq. (28). The tensors $\hat{\boldsymbol{\Omega}}_{\boldsymbol{k},pq}$ consist of the static–static (describing the influence of the static magnetization of the qth sublattice on the static magnetization of the pth sublattice), static–dynamic and dynamic–static, and dynamic–dynamic components. In the characteristic matrix $\left(\boldsymbol{\mu}_p \times \hat{\boldsymbol{\Omega}}_{\boldsymbol{k},pq}\right)$, which is effectively $2P \times 2P$-

dimensional, only the dynamic–dynamic components of the tensors $\hat{\boldsymbol{\Omega}}_{k,pq}$ are directly present. If all the off-diagonal dynamic–dynamic components of the tensors $\hat{\boldsymbol{\Omega}}_{k,pq}$ vanish, the characteristic matrix contains $2P^2$ components that are identically equal to zero: $\left(\boldsymbol{\mu}_p \times \hat{\boldsymbol{\Omega}}_{k,pq}\right)_{j,j+2n} = 0,\ n \in \mathbb{Z}$. As one can prove using the definition of a matrix determinant [57], in this case the characteristic equation of the eigenvalue problem Eq. (28) is bipolynomial with respect to the SW eigenfrequency ω_k. Therefore, the solution of this characteristic equation consists of P pairs of eigenvalues $\omega_{k,\nu}$ and $(-\omega_{k,\nu})$, which means that the collective SWs in the array are reciprocal.

Such a case is realized in an array of flat dots with the height being the same for all the dots, if the directions of static magnetization of the dots lie in the array's plane (x–y plane). Indeed, the dynamical components of the tensor $\hat{\boldsymbol{\Omega}}_{k,pq}$ are proportional to the dynamical components of the mutual demagnetization tensor $\hat{N}(\boldsymbol{r})$ (see Eqs. (29) and (30)), the xz and yz components of which are identically zero (see Eq. (34)). Thus, for an in-plane magnetized dot the cross-interaction between the dynamical m_z component of the dot magnetization and all the in-plane magnetization components vanishes identically. This means that in any array of flat in-plane magnetized dots the off-diagonal dynamic–dynamic components of the Hamiltonian matrix $\hat{\boldsymbol{\Omega}}_{k,pq}$ also identically vanish, independently of the lattice structure and the static state of the array. Therefore, the collective SWs in an array of in-plane magnetized flat magnetic dots are always reciprocal and, if a *nonreciprocity* is needed, there should exist a nonzero *out-of-plane component of the dot magnetization* $\mu_{p,z} \neq 0$. This can be achieved, for example, by using vertically elongated magnetic pillars.

Another necessary condition of the SW nonreciprocity follows from the general symmetry property $\hat{\boldsymbol{\Omega}}_{k,pq} = \hat{\boldsymbol{\Omega}}^*_{-k,pq}$ (see Eq. (31)). If the tensor $\hat{\boldsymbol{\Omega}}_{k,pq}$ is real, the eigenvalue problem Eq. (28) does not change with the reversal of the SW wave vector $\boldsymbol{k}$, meaning that the collective SWs are reciprocal. Whether the tensor $\hat{G}_k(\boldsymbol{\delta}_{pq})$ and, consequently, $\hat{\boldsymbol{\Omega}}_{k,pq}$ are real or complex depends only on the structure of the array's lattice (see Eq. (30)). In particular, $\hat{\boldsymbol{\Omega}}_k$ is real if all the dots in an array are almost identical (in shape, material parameters and static magnetization) *and* the dots are arranged in a simple lattice (having one dot per primitive cell), because there is only one sublattice with $\boldsymbol{\delta}_{pp} \equiv 0$ and, as follows from Eq. (33), the tensor $\hat{G}_k(0)$ is real. At the same time, the tensors $\hat{\boldsymbol{\Omega}}_{k,pq}$ for $p \neq q$ can be complex. Therefore, the collective SWs in a dipolarly coupled dot array can be *nonreciprocal* only if the *primitive cell of the array contains several magnetic dots*.

The simplest example of an array existing in such a complex spatially periodic ground state is an array having $P = 2$ dots in a primitive cell. This case can be easily analyzed analytically. If an array exists in an FM ground state, one can easily derive an explicit characteristic equation of the eigenvalue problem Eq. (28): $\omega_k^4 + C_2\omega_k^2 + C_1\omega_k + C_0 = 0$. It is evident that the nonreciprocity appears only when the coefficient $C_1 \neq 0$. In all the other cases, the characteristic equation has two pairs of solutions $[\omega_{k,\nu'}, -\omega_{k,\nu}]$, which, as has been explained above, means that collective SWs are reciprocal. The magnitude of the coefficient C_1 can be evaluated as:

$$C_1 = \left(\Omega_{11}^{(x'x')} - \Omega_{22}^{(x'x')}\right)\left(\Omega_{12}^{(x'y')} - \Omega_{12}^{(y'y')} - \text{c.c.}\right) + \ldots, \tag{38}$$

where the $x'x'$ term is shown explicitly and the two other terms can be obtained by a cyclic permutation of the indices $(x'x') \to (x'y') \to (y'y')$. The direction of the z'-axis is chosen to be parallel to the direction of the elements' static magnetization, $\boldsymbol{\mu}_p = \boldsymbol{e}_{z'}$. As is evident from the equation above, the necessary condition for the SW nonreciprocity in the considered case is the inequality $\hat{\boldsymbol{\Omega}}_{k,11} \neq \hat{\boldsymbol{\Omega}}_{k,22}$. This can be achieved if the dots, which belong to different sublattices, differ from one another by their geometric or/and material parameters, i.e., if there are two groups of different dots in the array.

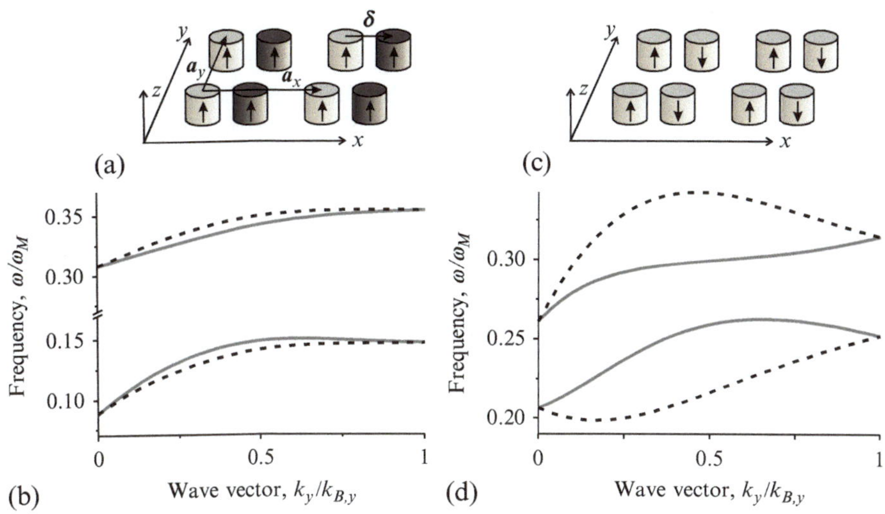

FIGURE 6

(a) Sketch of an array of two different types of nanodots existing in a perpendicular FM ground state. Dark gray dots have an additional uniaxial anisotropy $H_{an} = 0.2M_s$ in the z-direction. (b) Corresponding SW spectra at zero external field. Solid lines correspond to the "positive" propagation direction, dashed red lines to the opposite (negative) propagation direction. (c and d) Sketch of an array of identical nanodots in the antiferromagnetic ground state and corresponding spectra of SWs respectively. Parameters of calculations: dot aspect ratio $h/R = 5$, lattice constants $a_x = 10R$, $a_y = 3.3R$, shift between the sublattices $\delta = 3.3R$. After [46].

To illustrate the case of an array with two sublattices, we calculated the SW spectra in an array of circular nanodots arranged in a rectangular lattice (see Fig. 6a), assuming that one group of dots has additional out-of-plane anisotropy H_{an} (a qualitatively similar result can be obtained for an array in the FM state, when the dots differ by their radius). As expected, the SWs in such an array are nonreciprocal: for all the directions of the SW wave vector $\boldsymbol{k}$, except $\boldsymbol{k}=k_x\boldsymbol{e}_x$ (for this particular geometry) the waves with opposite wave vectors have different eigenfrequencies (Fig. 6b). Also, the SW eigenfrequency does not depend on the sign of the wave vector in all the symmetric points of the first Brillouin zone (e.g., $\boldsymbol{k}=0$, $\boldsymbol{k}=\pi\boldsymbol{e}_y/a_y$). This property is general and is related to the periodicity of the SW dispersion relation: $\omega_{\boldsymbol{k}}=\omega_{\boldsymbol{k}+\boldsymbol{K}}$ for all the vectors $\boldsymbol{K}$ of the reciprocal sublattice $\mathcal{SL}^*$. While the difference in the SW frequencies (nonreciprocal frequency splitting) is relatively small, the difference in group velocities of the oppositely directed SWs can be substantially larger. For the considered parameters, the maximum difference in group velocity is about 30%. This difference can be easily detected experimentally and potentially, used in applications.

Calculating the profiles of the SW modes we found that they change when the propagation direction is reversed. In particular, the ratio of the mode amplitude in different sublattices $|\boldsymbol{m}_{\boldsymbol{k},1}|/|\boldsymbol{m}_{\boldsymbol{k},2}|$ for both SW branches is greater for one propagation direction than for the opposite. Note that such profile asymmetry is present even if the dots in the array are identical, but naturally in this case it does not lead to frequency nonreciprocity. Similar features are well known for the surface magnetostatic waves in ferromagnetic films, the dispersion relation of which is also nonreciprocal if the boundary conditions at the opposite film surfaces are different [50,58].

The situation becomes more interesting when the ground state of an array is not ferromagnetic. Analysis of the eigenvalue problem Eq. (28) shows that in this case the condition $\hat{\boldsymbol{\Omega}}_{\boldsymbol{k},11} \neq \hat{\boldsymbol{\Omega}}_{\boldsymbol{k},22}$ is no longer necessary for the appearance of SW nonreciprocity. Thus, the nonreciprocal branches of the SWs can appear in an array of *nanopillars identical by shape and material*. For example, Fig. 6d shows the SW spectra of an array of identical dots existing in a stripe-like antiferromagnetic (SAFM) ground state. Moreover, in the case of a more complex dot lattice, with at least $P=3$ dots per primitive cell, the condition $\hat{\boldsymbol{\Omega}}_{\boldsymbol{k},pp} \neq \hat{\boldsymbol{\Omega}}_{\boldsymbol{k},qq}$ is not necessary even for an FM ground state of an array: the collective SWs in such dot arrays can be nonreciprocal even in an array consisting of *identical* nanodots in an *FM ground state*.

Comparing Fig. 6b and d, one can see that the nonreciprocal frequency splitting in the SAFM state is significantly larger than in the FM state, while the geometrical parameters are the same. Calculations performed using a well-known stationary perturbation theory [49] show that the nonreciprocal frequency splitting $(\omega_{\boldsymbol{k}}-\omega_{-\boldsymbol{k}})$ is proportional to the difference in polarization ellipticity of the SW eigenmodes $\boldsymbol{m}_p$ corresponding to two different sublattices. Thus, the nonreciprocal frequency splitting has a maximum magnitude when the eigenmodes of different sublattices are orthogonal, which is realized in an AFM state of an array of identical dots. A similar enhancement of the nonreciprocal frequency splitting can be achieved in the FM ground state by using two different types of dots having in-plane shape or material anisotropy with different directions of anisotropy axes.

Finally, it should be noted that the arrays shown in Fig. 6a and c have a *complex lattice structure*, i.e., if one neglects the difference in geometric and material parameters of the dots and considers only the positions of the dot's centers, the resulting geometric lattice will still be nonsimple—with two dots per primitive cell. The lattice becomes simple only when $\delta = a_x/2$. Using Eq. (33) one can show that in this case the tensor $\hat{G}_k(\boldsymbol{\delta})$ becomes real valued. Therefore, the collective SWs in any two-dimensional array with $P=2$ types of different dots arranged in a *simple* lattice are *always reciprocal*, since the coefficients of the inter-sublattice interaction in such an array are real. However, this property is valid only for the case $P=2$. For more complex ground states, having at least three different (in geometry, material parameters, or static magnetization) groups of dots, the collective SWs can become nonreciprocal in arrays with a simple structure of a *dot lattice*.

4 CONCLUSIONS

In this chapter, the collective SW excitations in an array of dipolarly coupled magnetic nanodots are considered. We present a general formalism which allows one to calculate the frequency and the spatial structure of the collective SW modes both in a finite and an infinite periodic nanodot array. Knowing the eigenfrequencies and the mode's structure, one can easily calculate the mode's damping rate, its excitation efficiency by an external microwave field and other practically important characteristics, and also determine if the considered static magnetic configuration of an array is stable or not. In the above-presented calculations, we restricted our attention to the macrospin approach, which is valid for the collective SW modes formed by spatially uniform modes of individual dots existing in a spatially uniform ground state. However, both these restrictions can be relaxed by introducing several effective demagnetization tensors $\hat{N}(\boldsymbol{r})$, which take into account the nonuniform profiles of the static and dynamic magnetization inside a dot.

As an example of application of the presented theory, we have calculated the SW spectra of a dot array with square and hexagonal lattices existing in the out-of-plane FM state. It is shown that a change in the lattice symmetry may lead to significant changes in the array's properties, in particular to the different types of SW instability at the stability boundary of the FM ground state.

We have also demonstrated that a single point defect in a magnetic dot array may lead to the appearance of several localized modes—one DM, which is always present, and several "well" modes, which appear if the "defect" dot creates a sufficiently large spatial inhomogeneity of the internal static magnetic field in the array. The number and structure of all the localized modes strongly depend on the magnetostatic interaction between the dots. Therefore, the investigation of these "defect" modes can provide valuable information about the interdot magneto-dipolar interaction. Also, in the presence of a point defect of the magnetization order all the spectral range of the "volume" SWs becomes visible in the FMR absorption spectrum of the array.

Finally, it has been demonstrated that the collective SWs in an array of dipolarly coupled magnetic nanodots can be nonreciprocal. The main requirements for SW nonreciprocity are: (i) the presence of a nonzero out-of-plane component of the dots' static magnetization and (ii) the existence of a nonzero imaginary part of the tensors $\hat{\Omega}_{k,pq}$. The last condition is satisfied in arrays existing in a complex periodic ground state with at least $P=2$ dots per primitive cell if the dots belonging to different arrays' sublattices are different, or at least $P=3$ dots per primitive cell if all the dots in the array are identical.

References

[1] Prinz GA. Magnetoelectronics. Science 1998;282:1660.

[2] Sellmyer DJ, Skomski R, editors. Advanced magnetic nanostructures. New York, NY: Springer; 2006.

[3] Stohr J, Siegmann HC. Magnetism. From fundamentals to nanoscale dynamics. Berlin: Springer; 2006.

[4] Jorzick J, Demokritov SO, Hillebrands B, Bartenlian B, Chappert C, Decanini D, et al. Spin-wave quantization and dynamic coupling in micron-size circular magnetic dots. Appl Phys Lett 1999;75:3859. http://dx.doi.org/10.1063/1.125480.

[5] Vavassori P, Donzelli O, Metlushko V, Grimsditch M, Ilic B, Neuzil P, et al. Magnetic switching in submicron-scale periodic magnetic arrays. J Appl Phys 2000;88:999. http://dx.doi.org/10.1063/1.373768.

[6] Novosad V, Guslienko KY, Shima H, Otani Y, Kim SG, Fukamichi K, et al. Effect of interdot magnetostatic interaction on magnetization reversal in circular dot arrays. Phys Rev B 2002;65:060402. http://dx.doi.org/10.1103/PhysRevB.65.060402.

[7] Malkinski L, Camley RE, Celinski Z, Winningham TA, Whipple SG, Douglas K. Hexagonal lattice of 10-nm magnetic dots. J Appl Phys 2003;93:7325. http://dx.doi.org/10.1063/1.1543861.

[8] Neusser S, Grundler D. Magnonics: spin waves on the nanoscale. Adv Mater 2009;21:2927. http://dx.doi.org/10.1002/adma.200900809.

[9] Chumak AV, Serga AA, Hillebrands B, Kostylev MP. Scattering of backward spin waves in a one-dimensional magnonic crystal. Appl Phys Lett 2008;93:022508. http://dx.doi.org/10.1063/1.2963027.

[10] Chumak AV, Serga AA, Wolff S, Hillebrands B, Kostylev MP. Scattering of surface and volume spin waves in a magnonic crystal. Appl Phys Lett 2009;94:172511. http://dx.doi.org/10.1063/1.3127227.

[11] Neusser S, Botters B, Grundler D. Localization, confinement, and field-controlled propagation of spin waves in $Ni_{80}Fe_{20}$ antidot lattices. Phys Rev B 2008;78:054406. http://dx.doi.org/10.1103/PhysRevB.78.054406.

[12] Kruglyak VV, Keatley PS, Neudert A, Hicken RJ, Childress JR, Katine JA. Imaging collective magnonic modes in 2d arrays of magnetic nanoelements. Phys Rev Lett 2010;104:027201. http://dx.doi.org/10.1103/PhysRevLett.104.027201.

[13] Chumak AV, Tiberkevich VS, Karenowska AD, Serga AA, Gregg JF, Slavin AN, et al. All-linear time reversal by a dynamic artificial crystal. Nat Commun 2010;1:141. http://dx.doi.org/10.1038/ncomms1142.

[14] Jorzick J, Krämer C, Demokritov SO, Hillebrands B, Bartenlian B, Chappert C, et al. Spin wave quantization in laterally confined magnetic structures (invited). J Appl Phys 2001;89:7091. http://dx.doi.org/10.1063/1.1357153.

[15] Mathieu C, Hartmann C, Bauer M, Buettner O, Riedling S, Roos B, et al. Anisotropic magnetic coupling of permalloy micron dots forming a square lattice. Appl Phys Lett 1997;70:2912. http://dx.doi.org/10.1063/1.119051.

[16] Metlov KL, Guslienko KY. Stability of magnetic vortex in soft magnetic nano-sized circular cylinder. J Magn Magn Mater 2002;242–245:1015–7. http://dx.doi.org/10.1016/S0304-8853(01)01360-9.

[17] Metlov KL, Lee YP. Map of metastable states for thin circular magnetic nanocylinders. Appl Phys Lett 2008;92:112506. http://dx.doi.org/10.1063/1.2898888.

[18] Bondarenko PV, Galkin AY, Ivanov BA, Zaspel CE. Collective modes for an array of magnetic dots with perpendicular magnetization. Phys Rev B 2010;81:224415. http://dx.doi.org/10.1103/PhysRevB.81.224415.

[19] Verba R, Melkov G, Tiberkevich V, Slavin A. Collective spin-wave excitations in a two-dimensional array of coupled magnetic nanodots. Phys Rev B 2012;85:014427. http://dx.doi.org/10.1103/PhysRevB.85.014427.

[20] Bishop JEL, Galkin AY, Ivanov BA. Ground states of an array of magnetic dots with Ising-type anisotropy and subject to a normal magnetic field. Phys Rev B 2002;65:174403. http://dx.doi.org/10.1103/PhysRevB.65.174403.

[21] Galkin AY, Ivanov BA. Analogue of a spin flop phase transition for an array of magnetic moments with dipole interaction. JETP Lett 2006;83:383. http://dx.doi.org/10.1134/S0021364006090049.

[22] Verba R, Melkov G, Tiberkevich V, Slavin A. Fast switching of a ground state of a reconfigurable array of magnetic nano-dots. Appl Phys Lett 2012;100:192412. http://dx.doi.org/10.1063/1.4714772.

[23] Verba R, Tiberkevich V, Guslienko K, Melkov G, Slavin A. Theory of ground-state switching in an array of magnetic nanodots by application of a short external magnetic field pulse. Phys Rev B 2013;87:134419. http://dx.doi.org/10.1103/PhysRevB.87.134419.

[24] Topp J, Heitmann D, Kostylev MP, Grundler D. Making a reconfigurable artificial crystal by ordering bistable magnetic nanowires. Phys Rev Lett 2010;104:207205. http://dx.doi.org/10.1103/PhysRevLett.104.207205.

[25] Tacchi S, Madami M, Gubbiotti G, Carlotti G, Goolaup S, Adeyeye AO, et al. Analysis of collective spin-wave modes at different points within the hysteresis loop of a one-dimensional magnonic crystal comprising alternative-width nanostripes. Phys Rev B 2010;82:184408. http://dx.doi.org/10.1103/PhysRevB.82.184408.

[26] Klein O, de Loubens G, Naletov VV, Boust F, Guillet T, Hurdequint H, et al. Ferromagnetic resonance force spectroscopy of individual submicron-size samples. Phys Rev B 2008;78:144410. http://dx.doi.org/10.1103/PhysRevB.78.144410.

[27] Gubbiotti G, Carlotti G, Okuno T, Shinjo T, Nizzoli F, Zivieri R. Brillouin light scattering investigation of dynamic spin modes confined in cylindrical permalloy dots. Phys Rev B 2003;68:184409. http://dx.doi.org/10.1103/PhysRevB.68.184409.

[28] Park JP, Eames P, Engebretson DM, Berezovsky J, Crowell PA. Imaging of spin dynamics in closure domain and vortex structures. Phys Rev B 2003;67:020403. http://dx.doi.org/10.1103/PhysRevB.67.020403.

[29] Francardi M, Sepioni M, Gerardino A, Sansone F, Gubbiotti G, Madami M, et al. Magnetic dot clusters for application in magneto-electronics. Microelectron Eng 2010;87:1614. http://dx.doi.org/10.1016/j.mee.2009.10.039.

[30] Zivieri R, Montoncello F, Giovannini L, Nizzoli F, Tacchi S, Madami M, et al. Collective spin modes in chains of dipolarly interacting rectangular magnetic dots. Phys Rev B 2011;83:054431. http://dx.doi.org/10.1103/PhysRevB.83.054431.

[31] Gubbiotti G, Madami M, Tacchi S, Carlotti G, Okuno T. Normal mode splitting in interacting arrays of cylindrical permalloy dots. J Appl Phys 2006;99:08C701.

[32] Giovannini L, Montoncello F, Nizzoli F. Effect of interdot coupling on spin-wave modes in nanoparticle arrays. Phys Rev B 2007;75:024416. http://dx.doi.org/10.1103/PhysRevB.75.024416.

[33] Gubbiotti G, Tacchi S, Madami M, Carlotti G, Adeyeye AO, Kostylev M. Brillouin light scattering studies of planar metallic magnonic crystals. J Phys D Appl Phys 2010;43:264003. http://dx.doi.org/10.1088/0022-3727/43/26/264003.

[34] Tacchi S, Madami M, Gubbiotti G, Carlotti G, Tanigawa H, Ono T, et al. Anisotropic dynamical coupling for propagating collective modes in a two-dimensional magnonic crystal consisting of interacting squared nanodots. Phys Rev B 2010;82:024401. http://dx.doi.org/10.1103/PhysRevB.82.024401.

[35] Arias R, Mills DL. Theory of collective spin-wave modes of interacting ferromagnetic spheres. Phys Rev B 2004;70:104425. http://dx.doi.org/10.1103/PhysRevB.70.104425.

[36] Chu P, Mills DL, Arias R. Exchange/dipole collective spin-wave modes of ferromagnetic nanosphere arrays. Phys Rev B 2006;73:094405. http://dx.doi.org/10.1103/PhysRevB.73.094405.

[37] Tartakovskaya E, Kreuzpaintner W, Schreyer A. Spin wave dynamics in two- and three-dimensional superlattices of nanosized ferromagnetic spheres. J Appl Phys 2008;103:023913. http://dx.doi.org/10.1063/1.2832756.

[38] Dvornik M, Bondarenko PV, Ivanov BA, Kruglyak VV. Collective magnonic modes of pairs of closely spaced magnetic nano-elements. J Appl Phys 2011;109:07B912. http://dx.doi.org/10.1063/1.3562509.

[39] Galkin AY, Ivanov BA, Zaspel CE. Collective modes for an array of magnetic dots in the vortex state. Phys Rev B 2006;74:144419. http://dx.doi.org/10.1103/PhysRevB.74.144419.

[40] Shibata J, Otani Y. Magnetic vortex dynamics in a two-dimensional square lattice of ferromagnetic nanodisks. Phys Rev B 2004;70:012404. http://dx.doi.org/10.1103/PhysRevB.70.012404.

[41] Galkin A, Ivanov B, Zaspel C. Collective magnon modes for magnetic dot arrays. J Magn Magn Mater 2005;286:351. http://dx.doi.org/10.1016/j.jmmm.2004.09.092.

[42] Politi P, Pini MG. Dipolar interaction between two-dimensional magnetic particles. Phys Rev B 2002;66:214414. http://dx.doi.org/10.1103/PhysRevB.66.214414.

[43] Verba RV. Spin waves in arrays of magnetic nanodots with magnetodipolar coupling, Ukr J Phys 2013;58:758.http://ujp.bitp.kiev.ua/files/journals/58/8/580808p.pdf.

[44] Verba R. Microwave properties of hexagonal arrays of magnetic nanodots. Visnyk Kyiv Nat Univ Radiophys Electronics 2012;17:29.

[45] Verba R, Tiberkevich V, Bankowski E, Meitzler T, Melkov G, Slavin A. Localized defect modes in a two-dimensional array of magnetic nanodots. IEEE Magn Lett 2013;4:4000404. http://dx.doi.org/10.1109/LMAG.2013.2277995.

[46] Verba R, Tiberkevich V, Bankowski E, Meitzler T, Melkov G, Slavin A. Conditions for the spin wave nonreciprocity in an array of dipolarly coupled magnetic nanopillars. Appl Phys Lett 2013;103:082407. http://dx.doi.org/10.1063/1.4819435.

[47] Beleggia M, Tandon S, Zhu Y, De Graef M. On the magnetostatic interactions between nanoparticles of arbitrary shape. J Magn Magn Mater 2004;278:270. http://dx.doi.org/10.1016/j.jmmm.2003.12.1314.

[48] Dmytriiev O, Meitzler T, Bankowski E, Slavin A, Tiberkevich V. Spin wave excitations of a magnetic pillar with dipolar coupling between the layers. J Phys Condens Matter 2010;22:136001. http://dx.doi.org/10.1088/0953-8984/22/13/136001.

[49] Landau LD, Lifshitz EM. Quantum mechanics. Course of theoretical physics, vol. 3. Oxford: Pergamon Press; 1965.

[50] Gurevich AG, Melkov GA. Magnetization oscillations and waves. New York, NY: CRC Press; 1996.

[51] Guslienko KY. Magnetostatic interdot coupling in two-dimensional magnetic dot arrays. Appl Phys Lett 1999;75:394. http://dx.doi.org/10.1063/1.124386.

[52] Ashkroft NW, Mermin ND. Solid state physics. New York, NY: Holt, Rinehart, & Winston; 1976.

[53] Kruglyak VV, Sokolovskii ML, Tkachenko VS, Kuchko AN. Spin-wave spectrum of a magnonic crystal with an isolated defect. J Appl Phys 2006;99:08C906. http://dx.doi.org/10.1063/1.2164419.

[54] Ding J, Kostylev M, Adeyeye AO. Magnonic crystal as a medium with tunable disorder on a periodical lattice. Phys Rev Lett 2011;107:047205. http://dx.doi.org/10.1103/PhysRevLett.107.047205.

[55] Sakoda K. Optical properties of photonic crystals. Berlin: Springer; 2001.

[56] Jorzick J, Demokritov SO, Hillebrands B, Bailleul M, Fermon C, Guslienko KY, et al. Spin wave wells in nonellipsoidal micrometer size magnetic elements. Phys Rev Lett 2002;88:047204. http://dx.doi.org/10.1103/PhysRevLett.88.047204.

[57] Lancaster P, Tismenetsky M. The theory of matrices. London: Academic Press; 1985.

[58] Kostylev M. Non-reciprocity of dipole-exchange spin waves in thin ferromagnetic films. J Appl Phys 2013;113:053907. http://dx.doi.org/10.1063/1.4789962.

CHAPTER 6

Thermal Properties of Magnetic Multilayers and Nanostructures: Applications to Static and Dynamic Behavior

Robert E. Camley

Center for Magnetism and Magnetic Nanostructures, University of Colorado at Colorado Springs, Colorado Springs, CO, USA

1 INTRODUCTION

Why have magnetic multilayers and nanoparticles received so much attention in the last few years? There are many answers:

(1) Multilayers and nanoparticles are a new class of materials where surface and interface effects fundamentally influence the properties of the entire material. For example, in a typical bulk material there might be 10^{15} surface atoms and 10^{23} atoms in the interior. Thus, the ratio of surface to bulk atoms is 10^{-8}. In contrast, a multilayer made up of films of alternating materials, where each film is five atomic layers thick, has a ratio of surface or interface atoms to bulk atoms of 40%.

(2) A rich variety of structures and behaviors is possible. One can combine ferromagnets with nonmagnets, ferromagnets with ferromagnets, ferromagnets with antiferromagnets, ferromagnets with helical magnets, to name only a few possibilities. Each of these structures has unique and potentially useful properties. For example, the giant magnetoresistance effect [1–4] originally found in Fe/Cr multilayers is now being used in magnetic memory storage systems. In addition, one can obtain and design unusual behaviors including magnetization that increases with temperature and a *thermal* hysteresis that is tunable with a small magnetic field.

(3) The importance of surface and interface effects may be controlled by the layering pattern for multilayers. Adding more layers to each film emphasizes bulk effects and conversely thinner films as constituents in the multilayer emphasize interface effects. Similarly, larger nanoparticles emphasize bulk effects, while smaller particles are surface dominated.

Magnetism of Surfaces, Interfaces, and Nanoscale Materials. http://dx.doi.org/10.1016/B978-0-444-62634-9.00006-0

(4) The properties of magnetic multilayers can be substantially modified by small external fields. Because of the tunability discussed above, one can design multilayers that are sensitive to small, convenient magnetic fields. For example, spin-flop fields in antiferromagnets are often in the 100 kOe range. In contrast, a spin flop in a magnetic multilayer can be designed to occur in the 100 Oe range.

In this chapter, we will explore simple calculational methods for determining the equilibrium spin structures of multilayers and nanostructures at any magnetic field and at any temperature. The results of these calculations will then be compared with experimental measurements. In addition, we will develop simple methods for finding the dynamic behavior of magnetic multilayer structures and again compare the results with available experiments.

2 THEORETICAL TREATMENT OF MAGNETIC MULTILAYERS

The magnetic multilayer may often be treated as an effective one-dimensional system if the spins in planes parallel to the surface are all equivalent. A spin in layer n, S_n, feels a number of fields which influence its motion. These include:

(1) Local exchange fields from within a layer and exchange fields from neighboring layers
(2) Local anisotropic fields
(3) The external field, H_0, and
(4) Demagnetizing fields caused by dipole motion.

We note that dealing with the system in a layer-by-layer fashion is particularly helpful in multilayer systems because surface/interface variations in exchange and anisotropy may easily be taken into account. A schematic illustration of the layer-by-layer structure is shown in Fig. 1.

We first describe an iterative energy minimization method for the $T=0$ case. One starts with an assumed set of values for the angular position of each layer of spins. In many cases the spins will lie in a plane parallel to the interfaces (this reduces the demagnetization energy) and for simplicity we will use this assumption. It is also assumed that all spins in each layer are equivalent. Due to exchange coupling, the energy of the spins in any given layer depends on the orientation and magnitude of the spins in its own layer and in the nearby layers. An arbitrary layer n, of spins, can be regarded as having an energy E_n, where

$$E_n = -\sum_{m=0,\pm1,\pm2} J_{nm} S_n \cdot S_{n+m} - g_n \mu_B H_0 \cdot S_n, \qquad (1)$$

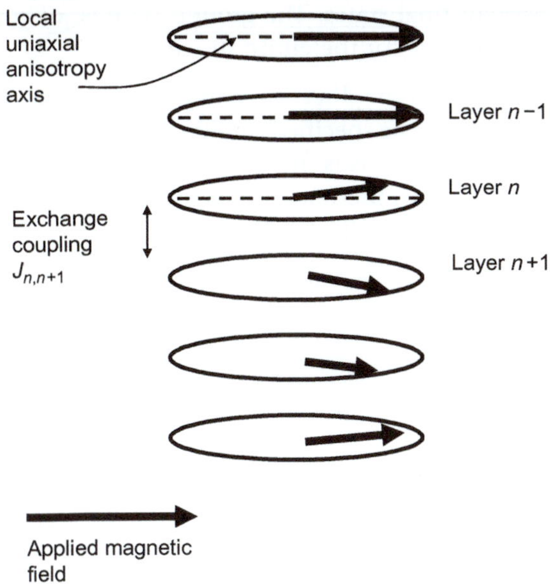

FIGURE 1
Schematic illustration of the layered structure and interactions for a magnetic multilayer. The top layers have uniaxial anisotropy. All the layers experience an external field and the exchange fields of the neighboring layers.

or in terms of the angle θ_n that the spins in layer n make with the applied field:

$$E_n = -\sum_{m=0_1\pm1_1\pm2} J_{nm} S_n S_{n+m} \cos(\theta_n - \theta_{n+m}) - g_n \mu_B S_n H_0 \cos(\theta_n). \tag{2}$$

For simplicity, we have neglected anisotropic energy in this example, but it can be easily included, as we will see later. Thus, at $T=0$ a single variable for each layer, θ_n, characterizes the magnetic structure. This energy in an individual layer is minimized by choosing the angle θ_n generated by the minimization condition:

$$\frac{\partial E_n}{\partial \theta_n} = 0. \tag{3}$$

This procedure can be viewed as rotating the spins in the nth layer to lie parallel to their effective field, a process which reduces the layer's energy. We can then find a configuration as follows: a different layer is then randomly chosen and the spins in that layer are rotated to lie in the direction of the local effective field. The process is continued until one has a self-consistent, stable state where all the spins are aligned with the local effective fields. Since the method only minimizes the energy locally, different initial configurations may lead to different self-consistent final states. The ground state is, of course, the stable state with the lowest energy for the entire structure.

At finite temperatures, both the direction and thermal-averaged magnitude of the spins in each layer must be specified. In this case, the iteration procedure is slightly different. A layer of spins is first rotated into the direction of the effective field, and then the spin's thermal-averaged magnitude in that direction is found through the use of the Brillouin function:

$$\langle S_n \rangle = S_n B_{S_n}(x), \tag{4}$$

where

$$x = \frac{g_n \mu_B S_n H_n}{kT}. \tag{5}$$

The Brillouin factor is given by

$$B_s(x) = \left(\frac{(2S+1)}{2S}\right) \coth\left(\frac{(2S+1)x}{2S}\right) - \left(\frac{1}{2S}\right) \coth\left(\frac{x}{2S}\right). \tag{6}$$

Here $\langle S_n \rangle$ is the thermal average of the spins in the nth layer in the direction of the effective field. H_n is the effective field acting on layer n. The effective field is now given by

$$H_n = H_0 + \sum_{m=0_1 \pm 1_1 \pm 2} \frac{J_{nm}}{g_n \mu_B} \langle S_{n+m} \rangle. \tag{7}$$

Note that in the effective field the spins in the neighboring layers are replaced by their thermal-averaged magnitudes. Again the entire operation is iterated through all spins until a self-consistent state emerges.

For a particular temperature and applied magnetic field, the iteration procedure requires of the order of 10^3–10^7 iterations per layer to reach a converged state. It generally takes a larger number of iterations to find a converged state near a phase transition. To speed up the calculation, one often uses the final state at one temperature to be the initial guess for the spin configuration at a nearby temperature. Of course, the final state may only be a local minimum and it is not guaranteed that the true ground state will always be found by this method. Nonetheless, if one takes a sufficient number of different initial sets of angles, the method seems to discover all the low-energy stable states.

The stability of a particular state can be checked in a direct way. Once a final state is found, one picks a new spin configuration slightly perturbed from the final state. Variations of a few degrees in each layer are typical. The iterative process then begins with this new configuration and one checks whether the system relaxes back toward the original state or diverges away from it.

We note in passing that this mean-field theory is really the simplest possible approximation that takes into account a layer-by-layer variation spin orientation and thermal-averaged magnitude. Such methods have been used previously in thin-film calculations [5]. Improvements would include Bethe–Peirels–Weiss methods or Monte Carlo techniques. However, these methods involve significantly larger computational investment.

3 EXAMPLES OF MAGNETIC MULTILAYER STRUCTURES

Among the most interesting of the magnetic superlattice structures are those that involve some antiferromagnetic coupling. In this case, the exchange energy favors some kind of antiparallel alignment while the Zeeman energy due to an external field is minimized if the magnetic moments are all parallel to the field. As a result, these structures can display a wide set of phase transitions, with the transition temperatures and fields controlled by changing the layering pattern. We therefore concentrate on multilayers with antiferromagnetic coupling.

3.1 Gd/Y Multilayers

One of the "simplest" layered systems with an interesting magnetic structure and a fascinating dependence on layering pattern is the Gd/Y superlattice. Although bulk Y is normally nonmagnetic, in the Gd/Y superlattice the Y layers mediate an effective exchange interaction between neighboring Gd films [6,7]. The application of a magnetic field in this system produces some interesting results. Because of the competition of interface exchange and Zeeman energies, the ground state is not a macroscopic antiferromagnetic state where the ferromagnetic films are strictly antiparallel. Instead the equilibrium structure resembles a macroscopic spin-flop state that represents a compromise between exchange and Zeeman energies. The resulting configuration is illustrated in Fig. 2. An increase in the external field reduces the canting angle θ and changes the net magnetic moment of the structure. The magnetization is thus very sensitive to applied fields and also to temperature, which plays a strong role in determining the average interlayer exchange energy.

In Fig. 3, we present theoretical and experimental results [7,8] for the magnetization as a function of temperature for Gd/Y superlattices. We see an unusual temperature dependence with a magnetization that *increases* as temperature is increased. This can be understood by looking at the exchange coupling of the Gd spins. The Gd spins on the interior are strongly coupled to each other compared to those on the exterior, which are only weakly coupled on the Y side via the RKKY interaction. When the temperature is increased, the Gd spins in the interior thus retain a larger thermal-averaged magnitude than those on the exterior since they see a larger effective field. As a result, the interface exchange

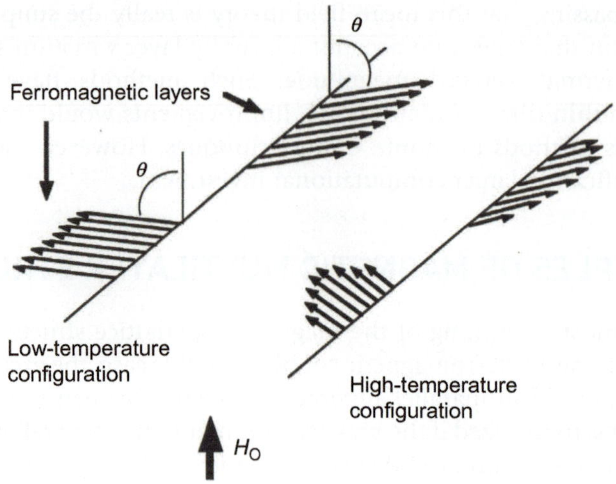

FIGURE 2
The spin configurations in the Gd/Y multilayer at low temperatures and high temperatures. Note that at high temperature the thermal-averaged magnetic moments in Gd are smaller at the surface than in the bulk. *(From Ref. [8].)*

energy (which should scale as $J_1\langle S_{\text{Gd-exterior}}\rangle^2$ is rapidly reduced and the competition between exchange and Zeeman energies now favors a state where the Gd moments point more closely along the external field. This is automatically included in the iterative calculation and the results are illustrated schematically in Fig. 2. Thus, even though $\langle S_{\text{Gd}}\rangle$ decreases as the temperature increases, the change in orientation of the Gd spins is sufficiently large that the magnetization increases. As T continues to increase, the $\langle S_{\text{Gd}}\rangle$ values begin to decrease very quickly and the magnetization decreases (more or less as single Gd films). The field dependence of the magnetization is also striking. Here we have an *ordered* state that displays a significant change in magnetization as a function of field.

3.2 Surface Effects in Multilayers

We now turn to the influence of the outermost surfaces of the magnetic structure of a finite multilayer system. The surface layer plays a special role in the magnetic structure simply because it is exchange coupled to other spins on only one side. Thus, in the competition between exchange and Zeeman energies, the influence of the exchange energy is reduced and it is easier for an external field to direct the surface spins. As the outermost layer of spins tries to turn toward the field direction, the remaining layers must adjust their orientations as well. Thus, surface-induced phase transitions can occur at magnetic field strengths well below that required for the equivalent bulk phase transition [9–12]. Also,

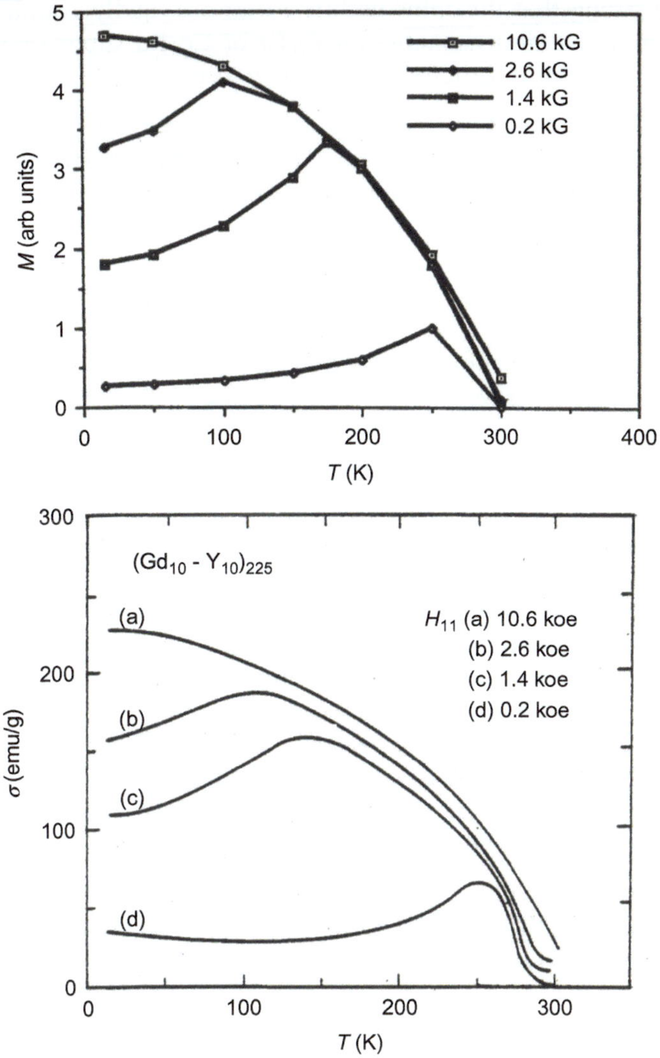

FIGURE 3
Magnetization as a function of temperature for different external fields for a Gd/Y multilayer. The top panel shows the theoretical calculation [8] and the bottom panel shows the experimental results from Ref. [7].

the reconstructed state can have surprisingly large penetration depths into the bulk of the superlattice.

As a first example of a surface phase in a superlattice we consider a finite Fe/Cr- or Gd/Y-type structure. We assume that the temperature is low enough that all spins have a thermal-averaged moment equal to their maximum value. Furthermore, we assume that the exchange coupling between atomic layers is sufficiently strong that the spins within a film are rigidly coupled together. These assumptions are reasonable for Fe/Cr or Co/Ru superlattices at room temperature, for film thicknesses under 10–20 nm.

For an infinitely extended Fe/Cr-type structure in an external field one expects a canted configuration, as discussed earlier. From energy considerations, it is easy to show that, in the absence of any anisotropy, the uniform canting angle is given by the equation:

$$\cos\alpha_0 = \frac{H_0 M}{4J}, \tag{8}$$

where M is the magnetization of a film and J is the interface exchange energy. In a finite multilayer, however, this uniformly canted state is not stable [12]. As mentioned above, the reason for this is that the outer layers of the finite structure experience only half the exchange coupling of the interior layers and thus are easier to turn toward the applied field.

Stable ground state configurations for finite Fe/Cr-type multilayers have been found [12] using the numerical method outlined above. These states are compared to the uniform canting state in Fig. 4. The structure in Fig. 4b, the low-field case, is quite complex. The outermost spins are twisted into the direction of the field as expected, but the spins of the second layer are actually turned farther away from the field than they would be in the bulk configuration. This alternation continues as one progresses into the bulk of the multilayer, but the amplitude of the deviation decreases with increasing distance from the surface. The main difference between the configurations of Fig. 4b and that found in Fig. 4c is that there is no alternation in the high-field configuration of Fig. 4c).

The surface twist states illustrated in Fig. 4b and c reduce the energy of the structure as compared to the uniform canting state by lowering the Zeeman energy. (As a result the surface twisted states have a slightly higher magnetization than does the uniformly canted state.) However, the twist also results in a small increase of exchange energy. The resulting structure is as usual a compromise between exchange and Zeeman energy.

It is worthwhile to compare these nonuniform canted structures discussed here to the very different case of nonuniform canting found in a domain wall. For example, the width of a domain wall in a ferromagnet is governed by the competition between the exchange energy and the anisotropic energy. The exchange energy can be minimized by spreading the canting out over many layers of spins and thus favors a very extended wall. In contrast, the anisotropic energy is minimized when spins point in an easy direction, thus favoring a narrow wall. In the surface twist problem, however, the competition is between exchange and Zeeman energies, and there is also the additional freedom of having two sublattices. This allows new kinds of states to occur when minimizing the total

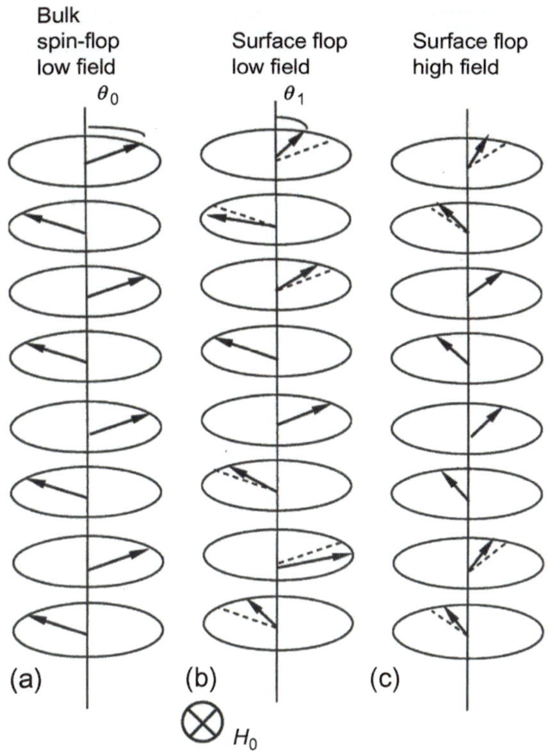

FIGURE 4
Illustration of bulk and surface spin-flop configurations in an antiferromagnet with no anisotropy.
From Ref. [12].

energy. The alternating spin configuration seen in Fig. 4b, for example, has the exchange energy between layers of spins alternately increase and decrease as one moves through the multilayer. This extra degree of freedom also allows the width of the nonuniform canting region to vary in an interesting manner as a function of applied field. At low fields, the nonuniform region is very large; for moderate fields, it is quite small and then for larger fields the width of the nonuniform regions is again large.

The considerations above apply to microscopic structures as well as multilayers. For example, one can use the same techniques as described above to examine the magnetic field and temperature behavior of ultra-thin antiferromagnetic films. A particularly interesting feature relates to the number of layers in the antiferromagnetic film. For an even number of layers, one of the outer layers must have its magnetization pointing opposite to the applied field and, as a result, a surface phase transition is nucleated at this surface. For an odd number of layers the magnetic moments at both surfaces will point parallel to the external field and no surface transition will be induced, and the change to a spin-flop structure must occur in the bulk. These configurations are illustrated in Fig. 5.

Calculations for the fields required to cause a spin flop have been done for several antiferromagnetic materials. An example is shown in Fig. 6, where the spin-flop field is shown as a function of the number of layers in an ultra-thin film of the antiferromagnet MnF_2 [13]. For an even number of layers the spin-flop field is nearly independent of the number of layers. The reason for this is that there is *always* a surface spin flop in this system. In contrast, when the number or atomic layers is odd the spin-flop field is significantly larger for thinner films, in part due to the pinning effect of the outer layers.

The spin-flop fields for MnF_2 calculated above are at $T=0$. Finite temperature calculations show that the odd/even behavior persists at finite temperature as

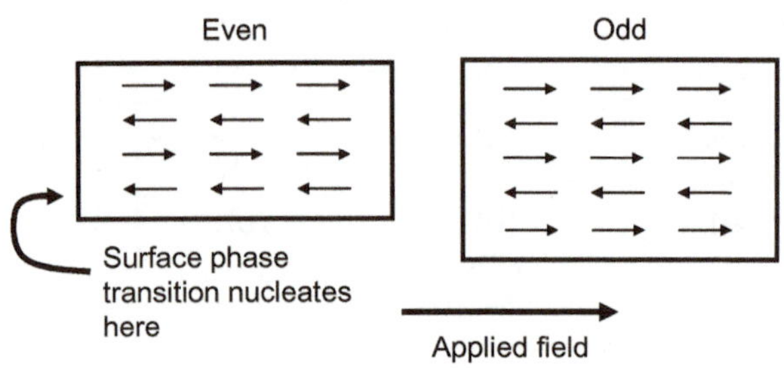

FIGURE 5
Illustration of the difference between an even and odd number of layers in an antiferromagnetic film. For an even number of layers one of the surface layers always has its moments pointing oppositely to the applied field.

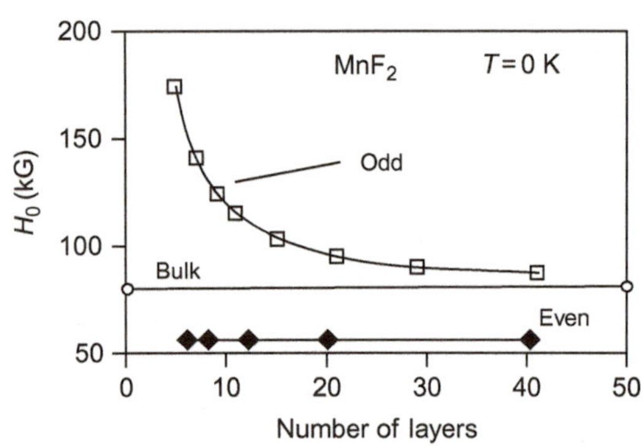

FIGURE 6
Spin-flop field in MnF_2 for an even and odd number of magnetic layers at $T=0$. When the number of layers is even the spin-flop field is smaller because a surface spin flop occurs. *From Ref. [13].*

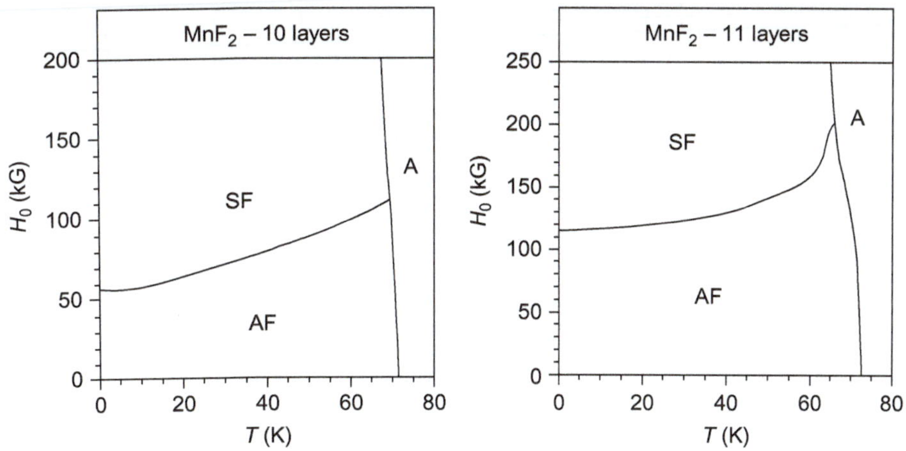

FIGURE 7
Phase diagrams for an ultrathin antiferromagnetic film of MnF_2 with an even ($N=10$) and odd ($N=11$) number of layers. *From Ref. [13].*

well, as can be seen in the phase diagrams of Fig. 7 [13] for an even number of atomic layers ($N=10$) and an odd number of layers ($N=11$). It is interesting to note that the difference between an odd number of atomic layers and an even number of atomic layers persists in temperature until the phase transition to a paramagnetic state.

The even/odd considerations in the microscopic antiferromagnet are all important in the multilayer as well. The details of surface phase transitions in an Fe/Cr(211) multilayer have been explored both theoretically and experimentally recently [14–16]. In these structures, the unit cell is rectangular in the plane of the surface and provides a uniaxial anisotropy.

The even/odd behavior was demonstrated most dramatically in measurements of the magnetization and magnetic susceptibility in the Fe/Cr(211) multilayer system [14], as shown in Fig. 8. As seen in this figure, the magnetization of the 20-layer Fe film structure shows two rapid changes, one at low field, associated with the surface spin flop, and one at higher field associated with a bulk spin flop. In contrast, the 21-layer Fe film structure displays only one transition close to the bulk transition of the 20-layer Fe film sample. Spin-flop states are generally associated with substantially increased susceptibility. The reason for this is simple. Susceptibility measures the ability of the spin configuration to change due to an external field. In the aligned states only the magnitudes of the magnetic moments can change. In contrast, in the spin-flop state both the magnitude and orientation can change, leading to a large susceptibility. We will see an enhanced susceptibility used as an indication of a canted structure in other systems as well.

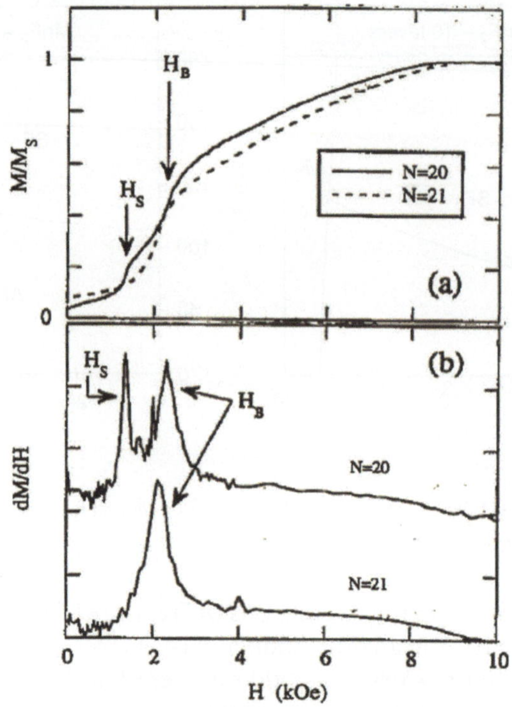

FIGURE 8

Experimental results on: (a) magnetization and (b) susceptibility in Fe/Cr(211) multilayers. For an even number of Fe films ($N=20$), the bulk and surface phase transitions each cause a sharp change in magnetization. This results in two peaks in the susceptibility. In contrast, there is only one peak in the susceptibility for an odd number of films. *From Ref. [14].*

The behavior of Fe/Cr(211) multilayers with an even number of Fe films has been shown to be particularly interesting [14–16]. If the system starts in the antiferromagnetic state at low field and the field is increased, a surface spin flop occurs at fields well below that required for the bulk spin flop. Essentially a domain wall is introduced near the surface. As the field is further increased, the domain wall migrates to the center of the film and then broadens in its spatial extent so that eventually the entire film is in a spin-flop configuration.

3.3 Fe/Gd Multilayers—Ferromagnetic Films with Antiferromagnetic Coupling

We now consider a multilayer system with alternating ferromagnetic films that are antiferromagnetically coupled at the interfaces. An example of such a system is the Fe/Gd multilayer [8,17–21]. There are some key differences between Fe and Gd. The Curie temperature of Fe is 1043 K, while Gd has a

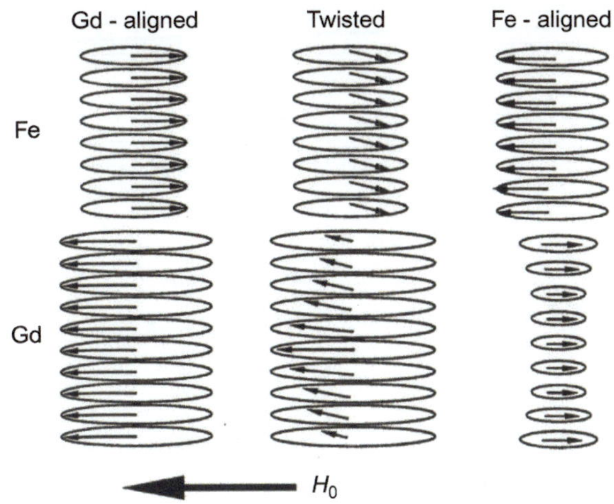

FIGURE 9
Illustration of the three main phases in the Fe/Gd superlattice. In the Gd-aligned phase the thermal-averaged moments of the Gd are large and aligned with the external field. In the Fe-aligned phase at higher temperature the Gd moments are small and the Fe moments align with the field. In the twisted phase the moments are nonuniform within a film and canted with respect to the magnetic field.

Curie temperature of 298 K. In addition, the magnetic moment of Fe is $2.2\mu_B$, while in Gd it is $7.2\ \mu_B$

Thus in the temperature range of 0 to about 300 K, the Gd thermal-averaged magnetic moment changes from near $7\mu_B$ to zero, while the Fe moment is nearly temperature independent. This has some interesting consequences, as we shall see below.

There are a number of known magnetic states for the Fe/Gd system that are illustrated in Fig. 9. There is a low-temperature-aligned Gd state where the Gd magnetization is aligned with the external field (and the Fe is opposite) and there is a high-temperature-aligned Fe state where the Fe magnetization is aligned with the external field (and Gd is opposite). These configurations are stable when the applied magnetic field is small; in higher magnetic fields there is a twisted or spin-flop-like state where the Fe and Gd moments are non-uniform within a film and both canted with respect to the applied field.

This system is essentially an artificial ferrimagnet where the effective concentrations of the two magnetic components are controllable by changing the layering pattern. The properties of the different phases and the phase diagram of Fe/Gd have been extensively studied experimentally and theoretically.

The different phases can be easily identified by an M versus H measurement, as seen in Fig. 10 [8,22]. The aligned phases are characterized by small changes in

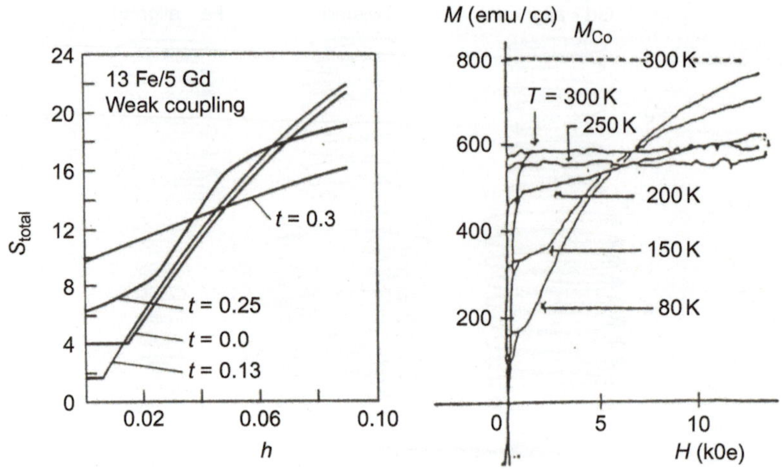

FIGURE 10

S_{total} versus applied field for an Fe/Gd multilayer structure. The field is measured in dimensionless units given by $h = g\mu_B H/J_{Fe}S$ and the temperature is given by $t = T/T_{Curie\text{-}Fe}$. At $t = 0.25$, for example, one sees a slow change in S_{total} versus h at low fields (Gd-aligned) then a rapid change at higher h (twisted phase), and then a slow variation at even higher h (the Fe-aligned phase). The right panel shows experimental results for Co/Gd, a similar system. *From Refs. [17,22].*

M as H is increased. In contrast, the spin-flop or twisted phase shows large changes in M as H is increased. Thus, the susceptibility dM/dH is a sensitive tool to characterize the phase with a large susceptibility indicating a twisted or canted state.

Figure 11 shows the experimental and theoretical phase diagram [18] for three different layering patterns. In (a) and (b) one sees similar behaviors—at low fields and low temperatures the system is in a Gd-aligned state; as the magnetic field is increased there is a transition to the spin-flop or twisted state, and at high temperatures the structure is in the Fe-aligned state. The critical magnetic field required for the transition from the Gd-aligned state to the twisted state depends on the layering pattern. In (a) the individual films are thinner—thus emphasizing interfacial exchange energy in the balancing of Zeeman and exchange energies. As a result, the critical field required for the transition to the twisted state is large. In (b) the films are thicker; the interfacial exchange becomes less important, and the transition field is smaller by about a factor of 2 when compared to (a). Finally, in (c) the thickness of the Gd films has been reduced sufficiently so that the Gd-aligned state is never stable, i.e., the net magnetic moment in the Gd never becomes larger that the net moment in the Fe.

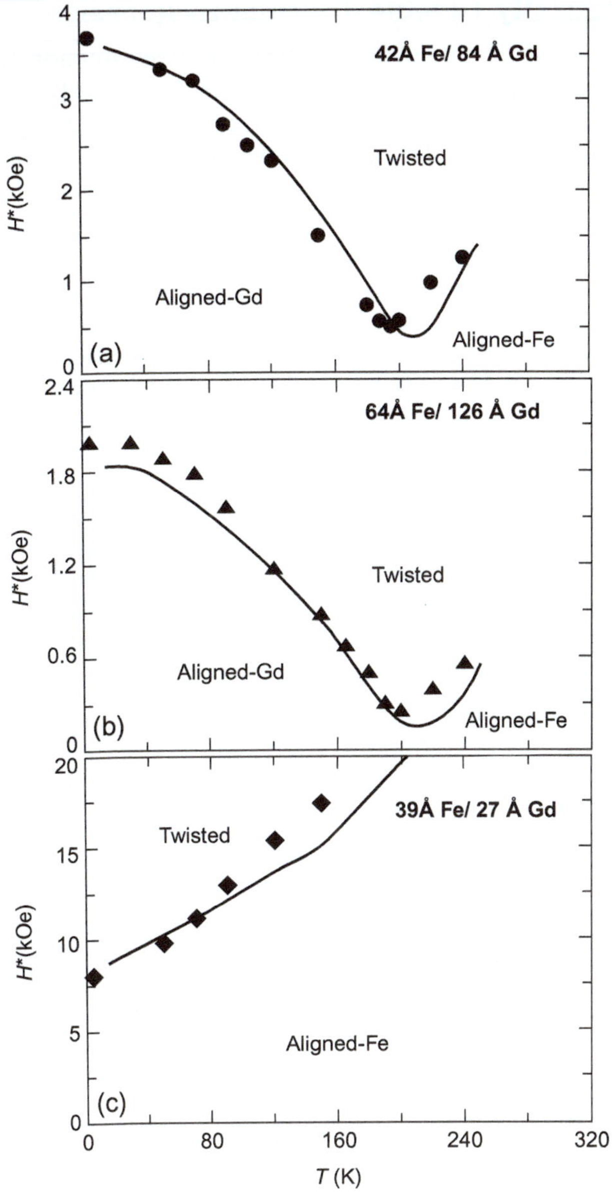

FIGURE 11

Theoretical and experimental phase diagrams for Fe/Gd multilayers with different layering patterns.
From Ref. [18].

3.4 Magnetically Tunable Thermal Hysteresis

We outline a recent treatment of an interesting phenomenon that occurs in Fe/Gd multilayers—magnetically tunable *thermal* hysteresis [23]. Thermal hysteresis is well known in physical changes of state such as melting or crystallization. The temperature width for the thermal hysteresis is often only a few kelvin or less. Many of these systems have a requirement of ultrapure samples since impurities, often at a surface, can act as nucleation sites and initiate the phase transition. In contrast, the Fe/Gd structure is a simple magnetic multilayer system that can be *designed* to show a thermal hysteresis curve. Calculations show the width, in temperature, of this hysteresis curve can be controlled by an external magnetic field with small fields (200–300 Oe) leading to thermal hysteresis spanning about 100 K. Larger fields (600 Oe) can reduce the width of the hysteresis curve to 20 K, and even larger fields can completely eliminate this effect.

The physical origin for the thermal hysteresis is simple. One must have two states that are stable at the same temperature. The two states are the Fe-aligned state and the Gd-aligned state. At high temperatures the system is in the Fe-aligned state; the Gd moments, which are small, are oppositely directed to the field. As the temperature decreases the Gd moments increase and the net magnetic moment can actually be opposite to the external field if an anisotropy in Fe holds the Fe spins in place. Eventually, however, this configuration becomes unstable and the structure reverses with the Gd moments along the field and the Fe moments antiparallel to the applied field. A similar situation occurs on heating the system from low temperatures.

The general situation of an Fe/Gd multilayer is quite complicated and, in fact, shows significant surface effects, as we shall see. However, the general features discussed here should persist if the anisotropy is found in the Fe, in both materials, or only in the Gd. The calculation proceeds in the usual way. A spin in monolayer *i* experiences an effective field composed of the exchange fields from the layers above and below, external fields and anisotropic fields:

$$\vec{H}_i = \sum_{nn} \left(J_{i,i+1} \left\langle \vec{S}_{i+1} \right\rangle + J_{i,i-1} \left\langle \vec{S}_{i-1} \right\rangle \right) + \hat{z} + H_a \langle S_i^z \rangle / S_i. \tag{9}$$

We use an anisotropy of 50 G in Fe and H_a is taken to be zero in Gd. All other parameters are given in Ref. [18]. (The reduction parameter in Ref. [18] is 0.75.) The minimum energy state is found using the iterative method described earlier.

The results for the Fe magnetization as a function of temperature for an infinite Fe/Gd superlattice are given in Fig. 12. This curve is calculated for a structure with 17 monolayers of Gd and 23 monolayers of Fe in each unit cell. This choice of values is intended to match the experimental parameters of Gd 50 Å/Fe 35 Å. Periodic boundary conditions are used to simulate the infinite

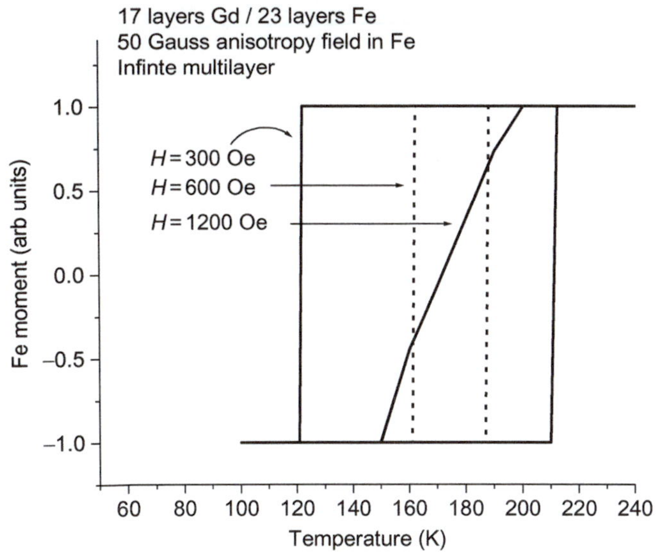

FIGURE 12

Theoretical thermal hysteresis curves for an infinite Fe/Gd multilayer with a unit cell of a 17 ML Gd/23 ML Fe superlattice for different applied magnetic fields. The hysteresis curves initially narrow as the applied field is increased. At higher fields (1200 Oe) the system is in a spin-flop phase and does not show substantial hysteresis. *From Ref. [23].*

superlattice. A thermal hysteresis spanning 90 K is seen for the lowest field value. This is reduced to about a 30 K span for a field of 600 Oe. For even higher fields, the system does not display hysteresis but makes a transition from the aligned phases to the spin-flop phase, and the magnetization essentially increases linearly as a function of temperature in the transition region.

It is known that the finite size of a Fe/Gd superlattice can have a significant influence on the phase transitions. For example, at low temperatures and low fields the system is typically in the Gd-aligned state. If the magnetic field is increased the magnetic structure will eventually change to a spin-flop-like state. However, the critical field at which this phase transition takes place depends dramatically on whether the outermost surface is Gd terminated or Fe terminated. If the structure has Fe spins on the outside, then in the Gd-aligned state the Fe spins are opposite to the external field and a surface phase transition to the spin-flop state nucleates at the Fe surface [11,24,25]. The critical field for this surface transition is about a factor of 5 lower than that for the bulk transition.

The thermal hysteresis effect discussed here also shows substantial finite size effects. In Fig. 13, we compare results for an infinitely extended multilayer with a unit cell 17 ML Gd/23 ML Fe with those of a finite multilayer with the same

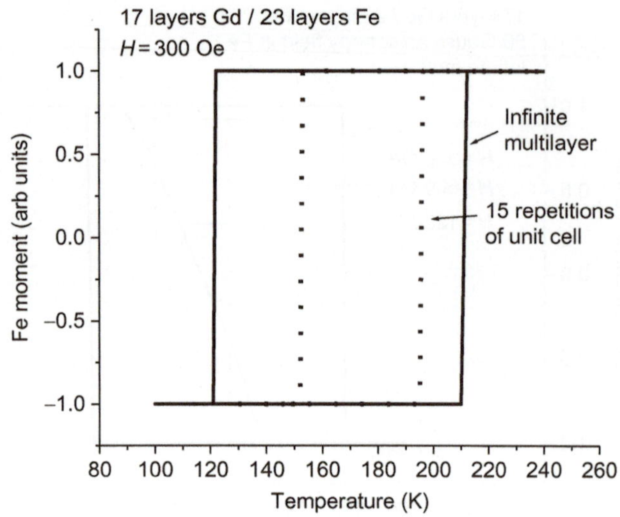

FIGURE 13

Theoretical thermal hysteresis curves for a finite and infinite Fe/Gd multilayer. The unit cell is 17 ML Gd/23 ML Fe in both cases. The applied field is 300 Oe. Note that the finite multilayer has a substantially narrower thermal hysteresis curve. *From Ref. [23].*

unit cell but with a total of 15 repetitions. The thermal hysteresis for an infinite structure is approximately twice as wide as that for the finite structure. This is due to the surface phase transition discussed above. In a finite structure composed simply of a repetition of the Fe/Gd unit cell, the spins at one surface will always be opposite to the applied field in an aligned state, and as a result a surface phase transition will nucleate at that surface, reducing the temperature range over which two states are stable.

Nonetheless, the main features of the thermal hysteresis remain qualitatively the same regardless of the number of unit cells, and it is important to compare the theoretical results with the results of an experiment. Neutron scattering measurements were taken on a multilayer with composition [Gd 50 Å/Fe 35 Å] × 15 deposited on an Si wafer with Fe as the top layer of the multilayered stack. The magnetic properties were studied with polarized neutron reflectometry at the Intense Pulsed Neutron Source of Argonne National Laboratory. In these experiments a neutron is incident on the sample. The scattered neutrons are collected and analyzed and I^+ represents the intensity of the scattered beam when the magnetic moment of the incident neutrons is parallel to the applied field. I^- denotes the similar case except that the magnetic moment of the incident neutrons is antiparallel. A convenient quantity is the spin asymmetry $P=(I^+ - I^-)/(I^+ + I^-)$, which is basically proportional to the magnetization of the Fe sublattice, and thus it can be immediately compared with the theoretical predictions. Figure 14 shows the spin asymmetry when the sample is cycled

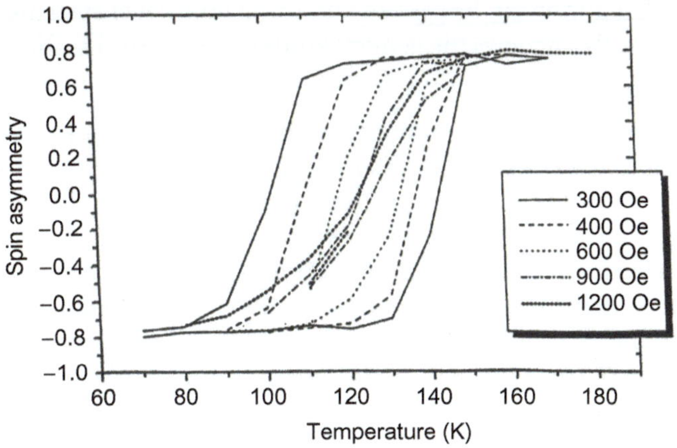

FIGURE 14

Experimental neutron scattering results for the thermal hysteresis of the asymmetry *P*. The vertical axis is proportional to the Fe magnetization. The experiments were done on a [Gd 50 Å/Fe 35 Å]$_{15}$ sample. *From Ref. [23].*

in temperature at different fields. For $H < 1200$ Oe, the loop shows hysteresis. The area of the loop decreases as the field is increased, yet the slope of the segments uniting the lower to the upper branch remains the same. Above 900 Oe, the loop has closed down almost completely, but with a smoother transition from the lower to the upper branch.

The results presented here show the excellent qualitative agreement of theory and experiments, particularly in that the width of the thermal hysteresis decreases as the external field is increased, and in the transition to the spin-flop state at higher fields, a state where the magnetization increases approximately linearly with temperature.

It is clear from these examples that the simple self-consistent local mean field theory does a good job of qualitative and quantitative description for magnetic multilayer systems of many different kinds. Additional examples can be found in Ref. [8].

4 DYNAMIC MODES IN MULTILAYERS

The thermal information obtained by the methods outlined above, can be used to estimate temperature-dependent dynamic properties in thin films, multilayers, and nanostructures. For example, we have seen that the thermal averaged magnetization in an individual layer depends on the position of the layer. As a simple approximation, one can use this thermal, spatially dependent, magnetization in calculations for the temperature-dependent dynamic behavior of the spin system. The following section discusses how such spatially dependent calculations can be carried out, using multilayer systems as an example. This section also reviews the general kinds of spin motions that can occur in multilayers with ferromagnetic and antiferromagnetic couplings.

As we have seen, much of the interesting physics that occurs in magnetic multilayers is due to the coupling across interfaces and the effects of having a surface. Unfortunately the nature and strength of interactions across an interface are difficult to measure directly. Even with techniques such as spin polarized low energy electron diffraction (SPLEED) and neutron scattering [26,27], it is difficult to obtain unambiguous information on the magnetic structure of surfaces and interfaces. Measurements of the static magnetization of magnetic multilayers give data on the averaged magnetic moment from a relatively large volume of sample, rather than providing detailed information on any of the individual films within the multilayer. Another problem, for example, is that ferromagnetic interlayer exchange coupling cannot always be inferred from static magnetization measurements.

One method of investigating surface and interface magnetism, which has proven useful in studies of ferromagnetic films and multilayers, is to probe the spin-wave excitations of a magnetic system [28–31]. In a spin wave the magnetic moments at each site precess about their individual equilibrium directions. Because the spins are coupled with one another through exchange and dipolar interactions, spin-wave excitations are the eigenmodes of the magnetic system and are characterized by frequency and wavelength. Thus, the frequency of a spin wave may depend quite sensitively on the exchange coupling between spins, as well as other effective fields caused by, for example, bulk and surface anisotropies [30] and magnetoelastic effects. These interactions will not only affect the frequency of precession but also the relative phase of precession between spins at neighboring lattice sites.

In ferromagnetic systems such as Fe and Co, the lowest spin-wave frequencies are typically of the order of 5–10 GHz. These are long-wavelength excitations and can be studied using ferromagnetic resonance [31] and Brillouin light scattering techniques [28–30] (see chapter 3 on Magnetization Dynamics in this book). Higher energy excitations can be observed by magnetic neutron scattering [32,33]. Spin-wave frequencies in antiferromagnets are typically much higher, existing at several hundred gigahertz and into the terahertz range, and thus also influence the optical properties in the far-infrared region [34,35]. Excitations in antiferromagnets can be observed with infrared reflectivity [36–38], Raman scattering as well as neutron scattering [33]. In magnetic multilayers, the energies of long-wavelength spin-wave excitations will also range from a few to several hundred gigahertz depending on the exchange interactions between the ferromagnetic films.

In this section, we will explore how the dynamic spin behavior reflects the static configurations, and how the frequencies of the excitations can be used to determine the important magnetic parameters that govern the spin system. Most of the discussion which follows will relate to excitations that can be measured in ferromagnetic resonance. In this case the spin-excitation wavevector parallel to the surface and interfaces is usually zero. This allows a substantial simplification of the calculations, particularly for the dipole fields. For nonzero wavevectors, one must include the dipole fields in a careful way, and a variety of interesting behaviors result [8,39].

We begin by presenting a simple calculation which illustrates the spin motion in two ferromagnetically coupled ferromagnetic films. As noted earlier, one cannot find the exchange coupling constant for this magnetic structure through static measurements. However, this is possible for dynamic measurements. The primary goal of this calculation is to obtain the excitation frequencies of the normal modes of the coupled ferromagnetic films and connect these frequencies to the interfacial coupling between the two films.

The energy per unit area in such a system is given by

$$\frac{E}{\text{area}} = -\left(\vec{M}_1 \cdot \vec{H} + \frac{1}{2}\vec{M}_1 \cdot \vec{H}_{D1}\right)t_1 - \left(\vec{M}_2 \cdot \vec{H} + \frac{1}{2}\vec{M}_2 \cdot \vec{H}_{D2}\right)t_2 - J_I\vec{M}_1 \cdot \vec{M}_2. \tag{10}$$

Here, t_1 is the thickness of one film, $\vec{H}_{D1}$ the demagnetizing field acting on the film, and $\vec{M}_1$ is its magnetization. Similar definitions hold for the second film. H is the applied field and J_I is the interfacial exchange coupling constant. This equation contains the Zeeman energy and demagnetizing energy for each film, and the interface coupling energy. Because we are using a total energy/area at this point, the Zeeman and dynamic demagnetizing energy densities have to be multiplied by the thickness of the appropriate film. For a spatially uniform mode the demagnetizing field in film 1 is given by

$$\vec{H}_{D1} = -4\pi M_{1y}\hat{y}, \tag{11}$$

which is due to the uniform distribution of magnetic surface charges at the top and bottom of each film. Similarly

$$\vec{H}_{D2} = -4\pi M_{2y}\hat{y}. \tag{12}$$

The energy density in film 1 is thus

$$U_1 = -\left(\vec{M}_1 \cdot \vec{H} - 2\pi M_{1y}^2\right) - J_I\frac{\vec{M}_1 \cdot \vec{M}_2}{t_1}. \tag{13}$$

From the last equation, we can find the effective field acting on $\vec{M}_1$:

$$\vec{H}_{\text{eff}-1} = -\frac{\partial U_1}{\partial \vec{M}_1} = \vec{H} - 4\pi M_{1y}\hat{y} + \frac{J_I\vec{M}_2}{t_1}. \tag{14}$$

Note that the last term on the right-hand side looks like a thickness-dependent exchange field.

A similar equation can be written for the effective field acting on the magnetization in the second film.

One now uses the torque equation of motion for each film:

$$\frac{d\vec{M}_1}{dt} = \gamma\left(\vec{M}_1 \times \vec{H}_{\text{eff}-1}\right) \tag{15}$$

$$\frac{d\vec{M}_2}{dt} = \gamma\left(\vec{M}_2 \times \vec{H}_{\text{eff}-2}\right), \tag{16}$$

where γ is the gyromagnetic ratio. The value of γ varies from material to material but is typically around 1.82×10^7 rad/Oe or equivalently 2.9 GHz/kOe for the ferromagnetic metals. For the simple problem of two ferromagnets with ferromagnetic coupling the calculation is relatively straightforward. The applied field is set in the $\hat{z}$ direction so we expect the magnetizations in each film to point in this direction as well, with small oscillations in the x and y directions. So:

$$\vec{M}_1 = M_{1x}\hat{x} + M_{1y}\hat{y} + M_1\hat{z} \tag{17}$$

$$\vec{M}_2 = M_{2x}\hat{x} + M_{2y}\hat{y} + M_2\hat{z}, \tag{18}$$

where the M_x and M_y terms are small. When all the substitutions are made into the equations of motion, we obtain a matrix equation that has the form of an eigenvalue equation. This equation is not, however, Hermitian!

$$\begin{pmatrix} -i\frac{\omega}{\gamma} & -\left(H + \frac{J_I}{t_1}M_2 + 4\pi M_1\right) & 0 & +M_1\frac{J_I}{t_1} \\ +\left(H + \frac{J_I}{t_1}M_2\right) & -i\frac{\omega}{\gamma} & -\frac{J_I}{t_1}M_1 & 0 \\ 0 & \frac{M_2 J_I}{t_2} & -i\frac{\omega}{\gamma} & -\left(H + \frac{J_I}{t_2}M_1 + 4\pi M_2\right) \\ -\frac{M_2 J_I}{t_2} & 0 & \left(H + \frac{J_I}{t_2}M_2\right) & -i\frac{\omega}{\gamma} \end{pmatrix} \begin{pmatrix} M_{1x} \\ M_{1y} \\ M_{2x} \\ M_{2y} \end{pmatrix} = 0. \tag{19}$$

In order to obtain a simple analytic expression, the problem is simplified further. We set $M_1 = M_2 = M$, and $t_1 = t_2 = t$. Also, we introduce the exchange field $H_{ex} = J_I M/t$. With these approximations, two frequencies are found:

$$\omega = \gamma\sqrt{H(H + 4\pi M)} \tag{20}$$

and

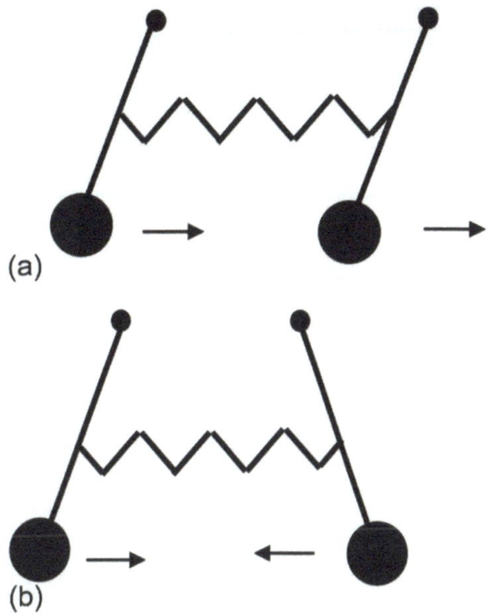

FIGURE 15
Two coupled pendula. (a) The two pendula move in-phase and the spring is never compressed or stretched, so the frequency of this motion is independent of the spring. (b) The motion of the two pendula is 180° out of phase and the spring influences the frequency of the mode.

$$\omega = \gamma\sqrt{4H_{ex}^2 + (4H + 8\pi M)H_{ex} + H(H + 4\pi M)}. \tag{21}$$

Obviously one mode has a frequency which is independent of the exchange coupling and the other mode frequency depends on this coupling. This is easy to understand in terms of a simple model of coupled pendula, as seen in Fig. 15. If we consider two identical independent pendula, they each oscillate with the same frequency. If we couple the two together using a spring, there are two normal modes. One occurs when the two pendula move in phase. In this case the spring is neither stretched nor compressed and the frequency remains unchanged. The other normal mode has the two pendula moving 180° out of phase with each other and in this case the spring modifies the frequency of oscillation. The spin motion in the two films also decomposes into an in-phase mode (acoustic mode) and an out-of-phase mode (optic mode). The in-phase mode has a frequency that is insensitive to the exchange coupling, but the optic mode frequency can be used to find the interface exchange coupling.

As we have seen above, one may obtain the interfacial exchange coupling between two ferromagnetically coupled films by measuring the frequencies of the dynamic modes of the entire structure. This has proven quite useful because it is often difficult to measure the exchange coupling using static

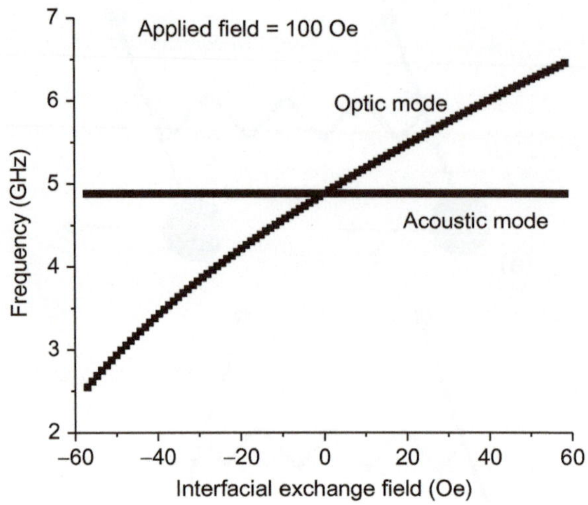

FIGURE 16

Frequencies for two exchange coupled ferromagnetic Co films as a function of the coupling strength. One mode is independent of the coupling, while the other changes smoothly as a function of the coupling.

measurements alone such as with magnetometry. The reason for this is that in a static measurement, the two films always act as a single unit as long as the coupling is ferromagnetic and anisotropies are small. With dynamic measurements one can determine the value of the exchange coupling, independent of the sign of the coupling [40,41]. We show in Fig. 16 an example of how the dynamic frequencies depend on the interface coupling constant for two coupled Co films. More complex systems with many layers and arbitrary couplings are treated in a similar manner. It is necessary in this case to find the complex equilibrium positions before obtaining the linearized eigenvalue equation. Nonetheless, this can be done in a straightforward manner, although the algebra can be painful. We will discuss an alternative method in the next section, but first we present some examples using the method outlined above.

We now consider two identical ferromagnetic films that are coupled together antiferromagnetically. Each film has an in-plane uniaxial anisotropy associated with the z-axis and an external field is applied along z with the y-axis normal to the films. Even this simple system can have a rich variety of behaviors, as we will see. Clearly if the external field is very large, both magnetizations will align with the external field. For small external fields, the antiferromagnetic exchange will dominate and the magnetizations will be antiparallel. For intermediate fields, the two magnetizations will have a spin-flop configuration. The energy density for the system includes the Zeeman energy density, the interface exchange coupling energy density, and the anisotropic energy density:

$$U = -\vec{M}_1 \cdot \vec{H} - \vec{M}_2 \cdot \vec{H} - H_{ex}\vec{M}_1 \cdot \vec{M}_2/M - H_a(M_{1z})^2/M - H_a(M_{2z})^2/M. \quad (22)$$

Here $\vec{M}_1$ is the magnetization in film 1, $\vec{M}_2$ defined similarly, and M is the magnitude of the total magnetization. If we assume a spin-flop like configuration where M_1 is canted away from the z-axis by an angle θ_1 and M_2 is canted away from the z-axis by an angle $-\theta_2$, the energy density of film 1 can be written as

$$U = -MH\cos\theta_1 - H_{ex}M\cos(\theta_1+\theta_2) - H_aM\cos^2\theta_1. \tag{23}$$

The equilibrium angle θ_1 is found by minimizing the energy density:

$$\frac{dU}{d\theta_1} = 0 = MH\sin\theta_1 + H_{ex}M\sin(\theta_1+\theta_2) + H_aM\cos\theta_1\sin\theta_1. \tag{24}$$

We can now assume, by symmetry, that the canting angles for the two magnetizations are equal in magnitude, and solve for the canting angle:

$$\frac{-H}{2H_{ex}+H_a} = \cos\theta_1. \tag{25}$$

Note that for antiferromagnetic coupling H_{ex} is a negative quantity in this example. If the external field is large (so that the magnetizations are parallel) and then is reduced, the field at which canting can begin occurs when $\cos\theta_1 = 1$ and we obtain

$$H_c = -(2H_{ex}+H_a). \tag{26}$$

As the external field is further lowered, a second transition from the spin-flop state to the antiparallel state occurs.

In Fig. 17, we show the net magnetic moment as a function of applied field for two antiferromagnetically coupled Co films. The parameters for the system are $M = 1.4$ kG, $H_a = 0.057$ kOe, and the thickness-dependent $H_{ex} = -0.058$ kOe. If we scan from large applied fields to small fields, we see the two transitions discussed above. However, if we start at a small external field in the antiferromagnetic state and increase the field, the system can remain stable in the antiparallel state until a transition to the parallel state occurs at a much higher field. Thus, an effective hysteresis is introduced into the system. Similar behaviors have been noted in more complex systems.

The dynamical modes of the two coupled ferromagnetic films may now be calculated as indicated above. We present the results of this calculation in Fig. 18, where we show the dynamic modes which exist when the system is started at a high external field and the field is subsequently reduced. The change of state from one configuration to another is very clearly indicated in the dynamical modes. It is interesting to note that the continuous, second-order transition from the parallel state to the spin-flop state is associated with a clear "soft mode" where the frequency is driven toward zero. In contrast, the first-order

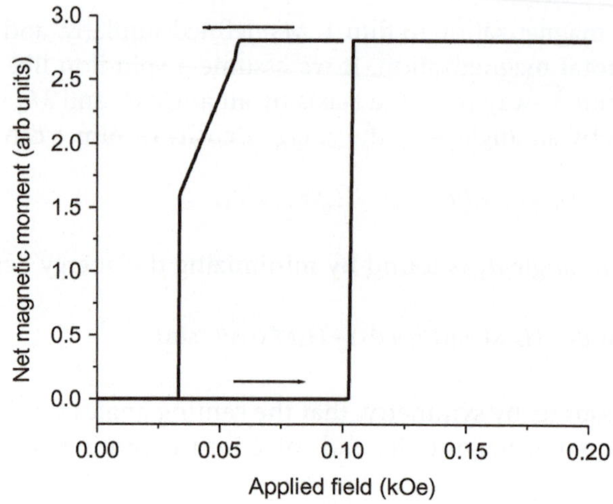

FIGURE 17
Magnetization versus applied field for two antiferromagnetically coupled Co films. The hysteretic behavior occurs because the system goes through a spin-flop phase as the magnetic field is reduced from large values, but when the system is started at zero field and the field is increased, the system goes directly from an antiparallel state to a parallel state.

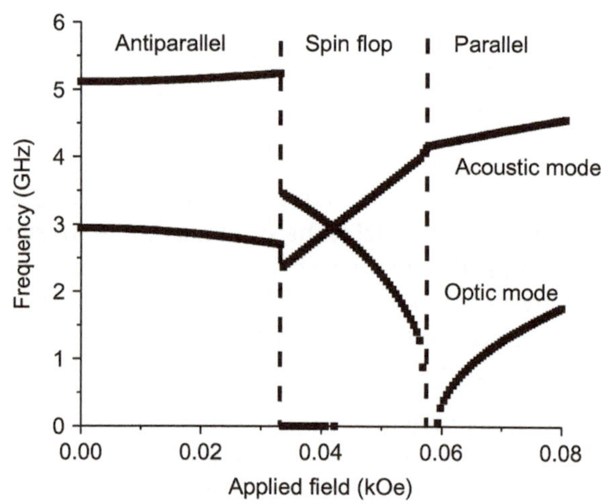

FIGURE 18
Frequency of spin excitations versus applied magnetic field for two antiferromagnetically coupled Co films. The field is started at a large value and reduced to zero. The parameters are the same as those used for Fig. 17.

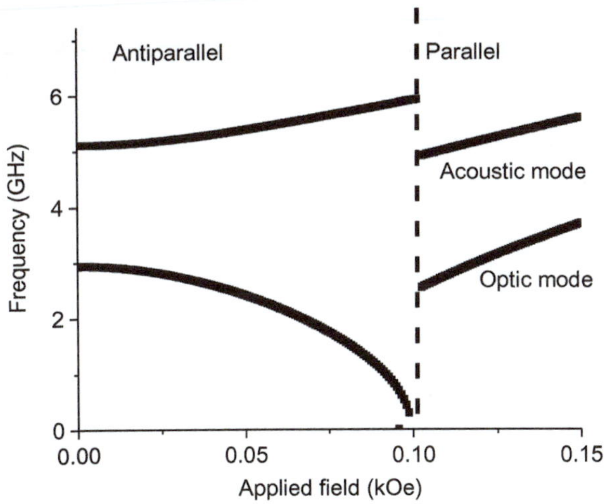

FIGURE 19
Frequency of spin excitations versus applied magnetic field for two antiferromagnetically coupled Co films. The field is started at zero and increased to a large value. The parameters are the same as those used for Fig. 17.

transition from the spin-flop state to the antiparallel state does not show a soft mode, but instead shows a discontinuous change in the frequency. Theoretical [41] and experimental [42] results for cases where an in-plane wave vector is not zero and the dipole fields need to be included carefully are quite similar to the ones shown in Fig. 18.

As seen in Fig. 17, the magnetic structure is different if the magnetic field is increased from zero rather than reduced from a high value. The dynamic modes also show this difference in a very clear way. Figure 19 presents the frequency of the two dynamic modes as a function of field when the field is increased from zero. In contrast to Fig. 18, there is only one phase transition which leads to a discontinuous jump in the frequency at $H = 100$ Oe.

It is interesting to see the information that is available in dynamic measurements for multilayers with larger numbers of films as well. Figure 20 shows theoretical and experimental results for an Fe/Cr multilayer with a total of 16 Fe films [43]. One expects 16 closely spaced modes, but from Fig. 20 it is seen that only four modes are resolvable in the low-field region where the spin structure is canted. Nonetheless, it is clear that the main features of the dynamic modes are measurable.

In Fig. 21, we present the theoretical results for magnetization as a function of field (scan from high field to low field) for a Co/Cu multilayer with 19 Co films. In this example, the outermost Co film is assumed to have an exchange coupling that is three times larger than the ferromagnetic coupling between all

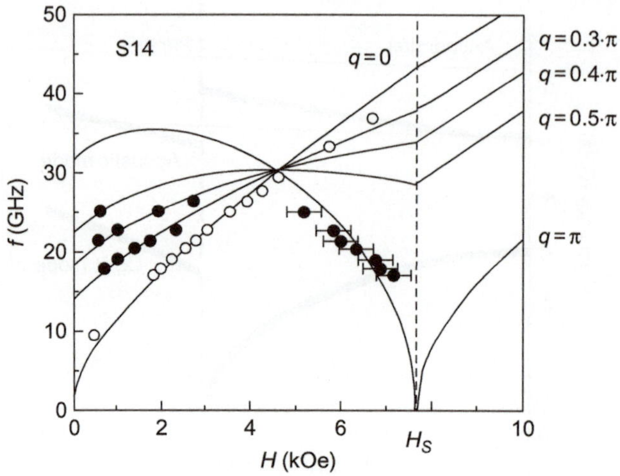

FIGURE 20

Experimental and theoretical results for frequency versus magnetic field for an Fe/Cr multilayer. *q* gives the phase shift between neighboring magnetic layers in an infinite multilayer system. *From Ref. [43].*

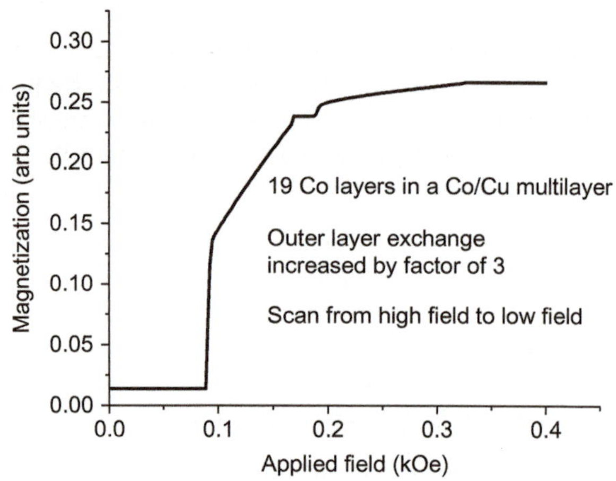

FIGURE 21

Theoretical results for magnetization versus field for a Co/Cu multilayer with 19 Co films. The thickness of the Cu is chosen so the Co layers have antiferromagnetic exchange coupling between each other. The outer Co layer experiences an antiferromagnetic exchange three times larger than the ferromagnetic coupling in the bulk. Some phase changes are easy to see as evidenced by sharp changes in the magnetization.

other Co films. Such a situation could be easily arranged by changing the thickness of the Cu spacer layer next to the outermost Co film. Figure 21, which plots magnetization as a function of field, shows a behavior that indicates a number of different phases. This is evident in a theoretical plot, but noise might easily

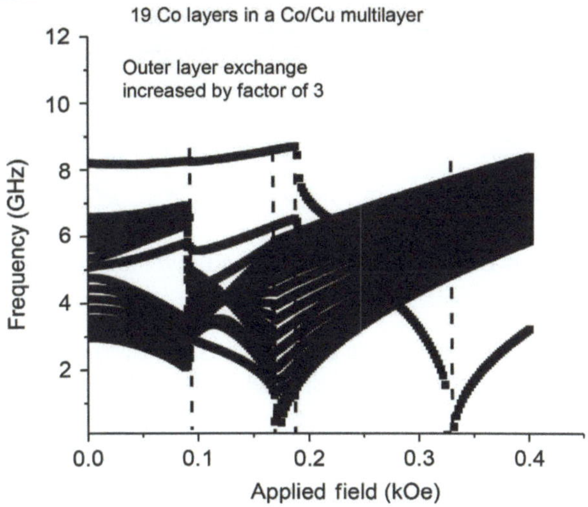

FIGURE 22
Mode frequencies as a function of field for the Co/Cu multilayer in Fig. 21. The soft mode near $H = 0.33$ kOe is due to a surface spin flop of the outermost Co moment. The different phases are separated by the vertical dashed lines. The bulk spin-flop region exists in the range $0.09 < H < 0.17$.

obscure the existence of the different phases in an experimental measurement. Figure 22 shows the behavior of the dynamic modes in the same structure. Even at high field, where the magnetic moments in all the films are aligned with the external field, one can see that one mode is far away from the continuum of bulk modes. This indicates an impurity-type mode, in this case associated with the outermost film. As the external field is reduced, the impurity mode goes soft ($H = 0.33$ kOe) as a surface phase transition takes place. This transition is virtually undetectable in the magnetic moment calculation of Fig. 21 because the outer Co film provides only a small portion of the net magnetic moment of the structure.

5 A SINGLE METHOD THAT PROVIDES STATIC AND DYNAMIC RESULTS

Up to now, we have primarily discussed equilibrium configurations obtained by the energy minimization method. This method is easy to implement and generally gives good results. However, some of the assumptions that are often used in association with this method can produce incorrect results. An example of this occurred in the study of the behavior of ferromagnetic/antiferromagnetic structures. If one grows a thin ferromagnetic film on an antiferromagnet and cools the system in the presence of a magnetic field, one often finds that the

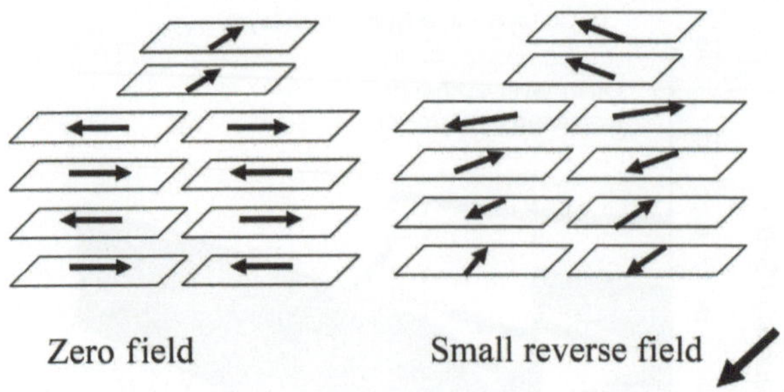

FIGURE 23
Schematic illustration of spin configurations in a ferromagnet/antiferromagnet structure with a spin-compensated antiferromagnetic surface. The ferromagnet is represented by the top two layers.

hysteresis curve associated with the ferromagnet is both shifted in field and the coercivity is substantially widened in field. The shift of the hysteresis curve is known as the exchange bias effect and has been studied for many years [44].

A recent calculation, based on energy minimization, indicated a mechanism for the exchange bias [45]. This calculation dealt with a ferromagnet that was coupled to an antiferromagnet with a compensated surface, i.e., the surface had an equal number of spins from each sublattice. The magnetic structure, in low field, is shown schematically in Fig. 23. When the field is reversed, as is shown in Fig. 23, a domain wall or twist is built up in the antiferromagnet. The building up of the domain wall prevents the ferromagnet from simply reversing at low reverse field. This produces the exchange bias shift. If the external magnetic field is returned to its original direction, the twist is expected to simply unwind and the system proceeds reversibly back to its original state.

The mechanism above would explain the exchange bias effect, but not the enhanced coercivity. Unfortunately, some limitations of the energy minimization calculation obscured some of the physics. In the energy minimization calculation, it is assumed that the magnetic moments in the ferromagnet and antiferromagnet lie in the plane of the sample. Later calculations [46,47] showed that as the domain wall was "wound up" in the reversal process, the spin system would become unstable due to out-of-plane fluctuations. The spin moments would briefly rise up out of plane, and then settle down in a new configuration—effectively unwinding the wall a little. In a compensated system, this leads to enhanced coercivity, but not to any exchange bias. In a system with some lack of compensation, one can find both exchange bias and enhanced coercivity.

The calculations that indicated the instability were based on the Landau–Lifshitz–Gilbert (LLG) equation. This equation is an analog of the classical equation $dL/dt = \text{torque}$ and is given by

$$\frac{d\vec{M}}{dt} = \gamma\left(\vec{M} \times \vec{H}_{\mathrm{eff}}\right) - \frac{\alpha}{M}\left(\vec{M} \times \frac{d\vec{M}}{dt}\right), \tag{27}$$

where the first term on the right-hand side represents the torque on a magnetic moment due to an effective field H_{eff}, and the second term is a phenomenological damping term which allows relaxation of the magnetization toward the equilibrium direction (H_{eff}) and keeps the length of the magnetization vector a constant in time.

The use of the LLG equations can give information on both the static and dynamic properties of magnetic multilayers. In this method, one writes a set of LLG equations for each layer in the multilayer:

$$\frac{d\vec{M}_i}{dt} = \gamma\left(\vec{M}_i \times \vec{H}_i\right) - \frac{\alpha}{M_i}\left(\vec{M}_i \times \frac{d\vec{M}_i}{dt}\right). \tag{28}$$

Layer i is coupled to the other layers through the effective field, H_i, acting on layer i, as we have seen earlier. One picks an initial spin configuration and integrates the coupled set of equations forward in time using a differential equation solver such as the fourth-order Runge–Kutta method. To find the equilibrium position a large value of the damping constant is used and one integrates forward in time until the M_i values do not change significantly. Again, there is no guarantee that this is the ground state rather than some metastable state, so care must be taken to choose an initial configuration that is close to the expected ground state.

In contrast to the energy minimization method, where it is often assumed that the spin moments lie in the plane of the film, the LLG calculation allows for complete three-dimensional motion of the spins. It is important, therefore, that the demagnetizing fields acting on each film be included in the effective field, because these fields provide the physical mechanism that encourages the equilibrium spin positions to be in the plane of the film.

To find the dynamic modes, the damping is then reduced substantially and the system is given a small perturbation. Again, the system is allowed to evolve in time and one finds $M_x(t)$, $M_y(t)$, and $M_z(t)$ for each layer. The Fourier transform of the time-dependent magnetizations then gives the spin-wave frequencies. The total effective field is a sum of exchange, anisotropic, dipolar, and external fields. We assume that the main contribution to the dipole field is the demagnetizing field of a thin layer acting on itself; thus, $-4\pi M_y$, where y is the direction perpendicular to the film surface. This approximation works well in the thin-film or long-wavelength limit.

The method outlined above has been used to study a number of different systems. These include ferromagnetic/antiferromagnetic structures, coupled ferromagnetic films, and magnetic nanostructures. In addition to finding spin wave frequencies, this method can also find the response of the structure to a driving field. If the driving field is oscillatory in time, one essentially models ferromagnetic resonance, but this method can also find the response of the spin system to a single field pulse.

In principle one can combine the thermodynamic treatment discussed earlier with the method illustrated in this section, i.e., one can use position-dependent magnetization values which depend on the temperature, and then one can find the temperature-dependent static and dynamic behaviors. If one want to do this in a self-consistent manner, all within one program, it does call for a certain amount of caution. Standard dynamical calculations often ignore the portion of the exchange field which produces zero torque. The thermal behavior, in contrast, depends on the entire effective field and not just the portion of the field which produces a torque. So a self-consistent thermal calculation must also use the entre effective field at each step. However because exchange fields are so large, this often means that the time step in the time integration has to be quite small, especially in the early portions of the calculation where the system is evolving toward an equilibrium configuration.

5.1 Exchange Spring Structures

As an example, we review the behavior of the exchange-spring system where a ferromagnet with a large anisotropy SmCo is coupled to a ferromagnet with a weak anisotropy, Fe. We emphasize that a single calculational scheme and a single set of parameters is able to describe both the static properties (hysteresis curves) and the dynamic properties (spin-wave frequencies) with reasonable accuracy [48]. The parameters for the calculation involve the exchange coupling between layers in the Fe and between layers in the SmCo and at the interface between the two materials. In addition, one needs the magnetization and anisotropy for each material. Except for the interface coupling, all of these parameters are well established for single films of each material.

The equilibrium spin structure depends on the direction of the applied field, as is illustrated in Fig. 24. For fields aligned with the anisotropic axis one has a simple configuration where the moments in both ferromagnets line up with the external field. When the field is reversed (Fig. 24b) the moments in the SmCo are held in place by the strong anisotropy in that material. The exchange coupling between the Fe and the SmCo then fixes the Fe moments as well. As the reversed field is increased the outermost Fe moments try to line up with the external field, creating a domain wall in the Fe (Fig. 24c). Eventually the external field is so large that the SmCo moments also switches and lines up with the applied field.

The hysteresis curve with the field along the easy axis for a 100 Å Fe film on SmCo is shown in Fig. 25. The symbols are experimental data and the full line

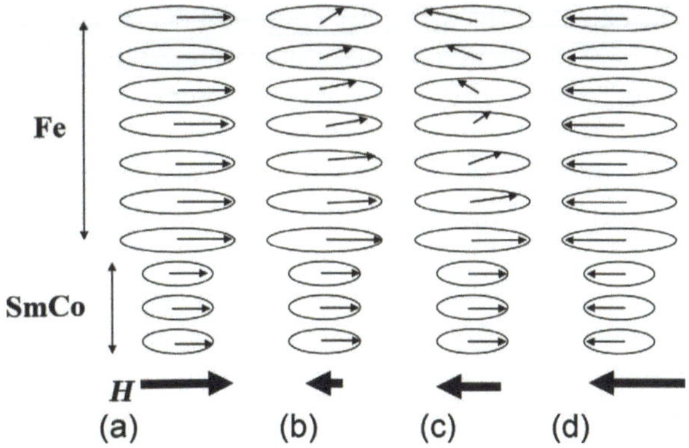

FIGURE 24

Schematic illustration of spin configurations in the Fe/SmCo exchange-spring system. (a) The moments are all aligned with the external field *H*. (b) The external field has been reversed but is small. The outermost Fe spins begin to turn toward the field. (c) The external field is reversed and larger, and the Fe moments are nearly completely reversed near the top of the film. (d) The external field is now so large that all the moments, both in Fe and SmCo, line up with the reversed field.

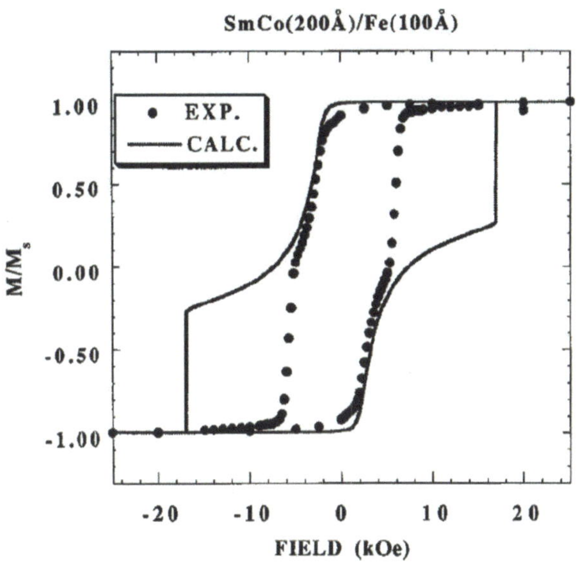

FIGURE 25

Experimental and theoretical results for magnetization in an applied field for an Fe/SmCo structure. The main difference between the calculated and measured behavior is in the reversal field of the SmCo. If this reversal is aided by domain wall motion the calculation will overestimate the field required for reversal. *From Ref. [48].*

is the result of the calculations using the coupled LLG equations. As the field is reversed from positive to negative values, the magnetization stays constant until about 1.5 kOe due to the pinning of the Fe moments. As the reversed field is further increased in magnitude, the twist seen in Fig. 24c occurs and the magnetization is gradually reduced. In the theoretical calculations this twist extends out to a value of about 17 kOe; however, the experiments show the SmCo film switches at a much lower field of 6 kOe. It is assumed that the reason for this is that the model only allows switching to occur by coherent rotation, but that in the sample domain nucleation and domain wall motion may contribute substantially to the switching field for the SmCo. Except for the switching field of the SmCo film, excellent agreement is obtained, and this allows one to obtain a reasonable value for the interface coupling constant.

Using the same numerical method, one can now find the frequencies of the dynamic modes as described above. In Fig. 26, we present frequency as a function of applied field for the 100 Å Fe/200 Å SmCo sample. Having found the interface coupling from the static measurements, there are no adjustable parameters in this calculation. The agreement between experiment and theory is quite good, with agreement in the region of -5 kOe $<H<0$ kOe, confirming the picture of a domain wall-like twist in the Fe.

The model discussed here has substantial predictive power. One can use the same parameters to calculate the excitation frequencies for structures with different Fe thicknesses. Figure 27 shows the theoretical and experimental results for frequency at $H=0$ as a function of the thickness of the Fe film. Again, the agreement between theory and experiment is excellent and shows a dramatic increase in frequency as the thickness of the Fe is reduced. Similar structures

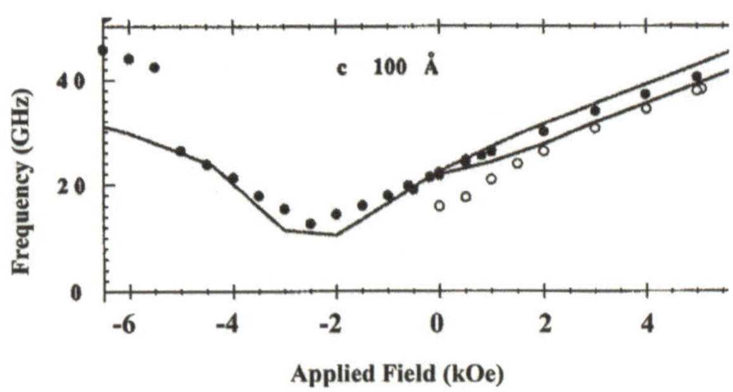

FIGURE 26

Frequency versus applied field for a 10-nm Fe film on a 20-nm SmCo film. The region -5 kOe $<H<0$ spans the twisted state seen in Fig. 23c and d. The filled circles are for measurements with the magnetic field parallel to the easy axis and the open circles represent measurements with the field perpendicular to the easy axis. *From Ref. [48].*

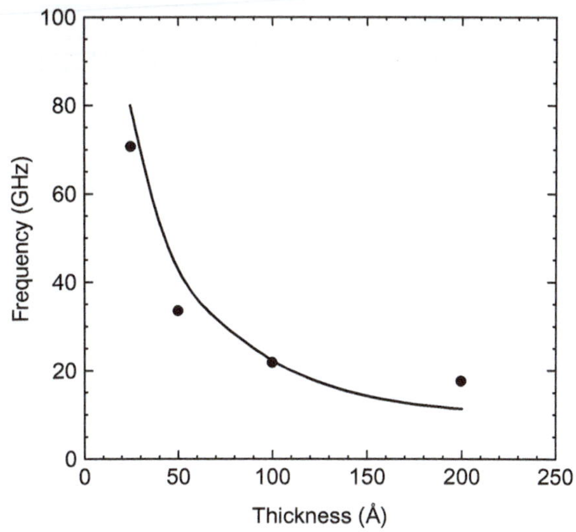

FIGURE 27
Theoretical (solid line) and experimental (dots) frequencies found at $H=0$ as a function of Fe film thickness for Fe/SmCo structures. *From Ref. [48].*

and the resulting frequency range of 20–70 GHz may be used to create new signal processing devices at high frequencies [49].

5.2 Other Examples

This same method has been employed in a number of calculations for the static and dynamic properties of nanostructures. A recent study, for example, examined the properties of permalloy dots with a diameter of 800 nm and a thickness of 60 nm [50]. For this size of dot, both a vortex magnetization state and a saturated state are possible, with the vortex state favored at low applied fields and the saturated state occurring in higher applied fields. The dynamics of the structure are predicted by using the LLG equations to first find the equilibrium spin structure. This is done with a large value of α to ensure relaxation to the ground state. The value of α is then reduced, the system is allowed to evolve in time, and the Fourier transform is calculated as discussed above. The results are shown in Fig. 28. It is immediately obvious that the vortex state has excitations at two frequencies, while the saturated state shows an excitation at only one frequency. These calculations were complemented by Brillouin light scattering measurements. As seen in Fig. 29, these measurements find two modes at low fields where the vortex state is stable and only one mode at high fields where the saturated state is stable.

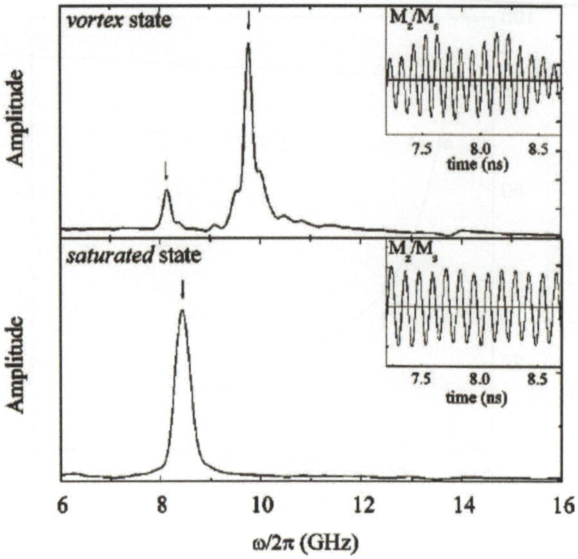

FIGURE 28

Fourier transform of the time evolution of one component of the magnetization for a nanodot in the vortex state and in a saturated state. The time evolution is shown in the inset for each case. *From Ref. [50].*

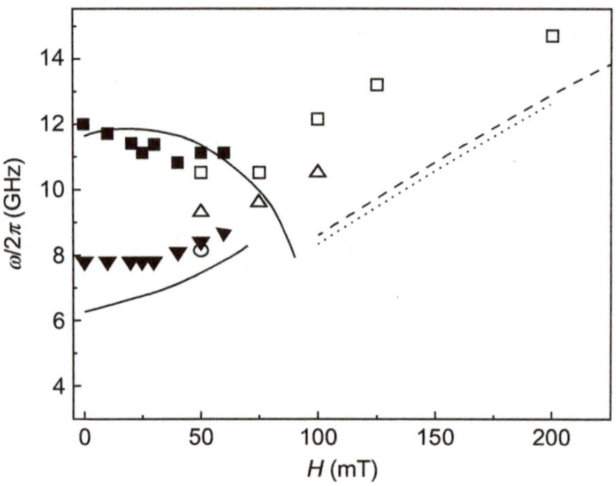

FIGURE 29

Theoretical calculation (solid lines and dashed lines) and experimental measurements (symbols) for dynamic mode frequencies in a nanodot. At low fields, two modes are observed, a result found in the vortex state. At high fields, only one mode is found in the saturated state. *From Ref. [50].*

A recent study points out the possibility of studying localized modes in nonuniform magnetic structures. The static and dynamic behavior in a rectangular bar of a ferromagnetic material was explored in Refs. [51,52]. Because of demagnetizing fields the entire structure is not in the saturated state even for large applied fields along the long direction of the bar. Nonetheless, the excitations of the system may be found using the methods outlined above. This calculation used a sophisticated method to calculate the dipole fields and as a result one may find the normal modes of the system without any artificial imposition of "pinned" or "unpinned" spins as a boundary condition at the surface of the nanostructures. As an example, a "snapshot" of one component of the magnetization is shown in Fig. 30. This end-mode excitation in this example has a frequency of 31 GHz, which is considerably lower than the lowest bulk mode at 52 GHz. The frequency difference is due, in part, to the nonuniform demagnetizing field that is largest near the ends of the bar, thus showing the ability of these calculations to allow for local variations.

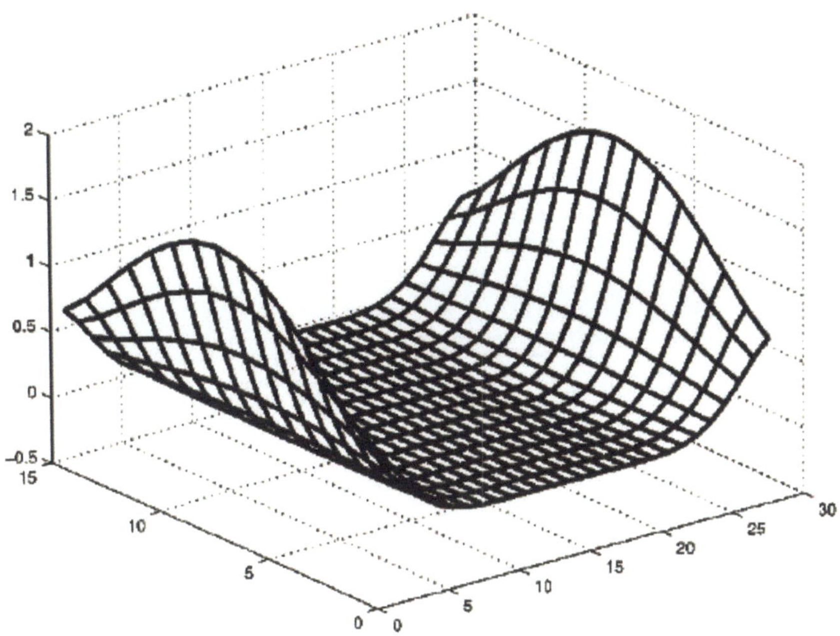

FIGURE 30
A "snapshot" of the behavior of one component of magnetization in a rectangular nanobar. The picture shows a well-defined end-mode. *From Ref. [52].*

6 NANOPARTICLES AND NANOSTRUCTURES

Magnetic nanoparticles have a number of important applications. Perhaps the two most developed directions concern magnetic recording [53] and biological applications. In magnetic recording one wants small particles which retain their magnetic state in zero field, even at elevated temperatures. For biological applications [54,55], magnetic nanoparticles are used for drug delivery, sensors, lab-on-a-chip applications, and for contrast agents in magnetic resonance imaging (MRI). All of this work is fundamentally influenced by the thermal behavior of the magnetic nanoparticles.

The calculation of the thermal behavior of nanoparticles is similar to that described earlier; however, there are two important differences in the outcomes:

(1) Spins near surfaces and edges have significantly reduced coordination numbers compared to those in the interior.
(2) The local magnetic structure (i.e., ferromagnetic, antiferromagnetic, etc.) can be thermally stable, but the orientation of the total magnetic moment with respect to an external field may not be stable due to fluctuations caused by thermal energy. This is the so-called superparamagnetic effect.

6.1 Reduced Coordination Effects in Nanoparticles

We first concentrate on reduced coordination issues. In Fig. 31, we give a simple example indicating the reduced coordination at different sites in a nanostructure. Here we look at a square two-dimensional structure and highlight the nearest neighbors for corner spins, edge spins, and interior spins. In this limit,

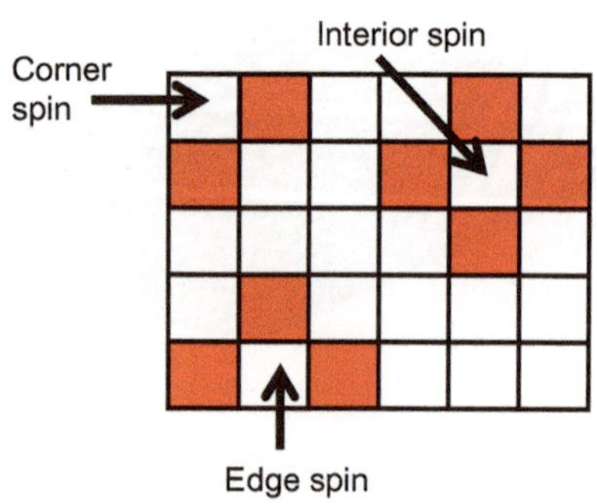

FIGURE 31
In a nanostructure the number of nearest neighbors for a spin depends on its position. For nearest neighbor coupling, corner spins are stabilized against thermal fluctuations by exchange fields from only two nearby spins. Edge spins have exchange coupling to three spins, and interior spins are exchange-coupled to four spins.

the corner spins have half as many nearest neighbors as interior spins. Thus, one could guess that the thermal averaged magnetization for the corner spins will vanish at about half the bulk transition temperature. Of course real systems are three-dimensional and are generally not simple cubic. However, the basic ideas will still hold—atoms on a surface will have reduced coordination numbers compared to those in the bulk and in structures with sharp edges and corners the coordination number will be further reduced.

We illustrate the consequences of these ideas with some examples. Figures 32 and 33 show results for the thermal-averaged spin values in a nanosized Gd cube with 10 lattice sites in each direction. For simplicity, we have taken the structure to be simple cubic. The spin value of Gd is taken as $S=3.5$ and the effective exchange constant is found by setting $T_{\text{C-bulk}}=292$ K. We use the self-consistent mean field theory to calculate the value of $\langle S(T)\rangle$ at each

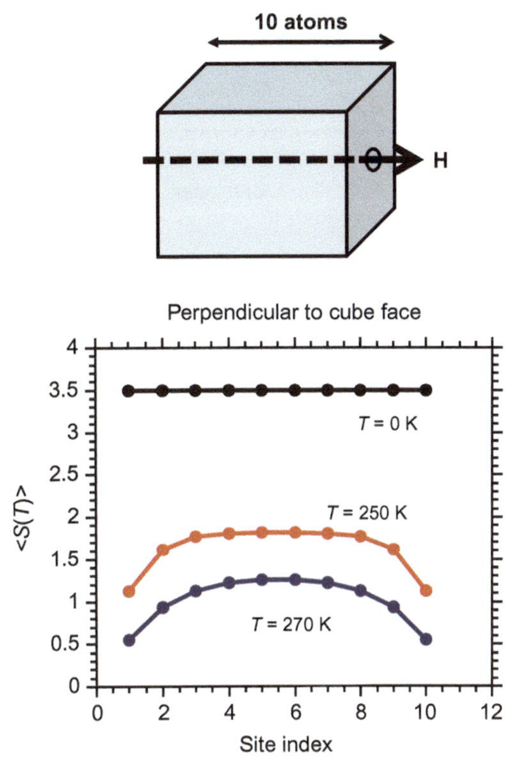

FIGURE 32

Calculated values for $\langle S(T)\rangle$ sites along the dashed line in a $10\times10\times10$ Gd cube. The applied field is 1 kOe and the bulk transition temperature is $T=292$ K. Well below the bulk transition temperature, one can already see significant reductions of the $\langle S\rangle$ values near the faces of the cube due to the reduced number of nearest neighbors at the faces.

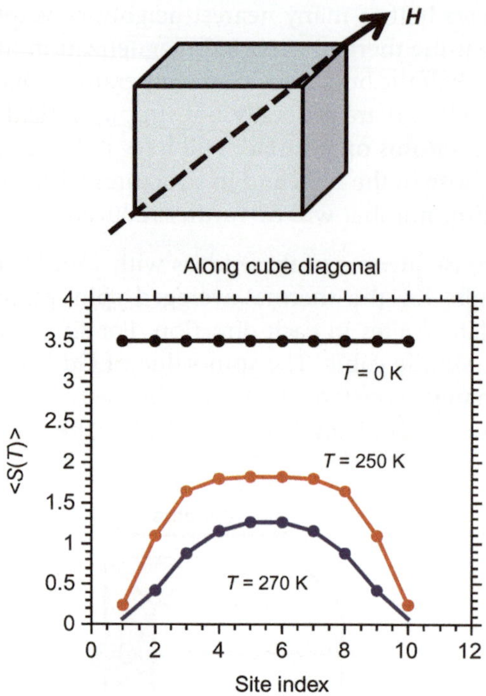

FIGURE 33

Calculated values for $\langle S(T)\rangle$ for sites along the dashed line in a $10 \times 10 \times 10$ Gd cube. The applied field is 1 kOe, and the bulk transition temperature is $T=292$ K. Well below the bulk transition temperature, one can already see significant reductions of the $\langle S\rangle$ values near the corners of the cube due to the reduced number of nearest neighbors at the corners. The reduction in this case is even larger than that seen for the faces in Fig. 32.

site as described earlier. (We note this is quite fast; one gets results for all temperatures from 0 to 350 K, with a 10 K step, in just a few minutes.) An applied field of 1 kOe is used to help stabilize the structure in both cases.

In Fig. 32, the external magnetic field is set perpendicular to a cube face and we measure $\langle S(T)\rangle$ for sites along the dashed line. At $T=0$, all the spins have their maximum value, $S=3.5$. As the temperature is increased, the outer spins are less stable against thermal fluctuations (having fewer neighbors) and the averaged value is reduced compared to the interior value. The effect is significant at higher temperatures. At $T=270$, the $\langle S\rangle$ value for the outer spins is only about half of the value found for the interior spins.

Based on the earlier discussion, one would expect an even larger effect found if one examined spins along a diagonal because this involves corner spins. This is demonstrated in Fig. 33, where the applied field and the $\langle S(T)\rangle$ values along the

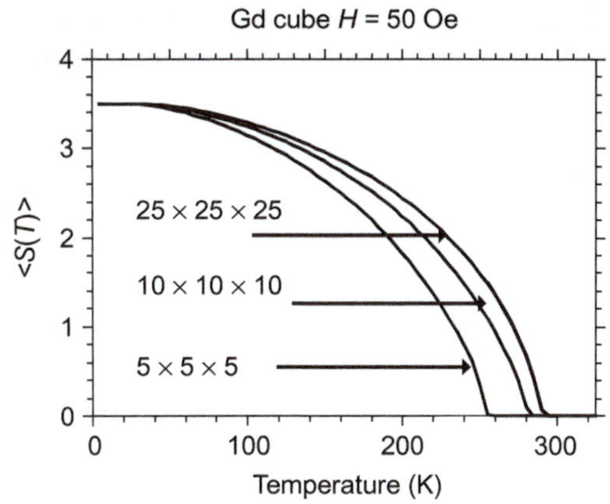

FIGURE 34
A plot of the thermal-averaged spin value (per atom) for different-sized magnetic structures. The largest nanoparticle, a Gd cube with 25 sites on a side, has results close to that of bulk Gd. The smallest structure, a Gd cube with only five sites on a side, has a substantially reduced transition temperature (graph in Gd percolation).

diagonal are presented. Even at 250 K, the $\langle S \rangle$ value for the corner spins is already close to zero. The variation in $\langle S \rangle$ with site index is quite large, both at $T=250$ and $T=270$ K. One sees that the system is truly surface dominated in that the interior value of $\langle S \rangle$ only holds for a small number of spins; the majority of spins are influenced by the reduced coordination at the surface.

The surface effects just outlined can have a significant impact on the overall magnetization of a nanostructure, particularly if the structure is quite small. Figure 34 shows the thermal-averaged spin value per atom as a function of temperature for three different-sized nanocubes. We continue with the Gd example and consider cubes with 5, 10, and 25 lattice sites on a side. Again, a small magnetic field is applied to speed numerical convergence. One sees that the average $\langle S \rangle$ values are smaller for the smaller structures at a given temperature. The transition temperature for the $25 \times 25 \times 25$ structure is very close to the bulk value of 292 K, but the transition temperature for the $5 \times 5 \times 5$ structure is reduced by about 13% compared to the bulk value.

6.2 Superparamagnetism and Magnetic Nanoparticles

All of the examples above examined the local magnetic structure. As indicated earlier, there is a second issue—whether the net magnetic moment of a structure can be aligned with an external field, or whether the fluctuations caused by thermal energy will reduce or eliminate this alignment. In the simplest version, we assume that the (local) magnetic moment of the nanoparticle is independent of the field and is a fixed value at a given temperature. We will also assume that there is no anisotropy in the problem at this stage. In essence the particle acts like a giant single magnetic moment. This is exactly what one uses in calculating the alignment of a group of noninteracting atomic moments, i.e., a paramagnetic material. Here, however, the moment is large compared to a single atomic moment, and this gives it the name superparamagnetism.

The problem then becomes a simple statistical mechanics problem—calculating the thermal-averaged value of a classical magnetic moment along the direction of the applied field. This is what would be measured in a typical magnetization system such as a SQUID or a vibrating sample magnetometer. We assume the applied field H is in the z direction and that the magnetic moment, m, can be at an angle θ with respect to the z-axis. One creates the ratio:

$$\frac{\text{Magnetic energy}}{\text{Thermal energy}} = \frac{-mH\cos\theta}{kT}. \tag{29}$$

The moment, in this case, is macroscopic and the magnetic energy is a continuous function of the angle between the applied field and the moment. This is in contrast to the quantum case of a single spin, where there is only a set of discrete values for the z component of the magnetic moment. The z component of the magnetic moment is then given by

$$< m_z > = \frac{\int_0^{\pi} m\cos\theta \exp(-mH\cos\theta/kT)\sin\theta d\theta}{\int_0^{\pi} \exp(-mH\cos\theta/kT)\sin\theta d\theta}. \tag{30}$$

For the macroscopic case, and using the Boltzmann occupation factor, one has to integrate over all solid angles (θ and ϕ variables) but the ϕ integration cancels above. After some calculation, one finds

$$< m_z > = mL(x). \tag{31}$$

Here $L(x)$ is the Langevin function, given by

$$L(x) = \coth(x) - 1/x, \tag{32}$$

where $x = mH/kT$.

The key issue is that as the particle becomes larger, there will be a larger net magnetic moment m. The factor x becomes larger and $L(x)$ approaches 1; the z component of the magnetic nanoparticle becomes fully aligned with the external field.

It is instructive to see some numerical examples. Figure 35 gives results for a small Gd cube with five atoms on a side. The dashed lines show what would be measured in a typical magnetization experiment: $\langle S_z \rangle$, the thermal averaged

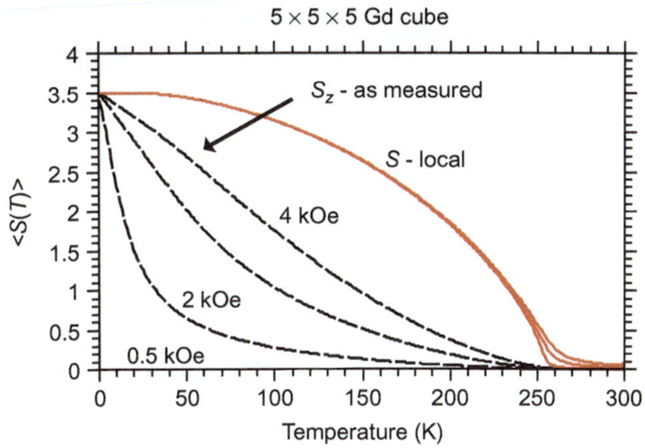

FIGURE 35

A comparison of the thermal-averaged spin moment as a function of temperature for a small Gd cube. The solid lines represent the magnitude of the moment and the dashed lines show what would be measured in a typical magnetization experiment—the component of $\langle S \rangle$ which lines up with the external field.

z component (per atom) of the spin oriented by the external magnetic field which is also along z. The solid lines show the thermal-averaged value of the spin magnitude $\langle S \rangle$ as calculated earlier in this section. The differences are dramatic. For such a small particle, the thermal energy is large compared to the total magnetic energy in the applied field. As a result, the measured magnetic moment will change dramatically depending on the applied field.

As the particle size is made larger, the superparamagnetic effect becomes less important. Figures 36 and 37 show results for Gd cubes with side lengths of 10 and 20 atoms. We note that the results depend strongly on the applied field for the $10 \times 10 \times 10$ structure. Even at 2 kOe there is a difference between the measured moment and the local moment. For the larger cluster, one finds reasonably close agreement between the measured results and the local results at a field of 0.5 kOe. To put this in better perspective, the largest structure has 8000 Gd atoms, which is equivalent to a Gd cube with a side length of 6.4 nm or a sphere with a radius of about 4 nm. (This is calculated using the density of Gd, which is 7.9 g/cm^3.) Nanoparticles, such as iron oxide, where the magnetic moments are less tightly packed, require even larger sizes to be above the superparamagnetic limit.

For biological applications, there is often a "sweet spot" for the size of the nanoparticles ranging from about 30 to 80 nm. Larger particles tend to be trapped and remain in the body; smaller particles are often rapidly excreted from the body. The size and range of the sweet spot may depend on the particular application. For example, a recent study of magnetic nanoparticles for therapy in the eyes and the ears indicated a wider range, from 60 to about 800 nm [56].

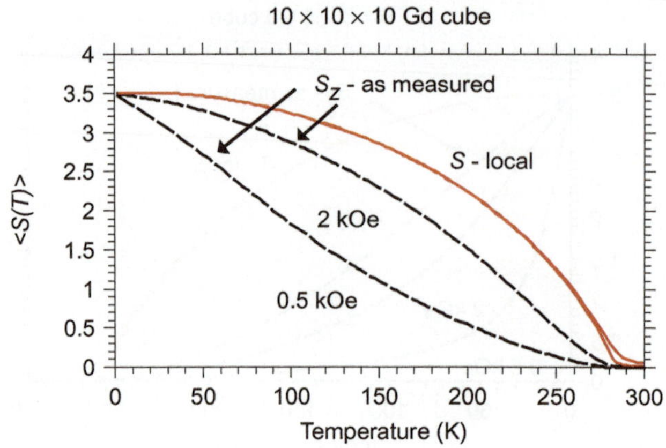

FIGURE 36

A comparison of the thermal-averaged spin moments as a function of temperature for a small Gd cube. The solid lines represent the magnitude of the moment and the dashed lines show what would be measured in a typical magnetization experiment—the component of $\langle S \rangle$ which lines up with the external magnetic field directed along *z*.

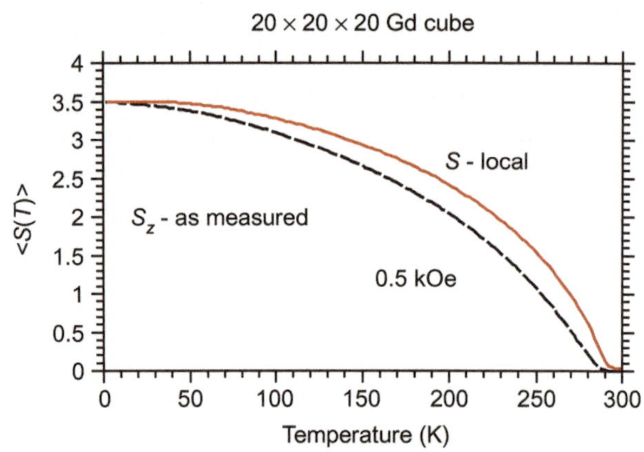

FIGURE 37

A comparison of the thermal-averaged spin moments as a function of temperature for a small Gd cube. The solid lines represent the magnitude of the moment and the dashed lines show what would be measured in a typical magnetization experiment—the component of $\langle S \rangle$ which lines up with the external magnetic field directed along *z*. For this size particle, the measured magnetization will be close to the total moment at an external field value of 0.5 kOe. This is in contrast to the smaller nanoparticles investigated in the previous two figures.

As a result of this size requirement, most iron oxide nanoparticle studies (a material allowed in human bodies by the FDA) for biological applications are done with superparamagnetic particles [57]. If one is using large magnetic fields, as is often the case for drug delivery or for MRI [58], the behavior of these particles will be similar to that of ferromagnetic particles except that there is no hysteresis. Ferromagnetic nanoparticles made of Fe, Ni, Co, and other materials had disadvantages in that they could be toxic or be highly reactive. Nonetheless, coated ferromagnetic nanoparticles may have some advantages [59]. One immediately seen advantage occurs in trying to direct the nanoparticles with external magnetic fields. As is well known, the force on a magnetic dipole depends on the gradient of the applied magnetic field and is given by

$$\vec{f} = \left(\vec{m} \cdot \nabla\right) \vec{H}. \tag{33}$$

For superparamagnetic particles, the value of the magnetic moment can be reduced from its maximum depending on temperature and applied field, as we have seen. However, for ferromagnetic particles the magnetic moment can have its maximum value and one can obtain the largest force. Ferromagnetic nanoparticles have also been studied as possible contrast agents for MRI applications [60].

In contrast, the difference between superparamagnetic and ferromagnetic nanoparticles is very important for the recording industry. In magnetic recording a stored state must be stable for a very long time (typically the requirement is 10–25 years) in the absence of any applied field. For magnetic recording the magnetization is held in place by an anisotropic field. In this case the key parameter is the ratio of the total anisotropic energy to the thermal energy, i.e.,

$$x = \frac{KV}{k_B T}, \tag{34}$$

where K represents the magnetic anisotropic energy density and V represents the volume. A standard measure for the stability time is given by an Arrhenius law relating to the probability of exceeding an energy barrier:

$$T = T_o e^x. \tag{35}$$

In this formula, T represents the desired stability time and T_o represents an attempt time (typically of the order of about 1 ns). With a desired stability time of 25 years, one needs a value of $x > 41$. This can put strict limits on the minimum volume.

The methods developed here have been applied to more complex nanostructures composed of multiple nanoparticles. Dots [61,62], dot arrays [63], and antidot arrays have been examined, and the influence of magnetic pinholes in the spacer layer between ferromagnetic films on the exchange coupling between the films as a function of temperature has been studied [64]. Differences in the dynamics between granular systems and multilayers have been examined [65]. In addition artificial, complex structures of elongated nano-objects, known as spin-ice structures, have had their thermal properties studied using the methods developed here [66].

7 OTHER METHODS

The method outlined in this chapter is simple, easy to implement, and can give quantitative agreement with experiment. It is, however, not as rigorous as some other methods. We will give a short discussion of two other methods:

(1) Micromagnetics with random fields representing the effect of the thermal bath
(2) Monte Carlo methods.

A nice review of both these methods can be found in Ref. [67].

7.1 Micromagnetics Methods

One fundamental method for calculating thermal magnetic behavior uses the coupled equations of motion for each atomic spin in a nanostructure. This is given by

$$\frac{d\vec{S}}{dt} = \gamma\left(\vec{S}\times\vec{H}_{\text{eff}} + \alpha\frac{\vec{S}\times\left(\vec{S}\times\vec{H}_{\text{eff}}\right)}{S}\right) + \gamma\left(\vec{S}\times\vec{h}\,(t)\right). \tag{36}$$

This is similar to the equations of motion discussed before, but there is an additional field $\vec{h}\,(t)$ which represents fluctuating magnetic fields produced by thermal effects. This fluctuating field must obey certain restrictions:

(1) The time-averaged value of the field is zero:

$$\left\langle\vec{h}(t)\right\rangle_{\text{time}} = 0 \tag{37}$$

(2) There is no correlation between different sites i and j, or between different components of the field
(3) The average time correlation for the fluctuating magnetic fields is given by

$$\left\langle\vec{h}_i(0)\vec{h}_j(t)\right\rangle = \left(\frac{2\alpha kT}{\gamma}\right)\delta_{i,j}\delta(t), \tag{38}$$

where the constant preceding the delta functions is obtained through the use of the fluctuation-dissipation theorem.

This results in a well-defined procedure for calculating the dynamics of the motion at any temperature. The equilibrium behavior can then be found by appropriate averaging of the dynamic quantities over time.

However, there are a number of important issues that can limit the use of this method:

(1) The individual spins are atomic in nature—this would typically require over a million sites to simulate a cube with a side length of only 1 μm.
(2) It is difficult to simulate long times with micromagnetics. For example, in studying magnetic reversals in magnetic recording, one is interested in reversals that could happen over a period of years. In micromagnetics one is typically restricted to net times of 10–100 ms.
(3) The time step required is typically a fraction of a picosecond, and the repeated computations of the dipolar fields make the calculations numerically intensive.
(4) In nonequilibrium states, the exchange fields can be very large. This also requires very small time steps in the numerical integration of the equations of motion.

Despite this, this method has seen substantial use. A recent example involved simulating the switching of a small grain for a typical perpendicular recording material [68] and examining the effect of sweep rate on the coercive field.

There have also been a number of variations and alternatives of traditional micromagnetic calculations with stochastic fields. Notable among these is work which allows calculation of the coercive field over time scales up to years in length [69].

Another variation on thermal micromagnetic calculations has been proposed [70] to understand magnetic reversals in the presence of significant heating [71]. A key issue is that single-spin micomagnetics models conserve the spin length at each site. But because the temperature can be extremely high, this may not be true in some experiments. This approach may also be useful in understanding and modeling heat-assisted magnetic recording (HAMR) [72].

7.2 Monte Carlo Calculations

Unlike the micromagnetic simulations discussed above, Monte Carlo techniques do not give true dynamics; there is no precessional motion and time does not explicitly enter the problem. However, there are advantages to this method.

The most common Monte Carlo calculation uses the Metropolis method to select states that will be considered in a thermal average. One starts by creating a collection of interacting spins. The interactions can include exchange fields, dipole fields, anisotropic fields, and an external magnetic field. For a classical system, the orientation of the spins may take on any angle. One spin is chosen at random and a trial step, rotation to a different angle, is considered. The change in energy $\Delta E = (E_{\text{trial}} - E_{\text{old}})$ is then found. If the energy of the spin is lower in the trial position, the step is accepted and a different spin is chosen.

If the new energy of the spin is higher, the trial step is accepted with probability $\exp(-\Delta E/kT)$, where k is Boltzmann's constant and T is temperature. If the trial step is rejected, one uses the old value again in the sums that follow. Thus, larger temperatures lead to more high-energy states being accepted.

To find the thermal behavior, one performs sums over all steps. For example, the average spin at site i in direction z is given by

$$< S_{iz} >= \frac{1}{A} \sum_{\substack{\text{all steps} \\ a=1}}^{A} S_{iz}(a). \tag{39}$$

This method is relatively simple to implement because it deals only in energies and not in vector fields as is found in the micromagnetics method. Nonetheless, there are a number of issues that must be considered. First, one should wait until the system is close to equilibrium before starting the sum. In this way, the sum will represent states that are thermally allowed from an equilibrium configuration. Second, it may not be efficient to do this sum including values after every single step. One common method is to sum over Monte Carlo Steps (MCS), where one MCS corresponds to N trial steps being considered, where N is the number of spins in the structure. Third, the size of the trial step (i.e., the change in angle from one step to the next) is important. If it is too large, many trials are rejected, and the system does not sample phase space properly. If it is too small, the system typically does not get far enough away from the initial state, and again phase space is not properly sampled. One often hears that as a rule-of-thumb, one wants an acceptance rate of about 50%.

Because this method is relatively easy to implement, there have been many papers using this technique on a wide variety of topics. These include magnetic reversals in a wire [73], single-atom magnetic chains [74], and hysteretic behavior in a collection of nanoparticles for biological applications [75].

There have been a number of extensions and variations to the standard Monte Carlo method as well. A number of papers have appeared using Kinetic Monte Carlo methods to discuss thermally driven magnetic reversals of magnetic media where the reversal time can be hundreds of years [76–79].

8 SUMMARY

It is clear that to understand magnetic multilayers and nanostructures both the static configurations and the spin dynamics must be examined. The techniques outlined here give simple methods to obtain complex magnetic structures as a function of magnetic field and temperature. All the parameters that characterize the magnetic material—anisotropy, exchange coupling, magnetization, surface effects, impurities, dipolar interactions, and magnetic structure—play a role in determining the frequency of the allowed spin waves. As a result, measurements of the dynamic properties are very useful in determining the fundamental parameters of magnetic structures.

The techniques discussed here have already been applied to a large range of problems. In addition to the ones mentioned earlier, these methods have been used to study giant magnetocaloric effects in Ho films [80], thermal hysteresis in ferromagnetic/antiferromagnetic bilayers [81], surface-induced nucleation of the heli-fan phase in helical rare-earth structures [82], phase transition in FeRh films [83], and thermal hysteresis in FeCo alloys [84], just to mention a few examples. Using these same methods, one can also study the nonlinear dynamic and thermal behavior of multilayers and other systems [85]. This is covered in detail in chapter 4 by K. Livesey in this book.

Acknowledgments

Some of the research in this chapter was supported by the UCCS BioFrontiers Center, the University of Western Ontario, and the University of Central Florida. REC would like to thank T.S. Rahman and M.G. Cottam for helpful conversations.

References

[1] Binasch G, Gruenberg P, Saurenbach F, Zinn W. Enhanced magnetoresistance in layered magnetic structures with antiferromagnetic interlayer exchange. Phys Rev B 1989;39:4828–30.

[2] Baibich MN, Broto JM, Fert A, Van Dau NF, Petroff F, Etienne P, et al. Giant magnetoresistance of (001)Fe/(001)Cr magnetic superlattices. Phys Rev Lett 1988;61:2472.

[3] Camley RE, Barnas J. Theory of giant magnetoresistance effects in magnetic layered structures with antiferromagnetic coupling. Phys Rev Lett 1989;63:664.

[4] Parkin SSP, More N, Roche KP. Oscillations in exchange coupling and magnetoresistance in metallic superlattice structures: Co/Ru, Co/Cr, and Fe/Cr. Phys Rev Lett 1990;64:2304.

[5] Corciovei A, Costache G, Vamanu D. Ehrenreich H, Seitz F, Turnbull D, editors. Solid state physics, vol. 27. New York: Academic; 1972. p. 237.

[6] Majkrzak CF, Cable JW, Kwo J, Hong M, McWhan DB, Yafet Y, et al. Observation of a magnetic antiphase domain structure with long-range order in a synthetic Gd-Y superlattice. Phys Rev Lett 1986;56:2700.

[7] Kwo J, Hong M, Di Salvo FJ, Waszczak JV, Majkrzak CF. Modulated magnetic properties in synthetic rare-earth Gd-Y superlattices. Phys Rev B 1987;35:7925.

[8] Camley RE, Stamps RL. Magnetic multilayers: spin configurations, excitations and giant magnetoresistance. J Phys Condens Matter 1993;5:3727.

[9] Keffer F, Chow H. Dynamics of the antiferromagnetic spin-flop transition. Phys Rev Lett 1973;31:1061.

[10] Mills DL. Surface spin-flop state in a simple antiferromagnet. Phys Rev Lett 1968;20:18.

[11] Lepage JG, Camley RE. Surface phase transitions and spin-wave modes in semi-infinite magnetic superlattices with antiferromagnetic interfacial coupling. Phys Rev Lett 1990;65:1152.

[12] Noertemann FC, Stamps RL, Carriço AS, Camley RE. Finite-size effects on spin configurations in antiferromagnetically coupled multilayers. Phys Rev B 1992;46:10847.

[13] Carriço AS, Camley RE, Stamps RL. Phase diagram of thin antiferromagnetic films in strong magnetic fields. Phys Rev B 1994;50:13453.

[14] Wang RW, Mills DL, Fullerton EE, Mattson JE, Bader SD. Surface spin-flop transition in Fe/Cr(211) superlattices: experiment and theory. Phys Rev B 1994;72:920.

[15] Rakhmanova S, Mills DL, Fullerton EE. Low-frequency dynamic response and hysteresis in magnetic superlattices. Phys Rev B 1998;57:476.

[16] Wang RW, Mills DL, Fullerton EE, Kumar S, Grimsditch M. Magnons in antiferromagnetically coupled superlattices. Phys Rev B 1996;53:2627.

[17] Camley RE, Tilley DR. Phase transitions in magnetic superlattices. Phys Rev B 1988;37:3413. Camley RE. Properties of magnetic superlattices with antiferromagnetic interfacial coupling: magnetization, susceptibility, and compensation points. Phys Rev B 1989;39:12316.

[18] Sajieddine M, Bauer Ph, Cherifi K, Dufour C, Marchal G, Camley RE. Experimental and theoretical spin configurations in Fe/Gd multilayers. Phys Rev B 1994;49:8815.

[19] Bauer Ph, Sajieddine M, Dufour C, Cherifi K, Marchal G, Mangin Ph. Direct evidence of the twisted state in ferrimagnet Gd/Fe multilayers by Mössbauer spectroscopy. Europhys Lett 1991;16:307.

[20] Takanashi K, Kamiguchi Y, Fujimori H, Motokawa M. Magnetization and magnetoresistance of Fe/Gd ferrimagnetic multilayer films. J Phys Soc Japan 1992;61:3721.

[21] Cherifi K, Dufour C, Bauer Ph, Marchal G, Mangin Ph. Experimental magnetic phase diagram of a Gd/Fe multilayered ferrimagnet. Phys Rev B 1991;44:7733.

[22] Tsunashima S, Ichikawa T, Nawate M, Uchiyama S. Magnetization process of Gd/Co multilayer films. J Phys Colloques 1988;49, C8-1803.

[23] Camley RE, Lohstroh W, Felcher GP, Hosoito N, Hashizume H. Tunable thermal hysteresis in magnetic multilayers: magnetic superheating and supercooling. J Magn Magn Mater 2005;286:65–71. Demirtas S, Hossu MR, Camley RE, Mireles HC, Koymen AR. Tunable magnetic thermal hysteresis in transition metal (Fe, Co, CoNi)/rare earth (Gd) multilayers. Phys Rev B 2005;72:184433.

[24] Hahn W, Loewenhaupt M, Huang YY, Felcher GP, Parkin SSP. Experimental determination of the magnetic phase diagram of Gd/Fe multilayers. Phys Rev B 1995;52:16041.

[25] Haskel D, Choi Y, Lee DR, Lang JC, Srajer G, Jiang JS, et al. Hard x-ray magnetic circular dichroism study of a surface-driven twisted state in Gd/Fe multilayers. J Appl Phys 2003;93:6507.

[26] Prigge S, Roux H, Bauer E. Pd layers on a W (100) surface. Surf Sci 1981;107:101–12. Rau C, Jin C, Robert M. Ferromagnetic order at Tb surfaces above the bulk Curie temperature. J Appl Phys 1988;63:3667.

[27] Siegman HC, Mauri D, Scholl D, Kay E. Surface and thin film magnetism with spin-polarized electrons. J Phys Colloques 1988;49:C89.

[28] Grünberg P. In: Cardona M, Güntherodt G, editors. Light scattering in solids, vol. V. Berlin: Springer; 1989 [chapter 8]. Grünberg P. Light scattering from magnetic surface excitations. Prog Surf Sci 1985;18(1):1–58.

[29] Cochran JF. In: Heinrich B, Bland JAC, editors. Ultrathin magnetic structures II. New York: Springer; 1994. p. 222.

[30] Hillebrands B, Güntherodt G. In: Bland JAC, Heinrich B, editors. Ultrathin magnetic structures II. Berlin: Springer; 1993.

[31] Heinrich B. In: Heinrich B, Bland JAC, editors. Ultrathin magnetic structures II. New York: Springer; 1994. p. 195–290.

[32] Bland JAC, Johnson AD, Lauter HJ, Bateson RD, Blundell SJ, Shackleton C, et al. Spin polarized neutron studies of epitaxial films. J Magn Magn Mater 1991;93:513.

[33] Collins MF. Magnetic critical scattering. Oxford: Oxford University Press; 1989.

[34] Saunders RW, Belanger RM, Motokawa M, Jaccarino V. Far-infrared laser study of magnetic polaritons in FeF_2 and Mn impurity mode in FeF_2:Mn. Phys Rev B 1981;23:1190.

[35] Remer L, Luthi B, Sauer H, Geick R, Camley RE. Nonreciprocal optical reflection of the uniaxial antiferromagnet MnF_2. Phys Rev Lett 1986;56:2752.

[36] Jensen MRF, Parker TJ, Abraha K, Tilley DR. Experimental observation of magnetic surface polaritons in FeF_2 by attenuated total reflection. Phys Rev Lett 1995;75:3756.

[37] Jensen MRF, Feiven SA, Parker TJ, Camley RE. Experimental determination of magnetic polariton dispersion curves in FeF_2. Phys Rev B 1997;55:2745.

[38] Jensen MRF, Feiven SA, Parker TJ, Camley RE. Experimental observation and interpretation of magnetic polariton modes in FeF_2. J Phys Condens Matter 1997;9:7233.

[39] Camley RE. Magnetization dynamics in thin films and multilayers. J Magn Magn Mater 1999;200:583.

[40] Vohl M, Barnas J, Gruenberg P. Effect of interlayer exchange coupling on spin-wave spectra in magnetic double layers: Theory and experiment. Phys Rev B 1989;39:12003.

[41] Stamps RL. Spin configurations and spin-wave excitations in exchange-coupled bilayers. Phys Rev B 1994;49:339.

[42] Rezende SM, Lucena MA, deAguiar FM, Azevedo A, Chesman C, Kabos P, et al. High-resolution Brillouin light scattering and angle-dependent 9.4-GHz ferromagnetic resonance in MBE-grown Fe/Cr/Fe on GaAs. Phys Rev B 1997;55:807.

[43] Drovosekov AB, Kholin DI, Kolmogorov AN, Kreines NM, Mescheriakov VF, Miliayev MA, et al. The observation of non-homogeneous FMR modes in multilayer Fe/Cr structures. J Magn Magn Mater 1999;198–199:455.

[44] See the reviews Nogues J, Schuller IK. Exchange bias. J Magn Magn Mater 1999;192:203. Stamps RL. Mechanisms for exchange bias. J Phys D Appl Phys 2000;33:R247. Kiwi M. Exchange bias theory. J Magn Magn Mater 2001;234:584.

[45] Koon NC. Calculations of exchange bias in thin films with ferromagnetic/antiferromagnetic interfaces. Phys Rev Lett 1997;78:4865.

[46] Schulthess T, Butler W. Consequences of spin-flop coupling in exchange biased films. Phys Rev Lett 1998;81:4516.

[47] Camley RE, Astalos RJ. Probing the ferromagnet/antiferromagnet interface with spin waves. J Magn Magn Mater 1999;198–199:402.

[48] Grimsditch M, Camley R, Fullerton EE, Jiang S, Bader SD, Sowers CH. Exchange-spring systems: coupling of hard and soft ferromagnets as measured by magnetization and Brillouin light scattering. J Appl Phys 1999;85:5901.

[49] Astalos RJ, Camley RE. Magnetic permeability for exchange-spring magnets: application to Fe/Sm-Co. Phys Rev B 1998;58:8646.

[50] Novosad V, Grimsditch M, Guslienko KYu, Vavassori P, Otani Y, Bader SD. Spin excitations of magnetic vortices in ferromagnetic nanodots. Phys Rev B 2002;66:052407.

[51] Gérardin O, Le Gall H, Donahue MJ, Vukadinovic N. Micromagnetic calculation of the high frequency dynamics of nano-size rectangular ferromagnetic stripes. J Appl Phys 2001;89:7012.

[52] Grimsditch M, Leaf GK, Kaper HG, Karpeev DA, Camley RE. Normal modes of spin excitations in magnetic nanoparticles. Phys Rev B 2004;69:174428.

[53] Frey NA, Sun S. Magnetic nanoparticle for information storage applications. Boca Raton, FL: CRC Press; 2010. p. 33–68.

[54] Tran N, Webster TJ. Magnetic nanoparticles: biomedical applications and challenges. J Mater Chem 2010;20:8760–7.

[55] Pankhurst QA, Connolly J, Jones SK, Dobson J. Applications of magnetic nanoparticles in biomedicine. J Phys D Appl Phys 2003;36:R167–81.

[56] Shapiro B, Kulkarni S, Nacev A, Sarwar A, Preciado D, Depireux DA. Shaping magnetic fields to direct therapy to ears and eyes. Annu Rev Biomed Eng 2014;16:455–81.

[57] Laurent S, Forge D, Port M, Roch A, Robic C, Vander Elst L, et al. Magnetic iron oxide nanoparticles: synthesis, stabilization, vectorization, physicochemical characterizations, and biological applications. Chem Rev 2008;108:2064–110.

[58] Qiao R, Yang C, Gao M. Superparamagnetic iron oxide nanoparticles: from preparations to in vivo MRI applications. J Mater Chem 2009;19:6274–93.

[59] Vitol EA, Novosad V, Rozhkova EA. Microfabricated magnetic structures for future medicine: from sensors to cell actuators. Nanomedicine (London) 2012;7:1611–24.

[60] Hadjipanayis CG, Bonder MJ, Balakrishnan S, Wang X, Mao H, Hadjipanayis GC. Metallic iron nanoparticles for MRI contrast enhancement and local hyperthermia. Small 2008;4:1925–9.

[61] Buda LD, Prejbeanu IL, Ebels U, Ounadjela K. Micromagnetic simulations of magnetisation in circular cobalt dots. Comput Mater Sci (Elsevier) 2002;24:181–5.

[62] Hoellinger R, Killinger A, Krey U. Statics and fast dynamics of nanomagnets with vortex structure. J Magn Magn Mater 2003;261:178–89.

[63] Stamps RL, Camley RE. Magnetization processes and reorientation transition for small magnetic dots. Phys Rev B 1999;60:11694 and 11670.

[64] Fulghum DB, Camley RE. Magnetic behavior of antiferromagnetically coupled layers connected by ferromagnetic pinholes. Phys Rev B 1995;52:13436.

[65] Wongsam MA, Chantrell RW. Simulations of spin dynamics in cobalt-based magnetic multilayers. Phys Rev B 1998;58:12207.

[66] Budrikis Z, Livesey KL, Morgan JP, Akerman J, Stein A, Langridge S, et al. Domain dynamics and fluctuations in artificial square ice at finite temperatures. New J Phys 2012;14:035014.

[67] Nowak U. Thermally activated reversal in magnetic nanostructures. In: Stauffer D, editor. Annual reviews of computational physics IX. Singapore: World Scientific; 2001. p. 105.

[68] Plumer ML, Leblanc MD, Whitehead JP, van Ek J. Micromagnetic simulations of sweep-rate dependent coercivity in perpendicular recording media. J Appl Phys 2012;111:123905.

[69] Suess D, Breth L, Lee J, Fuger M, Vogler C, Bruckner F, et al. Calculation of coercivity of magnetic nanostructures at finite temperatures. Phys Rev B 2011;84:224421.

[70] Evans RFL, Hinzke D, Atxitia U, Nowak U, Chantrell RW, Chubykalo-Fesenko O. Stochastic form of the Landau-Lifshitz-Bloch equation. Phys Rev B 2012;85:014433.

[71] Stanciu CD, Hansteen F, Kimel AV, Kirilyuk A, Tsukamoto A, Itoh A, et al. All-optical magnetic recording with circularly polarized light. Phys Rev Lett 2007;99:047601.

[72] Kryder MH, Gage EC, McDaniel TW, Challener WA, Rottmayer RE, Ju G, et al. Heat assisted magnetic recording. Proc IEEE 2008;96:1800.

[73] Yamaguchi K, Suzuki K, Nittono O, Yamada K, Enokizono M, Takagi T. Monte Carlo simulation for magnetic domain wall displacements in magnetic nano-wires with local disorders. IEEE Trans Magn 2009;45:1622.

[74] Li J, Liu B-G. Monte Carlo simulated dynamical magnetization of single-chain magnets. J Magn Magn Mater 2015;378:186–9.

[75] Saville SL, Qi B, Baker J, Stone R, Camley RE, Livesey KL, et al. The formation of linear aggregates in magnetic hyperthermia: implications on specific absorption rate and magnetic anisotropy. J Colloid Interface Sci 2014;424:141.

[76] Plumer ML, Fal TJ, Mercer JI, Whitehead JP, van Ek J, Ajan A. Kinetic Monte Carlo simulations of M–H loops for HAMR recording media: comparison with MOKE data. IEEE Magn Trans 2014;50:3100805.

[77] Fal TJ, Plumer ML, Whitehead JP, Mercer JI, van Ek J, Srinivasan K. Simulations of magnetic hysteresis loops for dual layer recording media. Appl Phys Lett 2013;102:202404.

[78] Fal TJ, Mercer JI, Leblanc MD, Whitehead JP, Plumer ML, van Ek J. Kinetic Monte Carlo approach to modeling thermal decay in perpendicular recording media. Phys Rev B 2013;87:064405.

[79] Deskins WR, Brown G, Thompson SH, Rikvold PA. Kinetic Monte Carlo simulations of a model for heat-assisted magnetization reversal in ultrathin films. Phys Rev B 2011;84:094431.

[80] Medeiros Filho FC, Mello VD, Dantas AL, Sales FHS, Carrico AS. Giant magnetocaloric effect of thin Ho films. J Appl Phys 2011;109:07A914-1–07A914-3.

[81] Nascimento FIF, Dantas AL, Oliveira LL, Mello VD, Camley RE, Carrico AS. Thermal hysteresis of ferromagnetic/antiferromagnetic compensated bilayers. Phys Rev B 2009;80:144407.

[82] Mello VD, Carrico AS. Surface induced nucleation of the helifan phase. Surf Sci 2001;482–485:960.

[83] Maat S, Thiele J-U, Fullerton EE. Temperature and field hysteresis of the antiferromagnetic-to-ferromagnetic phase transition in epitaxial FeRh films. Phys Rev B 2005;72:214432.

[84] Demirtas S, Camley RE, Koymen AR. Tunable thermal hysteresis in CoGd alloys. Appl Phys Lett 2005;87:202111.

[85] Rezende SM, de Aguiar FM. Nonlinear dynamics in microwave driven coupled magnetic multilayer systems. J Appl Phys 1996;79:6309.

CHAPTER 7

Spintronic Oscillators Based on Spin-Transfer Torque and Spin-Orbit Torque

Mario Carpentieri*, Giovanni Finocchio†
*Department of Ingegneria Elettrica e dell'Informazione, Politecnico di Bari, Bari, Italy
†Department of Electronic Engineering, Industrial Chemistry and Engineering, University of Messina, Messina, Italy

1 INTRODUCTION

An oscillator is a system that exhibits a periodic output having constant inputs only. The harmonic oscillator, the most important class of oscillators, has an output mainly characterized by a single frequency (single mode). In contrast, the anharmonic oscillator is characterized by an output not oscillating in simple harmonic motion. Electronic oscillators are very important for a huge number of applications such as digital circuits, quartz digital watches, timing modules, remote control, satellite receivers, radio and video transmission, etc. The oscillators can be classified into two main categories: linear and nonlinear. Linear oscillators are implemented in a linear system by introducing a feedback from output to input; if the device is designed in order to satisfy the Barkhausen criterion [1], the intrinsic losses of the system are compensated, giving rise to dynamic behavior characterized by two imaginary eigenvalues. However, linear devices can also be implemented by using nonlinear elements biased in a region where the nonlinear response can be well approximated by a linear curve. Examples of this include ring oscillators, inductor–capacitor LC-tuned oscillators, crystal oscillators, relaxation oscillators, and voltage-controlled oscillators (VCOs). The key component in harmonic oscillators is the presence of a resonant element made up of two subsystems which periodically exchange energy; a simple example is the LC circuit. The resonant circuit is called a tuned circuit or tank circuit (because the action of a tuned circuit is similar to a pendulum swinging back and forth or water sloshing back and forth in a tank).

Nonlinear oscillators are characterized by the presence of a nonlinear element (tetrodes, tunnel diodes, transistors, etc.) and a bias source that fixes the working point of the nonlinear element at the right voltage and current configuration. Spintronic oscillators fall into this category. In particular, this review will

Magnetism of Surfaces, Interfaces, and Nanoscale Materials. http://dx.doi.org/10.1016/B978-0-444-62634-9.00007-2

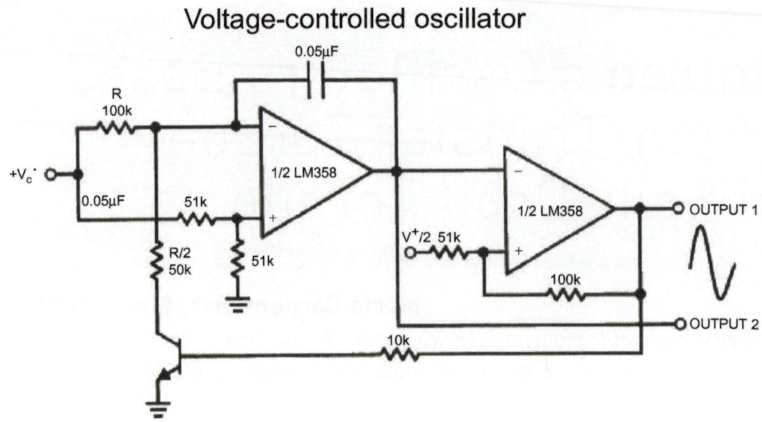

FIGURE 1
Example of electrical circuit of a typical VCO.

discuss the spin-transfer torque (STT) and the spin-Hall oscillators (SHOs) and how to combine them in a multiphysics scenario.

Oscillators are characterized by different technological properties, as listed below. An ideal oscillator should optimize all of these simultaneously.

Here we introduce the concepts that characterize an oscillator by considering the VCO as a reference; at the end of the chapter we will compare the VCO with the STT and SHE oscillators. The VCO is an electronic oscillator whose oscillation frequency is controlled by a voltage input V_c, which determines the instantaneous oscillation frequency. Figure 1 shows an example of an electrical circuit of a typical VCO, which contains a combination of an amplifier and a feedback network. It appears in many analog and RF signal processing systems used in several communication applications.

1.1 Frequency Tunability

In many oscillators, the frequency should be adjustable in operation over a specified frequency range, especially in radio circuits, where the radio is used to transmit or receive signals at different frequencies. There are also requirements on the speed with which the oscillation frequency can be adjusted. For example, in a typical VCO the oscillation frequency can be varied by applying a control voltage V_c to change the frequency of an amount $\omega_\Delta = K_V V_C$, where K_V is the frequency tuning constant. In this case, the output voltage of an oscillator is given by

$$v(t) = V_{out} \cos\left(\omega_C t + K_V \int_{-\infty}^{t} V_C dt\right), \qquad (1)$$

where V_{out} is the amplitude of the output voltage and ω_C is the center frequency. In reality, the frequency tuning is not a linear function of the tuning.

The most common method for tuning the frequency of an LC oscillator is to use nonlinear capacitors in the frequency-determining network to change the frequency. By adjusting the control voltage over these capacitors, the capacitance changes and thereby also the oscillation frequency. In traditional harmonic oscillators, the variable capacitance is conventionally a pn-junction diode in silicon circuits and a Schottky diode in GaAs. The capacitance tuning range for a pn-junction diode is quite narrow in a standard CMOS/BiCMOS process, and hence the resulting VCO tuning range is often unacceptable. Furthermore, the frequency tuning may cause excess phase noise.

1.2 Phase Noise, Q-Factor, and Linewidth

The term phase noise is widely used to describe short-term random frequency fluctuations of a signal. Frequency stability is a measure of the degree to which an oscillator maintains the same value of frequency over a given time. An ideal sine wave oscillator may be described by

$$v(t) = V_{out} \cos 2\pi f t, \tag{2}$$

where V_{out} is the nominal amplitude of the signal and f is the nominal oscillation frequency. However, when using a real oscillator, the amplitude and the phase are affected by noise and are time varying. The instantaneous output of an oscillator may be represented by

$$v(t) = V_{out}(1 + A(t)) \cos(2\pi f t + q(t)), \tag{3}$$

where $A(t)$ and $q(t)$ represent the amplitude and phase fluctuations of the signal respectively. While $A(t)$ introduces flutuations, $q(t)$ is the mechanism that fixes the minimum linewidth of the oscillator.

The phase term may be random or discrete and can be seen on the spectrum analyzer. Typically, there are two types of fluctuating phase terms. In the first one, discrete signals, referred to as spurious, appear as distinct components in the spectral density plot. The second type, random in nature, appears as random phase fluctuations and is commonly called phase noise. Many methods are used to characterize the phase noise of an oscillator. Essentially, all the methods measure the frequency or phase deviation of the source under test in either the frequency or time domain. Since frequency and phase are related to each other, all these terms are also related. One of the most common fundamental descriptions of phase noise is the one-sided spectral density of phase fluctuations per unit bandwidth. The term spectral density describes the energy distributions as a continuous function, expressed in units of energy per unit bandwidth. The phase noise of an oscillator is well described in the frequency domain where the spectral density is characterized by measuring the noise sidebands on either side of the output signal center frequency. When the energy is injected into the system, the LC-resonator circuit is capable of transferring that energy back and forth between its reactive components, i.e., the capacitance and inductance. The reactive elements store the energy for one half-cycle, while releasing it in the next half-cycle. Considering the case of finite tank resistance, the oscillation starts up and after a short time it eventually dies out due to the presence of dissipation that consumes the energy. This introduces the most important parameter of an LC tank, which is its quality factor (Q). Fundamentally, Q is defined as

$$Q = 2\pi f_0 \frac{\text{Energy stored}}{\text{Average power dissipated}}, \tag{4}$$

where f_0 is the resonance frequency. The above definition is not bound to any specific structure because there is no information about what and where the energy is stored and dissipated from the definition. A more useful approach for the oscillator Q is found by performing an AC analysis and extracting Q as

$$Q = \frac{f_0}{\text{FWHM}}, \tag{5}$$

where f_0 is the resonance frequency and FWHM is the full width at half-maximum or bandwidth of the oscillator spectrum.

Figure 2 shows a comparison between the power spectrum of an ideal and non-ideal (phase noise and amplitude noise not neglegible) harmonic oscillator, together with the indication of the FWHM.

Thus, for a fixed f_0, Q increases as the bandwidth decreases, corresponding to a sharpening of the peak in the magnitude response. Due to the losses and limitations in the reactive elements, practical oscillators need some sort of self-sustaining mechanism to ensure that they continue to generate these periodic signals for an indefinite period of time. In order to improve the quality factor Q it is necessary to obtain an oscillator with high oscillation frequency f_0 and narrow FWHM (or linewidth) of the frequency spectrum.

1.3 Power Dissipation

Low power consumption is necessary as an industry standard and for power-saving purposes. For example, mobile devices are required to have long standby times, indicating a need for low power consumption (typically only several tens of milliwatts). Minimum power dissipation is also important for devices that are connected to electric power grids, since the whole device has to fit into a cheap package with a certain thermal resistance. In this case, low power means

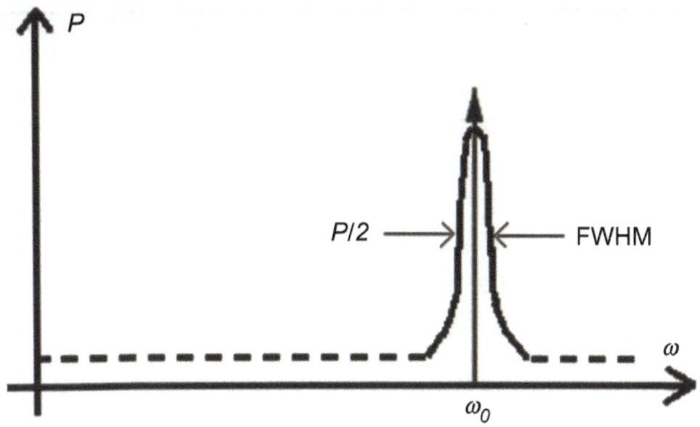

FIGURE 2
A comparison between the power spectrum of an ideal harmonic oscillator (solid arrow) with a real oscillator (dashed line); the spread of the dynamical response is linked to the phase noise, the reduction of the ampitude is linked to the power or amplitude noise. In correspondence with half power there is an indication of the linewidth (FWHM).

the possibility of a higher degree of integration. The maximum supply voltage can be dictated by the application or by the technology (e.g., breakdown voltages in CMOS). In a VCO design, it is difficult to have low phase noise and low power consumption simultaneously because the tank voltage amplitude is proportional to the current flowing [2]. In fact, as the current increases (consuming more power), the phase noise is reduced because the amplitude of the oscillating signal increases simultaneously. Therefore, a trade-off between phase noise and power consumption is necessary.

1.4 Supply Voltage and Output Power

A VCO has at least one tuning input and when the tuning voltage is varied, the frequency of the oscillator varies. Noise or spurious signals on the power supply line modulate the oscillator and cause phase noise or spurious components respectively. This phenomenon is called supply pushing. An oscillator design should work correctly within a certain supply voltage range, for example it can be 3 V nominal with a tolerance of ±10%. In general, the phase noise is inversely proportional to the power dissipated in the resistive part of the resonant LC circuit. Therefore, to obtain small phase noise it is necessary to increase the bias current of the resonator. However, any electronic part of the VCO (e.g., CMOS transistor) has a maximum voltage that cannot be exceeded without permanent damage; this aspect limits the maximum applied voltage. In addition, a larger supply voltage means larger power dissipation.

The primary task for the active network is to supply the power to keep the oscillations going. The supplied power must balance the power losses in the dissipative components in the feedback network. In general, the power efficiency can be approximated as

$$\eta = \frac{V_{\text{out1}}}{V_{\text{dc}}}, \tag{6}$$

where V_{dc} is the supply voltage and V_{out1} is the voltage amplitude at the fundamental frequency at the output.

1.5 Injection Locking

Injection locking may occur in oscillators when an input signal of sufficient magnitude is injected and the oscillation frequency changes from the free-running frequency to that of the injected signal. The frequency of the injected signal must be close enough to one of the multiples of the fundamental free-running frequency in order that the injection-locking phenomenon can occur. In some cases, injection locking is desired, such as in some radio receivers, but in other cases it can be a problem, such as when the oscillator locks to a disturbance from nearby circuitry. In other words, when the frequency of a large enough disturbance is close enough to one of the harmonics of the oscillator, the oscillation frequency may lock to the disturbing signal.

1.6 Immunity to Disturbances (Supply, Load Variations, and Substrate)

The fluctuations of the supply and substrate will induce perturbation at different stages of the oscillator. In general, integrated oscillators are implemented as differential circuits since the environment is noisy and the oscillator shares the substrate with other circuitry. Discrete oscillators are usually implemented as single-ended circuitry to keep the component count low and hence the price and size down.

1.7 Cost (Price, Size, and Design Time)

Technology plays an important role in the performance and cost of an oscillator. For low production costs, a VCO should occupy as small an area of silicon as possible. Therefore, some electronic devices, such as inductors and varactors, are strictly avoided because they are very expensive in terms of area. This fact leads to limited choices in design. Passive components (e.g., resistors, capacitors) need to be designed as small as possible for small area occupation.

2 STT AND MAGNETORESISTIVE EFFECTS

The word "spintronics" is related to two main milestones: (i) the discovery in 1988 of the giant magnetoresistance (GMR) effect due to the research groups of Peter Grunberg and Albert Fert [3,4] and (ii) the discovery in 1996 by John Slonczewski [5] and Luc Berger [6] that moving electrons can transfer spin angular momentum from one region to another, creating torque on the magnetization in a ferromagnet. This process has become known as STT.

A very simple structure to observe the GMR effect is composed of two ferromagnetic layers separated by a nonmagnetic layer; this is called a "spin valve". The nonmagnetic layer is used to magnetically decouple the ferromagnetic layers. The different thicknesses of the two ferromagnetic layers lead to different behavior in their response to an applied magnetic field. For example, the thicker film might have a larger anisotropy, tending to hold the magnetization in a particular direction. In this case, the magnetization of the thicker layer, the so-called "fixed" or "pinned" layer, is relatively insensitive to moderate applied fields, whereas the magnetization of the thinner film, the so-called "free" layer, easily aligns itself with the direction of the applied field. The electrical resistance of such a device depends not only on the electrical resistivity of the metallic layers in the device, but also on the angle between the magnetizations of the ferromagnetic layers. The fundamental reason behind this effect lies in the fact the electrons possess not only an electric charge, but also an intrinsic angular momentum called spin, leading to an intrinsic magnetic moment. Resistivity is caused by scattering processes, and electrons can undergo both spin-dependent and spin-independent scattering in a ferromagnet. The spin-dependent scattering depends on the angle of the magnetic moment of the electron with respect to the local magnetization direction. In such a configuration, the resistance of the magnetic structure depends strongly on the relative alignment of the ferromagnetic layers. Thus, the magnetic state of the junction influences the electric current that passes through it. The resistance has a minimum value when the layers are parallel and a maximum value when they are antiparallel (see Fig. 3), with an approximately cosine angular dependence.

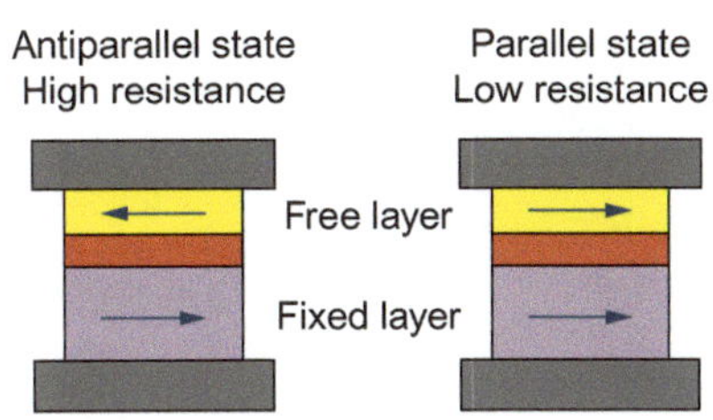

FIGURE 3
Parallel and antiparallel configurations of the magnetization of the two magnetic layers.

If the nonmagnetic interlayer is an insulator that is sufficiently thin to allow quantum mechanical tunneling of electrons the structure can exhibit so-called tunneling magnetoresistance (TMR). This phenomenon was discovered in 1975 [7], before GMR, but it did not receive much attention until higher quality barriers were developed in 1995 [8,9]. TMR is currently of great importance in the development of spin-based electronics, due in part to the fact it offers resistance changes that can be in excess of 500% [10] relative to the minimum value.

Although two stable zero-field resistance states are possible, corresponding to parallel and antiparallel magnetizations of the ferromagnetic layers, GMR and TMR effects have fundamentally different microscopic origins, as illustrated in Fig. 4. A TMR device consists of two ferromagnetic layers separated by an insulating spacer that, for a sufficiently thin spacer (less than about 2 nm), allows a tunnel current to flow between the ferromagnets. Because spin is conserved in the tunneling process, the current is larger for parallel than for antiparallel magnetizations, as shown in Fig. 5a. In a GMR device, the insulating spacer is replaced with a nonmagnetic metal and the electrons of a given spin orientation are typically scattered much more strongly when traveling through a ferromagnet with a majority of its electrons in the opposite spin state. For parallel magnetizations, one spin channel has almost no scattering, whereas the other is more strongly scattered. For antiparallel magnetizations, both channels have some scattering (see Fig. 5b). This again leads to differing currents and a resistance change of about 15% in this case.

The inverse process, theoretically demonstrated by Slonczewski in 1996, was based on the idea that follows the concept of the action–reaction principle [11]: If a ferromagnet can polarize the spin of an electric current (because of the spin-dependent scattering probability), this means that there is a transfer of angular momentum from the ferromagnet to the electrons that build up the current. Since angular momentum is conserved, there must be an equal amount of angular momentum from the electrons to the ferromagnet. This effect is called

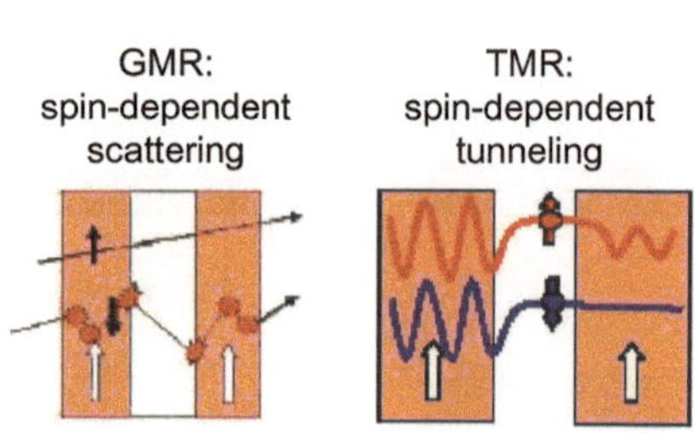

FIGURE 4
Sketch of the spin-dependent scattering and spin-dependent tunneling in the GMR and TMR cases.

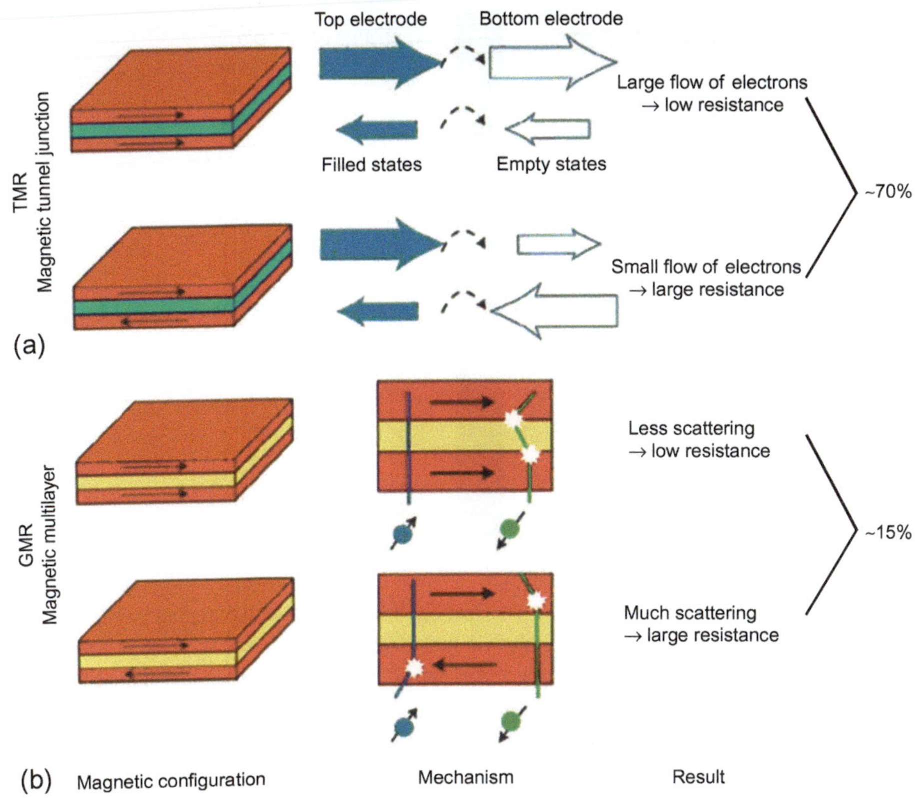

FIGURE 5
(a) TMR and (b) GMR mechanism in cases of low and high resistance.

spin transfer, because the spin-polarized electric current between the magnetic sublayers allows for a transfer of angular momentum, as shown in Fig. 6.

A change in angular momentum with time is equivalent to a torque, so there exists a STT from the electrons to the ferromagnet.

The idea of a current-controlled magnetoresistive random access memory (STT-MRAM) has been developed using this effect. This solution would not only decrease the power needed for operation, but also reduce the read and write times, which would make it a viable nonvolatile memory that does not wear out.

In fact, classical MRAM needs current methods for writing bits with magnetic fields generated by large (~10 mA) currents, leading to issues with power dissipation. In addition, bit density is limited due to issues with localization of the writing field. In STT-MRAM, spin-polarized currents localized to individual bits

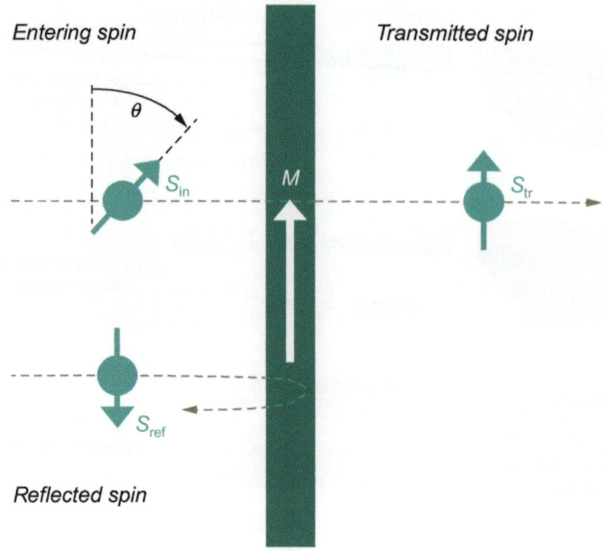

FIGURE 6

Representation of a single spin of angular momentum S_{in}, that suffers a torque exerted by the magnetization M of the magnetic thin film, which in turn is reoriented by the direction of M and is transmitted through the thin film with angular momentum S_{tr}. The amount of spin that is transmitted depends on the material properties and many are reflected with moment S_{ref}, which are mainly constituted by the spins of opposite polarization with respect to the transmitted ones.

are used. In this case, current-switched MRAM would require writing currents of the order of 0.1–0.2 mA to be compatible with CMOS technology, with the additional requirement that a bit maintain its state for >10 years (barrier between states >1 eV) to ensure nonvolatility. Typical devices take advantage of the small (of the order of 40 nm×100 nm) shapes possible.

Recently, in November 2012, the company Everspin launched the first commercial STT-MRAM chips [12], bringing forth what will most probably be a booming new era of spin-torque memories.

Additional devices have been proposed using STT. These are based on the propagation of spin currents by either conducting electrons or by spin waves. One of these devices, again in the area of magnetic storage, is the racetrack memory (Fig. 7) proposed by Parkin *et al.* [13], which again promises to be a more efficient and fast type of memory.

This type of memory is based on a moving domain wall (DW), which is a region where the magnetization gradually changes from one given direction to another, in long magnetic strips [14,15]. The domains are shifted up and down in the racetrack by nanosecond-long pulses of in-plane spin-polarized current using the phenomenon of spin momentum transfer [16].

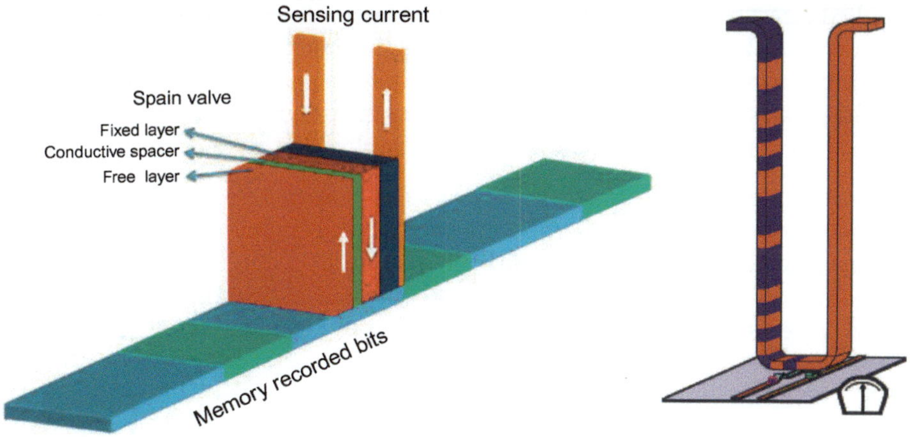

FIGURE 7
Racetrack memory proposed by Parkin. *Reproduced with permission from Ref. [11]. Copyright 2008 Science.*

Spin-torque nano-oscillators (STNOs) are another interesting prospect for such a device, since multiple devices that depend on oscillator principles could be further miniaturized.

As shown in Fig. 8, the idea for such a device came from the now well-established knowledge that a spin-polarized electric current injected into a ferromagnetic layer exerts a torque on the magnetization. This current can compensate the intrinsic damping, resulting in a steady precession of the magnetization rather than a decay of the magnetization back to its equilibrium position. The steady precession results in a resistance which is periodically varying, because of the GMR/TMR effect. Thus, due to Ohm's law, the output of an

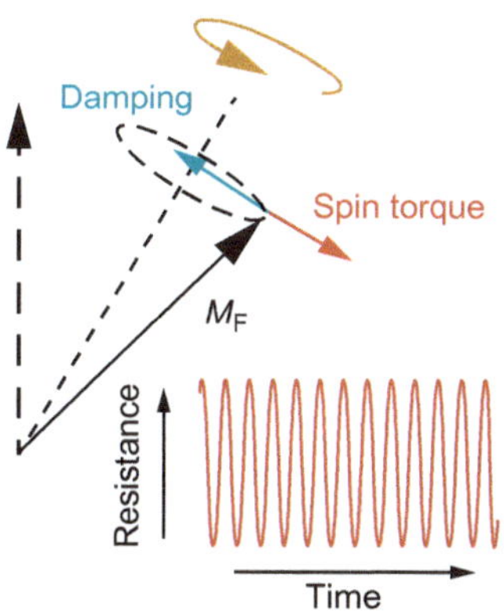

FIGURE 8
Magnetization precession obtained when the damping is compensated by the spin torque.

applied direct current is the generation of an alternating voltage at a microwave frequency (GHz). This generated voltage signal can be used in a wide variety of communications and signal processing applications. Since the signal is based on the precessional motion of magnetization, the frequency of the signal can be tuned by changing the current or the field applied to the nanomagnet [17].

Since the amplitude of such spin waves is small, researchers are trying to phase-lock a number of nano-oscillators in order to get a higher signal [18–21]. Up to now, this has proved to be a difficult task, in part due to finding the right combination of ferromagnetic materials that will allow for high-amplitude microwave oscillations using a direct current.

2.1 Micromagnetic Model

In order to study the physical phenomena discussed above, from either the fundamental or the experimental points of view, a bridge is needed to link the theoretical mathematical–physical models with the experimental devices. That link is achieved using numerical simulations. The latter type of investigation has been gaining more and more relevance in the last decades, as research into different physical properties becomes ever more complex. In order to justify the investment of expensive laboratory equipment that allows the scientific community to conduct research essential to advancements in physics and other sciences, simulations are used to predict, discover, and solve physical phenomena before experiments are attempted. Simulations provide a unique link between the theoretical research and the experimental findings, allowing both parties to see where they coincide and where the theories fail and need to be improved in order to better understand the experimental results. Micromagnetic simulations have played an important role because they have proved to be an efficient tool in the verification of the theoretical formalism [22–26], and subsequently in the interpretation and design of devices.

The micromagnetic model is based on the numerical solution of the Landau–-Lifshitz–Gilbert (LLG) equation, given by

$$\frac{d\mathbf{M}}{dt} = -\frac{\gamma_0}{1+\alpha^2}(\mathbf{M}\times\mathbf{H_{eff}}) - \frac{\gamma_0}{1+\alpha^2}\frac{\alpha}{M_S}\mathbf{M}\times(\mathbf{M}\times\mathbf{H_{eff}}), \tag{7}$$

which is used to solve the dynamics of the magnetization. Here, γ_0 is the gyromagnetic ratio and α is the Gilbert damping. The effective field $\mathbf{H_{eff}}$ takes into account both effective fields of a quantum nature such as the exchange and anisotropic fields, and fields that occur classically such as the magnetostatic, Oersted, and Zeeman fields. Furthermore, the effects of temperature are also included in $\mathbf{H_{eff}}$ as an additional thermal field.

The thermal field $\mathbf{H_{th}}$ is a random fluctuating three-dimensional vector quantity, given by $\mathbf{H_{th}} = \boldsymbol{\xi}(t)\sqrt{2\frac{\alpha}{1+\alpha^2}\frac{K_B T}{\mu_0\gamma' \Delta V M_S \Delta t}}$, where K_B is the Boltzmann constant, ΔV is the volume of the computational cubic cell, Δt is the simulation time step, T is the temperature of the sample [27,28], and $\xi(t)$ is a Gaussian stochastic process. The thermal field satisfies the following statistical properties:

$$\begin{cases} \left\langle H_{\text{th},k}\left(\vec{r},t\right)\right\rangle = 0 \\ \left\langle H_{\text{th},k}\left(\vec{r},t\right)H_{\text{th},l}\left(\vec{r}',t'\right)\right\rangle = D\delta_{kl}\delta(t-t')\delta\left(\vec{r}-\vec{r}'\right)' \end{cases} \quad (8)$$

where k and l represent the Cartesian coordinates x, y, z. Each component of $\mathbf{H}_{\text{th}} = (H_{\text{th},x}, H_{\text{th},y}, H_{\text{th},z})$ are space- and time-independent random Gaussian distributed numbers (Wiener process) with zero mean value. The constant D measures the strength of thermal fluctuations, and its value is obtained from the fluctuation dissipation theorem [29].

In order to take into account the STT contribution, an additional term has to be included in the LLG equation. In general, two different types of STT are used, one for devices in which the current flow is perpendicular to the plane (CPP), as in spin valves or MTJs (see Fig. 9), and the other when the current is in-plane, as is the case in DW dynamics along ferromagnetic strips. The case of interest for this discussion is the STT contribution with the current perpendicular to the plane.

For CPP, in the case of a spin valve, where a nonmagnetic metal spacer separates two magnetic layers, the STT in the thin magnetic layer has been given by Slonczewski [5]:

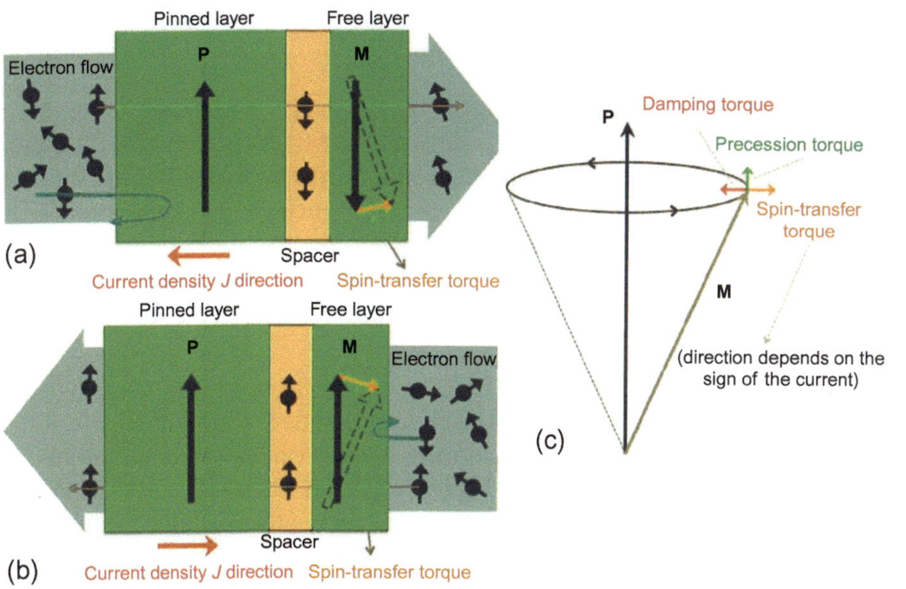

FIGURE 9
Spin-transfer torque in the case of current flow perpendicular to the plane.

$$\Gamma_{\text{STT-CPP}} = -\frac{g\mu_B J}{dM_s^3 e}\varepsilon(\mathbf{M},\mathbf{P})\,\mathbf{M}\times(\mathbf{M}\times\mathbf{P}), \tag{9}$$

where **M** and **P** are the magnetization of the free and fixed layer, respectively. J is the current density, g is the gyromagnetic splitting factor, μ_B is the Bohr magneton, d is the thickness of the layer undergoing the STT effect, e is the electron charge, and M_S is the saturation magnetization. ε(**M**,**P**) is a polarization function, which depends on the relative orientation between the magnetization of the assumed thin free layer **M** and that of the thick pinned layer **P** [5]:

$$\varepsilon(\mathbf{M},\mathbf{P}) = \frac{1}{-4 + (1+\eta^3)\dfrac{(3+(\mathbf{M}\cdot\mathbf{P}))}{4\eta^{3/2}}}, \tag{10}$$

where η is the spin polarization factor of the magnetic material under study. Other polarization functions can also be used for spin valves, which also take into account the giant magnetoresistance asymmetry χ, besides the polarization factor. This is given by [30]

$$\varepsilon(\mathbf{M},\mathbf{P}) = \frac{0.5\eta(\chi+1)}{2+\chi(1-(\mathbf{M}\cdot\mathbf{P}))}. \tag{11}$$

The previous expressions are valid for spin-valve devices; however, if considering an MTJ in which an insulator spacer is used instead of a nonmagnetic metal, the polarization function is [31]

$$\varepsilon(\mathbf{M},\mathbf{P}) = \frac{1}{2}\eta\frac{1}{1+\eta^2(\mathbf{M}\cdot\mathbf{P})}. \tag{12}$$

Adding the spin-torque term (9) to the dynamic LLG equation (7), applying $\mathbf{M}\times()$, using the identity $\mathbf{M}\times(\mathbf{M}\times \mathrm{d}\mathbf{M}/\mathrm{d}t) = -M_S^2(\mathrm{d}\mathbf{M}/\mathrm{d}t)$ and multiplying by α/M_S, it is possible to write the LLG equation with the STT effect for CPP devices, called the Landau–Lifshitz–Gilbert–Slonczewski equation (LLGS):

$$\begin{aligned}\frac{\mathrm{d}\mathbf{M}}{\mathrm{d}t} = &-\frac{\gamma_0}{1+\alpha^2}(\mathbf{M}\times\mathbf{H}_{\text{eff}}) - \frac{g\mu_B J}{(1+\alpha^2)dM_s^3 e}\varepsilon(\mathbf{M},\mathbf{P})\,\mathbf{M}\times(\mathbf{M}\times\mathbf{P})\\ &-\frac{\gamma_0}{(1+\alpha^2)}\frac{\alpha}{M_S}\mathbf{M}\times(\mathbf{M}\times\mathbf{H}_{\text{eff}}) + \frac{g\mu_B \alpha J}{(1+\alpha^2)dM_s^2 e}\varepsilon(\mathbf{M},\mathbf{P})\,\mathbf{M}\times\mathbf{P}.\end{aligned} \tag{13}$$

Usually, it is considered that the pinned layer magnetization **P** does not suffer the effects of the spin-torque interaction, since it is considered to be a pinned or fixed layer. Although this is a good approximation for many experiments it is not generally true, and in some cases the dynamics of the so-called pinned layer are relevant, as is the case of coupled device modes involving magnetization dynamics in both layers [32]. In such cases, both layers **M** and **P** are considered free to move and both polarize the conducting electrons depending on the direction of the current. Thus, when both layer magnetizations are considered dynamic, the names pinned and free are usually changed to thick and thin. When considering that both layers are free, the dynamic equation used (13) is the same but now has to be evaluated twice, once for each layer. In this case, the thin layer is acting in one of the expressions as the polarizing layer and in the other as the one suffering the effects of the spin-torque, and vice versa for the thick layer. This means that when evaluating, at each time step, the dynamics of the thin layer, an equation such as (13) is used , whereas for the thick layer the equation is

$$\begin{aligned}\frac{\mathrm{d}\mathbf{P}}{\mathrm{d}t} = & -\frac{\gamma_0}{1+\alpha^2}(\mathbf{P}\times\mathbf{H}_{\text{eff}}) - \frac{g\mu_B J}{(1+\alpha^2)dM_s^3 e}\varepsilon(\mathbf{M},\mathbf{P})\,\mathbf{P}\times(\mathbf{P}\times\mathbf{M}) \\ & -\frac{\gamma_0}{(1+\alpha^2)}\frac{\alpha}{M_S}\mathbf{P}\times(\mathbf{P}\times\mathbf{H}_{\text{eff}}) + \frac{g\mu_B\alpha J}{(1+\alpha^2)dM_s^2 e}\varepsilon(\mathbf{M},\mathbf{P})\,\mathbf{P}\times\mathbf{M}.\end{aligned} \tag{14}$$

This gives rise to what is usually called back-torque due to the fact that at each time step the STT effect of one layer over the other is evaluated.

In the following sections, it will be possible to note how full micromagnetic simulations of spin-torque-driven oscillations in magnetic multilayer devices are very useful for predicting the behavior of spintronics nano-oscillators and to confirm experimental data, providing information on the spatially non-uniform dynamics of the magnetization.

3 SPIN-HALL EFFECT AND ANISOTROPIC MAGNETORESISTANCE

Spin currents can be also generated by means of the spin-Hall effect (SHE) [33–35]. In the SHE, an applied charge current density J_e flowing through a heavy metal (HM) such as Pt generates a transverse-flowing spin current density $(\hbar/2)J_S/e$ with a magnitude described by $J_S = \theta_{SH}J_e$, where θ_{SH} is the spin-Hall angle that describes the magnitude of the SHE and the effectiveness of this spin-charge conversion. Figure 10 shows a sketch of a ferromagnet (FM)/HM device,

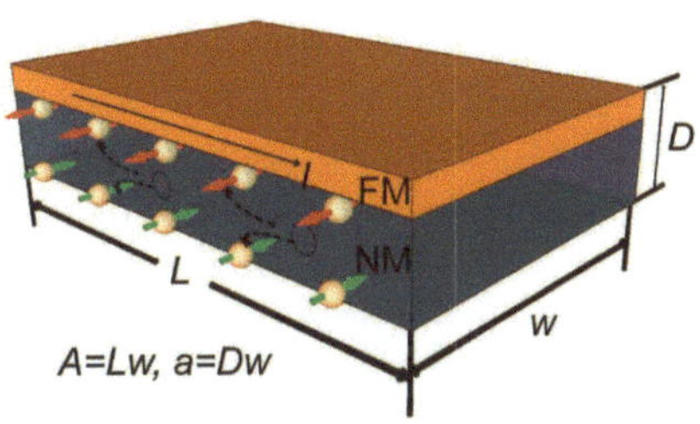

FIGURE 10

Gemotry of a typical spin-orbit-based device. The creation of a spin current in the normal metal transfers spin angular momentum to the ferromagnet, creating an excitation.

where the SHE can be observed. Because of the large spin-orbit coupling, the scattering in the HM is spin dependent and the spin-up (down) electrons flowing in it are separated towards the top (bottom) or vice versa depending on the sign of the spin-orbit coupling. At the interface between HM and FM there is spin accumulation due to the discontinuities in the spin current (zero in the FM), with a consequent spin-current drift from the HM to the FM. The spin current in the HM has an in-plane polarization perpendicular to the direction of the current flow, while in the FM the spin current should be parallel or antiparallel to the magnetization. Therefore, when the two polarizations do not coincide there is an STT from the spin current to the FM magnetization similar to that observed in spin valves.

Theoretically, it was predicted that an inverse effect could also exist, where a pure spin current can be converted into a charge current by the same spin-orbit scattering mechanism. In this case, the efficiency of this inverse spin-Hall effect (ISHE) was given by the same spin-Hall angle, $J_e = \theta_{SH} J_S$ [34]. Because in materials with the SHE the spin current can be generated for nonmagnetic materials, it is clear that the SHE has potential for use as a nonmagnetic source for spin current.

From an experimental point of view, the SHE in metallic systems was first verified using an all-electric measurement technique [36]. Following the initial discoveries of the SHE, many different techniques have been developed for the determination of the spin-Hall angle, both in the DC and in the RF regime. In particular, large spin-Hall angles were discovered in metals [37,38], making the SHE a promising way to generate or detect spin currents.

In relation to the spin-Hall angle, it is also useful to consider the spin-Hall conductivity σ_{SH}, which is connected to the spin-Hall angle by means of the expression $\sigma_{SH} = (\hbar/2e)\theta_{SH}\sigma$, where σ is the charge conductivity of the material. The SHE can arise from either intrinsic (band structure) or extrinsic (scattering) effects in materials, within which there is significant spin-orbit coupling [39,40]. In general, in the intrinsic limit it is expected that the spin-Hall conductivity is largely independent of the mean free path for scattering (and hence also of the electrical conductivity of the material), so that in the intrinsic limit for a given type of material the spin-Hall angle should scale as $\theta_{SH} \propto 1/\sigma$. In an extrinsic limit, where skew scattering processes are dominant, one should have $\theta_{SH} \propto \sigma$ and the spin-Hall angle θ_{SH} should be approximately independent of scattering within the material. In this picture, in order to correctly compute the spin-Hall angle, the spin diffusion length of the HM λ_{sf} has to be considered, because the SHE or the ISHE will only be significant within the range of λ_{sf} at the HM interface. When the thickness of the HM d is larger than λ_{sf}, the bottom part of the HM film will not contribute to the SHE or ISHE signal. If d is smaller than λ_{sf} then the spin accumulation at the HM/substrate interface will cause a backflow spin current, which can also reduce the real spin current.

In this scenario, considering a bilayer composed by normal metal (HM) and ferromagnet as shown in Fig. 10, it is possible to manipulate a ferromagnetic moment with a pure spin current generated from a nonmagnetic source by means of the SHE. This spin current flows towards the interfaces, where it can be absorbed as a spin-orbit torque (SOT) on the ferromagnet.

In these systems, it is therefore possible to excite magnetic dynamics using the spin current generated by the SHE. For example, by making an HM/FM bilayer and applying an oscillating charge current through the bilayer film, a transverse spin current will be generated inside the HM if there is an SHE and the spin current can be further injected into the adjacent FM, thereby exerting an oscillating spin torque on the FM that induces magnetization precession. Similar to the case of spin valves or MTJs, where the resistance oscillates due to the GMR or TMR, there will also be an oscillation in the resistance of the bilayer sample because of the existence of anisotropic magnetoresistance (AMR). This is a magnetoresistance phenomenon, where the electrical resistance depends on the angle between the current and the magnetization in metallic ferromagnets. The transverse component of the AMR is also called the planar Hall effect. The physical origin of AMR is due to the simultaneous action of magnetization and spin-orbit interaction and its detailed mechanism depends on the material. It can, for example, be due to a larger probability of *s-d* scattering of electrons in the direction of magnetization. The net effect (in most materials) is that the electrical resistance has a maximum value when the direction of current is parallel to the applied magnetic field. AMR contrasts with GMR, which reflects the change in resistance that accompanies the magnetic field-induced magnetic configuration in magnetic multilayers.

The SHE-driven spin accumulation at the HM/FM interface generates a Slonczeswki-like torque [41–43], which can be strong enough to switch uniformly magnetized films [44–46], or to obtain DW motion [47], or can induce persistent magnetization precession of the magnetization [48,49].

In the pioneering experiment of magnetization switching [42] the magnetization switching was attributed to the Rashba field due to the strong Rashba spin-orbit coupling. Indeed, the existence of an STT of the Slonczewski type was also connected in that experimental framework. However, in the experimental results achieved in Refs. [44,46,48], the switching was attributed to the STT from the spin-Hall effect (ST-SHE), because the measured Rashba field was negligible in those latter devices. In particular, comparison between the experimental data on different heavy-metal materials such as Pt [44], Ta [48], and W [46] clearly indicates that the ST-SHE is the main mechanism that drives the switching. In fact, this conclusion is derived from the experimental results that the spin-Hall angle in Pt has opposite sign compared to the one in Ta or W, and consequently the electric current to achieve the switching in the presence of Pt has opposite sign than the one used in the presence of Ta or W. On the other hand, the sign of the Rashba field torque should be independent on the HM used.

3.1 Micromagnetic Model

The ST-SHE is included in the LLG equation as additional torque of Slonczewski type [5,44]:

$$\frac{d\mathbf{M}}{dt} = -\gamma_0 \mathbf{M} \times \mathbf{H}_{\mathbf{eff}} + \frac{\alpha}{M_S} \mathbf{M} \times \frac{d\mathbf{M}}{dt} - \frac{d_J}{M_S} \mathbf{M} \times \mathbf{M} \times \boldsymbol{\sigma}. \tag{15}$$

Here $\mathbf{M}$ and $\mathbf{H}_{\mathbf{eff}}$ are the magnetization and the effective field. $\mathbf{H}_{\mathbf{eff}}$ takes into account the standard micromagnetic contributions, the exchange, the magnetocrystalline anisotropy, and the external and self-magnetostatic fields. α is the Gilbert damping, M_S is the saturation magnetization, and γ_0 is the gyromagnetic ratio. σ is the direction of the spin current in the HM. θ_{SH} is the spin-Hall angle given by the ratio between the amplitude of transverse spin current density generated in the HM and the charge current density flowing in it. The coefficient d_J is given by $d_J = (\mu_B \theta_{SH})/(eM_S t) \times j$, where j is the charge current density, μ_B is the Bohr Magneton, e is the electric charge, and t is the thickness of the magnetic layer. Equation (15) can also be rewritten to identify the contribution of a field-like torque term proportional to the Gilbert damping, and by considering the dimensionless form, it is

$$\begin{aligned}\frac{d\mathbf{m}}{\gamma_0 M_S dt} = &-\frac{1}{(1+\alpha^2)} \mathbf{m} \times \mathbf{h}_{\mathbf{eff}} - \frac{\alpha}{(1+\alpha^2)} \mathbf{m} \times \mathbf{m} \times \mathbf{h}_{\mathbf{eff}} \\ &- \frac{d_J}{(1+\alpha^2)\gamma_0 M_S} \mathbf{m} \times \mathbf{m} \times \boldsymbol{\sigma} + \frac{\alpha d_J}{(1+\alpha^2)\gamma_0 M_S} \mathbf{m} \times \boldsymbol{\sigma},\end{aligned} \tag{16}$$

with $\mathbf{m} = \mathbf{M}/M_S$ and $\mathbf{h}_{\mathbf{eff}} = \mathbf{H}_{\mathbf{eff}}/M_S$.

In a different structure, such as a three-terminal device consisting of an MTJ built over a layer of HM [44], the spintronics effect is efficiently coupled with the spin-orbitronics. In this case, it is possible to inject at the same time two current densities: the in-plane current in the HM and the perpendicular current flowing into the MTJ stack, achieving an additional degree of freedom in the control of the magnetization dynamics.

In this scenario, the magnetization dynamics are described by the following equation, which includes the STT from a spin-polarized current and the SOT driven by SHE:

$$\begin{aligned}\frac{d\mathbf{m}}{\gamma_0 M_S dt} = &-\frac{1}{(1+\alpha^2)} \mathbf{m} \times \mathbf{h}_{\mathbf{eff}} - \frac{\alpha}{(1+\alpha^2)} \mathbf{m} \times \mathbf{m} \times \mathbf{h}_{\mathbf{eff}} \\ &- \frac{d_J}{(1+\alpha^2)\gamma_0 M_S} \mathbf{m} \times \mathbf{m} \times \boldsymbol{\sigma} + \frac{\alpha d_J}{(1+\alpha^2)\gamma_0 M_S} \mathbf{m} \times \boldsymbol{\sigma} \\ &+ \frac{g|\mu_B| J_{MTJ}}{|e|\gamma_0 M_s^2 t} g_T(\mathbf{m}, \mathbf{m_p}) \left[\mathbf{m} \times (\mathbf{m} \times \mathbf{m_p}) - q(V)(\mathbf{m} \times \mathbf{m_p})\right],\end{aligned} \tag{17}$$

where $\mathbf{m}$, $\mathbf{h}_{\mathbf{eff}}$, and $\mathbf{m}_{\mathbf{p}}$ are the magnetization, the effective field of the free layer, and the magnetization of the polarizer, and g is the Landé factor. J_{MTJ} is the current density flowing through the MTJ stack, $g_{\mathrm{T}}(\mathbf{m}, \mathbf{m}_{\mathbf{p}}) = 2\eta_{\mathrm{T}}/(1+\eta_{\mathrm{T}}^2\mathbf{m}\cdot\mathbf{m}_{\mathbf{p}})$ characterizes the angular dependence of the spin-polarization function as computed by Slonczewski [31,50], and η_{T} is the polarization efficiency. $q(V)$ is a function which takes into account the voltage dependence of the field-like torque term in the MTJ.

4 STATE OF THE ART OF STT OSCILLATORS

The STT can be classified by considering three main aspects: the geometry (spin valves or magnetic tunnel junctions (MTJs)), the injection site of the current (uniform and nonuniform), and the ferromagnetic materials (in-plane and out-of-plane). Figure 11 summarizes the possible configurations we are going to discuss below in detail.

4.1 Spin Valves—One Free Layer, One Polarizer—Nonuniform Current Injection Via a Point Contact

The first experimental measure of microwave emission in those devices with a large out-of-plane bias field, which coincides with the first measure concerning this topic, was reported by Tsoi *et al.* [51,52]. Theoretically, Slonczewski identified those experimental excitations as exchange-dominated cylindrical spin-wave propagating modes (called "Slonczewski mode") developing a spatially nonuniform linear theory of spin-wave excitations [53]; the propagating nature was also demonstrated by Madami *et al.* [54] by means of optical

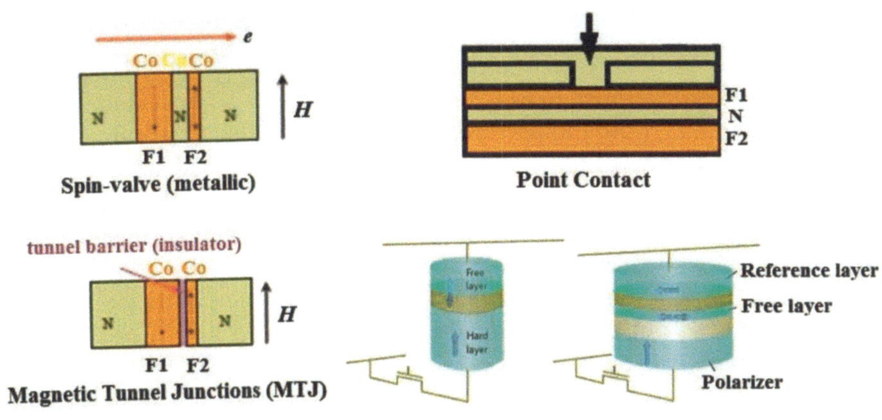

FIGURE 11
Different configurations of spin-transfer torque oscillators. F1 and F2 indicate the two ferromagnetic films. N denotes a normal metal.

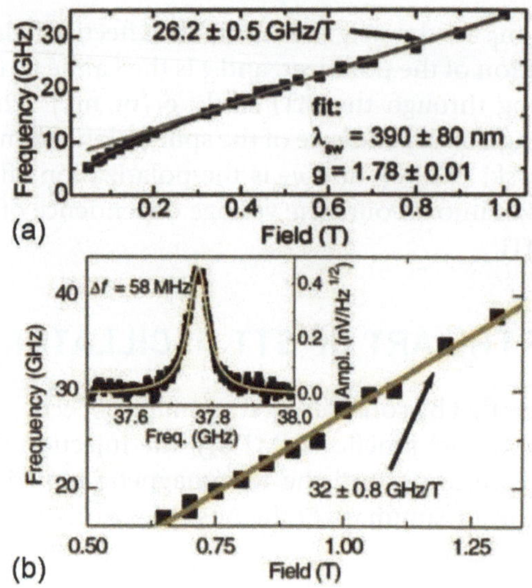

FIGURE 12

(a) In-plane and (b) out-of-plane frequency versus applied field dispersion curve. *Adapted with permission from Ref. [55]. Copyright 2004 American Physical Society.*

measurements. Those modes propagate radially out of the contact for hundreds of nanometers, giving rise to the possibility to achieve phase locking between the mode excited by point contacts localized at different positions in the thin-film nanomagnet [18,19]. Even if potentially this phase locking can be obtained for a large number of point contacts, it was demonstrated for a maximum of three point contacts. Experiments by Rippard *et al.* [55] pointed out that the excitation of magnetization precession could also occur for large in-plane bias fields (see Fig. 12 for the frequency tunability).

The linear theory developed in Ref. [53] was not able to also reproduce the later experimental data. Those modes have now been identified as nonlinear localized spin-wave modes (called "bullet modes") analytically [56], numerically by means of micromagnetic simulations [57], and experimentally [58]. The origin of these different kinds of excitations has been studied systematically as a function of the external magnetic field angle [57,59,60]. The key finding is that while the excitation threshold dependence on the field angle of the bullet mode (from in-plane to out-of-plane) increases, the threshold for the Slonczewski mode decreases. This property gives rise to excitation of a bullet mode below a critical angle and the excitation of a Slonczewski mode above that critical angle. In addition, it has been demonstrated that far above the critical current, these two modes can be excited in a nonstationary way for a certain range of field angles and currents [60]. Micromagnetic simulations demonstrated that the Oersted field is the key mechanism in the explanation of these latter experimental results [60].

As discussed in Section 1, the power spectrum of a nonlinear auto-oscillator in the presence of noise has a finite width, which is configurable with the linewidth generation. The measurement of the linewidth is related to the full width at half-maximum (FWHM) of the power spectrum. From a practical point of view, the generation of the FWHM is one of the most important parameters of a nano-oscillator. Since STNOs are strongly nonlinear oscillators, the amplitude fluctuations are transformed into phase fluctuations, and the nonlinear frequency shift generates a source of phase noise that implies a broadening of the linewidth [61].

An example of nanopillar STNO spectra showing a narrow linewidth at room temperature is shown in Fig. 13. The STNO geometry is the same as that in Ref. [62]. The main panel of Fig. 13 shows the operating frequency as a function of field angle and magnitude at a single current. The inset shows a typical signal observed in this kind of sample, with a high frequency of $f = 32.7$ GHz and very narrow linewidth FWHM$=4.5$ MHz, so that the quality factor $Q = f/\mathrm{FWHM}$ is about 7300.

The universal model of nonlinear oscillators indicates a minimum of the value N of the coupling between phase and power corresponding to the minimum of the linewidth [63]. Several procedures have been presented to experimentally

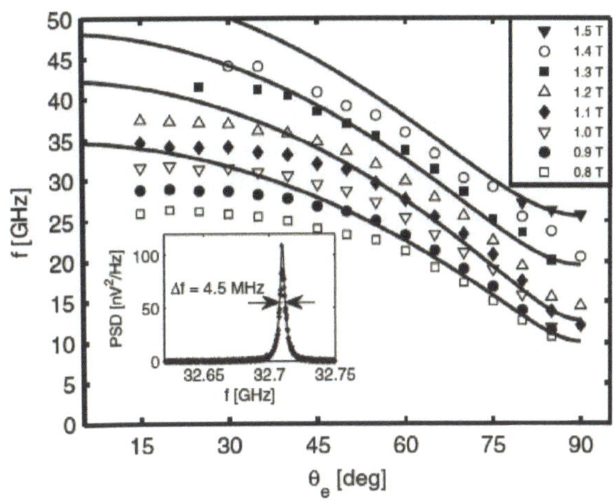

FIGURE 13
An example of frequency behavior as a function of the applied field angle (from in-plane 0° to out-of-plane 90°) and for a fixed perpendicular current of −15 mA. Inset: Typical power spectrum observed in these measurements, with high quality factor. *Reproduced with permission from Ref. [62]. Copyright 2009 American Physical Society.*

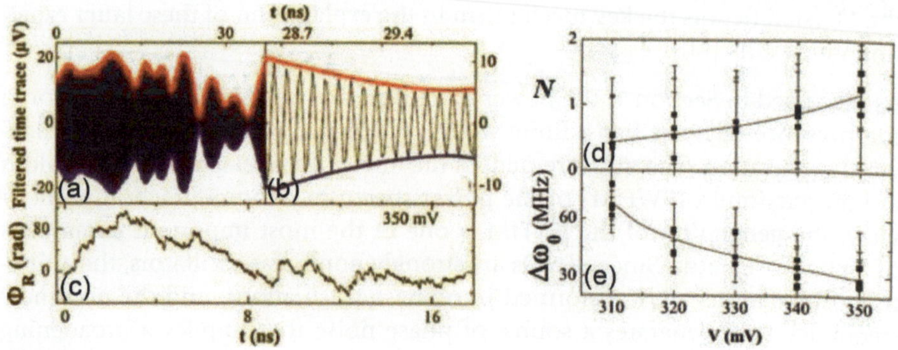

FIGURE 14

(a) Time-resolved voltage trace at $V=350$ mV and $H=-32$ mT in a 37-ns window with time-varying amplitude $A(t)$ of $x_a(t)$ and $-A(t)$ as the upper and lower envelope traces. (b) A 2.6-ns window zoom of the previous trace. (c) Deduced time-drift of the phase deviation $\Phi_R(t)$. (d) The nonlinear coefficient N and (e) the linear linewidth as a function of applied bias voltage. Lines indicate the mean values for each voltage. *Adapted with permission from Ref. [64]. Copyright 2010 American Institute of Physics.*

measure N. To this end, Bianchini *et al.* [64] developed a procedure for a direct experimental measurement of this coupling by means of time-domain data – see the plot of V and linewidth as a function of voltage in Fig. 14d and e respectively, as computed from the time-domain traces (see, for example, Fig. 14a, b). In general, the analysis of the time-domain measurements allows separation of the linear and nonlinear contributions to the phase noise, the nonlinear contribution being due to the power-phase coupling [65].

This particular category of devices has been used to study soliton dynamics, namely self-oscillations, of vortices, vortex–antivortex pairs, droplets, and skyrmions. In detail, at low bias fields (near zero field) the excited modes are related to a vortex motion with circular trajectory around the injection site of the current [66] or with elliptical trajectory coupled to continuous vortex polarity switching [67]. Rotating vortex–antivortex pairs and quadruplets have been observed in experiments and predicted micromagnetically [68,69]. In devices with an out-of-plane free layer, both at small and large out-of-plane fields, there can be excitation of droplets characterized by different topological charges [70–72]. For the sake of clarity, to characterize the topological complexity, it is possible to evaluate the topological density, which is defined at every point of the film and indicates how the magnetization vector **M** varies locally in the plane. Moreover, the integral of the topological density over the film plane, namely the topological charge, always turns out to be an integer.

A new interesting feature introduced by the droplet is the concept of self-modulation. Together with the main excited mode, there is excitation of two nonstationary low-frequency modes, which modulate the main frequency by introducing in the power spectrum two sidebands. The excitation of the self-modulation phenomena is strongly dependent on the physical parameters of the materials, as indicated by micromagnetic simulations [73].

A different idea to realize a nanocontact was proposed by Boone *et al.* [74]. Instead of using a two-dimensional extended film with a point contact, they used a ferromagnetic nanowire (extended only in one dimension) with the contact localized in the center. In this category of devices, for in-plane bias fields, a bullet-type localized mode is excited, while for out-of-plane fields it has been predicted, by means of micromagnetic simulations, that a propagating mode can be generated [75,76]. This solution seems very promising for improving the propagating distance of spin waves because the losses due to radiation are smaller than those in the 2D point contacts. In detail, for the nanowire the spin-wave amplitude scales proportional to $\exp(-\alpha r/2)$ while for the 2D point contact structure the amplitude scales as $\exp(-\alpha r/r)$ [75–77], where α is the damping and r is the distance between the center of the nanocontact and a generic point of the ferromagnet.

The study of microwave emission in spin valves with nonuniform current injection is very important from a fundamental point of view; they are an important building block for the realization of magnonic systems, in particular for the generation and detection of spin waves. In addition, it is possible to reach high output frequencies with narrow linewidths. On the other hand, their use as oscillators was limited by the low output power.

4.2 Spin Valves—One Free Layer, One Polarizer—Uniform Current Injection

In 2003, it was demonstrated experimentally that one could achieve microwave emission in nanopillar "spin valves" with the current density uniformly distributed in the whole free-layer section [78]. In a complete phase diagram of the magnetization dynamics as a function of the bias field and current, several kinds of nonuniform excited modes were reported [79]. Only 2 years later, the final confirmation of the Slonczewski prediction was obtained with time-domain measurements of the dynamics excited by a spin-polarized current in a spin valve with the polarizer exchange biased at 45° with respect to the easy axis of the free layer [80]. A key advance was the achievement of repeatable measurements. Complete analyses of these devices indicate the excitation of a uniform mode over a wide range of currents [81,82], and the possibility to achieve narrow linewidths for fields applied near the in-plane hard axis [83]. This latter result has been explained by means of micromagnetic simulations as a transition of the exited mode from nonuniform (field applied near the easy axis) to uniform (field applied near the hard in-plane axis) [84]. A different strategy to reduce the linewidth was the realization of a tapered spin valve where, for a moderate in-plane hard axis field (<60 mT), the excited modes show a sharp increase in power and a sharply narrowed linewidth (<10 MHz), which in this case was ascribed to a transition from incoherent to coherent dynamics [85].

In recent years, several studies on spin-valve-based STNOs have been carried out on perpendicular materials (the dynamics have been achieved experimentally only at very high fields and currents) [86] and half-metallic materials with large spin polarization to try to find the optimal configuration for different properties of STO at the same time [87,88].

One key issue addressed experimentally was the attainment of oscillation frequencies as large as possible while reducing or avoiding the need for an external field. One solution was the use of an in-plane free layer with two polarizers, one out-of-plane and one in-plane [89,90]. In this configuration, the out-of-plane polarizer is designed in order to generate a spin current with a larger polarization to excite a self-oscillation with an out-of-plane trajectory while the in-plane polarizer has the ultimate aim to convert the magnetization oscillation into a magnetoresistive signal. These devices are characterized by frequency oscillations around 6–8 GHz for applied magnetic fields of about 4–8 kA/m. Another solution was the use of spin valves with two in-plane free layers and two out-of-plane polarizers [91]. The main result achieved in this configuration is linked to the achievement of an oscillation frequency (>5 GHz) at zero bias field. In these devices, each free layer sees a direct STT from one out-of-plane polarizer and a back STT originating from the other free layer. These two actions excite a mode that is characterized by a nonuniform magnetization pattern that in each layer is related by a 180° rotation around the *z*-axis between the two time points, so that the relative configurations of the two layers are the same. The dynamics of the two free layers are phase locked and this is the cause of the frequency doubling (the magnetization of each layer oscillates at the same frequency f_0 while the magnetoresistive signal oscillates at $2f_0$). For the coupling mechanisms, both the dipolar field and the back torque are important.

One of the most promising configurations in terms of oscillation frequency has been proposed by Braganca *et al.* [92]. They measured microwave oscillations in spin valves with two in-plane free layers and without any polarizers. These devices exhibit modes that can reach frequencies >25 GHz in the absence of an applied magnetic field; to our knowledge, this is the largest frequency measured at zero field so far. Figure 15 shows the high-frequency oscillation of the magnetization for negative applied current of −0.8 mA. Samples are elliptical nanopillars with cross-section of 30 nm × 60 nm [92].

Spin valves have also been used to study complex dynamics. For instance, "wavy" torque and the first zero-field dynamics were obtained by choosing materials with different spin-diffusion lengths for the polarizer and for the free

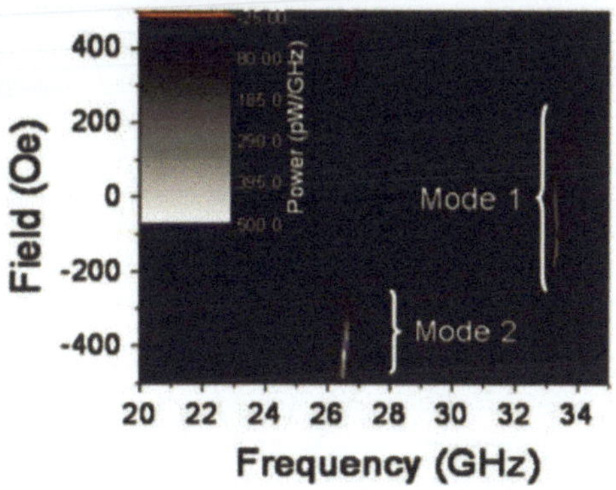

FIGURE 15
(a) High–frequency phase diagram for an STNO device. *Adapted with permission from Ref. [92]. Copyright 2013 American Institute of Physics.*

layers. The first excitation of vortex dynamics driven by STT [93] was seen in spin valves with one of the two ferromagnetic layers thick enough, greater than 50 nm. In that system, micromagnetic simulations demonstrated that both magnetic layers were involved in the magnetization dynamics, a single mode can be excited at large in-plane or out-of-plane fields [94,95], while two non-stationary modes have been measured experimentally [94].

In such devices, and in agreement with general synchronization theories, it is also possible to couple the auto-oscillation state of an STNO to a weak microwave field. This gives rise to several phenomena: frequency locking, frequency pulling, and frequency modulation [96]. P modes are excited in the injection-locking region where basically the self-oscillation mode is locked at the field frequency. In the frequency-pulling region, the dynamics are characterized by a quasi-periodic dynamical state (Q modes), where in the power spectrum it is observed that the weak mode is pulled from its original oscillation frequency due to the AC field and the self-oscillation mode [97]. A phase diagram for the parameter regions in these different behaviors is shown in Fig. 16.

4.3 Magnetic Tunnel Junction

The first report of measurements of persistent magnetization dynamic in MTJs with an AlO tunnel barrier was presented by Fuchs *et al.* in 2003 [98]. However, only after the discovery of a giant TMR at room temperature was achieved with a MgO tunnel barrier [99,100] it was possible to perform a systematic characterization of the dynamical properties of MTJ-based STNOs. Since 2006, microwave emissions in MTJs have been measured [101–104], including those arising from vortex dynamics [105]. In our view, another milestone in the design of STNOs was the discovery of perpendicular MTJs by Ikeda *et al.* [106].

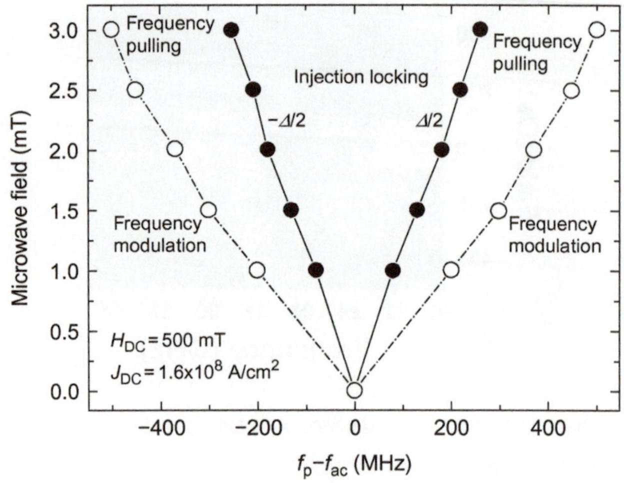

FIGURE 16

Stability diagram as a function of the amplitude of the microwave external field for a fixed current density. *Adapted with permission from Ref. [97]. Copyright 2010 American Physical Society.*

This gave rise to a number of experiments where, by controlling the trade-off between shape anisotropy and interfacial perpendicular anisotropy (IPA), it is possible to define the easy axis of either polarizer or free layer. In particular, for STNO applications, the IPA can be used to reduce the out-of-plane demagnetizing field while maintaining the orientation of both of the two magnetizations in the film plane [107,108], or by maintaining the polarizer in-plane and the free layer tilted [109] or out-of-plane [110]. In the former configuration, for an optimal field amplitude and direction, it was found that a large output power of the order of hundreds of nanowatts could be delivered over a load.

Generally, the oscillation characteristics of STNOs are determined from the microwave signals generated due to the MR effect. As described in Ref. [107], this oscillatory output was measured at room temperature for various field orientations. Figure 17 shows plots of the power spectral density, where a dependence on the bias current for two different field orientations (in-plane field and out-of-plane field) is illustrated. The latter configuration is the basis of possible excitation of microwave emission at ultralow current and with no external field.

As measured in Ref. [110], Fig. 18 shows a comparison between the oscillation power as a function of the current for the two thicknesses of the free layer 1.60 and 1.62 nm. In this case, the device is a magnetic stack consisting of a synthetic antiferromagnet (SAF) $Co_{70}Fe_{30}$ (2.3 nm)/Ru (0.85 nm)/$Co_{40}Fe_{40}B_{20}$ (2.4 nm) layer and a $Co_{20}Fe_{60}B_{20}$ free layer separated by an MgO insulator (0.8 nm). The MTJs are pillar-shaped devices with nominal dimensions of

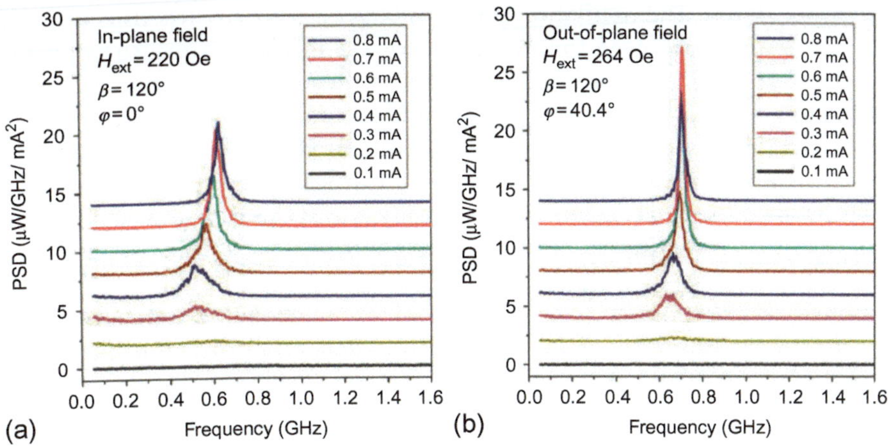

FIGURE 17

Measured microwave emission spectra for an ellipse-shaped pillar formed by the $Co_{70}Fe_{30}$ (2.3 nm)/Ru (0.85 nm)/$Co_{40}Fe_{40}B_{20}$ (2.4 nm) in-plane magnetized reference layer, MgO insulator (0.8 nm), and the $Co_{20}Fe_{60}B_{20}$ in-plane magnetized free layer with perpendicular anisotropy. The devices in this study have nominal dimensions of 170 nm × 60 nm and a free layer thickness of 1.7 nm. The power spectral densities (PSD) are measured for (a) the in-plane field and (b) for the out-of-plane field. *Reproduced with permission from Ref. [107]. Copyright 2012 American Chemical Society.*

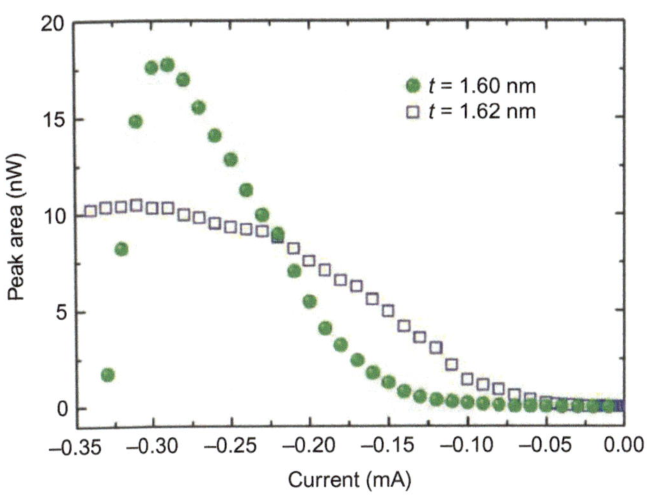

FIGURE 18

Integrated power of fundamental signals at zero applied magnetic field as a function of current for STNO. Olive circles are for 1.60 nm and blue squares for 1.62 nm free-layer thickness of the sample. *Reproduced with permission from Ref. [110]. Copyright 2012 Scientific Report Nature Group.*

150 nm × 70 nm. The oscillation power is larger for the smaller thickness and the best result is achieved for a current $I = -0.3$ mA with a measured output power of 18 nW. The main disadvantages of those configurations is the low oscillation frequency (around 1 GHz) and a larger linewidth compared to the spin-valve devices. A very recent experiment shows that it is possible to use microwave emission to improve the microwave detection properties of MTJs by means of injection locking. Basically, the MTJ is biased by a direct current that excites a self-oscillation while an additional weak microwave current drives the injection-locking mechanism. In the locking region, the magnetoresistive signal and the microwave current oscillate at the same frequency, and the output voltage exhibits a DC component which is proportional to the output power of the oscillators [111].

5 STATE OF THE ART OF SHOs

Experimental studies of a ferromagnet coupled with an HM with large spin-orbit coupling have opened a route to the development of a more efficient category of spintronic devices. Here the magnetic state can be changed by the effects connected to spin-orbit coupling, including the Dzyaloshinskii–Moriya interaction, and the Rashba and SHEs. Persistent magnetization precession can also be obtained in these systems [47–49,44,112,45,113–115,34,116,117]. For the systems we will discuss below, the two main spin-orbit effects are the interfacial Dzyaloshinskii–Moriya interaction [118,119] and the SHE; the latter can be seen as an antidamping which is able to compensate the intrinsic magnetic losses linked to the Gilbert damping (see Section 2). The persistent magnetization precession driven by the SHE was first measured by Demidov *et al.* [116] in a Pt/Py bilayer by the microfocused Brillouin light scattering (BLS) technique for an in-plane bias field. The current was injected via two contacts of gold in the central region of the ferromagnet. Figure 19 shows the device structure.

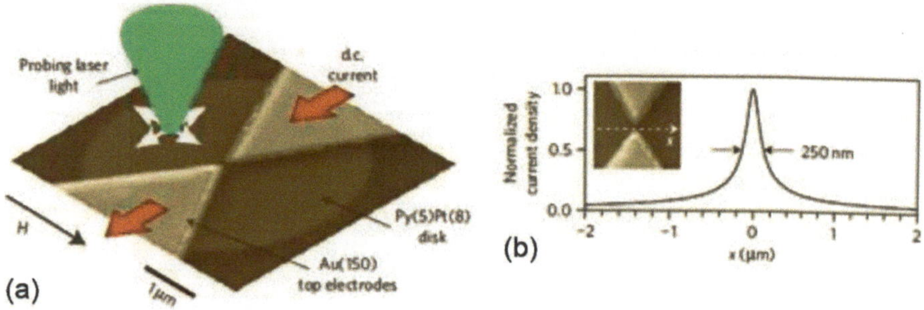

FIGURE 19

(a) Schematic device and (b) current distribution. *Adapted with permission from Ref. [116]. Copyright 2012 Nature Materials Nature Group.*

The current density is strongly nonuniform, resulting in a spin current which is also nonuniform. Due to the spatial resolution of the microfocused BLS, Demidov *et al.* [116] also demonstrated the nature of the excited mode to be a nonpropagating spin wave with a localization region of less than 100 nm. By analogy with the STNO, this can be seen as a point-contact SHO. Micromagnetic simulations have indicated the excitation of a bullet-like mode localized near the injection site of the current and also quantitatively reproduce the oscillation frequency versus current and field behavior (see Fig. 20).

In addition, as shown in Fig. 20e, micromagnetic computations predict the possibility to excite propagating spin waves with an asymmetric propagation pattern [120]. Demidov *et al.* [121] also demonstrated the possible synchronization of the microwave emission with an external microwave source. A similar system was also studied by Liu *et al.* [117]; in this experiment the microwave emission was measured via the conversion of the magnetization precession using the anisotropic magnetoresistive (AMR) signal. Measurements at lower fields indicate the possible excitation of two modes.

A different point-contact geometry was studied by Madami *et al.* [122], where the current is injected locally in a reduced region of a nanowire; this can be seen

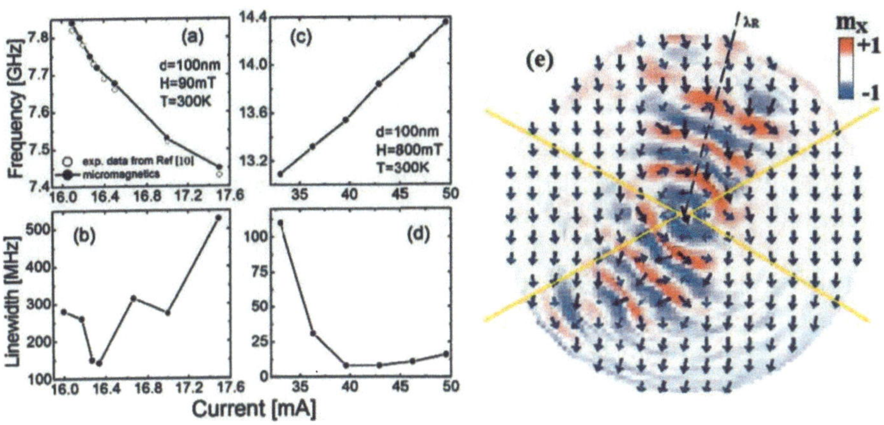

FIGURE 20

(a) A comparison between experimental oscillation frequencies from figure 2D of Ref. [116] (open circles) and the micromagnetic computations (solid line with circles), and (b) predicted micromagnetic linewidths (in-plane) as a function of the current. (c) Predicted oscillation frequencies and (d) linewidths as a function of the current for an out-of-plane field. (e) Example of spatial distribution of the magnetization as computed by means of micromagnetic simulations for an out-of-plane field (the color is related to the *x* component of the magnetization, the arrows indicate the in-plane component of the magnetization). The dashed line shows the ideal path for the estimation of the wavelength. *Adapted with permission from Ref. [120]. Copyright 2014 American Institute of Physics.*

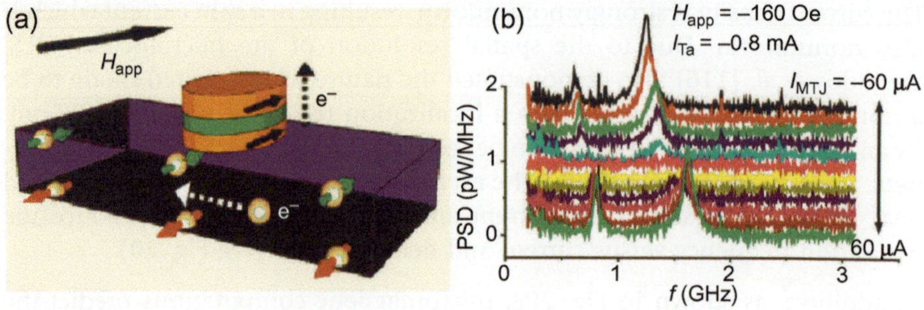

FIGURE 21

(a) Schematic of the three-terminal device and (b) measured microwave spectra. *Adapted with permission from Ref. [49]. Copyright 2010 American Physical Society.*

as a 1D point-contact SHOs. The dynamic behavior of these devices is more complex; two modes, and at high currents three modes, can be excited. Micromagnetic computations coupled with wavelet-based analyses identify those modes as nonstationary.

Persistent magnetization oscillations driven by the SHE have been also measured by Liu *et al.* [49] in three-terminal MTJs, where the torque from the SHE is due to a bias current flowing in a tantalum (Ta) strip (see Fig. 21).

In this case, the persistent magnetization precession is excited in a CoFeB free layer of an MTJ etched over the Ta strip itself. The main advantages of the device configurations in Refs. [116,117] over the ones studied in Ref. [49] are the simple fabrication process and the possibility to detect the signal using both optical and electrical measurements. On the other hand, for a fixed precessional angle of the magnetization the tunneling magnetoresistive signal in the three-terminal MTJs is much larger, more than one order of magnitude, than the AMR signal in bilayer devices. In our opinion, this system is very promising for applications and for coupling the STT and the SHE [123,124].

The last category of devices studied was introduced by Duan *et al.* [125]. This is an extended nanowire where the spin current is injected uniformly to the whole cross-section of the ferromagnet. It is the first example of a microwave oscillator based on STT that can reach a large size, in other words the magnetization dynamics can be excited coherently for regions of the order of tens of micrometers. This is possible because the Oersted field generated here is uniform while in STNOs it is not uniform and becomes the most important field at large sizes [125,126].

6 SUMMARY AND FUTURE PERSPECTIVES

In summary, we have discussed the main properties of typical oscillators. After a brief introduction about the magnetoresistance effect, the concept of spintronics and the associated micromagnetic models, we have presented some possible configuration of nano-oscillator devices based on STT and the SOT effect. In addition, we have given an indication of recent possible materials, geometry design, and features of STNOs. A summary and comparison of the STNO to classical semiconductor oscillators are shown in Table 1, where all the main properties of nano-oscillator devices are included.

There are still different challenges open for research in nanoscale spintronic oscillators. However, the most interesting one is the achievement of an output power of the order of milliwatts at reduced linewidth. This result can be achieved with a new design for a synchronization scheme, which involves, for instance, the use of a combination between the SHE and STT [126], or the realization of spin-Hall nanoscillators with multiple contacts. From a fundamental point of view, it will be interesting to find a skyrmion-based STT. An interesting challenge is also the use of MTJs in high-frequency oscillators using the experimental structure investigated by Braganca *et al.* [92]; in this way, it is possible to obtain a greater output power due to the large magnetoresistance signal typical of MTJs.

Table 1 Comparison Between the Main Properties of a Typical Voltage-Controlled Oscillator and Spin-Transfer Torque Oscillators

Device	Frequency Tunability	Frequency Operation (GHz)	Linewidth (MHz)	Q-Factor	External Magnetic Field
VCO—Ref. [127]	±22.9%	24.7	600	40	–
STNO—Ref. [128]	±25%	6–10	6	1500	Yes
STNO—Ref. [89]	±33%	2–4	–	–	Yes
STNO—Ref. [105]	–	~0.78	1.1	700	Yes
STNO—Ref. [110]	±40%	0.6–1.5	28	50	No
STNO—Ref. [94]	±15%	1.1–1.5	0.45	2600	No
STNO—Ref. [91]	–	6	74	60	No
STNO—Ref. [92]	-	26.6	140	190	No
SHO—Ref. [116]	±55%	5–10.5	<60	200	Yes
SHO—Ref. [49]	±30%	1.6	104	15	Yes

Acknowledgments

This work was supported by the Italian MIUR under Grant PRIN2010ECA8P3. The work of M. Carpentieri was also supported by the Fondazione Carical through the Project entitled "Progettazione e Sviluppo di Sensori Spintronici per Imaging a Microonde con Applicazione ai test Non Distruttivi e alla Caratterizzazione dei Materiali."

References

[1] Lindberg E. The Barkhausen criterion (observation?). In: Proceedings of 18th IEEE workshop on nonlinear dynamics of electronic systems, (NDES2010), Dresden, Germany: Inst. of Electrical and Electronic Engineers; 2010. p. 15–8.

[2] Samori C, Lacaita A, Villa F, Zappa F. Spectrum folding and phase noise in LC tuned oscillators. IEEE Trans Circuits Syst II, Analog Digit Signal Process 1998;45:781–90.

[3] Baibich MN, Broto JM, Fert A, Petroff F, Vandau FN, Eitenne P, et al. Giant magnetoresistance of (001) Fe/(001) Cr magnetic superlattices. Phys Rev Lett 1988;61(3):2472–5.

[4] Binasch G, Grunberg P, Saurenbach F, Zinn W. Enhanced magnetoresistance in layered magnetic-structures with antiferromagnetic interlayer exchange. Phys Rev B 1989;39(7):4828–30.

[5] Slonczewski J. Current-driven excitation of magnetic multilayers. J Magn Magn Mater 1996;159:L1–7.

[6] Berger L. Emission of spin waves by a magnetic multilayer traversed by a current. Phys Rev B Condens Matter Mater Phys 1996;54(13):9353–8.

[7] Jullière M. Tunneling between ferromagnetic films. Phys Lett A 1975;54:225–6.

[8] Moodera JS, Kinder LR, Wong TM, Meservey R. Large magnetoresistance at room temperature in ferromagnetic thin film tunnel junctions. Phys Rev Lett 1995;74:3273–6.

[9] Miyazaki T, Tezuka N. Giant magnetic tunneling effect in Fe/Al_2O_3/Fe junction. J Magn Magn Mater 1995;139:L231–4.

[10] Lee YM, Hayakawa J, Ikeda S, Matsukura F, Ohno H. Effect of electrode composition on the tunnel magnetoresistance of pseudo-spinvalve magnetic tunnel junction with a MgO tunnel barrier. Appl Phys Lett 2007;90:212507.

[11] Bürgler ED. Lecture notes of the 40th IFF Springschool 2009: spintronics—from GMR to quantum information, Jülich . Germany [chapter D-3]; 2009.

[12] Slaughter JM, Rizzo ND. ST-MRAM gets practical. ST-MRAM, Technical Brief; 2013. www.everspin.com.

[13] Parkin SSP, Hayashi M, Thomas L. Magnetic domain-wall racetrack memory. Science 2008;320:190.

[14] Thiaville A, Nakatani Y. Domain-wall dynamics in nanowires and nanostrips. In: Hillebrands B, Thiaville A, editors. Spin dynamics in confined magnetic structures III. Berlin: Springer; 2006.

[15] Thiaville A, Garcia JM, Miltat J. Domain wall dynamics in nanowires. J Magn Magn Mater 2002;242–245:1061–3.

[16] Martinez E, Lopez-Diaz L, Alejos O, Torres L, Carpentieri M. Domain-wall dynamics driven by short pulses along thin ferromagnetic strips: micromagnetic simulations and analytical description. Phys Rev B 2009;79:094430.

[17] Kiselev SI, Sankey JC, Krivorotov IN, Emley NC, Schoelkopf RJ, Buhrman RA, et al. Microwave oscillations of a nanomagnet driven by a spin-polarized current. Nature (London) 2003;425:6956.

[18] Kaka S, Pufall MR, Rippard WH, Silva TJ, Russek SE, Katine J. Mutual phase-locking of microwave spin torque nano-oscillators. Nature 2005;437:389.

[19] Mancoff FB, Rizzo ND, Engel BN, Tehrani S. Phase-locking in double-point-contact spin-transfer devices. Nature 2005;437:393.

[20] Grollier J, Cros V, Fert A. Synchronization of spin-transfer oscillators driven by stimulated microwave currents. Phys Rev B 2006;73:060409(R).

[21] Rippard WH, Pufall MR, Kaka S, Silva TJ, Russek SE, Katine JA. Injection locking and phase control of spin transfer nano-oscillators. Phys Rev Lett 2005;95:067203.

[22] Bertotti G. Hysteresis in magnetism—for physicists, materials scientists and engineers. USA: Academic Press: Elsevier; 1998.

[23] Hillebrands B, Thiaville A. Spin dynamics in confined magnetic structures III. In: Topics in applied physics, 101. Berlin: Springer; 2006.

[24] Hillebrands B, Ounadjela K. Spin dynamics in confined magnetic structures I. In: Topics in applied physics, 83. Berlin: Springer; 2002.

[25] Stancil D, Prabhakar A. Spin waves—theory and applications. Springer; 2009.

[26] Shinjo T, editor. Nanomagnetism and spintronics. Waltham: Elsevier; 2009.

[27] Krivorotov IN, Emley NC, Garcia AGF, Sankey JC, Kiselev SI, Ralph DC, et al. Temperature dependence of spin-transfer-induced switching of nanomagnets. Phys Rev Lett 2004;93:166603.

[28] Li Z, Zhang S. Thermally assisted magnetization reversal in the presence of a spin-transfer torque. Phys Rev B 2004;69:134416.

[29] Martínez E, Lopez-Diaz L, Torres L, Garcia-Cervera CJ. Minimizing cell size dependence in micromagnetics simulations with thermal noise. J Phys D Appl Phys 2007;40:942–8.

[30] Slonczewski J. Currents and torques in metallic magnetic multilayers. J Magn Magn Mater 2002;247:324–38.

[31] Slonczewski J. Currents, torques, and polarization factors in magnetic tunnel junctions. Phys Rev B 2005;71:024411.

[32] Carpentieri M, Moriyama T, Azzerboni B, Finocchio G. Injection locking at zero field in two free layer spin-valves. Appl Phys Lett 2005;71:024411.

[33] Dyakonov MI, Perel VI. Current-induced spin orientation of electrons in semiconductors. Phys Lett A 1971;35:459.

[34] Hirsch JE. Spin Hall effect. Phys Rev Lett 1999;83:1834.

[35] Zhang S. Spin Hall effect in the presence of spin diffusion. Phys Rev Lett 2000;85:393.

[36] Valenzuela SO, Tinkham M. Direct electronic measurement of the spin Hall effect. Nature 2006;442:176.

[37] Kimura T, Otani Y, Sato T, Takahashi S, Maekawa S. Room temperature reversible spin Hall effect. Phys Rev Lett 2007;98:156601.

[38] Ando K, Takahashi S, Harii K, Sasage K, Ieda J, Maekawa S, et al. Electric manipulation of spin relaxation using the spin Hall effect. Phys Rev Lett 2008;101:036601.

[39] Tanaka T, Kontani H, Naito M, Naito T, Hirashima DS, Yamada K, et al. Intrinsic spin Hall effect and orbital Hall effect in 3d and 5d transition metals. Phys Rev B 2008;77:165117.

[40] Nagaosa N, Sinova J, Onoda S, MacDonald AH, Ong NP. Anomalous Hall effect. Rev Mod Phys 2010;82:1539.

[41] Lavrijsen R, Haazen PPJ, Murè E, Franken JH, Kohlhepp JT, Swagten HJM, et al. Asymmetric Pt/Co/Pt-stack induced sign-control of current-induced magnetic domain-wall creep. Appl Phys Lett 2012;100:262408.

[42] Wang X, Manchon A. Diffusive spin dynamics in ferromagnetic thin films with a Rashba interaction. Phys Rev Lett 2012;108:117201.

[43] Kim K-W, Seo S-M, Ryu J, Lee K-J, Lee H-W. Magnetization dynamics induced by in-plane currents in ultrathin magnetic nanostructures with Rashba spin-orbit coupling. Phys Rev B 2012;85:180404.

[44] Liu L, Lee OJ, Gudmundsen TJ, Ralph DC, Buhrman RA. Current-induced switching of perpendicularly magnetized magnetic layers using spin torque from the spin Hall effect. Phys Rev Lett 2012;109:096602.

[45] Miron IM, Garello K, Gaudin G, Zermatten P-J, Costache MV, Auffret S, et al. Perpendicular switching of a single ferromagnetic layer induced by in-plane current injection. Nature 2011;476:189–93.

[46] Pai C-F, Liu L, Li Y, Tseng HW, Ralph DC, Buhrman RA. Spin transfer torque devices using the giant spin Hall effect of tungsten. Appl Phys Lett 2012;101:122404.

[47] Emori S, Bauer U, Ahn S-M, Martinez E, Beach GSD. Current-driven dynamics of chiral ferromagnetic domain walls. Nat Mater 2013;12:611–6.

[48] Liu L, Pai C-F, Li Y, Tseng HW, Ralph DC, Buhrman RA. Spin-torque switching with the giant spin Hall effect of tantalum. Science 2012;336:555–8.

[49] Liu L, Pai C-F, Ralph DC, Buhrman RA. Magnetic oscillations driven by the spin Hall effect in 3-terminal magnetic tunnel junction devices. Phys Rev Lett 2012;109:186602.

[50] Slonczewski JC, Sun JZ. Theory of voltage-driven current and torque in magnetic tunnel junctions. J Magn Magn Mater 2007;310:169–75.

[51] Tsoi M, Jansen AGM, Bass J, Chiang W-C, Seck M, Tsoi V, et al. Excitation of a magnetic multilayer by an electric current. Phys Rev Lett 1998;80:4281–4.

[52] Tsoi M, Jansen AGM, Bass J, Chiang W-C, Tsoi V, Wyder P. Generation and detection of phase-coherent current-driven magnons in magnetic multilayers. Nature 2000;406:46–8.

[53] Slonczewski JC. Excitation of spin waves by an electric current. J Magn Magn Mater 1999;195:L261–8.

[54] Madami M, Bonetti S, Consolo G, Tacchi S, Carlotti G, Gubbiotti G, et al. Direct observation of a propagating spin wave induced by spin-transfer torque. Nat Nanotechnol 2011;6:635–8.

[55] Rippard WH, Pufall MR, Kaka S, Russek SE, Silva TJ. Direct-current induced dynamics in $Co_{90}Fe_{10}/Ni_{80}Fe_{20}$ point contacts. Phys Rev Lett 2004;92:027201.

[56] Slavin A, Tiberkevich V. Spin wave mode excited by spin-polarized current in a magnetic nanocontact is a standing self-localized wave bullet. Phys Rev Lett 2005;95:237201.

[57] Consolo G, Azzerboni B, Lopez-Diaz L, Gerhart G, Bankowski E, Tiberkevich V, et al. Micromagnetic study of the above-threshold generation regime in a spin-torque oscillator based on a magnetic nanocontact magnetized at an arbitrary angle. Phys Rev B 2008;78:014420.

[58] Demidov VE, Urazhdin S, Demokritov SO. Direct observation and mapping of spin waves emitted by spin-torque nano-oscillators. Nat Mater 2010;9:984–8.

[59] Rippard WH, Pufall MR, Kaka S, Silva TJ, Russek SE. Current-driven microwave dynamics in magnetic point contacts as a function of applied field angle. Phys Rev B 2004;70:100406 (R).

[60] Bonetti S, Tiberkevich V, Consolo G, Finocchio G, Muduli P, Mancoff F, et al. Experimental evidence of self-localized and propagating spin wave modes in obliquely magnetized current-driven nanocontacts. Phys Rev Lett 2010;105:217204.

[61] Kim J, Tiberkevich VS, Slavin AN. Generation linewidth of an auto-oscillator with a nonlinear frequency shift: spin-torque nano-oscillator. Phys Rev Lett 2008;100:017207.

[62] Bonetti S, Muduli P, Mancoff F, Åkerman J. Spin torque oscillator frequency versus magnetic field angle: the prospect of operation beyond 65 GHz. Appl Phys Lett 2009;94:102507.

[63] Tiberkevich V, Krivorotov I, Gerhart G, Slavin A. Compensation of nonlinear phase noise in an in-plane-magnetized anisotropics spin-torque oscillator. J Magn Magn Mater 2009;321: L53–6.

[64] Bianchini L, Cornelissen S, Kim J-V, Devolder T, Chappert C. Direct experimental measurement of phase-amplitude coupling in spin torque oscillators. Appl Phys Lett 2010;97:032502.

[65] Quinsat M, Gusakova D, Sierra JF, Michel JP, Houssameddine D, Delaet B, et al. Amplitude and phase noise of magnetic tunnel junction oscillators. Appl Phys Lett 2010;97:182507.

[66] Mistral Q, van Kampen M, Hrkac G, Kim J-V, Devolder T, Crozat P, et al. Current-driven vortex oscillations in metallic nanocontacts. Phys Rev Lett 2008;100:257201.

[67] Watelot SP, Kim J-V, Ruotolo A, Otxoa RM, Bouzehouane K, Grollier J, et al. Commensurability and chaos in magnetic vortex oscillations. Nat Phys 2012;8:682–7.

[68] Finocchio G, Ozatay O, Torres L, Buhrman RA, Ralph DC, Azzerboni B. Spin-torque-induced rotational dynamics of a magnetic vortex dipole. Phys Rev B 2008;78:174408 (R).

[69] Giordano A, Puliafito V, Torres L, Carpentieri M, Azzerboni B, Finocchio G. Micromagnetic study of spin-transfer-driven vortex dipole and vortex quadrupole dynamics. IEEE Trans Magn 2014;50:4300404.

[70] Mohseni SM, Sani SR, Persson J, Anh Nguyen TN, Chung S, Pogoryelov Y, et al. High frequency operation of a spin-torque oscillator at low field. Status Solidi RRL 2011;5:432–4.

[71] Macià F, Backes D, Kent AD. Stable magnetic droplet solitons in spin-transfer nanocontacts. Nat Nanotechnol 2014;9:992–6.

[72] Puliafito V, Torres L, Ozatay O, Hauet T, Azzerboni B, Finocchio G. Micromagnetic analysis of dynamical bubble-like solitons based on the time domain evolution of the topological density. J Appl Phys 2014;115:17D139.

[73] Puliafito V, Siracusano G, Azzerboni B, Finocchio G. Self-modulated soliton modes excited in a nanocontact spin-torque oscillator. IEEE Magn Lett 2014;5:3000104.

[74] Boone C, Katine JA, Childress JR, Zhu J, Cheng X, Krivorotov IN. Experimental test of an analytical theory of spin-torque-oscillator dynamics. Phys Rev B 2009;79:140404 (R).

[75] Berkov DV. Synchronization of spin-torque-driven nano-oscillators for point contacts on a quasi-one-dimensional nanowire: micromagnetic simulations. Phys Rev B 2013;87:014406.

[76] Giordano A, Carpentieri M, Zivieri R, Siracusano G, Azzerboni B, Finocchio G. Nanowire spin-torque oscillator with non-uniform polarizer: a micromagnetic study. IEEE Trans Magn 2014;50:1402204.

[77] Finocchio G, Maugeri N, Torres L, Azzerboni B. Domain wall dynamics driven by a localized injection of a spin-polarized current. IEEE Trans Magn 2010;46:1523–6.

[78] Kiselev SI, Sankey JC, Krivorotov IN, Emley NC, Schoelkopf RJ, Buhrman RA, et al. Microwave oscillations of a nanomagnet driven by a spin-polarized current. Nature 2003;425:380–3.

[79] Lee K-J, Deac A, Redon O, Nozieres J-P, Dieny B. Excitations of incoherent spin-waves due to spin-transfer torque. Nat Mater 2005;3:877–81.

[80] Krivorotov IN, Emley NC, Sankey JC, Kiselev SI, Ralph DC, Buhrman RA. Time-domain measurements of nanomagnet dynamics driven by spin-transfer torques. Science 2005;307: 228–31.

[81] Krivorotov IN, Berkov DV, Gorn NL, Emley NC, Sankey JC, Ralph DC, et al. Large-amplitude coherent spin waves excited by spin-polarized current in nanoscale spin valves. Phys Rev B 2007;76:024418.

[82] Finocchio G, Siracusano G, Tiberkevich V, Krivorotov IN, Torres L, Azzerboni B. Time-domain study of frequency-power correlation in spin-torque oscillators. Phys Rev B 2010;81:184411.

[83] Thadani KV, Finocchio G, Li Z-P, Ozatay O, Sankey JC, Krivorotov IN, et al. Strong linewidth variation for spin-torque nano-oscillators as a function of in-plane magnetic field angle. Phys Rev B 2008;78:024409.

[84] Finocchio G, Prattella A, Consolo G, Martinez E, Giordano A, Azzerboni B. Hysteretic spin-wave excitation in spin-torque oscillators as a function of the in-plane field angle: a micromagnetic description. J Appl Phys 2011;110:123913.

[85] Braganca PM, Lee OJ, Ozatay O, Liu L, Finocchio G, Ralph DC, et al. Coherent and incoherent spin torque oscillations in a tapered nanopillar magnetic spin-valve. Appl Phys Lett 2013;102:2521402.

[86] Mangin S, Ravelosona D, Katine JA, Carey MJ, Terris BD, Fullerton EE. Current-induced magnetization reversal in nanopillars with perpendicular anisotropy. Nat Mater 2006;5:210–5.

[87] Okura R, Sakuraba Y, Seki T, Izumi K, Mizuguchi M, Takanashi K. High-power rf oscillation induced in half-metallic Co_2MnSi layer by spin-transfer torque. Appl Phys Lett 2011;99: 052510.

[88] Sinha J, Hayashi M, Takahashi YK, Taniguchi T, Drapeko M, Mitani S, et al. Large amplitude microwave emission and reduced nonlinear phase noise in $Co_2Fe(Ge_{0.5}Ga_{0.5})$ Heusler alloy based pseudo spin valve nanopillars. Appl Phys Lett 2011;99:162508.

[89] Houssameddine D, Ebels U, Delae B, Rodmacq B, Firastrau I, Ponthenier F, et al. Spin-torque oscillator using a perpendicular polarizer and a planar free layer. Nat Mater 2007;6:447–53.

[90] Ebels U, Houssameddine D, Firastrau I, Gusakova D, Thirion C, Dieny B, et al. Macrospin description of the perpendicular polarizer-planar free-layer spin-torque oscillator. Phys Rev B 2008;78:024436.

[91] Moriyama T, Finocchio G, Carpentieri M, Azzerboni B, Ralph DC, Buhrman RA. Phase locking and frequency doubling in spin-transfer-torque oscillators with two coupled free layers. Phys Rev B 2012;86:06041(R).

[92] Braganca PM, Pi K, Zakai R, Childress JR, Gurney BA. Zero field high frequency oscillations in dual free layer spin torque oscillators. Appl Phys Lett 2013;103:232407.

[93] Pribiag VS, Krivorotov IN, Fuchs GD, Braganca PM, Ozatay O, Sankey JC, et al. Magnetic vortex oscillator driven by d.c. spin-polarized current. Nat Phys 2007;3:498–503.

[94] Pribiag VS, Finocchio G, Williams BJ, Ralph DC, Buhrman RA. Long-timescale fluctuations in zero-field magnetic vortex oscillations driven by dc spin-polarized current. Phys Rev B 2009;80:180411 (R).

[95] Finocchio G, Pribiag VS, Torres L, Buhrman RA, Azzerboni B. Spin-torque driven magnetic vortex self-oscillations in perpendicular magnetic fields. Appl Phys Lett 2010;84:102508.

[96] Pikovsky A, Rosenblum M, Kurths J. Synchronization: a universal concept in nonlinear sciences. Cambridge, MA: Cambridge University Press; 2003.

[97] Carpentieri M, Finocchio G, Azzerboni B, Torres L. Spin-transfer-torque resonant switching and injection locking in the presence of a weak external microwave field for spin valves with perpendicular materials. Phys Rev B 2010;82:094434.

[98] Fuchs GD, Emley NC, Krivorotov IN, Braganca PM, Ryan EM, Kiselev SI, et al. Spin-transfer effects in nanoscale magnetic tunnel junctions. Appl Phys Lett 2004;85:1205–7.

[99] Parkin SSP, Kaiser C, Panchula A, Rice PM, Hughes B, Samant M, et al. Giant tunnelling magnetoresistance at room temperature with MgO (100) tunnel barriers. Nat Mater 2004;3:862–7.

[100] Yuasa S, Nagahama T, Fukushima A, Suzuki Y, Ando K. Giant room-temperature magnetoresistance in single-crystal Fe/MgO/Fe magnetic tunnel junctions. Nat Mater 2004;3:868–71.

[101] Nazarov AV, Olson HM, Cho H, Nikolaev K, Gao Z, Stokes S, et al. Spin transfer stimulated microwave emission in MgO magnetic tunnel junctions. Appl Phys Lett 2006;88:162504.

[102] Houssamedine D, Florez SH, Katine JA, Michel J-P, Ebels U, Mauri D, et al. Spin transfer induced coherent microwave emission with large power from nanoscale MgO tunnel junctions. Appl Phys Lett 2008;93:022505.

[103] Deac AM, Fukushima A, Kubota H, Maehara H, Suzuki Y, Yuasa S, et al. Bias-driven high-power microwave emission from MgO-based tunnel magnetoresistance devices. Nat Phys 2008;4:803–9.

[104] Zeng ZM, Upadhyaya P, Amiri PK, Cheung KH, Katine J, Langer J, et al. Enhancement of microwave emission in magnetic tunnel junction oscillators through in-plane field orientation. Appl Phys Lett 2011;99:032503.

[105] Dussaux A, Georges B, Grollier J, Cros V, Khvalkovskiy AV, Fukushima A, et al. Large microwave generation from current-driven magnetic vortex oscillators in magnetic tunnel junctions. Nat Commun 2010;1:1–6.

[106] Ikeda S, Miura K, Yamamoto H, Mizunuma K, Gan HD, Endo M, et al. A perpendicular-anisotropy CoFeB–MgO magnetic tunnel junction. Nat Mater 2010;9:721–4.

[107] Zeng ZM, Khalili Amiri P, Krivorotov IN, Zhao H, Finocchio G, Wang J-P, et al. High-power coherent microwave emission from magnetic tunnel junction nano-oscillators with perpendicular anisotropy. ACS Nano 2012;6:6115–21.

[108] Masugata Y, Ishibashi S, Tomita H, Seki T, Nozaki T, Suzuki Y, et al. Spin-torque induced rf oscillation in magnetic tunnel junctions with an Fe-rich CoFeB free layer. J Phys Conf Ser 2011;266:012098.

[109] Skowroński W, Stobiecki T, Wrona J, Reiss G, van Dijken S. Zero-field spin torque oscillator based on magnetic tunnel junctions with a tilted CoFeB free layer. Appl Phys Express 2012;5:063005.

[110] Zeng ZM, Finocchio G, Zhang BS, Katine JA, Krivorotov I, Huai Y, et al. Ultralow-current-density and bias-field-free spin-transfer nano-oscillator. Sci Rep 2013;3:1426.

[111] Fang B., Carpentieri M., Hao X., Jiang H., Katine JA, Krivorotov IN, et al. Giant spin-torque diode sensitivity at low input power in the absence of bias magnetic field. http://arxiv.org/abs/1410.4958. Submitted to Nature Nanotechnology.

[112] Avci CO, Garello K, Miron IM, Gaudin G, Auffret S, Boulle O, et al. Magnetization switching of an MgO/Co/Pt layer by in-plane current injection. Appl Phys Lett 2012;100:212404.

[113] Haazen PPJ, Murè E, Franken JH, Lavrijsen R, Swagten HJM, Koopmans B. Domain wall depinning governed by the spin Hall effect. Nat Mater 2013;12:299–303.

[114] Ryu K-S, Thomas L, Yang S-H, Parkin S. Chiral spin torque at magnetic domain walls. Nat Nanotechnol 2013;8:527–33.

[115] Bhowmik D, You L, Salahuddin S. Spin Hall effect clocking of nanomagnetic logic without a magnetic field. Nat Nanotechnol 2014;9:59–63.

[116] Demidov VE, Urazhdin S, Ulrichs H, Tiberkevich V, Slavin A, Baither D, et al. Magnetic nano-oscillator driven by pure spin current. Nat Mater 2012;11:1028–31.

[117] Liu RH, Lim WL, Urazhdin S. Spectral characteristics of the microwave emission by the spin Hall nano-oscillator. Phys Rev Lett 2013;110:147601.

[118] Moriya T. New mechanism of anisotropic superexchange interaction. Phys Rev Lett 1960;4:228.

[119] Dzyaloshinskii I. A thermodynamic theory of 'weak' ferromagnetism of antiferromagnetics. J Phys Chem Solids 1958;4:241–55.

[120] Giordano A, Carpentieri M, Laudani A, Gubbiotti G, Azzerboni B, Finocchio G. Spin-Hall nano-oscillator: a micromagnetic study. Appl Phys Lett 2014;105:042412.

[121] Demidov VE, Ulrichs H, Gurevich SV, Demokritov SO, Tiberkevich VS, Slavin AN, et al. Synchronization of spin Hall nano-oscillators to external microwave signals. Nat Commun 2014;5:3179.

[122] Madami M, Gubbiotti G, Moriyama T, Tanaka K, Siracusano G, Carpentieri M, et al. Micro-focused Brillouin light scattering study of the magnetization dynamics driven by Spin Hall effect in a transversely magnetized NiFe nanowire. J Appl Phys 2015;117:17D504.

[123] Tomasello R, Carpentieri M, Finocchio G. Dynamical properties of three terminal magnetic tunnel junctions: spintronics meets spin-orbitronics. Appl Phys Lett 2013;103:252408.

[124] Siracusano G, La Corte A, Puliafito V, Finocchio G. A generalized tool for accurate time-domain separation of excited modes in spin-torque oscillators. J Appl Phys 2014;115 (17):17D108.

[125] Duan Z, Smith A, Yang L, Youngblood B, Lindner J, Demidov VE, et al. Nanowire spin torque oscillator driven by spin orbit torques. Nat Commun 2014;5:5616.

[126] Siracusano G, Tomasello R, Puliafito V, Giordano A, Azzerboni B, La Corte A, et al. Intrinsic synchronization of an array of spin-torque oscillators driven by the spin-Hall effect. J Appl Phys 2015;117:17E504.

[127] Sadhu B, Ferriss MA, Natarajan AS, Yaldiz S, Plouchart JO, Rylyakov AV, et al. A linearized, low-phase-noise VCO-based 25-GHz PLL with autonomic biasing. IEEE J Solid State Circ 2013;48(5):1138–50.

[128] Rippard WH, Deac AM, Pufall MR, Shaw JM, Keller MW, Russek SE, et al. Spin-transfer dynamics in spin valves with out-of-plane magnetized CoNi free layers. Phys Rev B 2010;81:014426 (R).

CHAPTER 8

Domain Wall Motion in Nanostructures

João Sampaio*, Julie Grollier*, Peter J. Metaxas†
**Unité Mixte de Physique CNRS-Thales, Palaiseau, France*
†School of Physics, University of Western Australia, Crawley, WA, Australia

1 INTRODUCTION

Domain walls (DWs) are the magnetic structures of finite width that separate regions of uniform magnetization in a magnetic material (the domains). The structure of DWs depends on the fundamental interactions of magnetic materials: the exchange interaction, the dipolar interaction, and the magnetocrystalline anisotropy. The way DWs move is defined by the fundamentals of magnetic dynamics: the precession of spins around their equilibrium states, the torques induced by fields, the torques arising from charge currents, and the processes that damp the spin precession. The structure and dynamics of DWs are thus an invaluable window to study these magnetic phenomena. The understanding of other spin textures, such as magnetic bubbles, vortices, or skyrmions [1], is also deeply connected to DWs, their structure and dynamics.

In recent decades, improvements in deposition and patterning techniques have enabled the study of nano-sized magnets. These objects provide a way to study the magnetism and spintronic effects at length scales relevant to these phenomena: a DW is typically a few to hundreds of nanometers thick, and the characteristic spin-flip diffusion length of an electron in a transition metal is a few to tens of nanometers. Moreover, the shape and dimensions of the nanomagnets induce large changes to the structure and subsequently the dynamics of DWs in these objects, offering a way to control the creation, propagation, and pinning of DWs.

As magnetic reversal is important to many technologies, so is the study of the statics and dynamics of the DWs that cause it. In particular, the efficient manipulation of DWs in nanomagnets has also a strong technological interest, as it is at the core of many current and proposed devices, from storage and processing of digital and analogical information [2], memristors [3], tuneable RF signal sources [4] and detectors, and even precise mechanic manipulators of magnetic beads for fluidic devices [5].

Magnetism of Surfaces, Interfaces, and Nanoscale Materials. http://dx.doi.org/10.1016/B978-0-444-62634-9.00008-4

2 DOMAIN WALL STATICS

Experimentally, domain walls are typically either created via (stochastic) nucleation processes driven by local or globally applied magnetic fields, or by fabricating curved or zigzag strips where domain walls can be nucleated at the curves using transversally applied fields (e.g., Fig. 1). The internal magnetic structure of a created domain wall within a domain wall conduit or nanostrip depends not only on the conduit's physical composition (as in bulk materials) but also on its geometry (e.g., [7–9]). The dynamics of domain walls are intimately linked to this internal structure, which can undergo complex dynamic transformations during wall motion. Here we will focus predominantly on four domain wall structures observed in the flat, narrow nanostrips which are the main focus of the majority of current device-oriented studies: Bloch/Néel domain walls in systems with perpendicular anisotropy and transverse/vortex walls in systems with in-plane anisotropy, illustrated in Fig. 2a–d. There will be some mention of asymmetric transverse domain walls and antivortex domain walls (Fig. 2e and f).

The structure of the domain wall in isolated ferromagnetic nanostrips is determined by a balance between the exchange energy (A, measured in J/m) and the anisotropy energy (K, measured in J/m^3). Note that the latter encompasses demagnetizing effects, which depend upon the conduit's geometry and the saturation magnetization of the material making up the conduit (M_s, measured in A/m) [11].

We first consider a domain wall in a system with a uniaxial out-of-plane magnetic anisotropy separating two oppositely magnetized domains with their magnetizations lying out of the plane of the strip. The domain wall is the region over which the magnetization changes direction from one domain to another and the width of this region, defined below, is determined largely by a

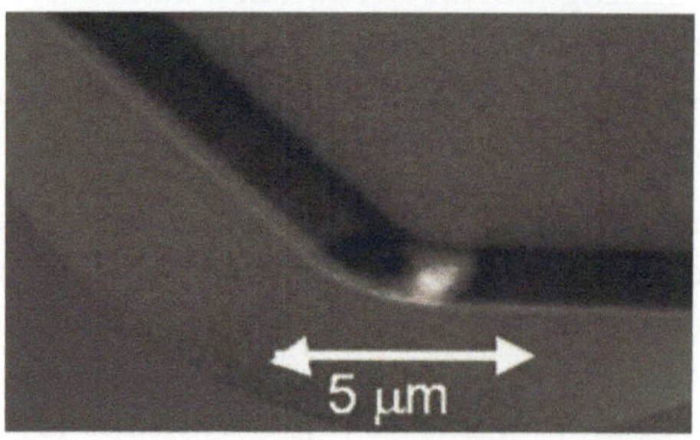

FIGURE 1
Vortex domain wall created at a bend in an NiFe nanostrip imaged via photoemission electron microscopy [6].

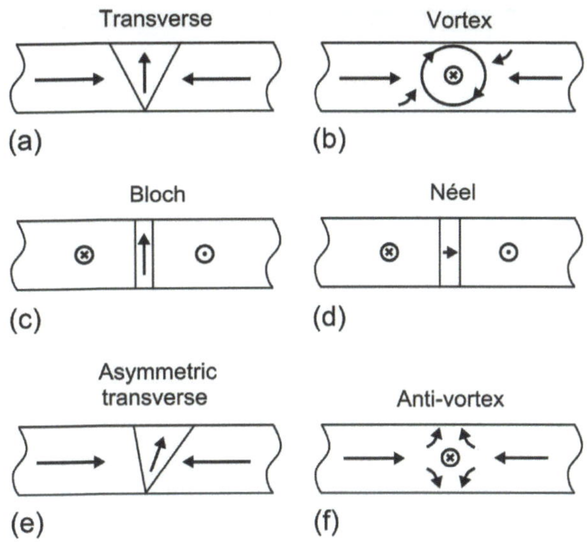

FIGURE 2
Sketches of domain wall structures in nanostrips (top-down view): (a) transverse domain wall, (b) vortex domain wall, (c) Bloch domain wall, (d) Néel domain wall, (e) asymmetric transverse domain wall, and (f) anti-vortex domain wall. The (asymmetric) transverse and (anti-)vortex domain walls are shown as head-to-head domain walls. The core of the (anti-)vortex structures is shown here as lying out of the plane. We use the Bloch and Néel nomenclature only for strips with perpendicular anisotropy (c and d). (a), (b), and (e) follow [10].

competition between exchange and anisotropy energies. The exchange energy favors parallel alignment of neighboring magnetic moments. As such, it tends to widen the domain wall since this minimizes the angle between adjacent spins within the domain wall. In contrast, the anisotropy energy favors alignment of the local moment, $\vec{m}\,(r)$, along a certain axis (or axes). This energy has the effect of narrowing the domain wall since this will reduce the number of spins misaligned with the anisotropy axis. Note that some degree of misalignment *must* occur within the domain wall region to enable the magnetization to change direction between two domains. Given this competition between A and K, it is perhaps not surprising that a simple expression for the domain wall width Δ can be derived by minimizing the energies associated with the domain wall structure itself (see, e.g., [12,13]) (Fig. 3):

$$\Delta = \sqrt{\frac{A}{K}}. \tag{1}$$

The domain wall width is generally small in systems with perpendicular magnetic anisotropy (PMA), typically of the order of 3–10 nm (Fig. 2c and d). This is because such systems have a large intrinsic out-of-plane anisotropy, which is

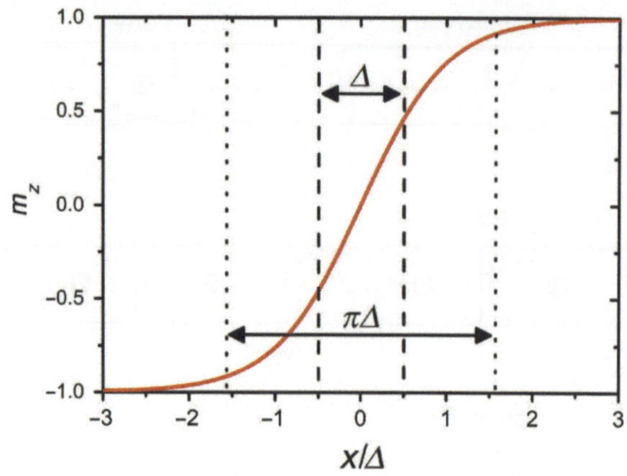

FIGURE 3

Domain wall width, Δ or $\pi\Delta$. The change in the normalized out-of-plane component of the magnetization is shown when moving across a Bloch domain wall in a system with PMA.

in fact a requirement to have the magnetization in the domains lie out of the film plane. In its absence, demagnetizing effects will dominate and will drive the domain magnetization in-plane. To give a physical feeling for Δ, we can consider a Bloch wall (Fig. 2c) located at $x=0$ in an ideal continuous film with PMA. As one moves across a domain wall, the out-of-plane component of the magnetization varies in a simple model as

$$m_z = \frac{M_z}{M_s} = \pm \tanh\left(\frac{x}{\Delta}\right) \tag{2}$$

($m_z \sim 0.46$ at $x = \Delta/2$ and $m_z \sim 0.92$ at $x = \pi\Delta/2$).

In nanostrips with low or negligible intrinsic anisotropy (e.g., NiFe, CoFeB), walls can be an order of magnitude wider than in systems with strong PMA. Here, anisotropy arises primarily due to demagnetizing effects via so-called shape anisotropy, with walls in these systems having what are generally more complex internal structures (Fig. 2a, b, e, and f). Nakatani *et al.* have provided a comprehensive geometrical phase diagram for head-to-head walls in in-plane magnetized nanostrips, which is shown in Fig. 4 [10]. Three main domain wall structures can be identified. In thin and/or narrow structures, the lowest energy wall is the transverse domain wall, which is quite a simple structure with an almost triangular region of magnetization that is perpendicular to the strip's long axis. As the strip thickness and/or width become larger, the vortex domain wall becomes the lowest energy wall. This wall is in many ways analogous to the vortex configuration observed in flat magnetic discs [14]. Although there is an increased exchange energy associated with the curled structure, this

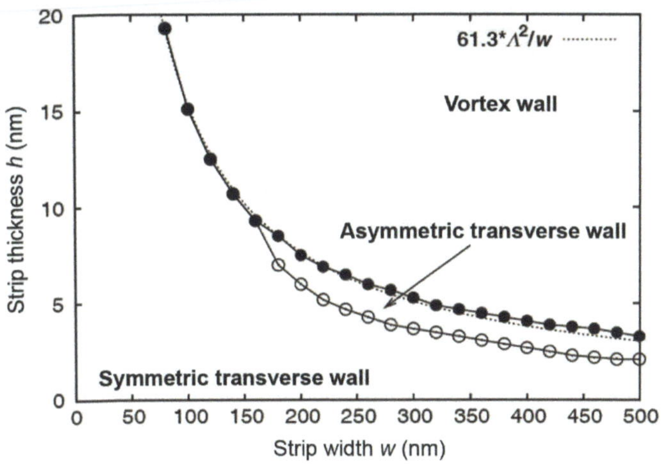

FIGURE 4
Phase diagram for head-to-head domain walls in an NiFe strip [10].

configuration minimizes the creation of magnetic charges on the strip edges, thereby reducing the demagnetizing energy of the wall. An intermediate asymmetric transverse domain wall will arise for a narrow regime of thickness/width values in between the vortex and transverse domain wall regions. Note that the phase diagram shown in Fig. 4 will be changed in the presence of additional in-plane anisotropies [15] and in multilayer spin valve structures due to inter-wall coupling [16] or the creation of quasi-walls in underlying magnetic layers [17]. The latter also modify wall dynamics [18]. It can be seen from the schematics in Fig. 2 that the domain wall width in in-plane magnetized systems is not so trivial to define. For example, a transverse wall structure has a width that varies across the strip. In such cases, one can employ the expression for the Thiele domain wall width, Δ_{T} [10,19]:

$$\frac{2}{\Delta_{\mathrm{T}}}=\frac{1}{wt}\int\left(\frac{\partial m}{\partial x}\right)^2 \mathrm{d}V, \tag{3}$$

where the integral is over the entire volume of a nanostrip with a width w and a thickness t lying along the x-axis.

Strip geometry also influences the domain wall structure in systems with PMA. Considering simple nanostrips with varying width, it has been found that the stable wall structure is a function of the strip width [8]. In Fig. 5, Koyama *et al.* used the fact that Neel and Bloch DWs produce different anisotropic magnetoresistance (AMR) signals to determine the type of DW present in strips of different widths. Néel walls are favored at small strip widths (below a critical width, w_c, typically of the order of 40–80 nm) [20–22] and Bloch walls favored

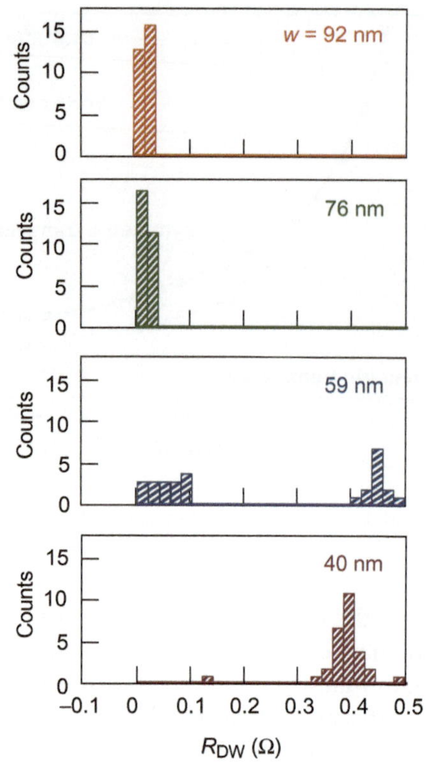

FIGURE 5

Evidence of strip-width-dependent domain wall structures in a nanostrip with PMA. The presence of Néel and Bloch domain walls in a nanostrip will generate different anisotropic magnetoresistance (AMR) signals, which enables their experimental identification. The data from Koyama *et al.* [8] shows histograms of resistance measurements of 30 domain walls prepared in a Co/Ni strip with PMA. When the strip width reaches 59 nm there is approximately equal chance of having a Néel or Bloch domain wall. Above and below this critical width, the prevalent wall structure is respectively Bloch and Néel.

for larger strip widths. The Néel wall configuration minimizes the charges that are created at the edges of the strip in the Bloch structure, which would lead to a large demagnetizing field across the strip. On the other hand, the presence of a Néel wall will increase the demagnetizing field across the wall itself.

In this section, we have presented an overview of the domain wall types observed in nanostrips as well as the physical factors that determine the structure itself. In the following, the dynamics of these structures will be considered. As will be demonstrated, this motion can be relatively complex with periodic or turbulent transitions between wall types occurring during domain wall motion.

3 DOMAIN WALL DYNAMICS

Although current-driven domain wall motion is critical for the success of scalable nanoscale domain wall devices, field-driven domain wall motion still provides an excellent arena in which to study the fundamentals of domain wall dynamics. This section will focus on the dynamics under a magnetic field firstly in perfect, defect-free films. A number of other reviews have been written on domain wall dynamics in recent years focusing both on theory [23] and experiment [6]. Secondly, we will discuss the dynamics in films containing pinning sites where motion is limited by engineered domain wall traps or intrinsic disorder and/or inhomogeneities. Finally, we will consider domain wall motion driven by electrical currents. Spin wave-assisted domain wall motion in nanostrips has also been proposed but is not discussed here (see [24] and references therein).

3.1 Domain Wall Motion in Defect-Free Nanostrips

An applied field will favor the growth of one domain over the other via the Zeeman energy. This leads to domain expansion, which occurs via domain wall motion. For fields applied collinearly with the domain magnetizations, the torques that drive domain wall motion are nonzero only inside the domain wall where the magnetization and the applied field are misaligned. Although velocity is often discussed as a function of the driving torque (as is done here), it is important to remember that the domain wall velocity is also a function of time since the wall undergoes transient dynamics before reaching a steady-state velocity. Furthermore, just as there are transient dynamics at the onset of domain wall motion, there can be transient dynamics at the end of an excitation as measured in in-plane magnetized strips [25,26]. In contrast, measurements on ultrathin strips with PMA suggested negligible measurable inertial effects which is thought to stem in part from the intrinsically high damping in these systems [27].

The *steady-state* domain wall velocity v, in the absence of pinning, is predicted to vary linearly with the applied field as [28]:

$$v=\frac{\gamma\Delta}{\alpha}H, \tag{4}$$

where H is the applied field, Δ is the domain wall width (described above), α is the magnetic damping parameter, and $\gamma>0$ is the gyromagnetic ratio. $\gamma\Delta/\alpha$ is referred to as the domain wall mobility. From the above equation, one sees that large domain wall widths and low damping will lead to high domain wall velocities, which are linear in H. However, this linearity disappears above

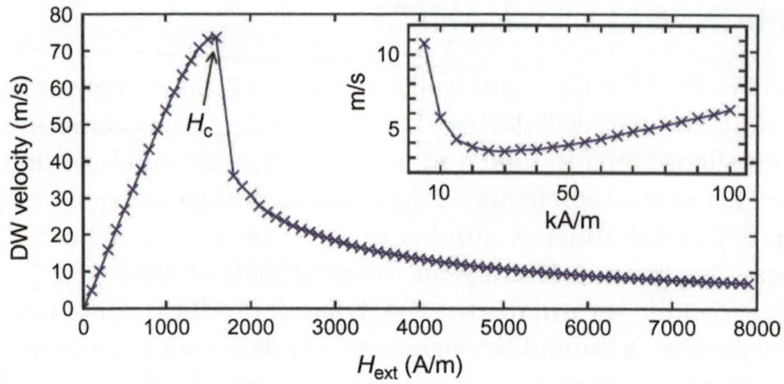

FIGURE 6

Walker breakdown dynamics. Simulated domain wall velocity as a function of external field in a nanostrip with PMA showing a drop in the domain wall velocity at the Walker breakdown field followed by an eventual continued increase in velocity at high fields (inset) [20].

(or just below) a critical field H_c known as the Walker breakdown field [28]. Domain wall motion above this field is characterized by a region of negative mobility (as seen in Fig. 6) followed by an eventual continued increase in the domain wall velocity (see the inset of Fig. 6). Above the Walker breakdown field, domain wall motion changes from steady forward motion to precessional and at times turbulent motion, as discussed below.

The dynamic response of magnetic moments, $\boldsymbol{M}$, to an applied field can be described using the Landau–Lifschitz–Gilbert (LLG) equation:

$$\frac{\mathrm{d}\boldsymbol{M}}{\mathrm{d}t} = -\gamma \boldsymbol{M} \times \boldsymbol{H}_{\mathrm{eff}} + \frac{\alpha}{M_s} \boldsymbol{M} \times \frac{\mathrm{d}\boldsymbol{M}}{\mathrm{d}t}. \tag{5}$$

$\boldsymbol{H}_{\mathrm{eff}}$ contains all (effective) magnetic fields including external magnetic fields, demagnetizing fields, effective anisotropy fields, and the exchange field. The first term on the right-hand side of the equation describes the torque on the local magnetic moment due to all the effective fields in the system. The second term describes the effect of the magnetic damping on the $\boldsymbol{H}_{\mathrm{eff}}$-driven precession of the moment.

Let us return to the case of a Bloch wall in a system with PMA in the presence of a simple out-of-plane field: $\boldsymbol{H} = -H\hat{\boldsymbol{z}}$. The only place that the torque is nonzero is within the domain wall since $\boldsymbol{H}$ is aligned along the same axis as $\boldsymbol{M} = \pm M_s\hat{\boldsymbol{z}}$ within the domains. Bloch walls can have one of two chiralities. If we assume the configuration shown in Fig. 2c, we find that the torque on the central part of the domain wall is towards the right of the page ($+\hat{x}$) direction. This generates a slight canting of the magnetization within the domain wall and this canting subsequently generates a magnetic field across the wall. This can be understood in a simple way by picturing magnetic charges appearing on either side of the wall due to the canting. The resultant internal demagnetizing field is in the $-\hat{x}$ direction. This demagnetizing field is critical to understanding the domain wall motion. The torque produced by this field on the magnetization in the domain wall is in the $-\hat{z}$ direction, thereby canting the spins within the domain wall downwards, which translates the wall in the $+\hat{x}$ direction. The direction of motion will be the same for a wall with the opposite chirality.

At fields below the Walker breakdown field and ignoring the transient motion during the initial wall canting, a Bloch wall in a system with PMA will move at steady velocity with an internal canting of the magnetization at an angle which depends upon H. Above the Walker breakdown field, the torque on the system is dominated by that of the applied field and the canting angle oscillates, leading to an oscillatory transition between a Bloch and Néel wall structure accompanied by back and forth motion, albeit with a (reduced) net forward velocity (see Fig. 7 for simulated domain wall images above Walker breakdown).

FIGURE 7

Precessional domain wall dynamics above the Walker breakdown field. Simulated snapshots of precessional, back and forth domain wall dynamics in a nanostrip with PMA under the action of a constant out-of-plane field with $H > H_W$ [20].

For systems with in-plane magnetization we can consider the case of a transverse wall under the action of a field applied along the strip. Similar calculations can be made as for the Bloch wall, which indicate that the magnetization within the transverse domain wall cants out of the plane of the film. Domain wall motion below the Walker field thus corresponds to a translation of the transverse wall with an out-of-plane magnetization component. In contrast, dynamics above the Walker breakdown are typically more complex than in films with PMA, characterized by (turbulent) periodic transitions between the transverse, anti-vortex, and vortex domain wall structures (e.g., [29,30]).

Although Walker breakdown was first detailed in the mid twentieth century for systems with PMA (see [28] and references therein), its first clear experimental observation in a magnetic nanostrip was by Beach *et al.*, who observed the characteristic change in velocity in a nanostrip with in-plane anisotropy (Fig. 8a) [31]. Later, the oscillatory motion which occurs above the Walker breakdown was directly measured [33,34]. One of the few observations of Walker breakdown-like dynamics in a system with PMA has been measured in the dilute magnetic semiconductor, GaMnAs (Fig. 8b) [32]. However, in most measurements of systems with PMA, the Walker breakdown is not directly observable in the velocity-field/current response, which is thought to be due to masking of this velocity drop by pinning effects [35,36].

In order to facilitate the design of *high-speed* domain wall-based devices, a number of methods to suppress Walker breakdown (and the slow, turbulent dynamics associated with it) have been investigated. Many works have studied the

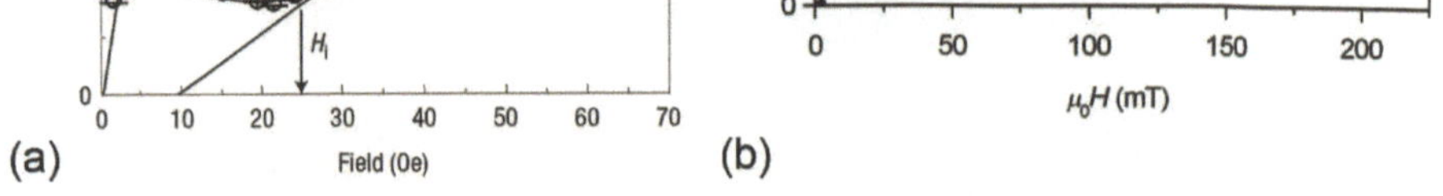

FIGURE 8

Experimental signatures of Walker breakdown in in-plane and out-of-plane magnetized materials. (a) Domain wall velocity versus field in a 600-nm-wide, 20-nm-thick in-plane magnetized NiFe nanostrip where the field is applied along the nanostrip's long axis [31]. (b) Domain wall velocity versus field in a GaMnAs film with PMA under the action of a field applied out of the film plane [32].

use of effective [37] or external [33,38] transverse fields to stabilize the wall structure and suppress domain wall deformation at $H > H_W$. Both edge roughness [30] and engineered comb-like structures [39] can also inhibit the formation of anti-vortex walls which accompany Walker breakdown in in-plane magnetized strips. Another approach simply reduces the domain wall travel distance below the distance required for the oscillatory motion to occur, which is well suited to devices with short travel distances such as magnetic tunnel junctions (MTJs) [40]. Finally, it should be noted that the Walker breakdown is also expected to be suppressed in nanowires with a circular cross-section (not discussed here; see, e.g., [41] and references therein).

The above text has focused squarely on the motion of domain walls under the assumption that only damping and internal structural transformations are limiting the domain wall velocity. In practice, however, pinning centers often play a major role in real systems, which can lead to a slowing of the domain wall velocity, increased domain wall propagation fields/currents, and the appearance of thermally activated pinning-limited regimes of motion. This is discussed in the next section.

3.2 Domain Wall Motion and Dynamics in the Presence of Pinning

A pinning center is a localized region in a material where domain wall propagation through or beyond the region is impeded. This may be because the domain wall energy is modified at that point or there may be an (effective) field acting at that point, which favors one domain magnetization direction over another. The latter locally impedes domain expansion and thus the torques that drive wall motion. Depending on its source and its consequences, pinning can be either detrimental or critical to device operation (e.g., unwanted pinning due to edge roughness induced during lithography vs. engineered pinning sites for positional stability or bit length definition [42]). When critical for positional stability in devices, it is important to have the ability to correctly balance domain wall stability and low power operation: the pinning sites must be strong enough to reliably pin the wall but not so strong that prohibitively large fields or currents have to be employed to move the wall (e.g., [42]).

Many types of engineered pinning sites are based on localized structural modifications of the nanostrip. Although it is possible to pin domain walls via changes to the thickness of a nanostrip [43], the majority of studies focus on localized modifications to the width of a nanostrip via lithographically defined notches (Fig. 9), constrictions, protrusions, or holes (e.g., [42,44–47]). Such pinning centers have been used in applications ranging from the creation of domain wall diodes [44] and ratchets [48] to the experimental quantification

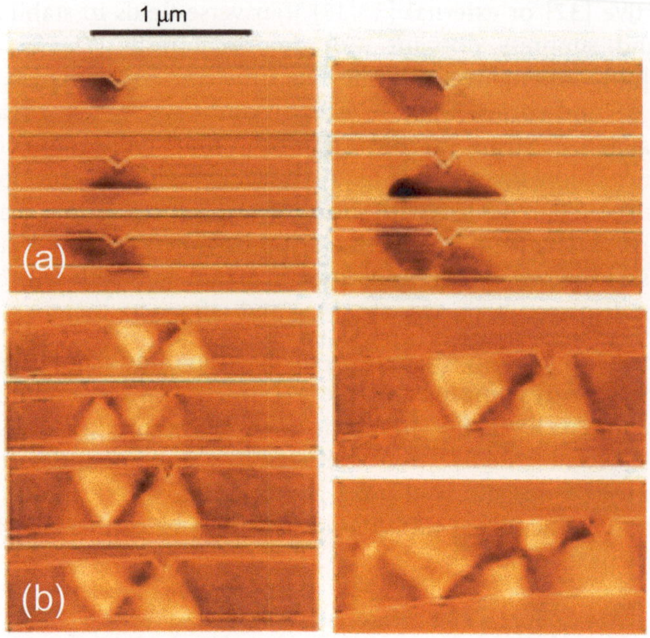

FIGURE 9

Transverse and vortex domain walls pinned at notches in an in-plane magnetized nanostrip and imaged using magnetic force microscopy [42].

of spin torques [49]. However, pinning sites' effects on domain walls are not always simple. Pinning effects will not only depend upon the pinning site itself but also on the domain wall type and its deformation at the pinning site (e.g., [50–54]). Pinning effects can also depend on whether a wall is moving or stationary when it is at a pinning site, highlighting the need to separate both dynamic and static pinning effects [55,56]. Given the focus on dynamics, it is also worth mentioning resonant domain wall motion *within* a pinning site. When pinned, walls can undergo a high-frequency resonant motion or deformation while exhibiting no average net displacement away from the pinning site over time [57–59]. Pinning may be generated by a structural pinning site or by another domain wall [60] and there have been a number of significant outcomes coming from such studies, ranging from domain wall mass determinations [61] to low-power domain wall depinning [62–64] and domain wall-based oscillators (e.g., [65]).

Structural pinning methods which are less "severe" than constrictions/protrusions include spatially localized, ion irradiation-induced changes to anisotropy [66] or the introduction of bends [67]. However, it is also possible to create pinning sites without making any structural modifications to the nanostrip itself. One such route uses stray magnetic fields generated by neighboring reconfigurable magnetic elements that are fabricated either above [68–70] or next to [71] a domain wall conduit, which acts to locally impede domain wall motion. Since the stray field and thus the pinning effects depend on the magnetic configuration in the nano-element, there is a possibility to magnetically tune the pinning site. One can also use neighboring or overlaying domain walls to pin one another [72,73]. Another way to controllably tune domain wall velocities (and also domain nucleation) is via electric field control of anisotropy using (localized) electric field gating [74–78]. This is promising for a number of applications since it offers the ability to reversibly change the strength of pinning sites via modulation of a gating voltage. Perhaps unsurprisingly, such methods have been shown to be most efficient for domain wall motion regimes such as creep (see below), where the domain wall velocity is strongly affected by energy barriers linked to anisotropy [79]. Ferroelectric gates have also been successfully employed to locally control magnetic reversal [80].

The above examples concentrate predominantly on systems where dynamics occur within a pinning site or where pinning sites are used to discretely position domain walls (i.e., preventing the smooth passage of a domain wall past a given point in a structure). An interesting example of how pinning sites can dramatically change *long-range domain wall dynamics* occurs in ultrathin Co films with PMA, which are good experimental realizations of quasi-two-dimensional Ising systems containing narrow (and thus quasi-one-dimensional) domain walls. A number of studies have demonstrated that collective interactions between these domain walls and *intrinsic* weak defects in such films can be interpreted using general theories for elastic interface motion in disordered media (see [81] and references therein). As a result, such systems have been studied intensely to test theories that are also applicable to numerous other physical systems and phenomena ranging from wetting to vortex motion in high-temperature superconductors (e.g., [82]). These theories successfully predict three regimes of domain wall (interface) motion at finite temperature (Fig. 10a): a thermally activated low-field creep regime for which finite wall velocities are observed for all values of $H>0$, an intermediate depinning regime and a high-field viscous or flow regime in which the velocity approaches that expected for defect-free films. All three regimes have been experimentally identified in ultrathin Co layers with PMA (Fig. 10b) that are typically sandwiched between Pt layers or Pt and AlO_x layers [36,81,84–87]. Creep motion, characterized by avalanche-like wall motion, leads to a $H^{-\mu}$ scaling for the domain wall pinning-energy barrier with a predicted (and observed) scaling exponent $\mu=1/4$ for a one-dimensional interface moving in a two-dimensional system with weak disorder. Notably, μ is modified in very narrow strips due to a change in the system dimensionality [88]. Universal creep scaling is observed only for fields below

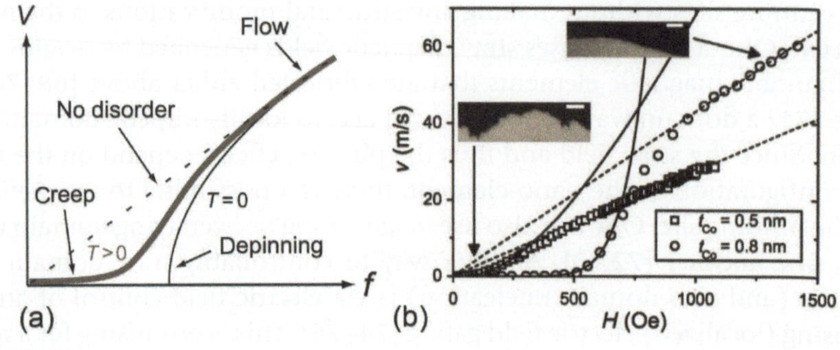

FIGURE 10
Universal domain wall (interface) dynamics. (a) Predicted velocity–force curve for an elastic interface driven through a weakler disordered medium at finite temperature and zero temperature [83]. (b) Measured velocity-field responses of domain walls in 0.5-nm- and 0.8-nm-thick Co films with PMA together with magneto-optical images of the domain structure in the 0.8 nm film when moving in the creep and flow regimes [36].

the so-called depinning field, at which point there would be a critical depinning transition in the case of $T=0$, below which the domain wall is predicted to be completely blocked. Although systems with PMA have been at the focus of recent studies on disorder-induced pinning, a number of simulation studies have also been carried out looking at the effects of disorder on the dynamics of wider transverse and vortex domain walls in in-plane magnetized systems [89–91].

3.3 Domain Wall Dynamics Under Current

The possibility of propagating DWs using currents instead of magnetic fields, first proposed and experimentally demonstrated by Berger *et al.* in the 1980s [92,93], is of enormous importance to the technologies described in the next section, as well as to the understanding of magnetotransport. Though current-driven DW motion has been demonstrated in several material systems, most studies focus on transition metals and their alloys, especially on Permalloy, due to their technologically convenient properties: high Curie temperature (T_c), low magnetocrystalline anisotropies and magnetostriction, low DW propagation field, and ease of deposition and fabrication.

Current-driven domain wall propagation typically results from the injection of a charge current in a metallic ferromagnetic film, with the partially spin-polarized electrons traversing the DWs. The resultant motion depends both on the structure and dynamics of the DW and on the details of the current-induced spin-transfer torques. In transition metal ferromagnets, the largest current-induced torque comes from the transfer of angular momentum from the polarized conduction electron to the local spins as the electron relaxes and aligns its spin parallel to the local magnetization. This torque is known as *adiabatic torque* and has the form [92]:

$$\dot{\mathbf{m}}_{\text{adiab.}} = u\mathbf{m} \times \left(\frac{\partial \mathbf{m}}{\partial x} \times \mathbf{m} \right) = u\frac{\partial \mathbf{m}}{\partial x}, \quad u = \frac{j\hbar\gamma P}{2eM_s} \tag{6}$$

where $\mathbf{m}$ is the local magnetization direction, x is the direction of the current, P is the ratio of current polarization, j is the current density, γ is the gyromagnetic ratio, and M_s is the saturation magnetization. The dot over the $\mathbf{m}$ on the left-hand side indicates a time derivative.

It was quickly seen, however, that this torque alone could not explain the experimental observations of current-induced DW propagation [94]. A phenomenological perpendicular torque was then introduced, labeled (imprecisely) as *non-adiabatic torque*, with the form [95]:

$$\dot{\mathbf{m}}_{\text{non-adiab.}} = \beta u\mathbf{m} \times \frac{\partial \mathbf{m}}{\partial x}, \tag{7}$$

where β quantifies the strength of this torque (in Permalloy, $\beta \sim \alpha \sim 10^{-2}$). The physical origin of so-called non-adiabatic torque is still subject to much debate, and several origins have been suggested for different systems: linear momentum transfer in extremely thin DWs, linear momentum transfer caused by the strong spin-linear momentum coupling (in GaMnAs), spin mistracking (true *non-adiabaticity*), spin-flip scattering, or micromagnetic processes such as magnon excitation in fast DWs (see, e.g., [96] and references therein). Although in some systems, such as in Permalloy and other transition metals, the present microscopic models predict well the experimentally determined β [94], in other systems, such as metallic multilayered films or the semiconductor GaMnAs, this is not yet the case.

The current-driven DWs under these two torques exhibit two propagation regimes. For low currents, there is the *steady-state regime* where the DW velocity follows a linear relation with the current density, $v_{\text{DW}} = u\beta/\alpha$. Note that in the absence of the non-adiabatic torque there is no steady DW motion at low current density. Above a threshold current, the *intrinsic threshold current* (sometimes called the Walker current after its field-driven analogue), the DW shows a *precessional propagation*. In this regime, the DW exhibits periodic oscillations of its instantaneous velocity and deformation of its structure, and its average velocity tends to u asymptotically (see Fig. 11). In both regimes the DW propagates along the direction of the electron flow. If there are significant pinning defects, there is an additional threshold current needed to depin the DW and start the current-induced motion, the *extrinsic threshold current*. As with field-driven motion, below this threshold the DW propagates by a series of thermally activated jumps between weak pinning sites (the creep regime, discussed above).

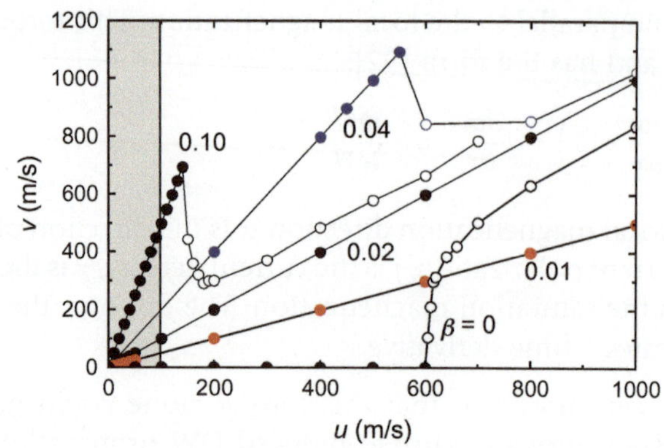

FIGURE 11

Current-induced DW velocity versus the parameter u, which is a measure of the current density, for various values of β (micromagnetic study of in-plane magnetized tracks, $\alpha = 0.02$). The open symbols indicate precessional propagation. *From Thiaville* et al. *[94].*

As with the Walker field, the value of the intrinsic threshold current is determined by the DW structure and its energy landscape, specifically the energy barrier holding the DW structure, in many cases originating from the demagnetization energy [94]. In tracks with in-plane magnetization, this barrier can correspond to the out-of-plane tilting of the DW, or to the injection of a magnetic vortex, while for PMA films it is the in-plane rotation of a Bloch DW to a Néel DW. The intrinsic threshold can be suppressed if the energy barrier to precession is eliminated. This was observed in studies of the PMA system Co/Ni, where the energy difference between Néel and Bloch structures was suppressed by tuning the cross-section of a track [8,97]. Koyama *et al.* found that the DW rotated from the Bloch configuration to the Néel configuration for narrower tracks, due to increasing transverse demagnetizing anisotropy. A strong reduction of the threshold current was observed in tracks of intermediate width, where the energy difference between Bloch and Néel was minimal.

3.3.1 Currents Injected Perpendicular to the Magnetic Film

In systems where a polarized spin current propagates perpendicularly to the DW-containing film, a more efficient current-induced domain wall motion may be observed [98]. Similarly to the previous case, two torques have been identified, often labeled *Slonczewski* and *Field-like* torques [99]:

$$\dot{\mathbf{m}}_{\text{Sloncz.}} = {}^{u'}\!/_{t}\, \mathbf{m} \times (\mathbf{m}_{\text{p}} \times \mathbf{m}) \qquad \dot{\mathbf{m}}_{\text{field-like}} = {}^{\xi u'}\!/_{t}\, \mathbf{m} \times \mathbf{m}_{\text{p}}, \tag{8}$$

where t is the film thickness, $\mathbf{m}_{\text{p}}$ is the direction of current polarization, ξ is a magnitude factor (typically ~0.1 in transition metal ferromagnets), and u' is a normalized current density, identical to u in the case of MTJs. As for the current in-plane case, only one of the torques produces a steady-state propagation (the field-like torque when $\mathbf{m}_p /\!/ \mathbf{m}_{domain}$, or the Slonczewski torque when $\mathbf{m}_p /\!/ \mathbf{m}_{DW} \times \mathbf{m}_{domain}$). In this geometry, an important efficiency gain can be obtained as $\xi > \beta$ and the film thickness can be reduced. This effect was first demonstrated in studies of DW propagation in MTJs [40,100,101].

The alignment of the polarization of the current and the DW magnetization drastically changes the observed effects, with steady-state DW propagation occurring only for some of the orientations [98,102]. This has been highlighted by the recent studies of current-induced DW motion in asymmetric film stacks of type NM/FM, where NM is a metal with strong spin–orbit coupling (SOC) and FM stands for a ferromagnetic layer. Experiments by Miron *et al.* on Pt/Co/AlO_x [103], and later by other groups (e.g., [104,105]), revealed a current-induced DW motion orders of magnitude more efficient than that observed in symmetric Co films, with a critical dependence on the materials of the under- and over-layers. Which effects are responsible is still a hotly debated question. A mechanism proposed by Thiaville *et al.* [106] implicates both the perpendicular diffusion of spin currents, generated by the spin Hall effect (SHE) in the adjacent Pt layer, and the Pt-induced interfacial Dzyaloshinskii–Moriya exchange interaction (DMI), which changes the stable DW structure. The SHE, a consequence of the spin-dependent scattering of electrons in metals with strong SOC, consists of the appearance of a spin current in the transverse direction to an applied charge current, polarized perpendicularly to the propagation directions of both the charge and spin currents. In a stack such as Pt/Co, the SHE spin current generated in the Pt will diffuse into the Co layer. In order not to cancel the effect, the magnetic film must be in an asymmetric stack. The DMI asymmetric exchange is responsible for the stabilization of Néel DWs where Bloch DWs would otherwise be expected. The Néel DWs have the correct orientation to be propagated in the translational regime by the perpendicularly propagating SHE spin current. Further studies where an external field was used to control the Néel/Bloch configuration confirmed the importance of the DW structure [104].

4 DOMAIN WALL-BASED DEVICES

Magnetic nanotracks are the basis of several new proposed devices, such as the racetrack memory [107], the domain wall shift register [108], the spintronic memristor [3,109], or the digital turn counter [110]. These devices take advantage of the binary stable domains of soft magnetic nanotracks and the possibility of moving them with magnetic fields or currents, to achieve high-density nonvolatile encoding of digital information, which can be propagated and even processed in complex logic circuits. We will see now how this can be done, and how the magnetic nanotrack is an ideal system to implement digital devices.

In this brief text, we have left out some related systems that explore other functionalities beyond digital applications. For example, the stray or fringing magnetic field generated by a DW in a nanotrack can be used to trap and displace (superpara)magnetic beads. These beads can be chemically functionalized to attach to specific biological targets (e.g., specific human cells or bacteria), allowing the precise detection and transport via domain wall displacement of these biological targets [5,111,112]. Domain wall-based electronic oscillators, which have been predicted to be capable of generating a.c. output signals under d.c. driving currents, are also of interest [4,22,113,114].

In order to have a domain wall-based logic circuit or a memory device, it is necessary to displace the domain walls in magnetic tracks without causing unwanted changes to the domain structure. Central to this process is the notion of a *DW conduit*, i.e., a track where it is possible to drive the DWs by a field or current much smaller than the field or current that nucleates new domains. The use of nanoscale magnetic objects, enabled by improvements in nanofabrication in the last decades, has been essential to the creation of good DW conduits. Planar Permalloy nanotracks fabricated by electron beam lithography typically show DW propagation to domain nucleation field ratios of about 1:20 [115]. Nanotracks integrated into multilayered films, such as giant magnetoresistance (GMR) stacks, often present larger propagation/nucleation ratios (e.g., ~1:10 in [116]). Vertical DW conduits, i.e., wires perpendicular to the substrate, can be made by electrodeposition, although they typically show very large propagation/nucleation ratios (e.g., ~1:1 for Ni nanowires [117]). Natural defects, such as border roughness or material inhomogeneities, pin the DWs and increase the propagation field. Likewise, artificial structures, such as artificial pinning sites or DW logic gates, or even track bends [67], are also responsible for larger propagation fields (or currents). On the nucleation side, the anisotropy of a track, defined by its shape, determines a maximum nucleation field: wider tracks have smaller nucleation fields. However, nucleation sites, such as large artificial structures, can lower the nucleation field. The track ends are also crucial: they present a much lower local shape anisotropy, and it is often at the ends that new domains are nucleated.

4.1 Domain Wall-Switched MRAM and the Spintronic Memristor

Magnetic random access memory (MRAM) is a nonvolatile data storage device that uses an array of MTJs to store and read information in bi-stable nanomagnets (see, e.g., [118,119]). Though several different technologies have been proposed (and commercialized), a typical spin transfer torque MRAM device works as follows. An array of elliptical MTJs is fabricated so that each can be electrically addressed individually. The shape anisotropy makes the free layer of the MTJ a bi-stable system, the state of which can be readily read by measuring the MTJ resistance. To write a bit, a large current is applied that sets the MTJ free layer in the desired state, through the spin transfer torque. This technology suffers from the need of a high writing current, which impedes its downscaling. The high writing current is needed because the spin transfer torque is a small effect, and one that vanishes when the free and reference layers are parallel, as is the case in the equilibrium states. The switching process depends then on the thermal fluctuations to create a finite angle between the reference and free layers (and thus a finite spin transfer torque), making the magnetic reversal a stochastic process at lower current densities. A DW-switched nanomagnet presents a solution to this problem [120]. DW switching, instead of the reversal by quasi-uniform rotation of the magnetization, presents an important advantage to the implementation of MRAMs: it can be a deterministic process at much lower current densities. The working principle is as follows. A nanotrack is integrated in an MTJ, with two stable DW pinning positions defined by artificial defects. As the two DW positions correspond to different sizes of the parallel (low-resistance) and anti-parallel (high-resistance) domains, the DW position is read using the tunneling magnetoresistance (TMR) effect. To write, a current is applied that produces a spin-transfer torque on the DW. Above a current threshold, which is lower than the current needed to reverse a uniform domain, the DW depins and propagates in a deterministic process. Furthermore, the small track length makes this process extremely fast, as it avoids the Walker breakdown process [40].

Though not in the scope of digital devices, DW-reversed nanotracks are also the basis of a similar device with applications to analogue data storage and neuromorphic computation, the spintronic memristor [3,121]. In this device, no artificial pinning sites are fabricated and the DW can occupy any position in the nanotrack. The resistance is linear with the DW position, which now takes a continuum of values. The properties of current-induced DW propagation are such that when a writing current pulse is applied, i.e., with a current greater than the residual depinning threshold, the DW will move with a velocity roughly linear with the applied current density and as long as the current pulse lasts. It is then an analogue device with a resistance proportional to the product of the applied current and the pulse duration, i.e., the total charge that was injected through the device. This behavior of the voltage–current relation, $V=M(q)I$, where q is the injected charge and M is a nonconstant function, is called the memristive effect.

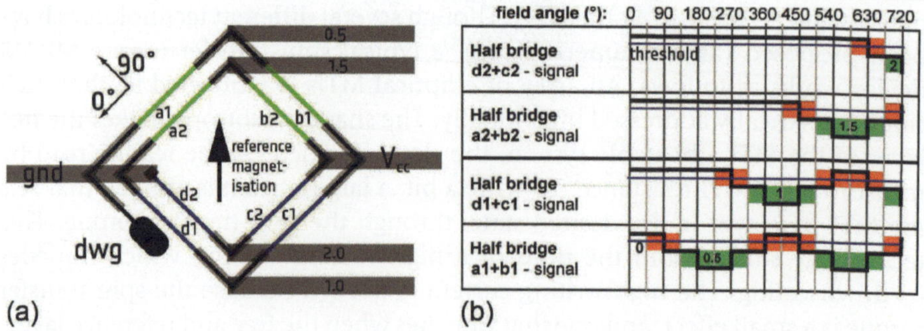

FIGURE 12

Domain wall turn counter. (a) Schematic of a 720° turn counter, with the magnetic nanotrack spiral with a nucleation pad and the electric contacts. (b) Resistance levels at the different contacts versus the field angle. *Adapted from Diegel* et al. *[122].*

4.2 Domain Wall Turn Counter

The role of the track shape and of the applied field in the controlled propagation of DWs can be seen in a very simple device, the DW turn counter. It is a device proposed and developed by Mattheis and colleagues, and is used in the contactless, nonvolatile digital turn sensors commercialized by Novotechnik [110,122,123]. It consists of a square spiral nanotrack with a large pad on one end (see Fig. 12a). The nanotrack is made of a GMR film stack, which enables the electric probing of the magnetic state. The pad serves as a nucleation site from which new domains (and DWs) are easily injected into the track. Starting from a state with no DWs in the spiral, a rotating homogeneous magnetic field is applied to the whole system. As the field turns in the sense of the spiral, new DWs are injected from the nucleation pad each half turn. All the DWs will wind in the spiral in discrete steps, reversing the straight segments one each quarter turn. If the field turns in reverse, DWs are annihilated in the pad and no new DWs are injected. The binary states of the spiral segments correspond unambiguously to a number of executed quarter turns, up to a maximum defined by the number of spiral loops. Using the GMR effect, the binary magnetic state of the different straight segments is measured (Fig. 12b) and, from this state, the exact number of quarter turns that the field has executed can be determined. The use of magnetic nanotracks gives this rotation sensor several interesting technological features. As the system state is nonvolatile, power is required only at the moment of the reading, making it a low-power, nanoscale sensor which is perfect for embedded systems. Its outputs are digital signals that encode the absolute rotated angle: no drift can occur. Finally, the magnetic coupling between the rotator and the sensor make it almost frictionless and robust.

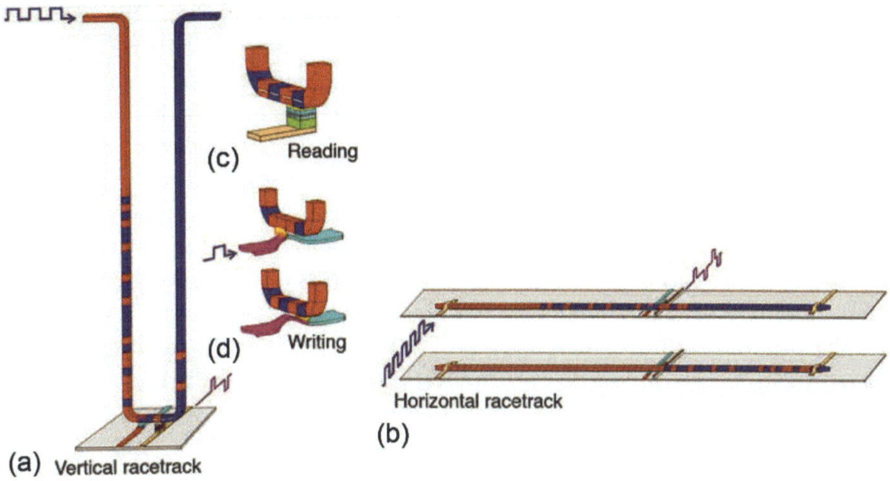

FIGURE 13

Domain wall racetrack, composed of a vertical (a) or planar (b) magnetic nanotrack, containing multiple magnetic domains (red/blue) which encode binary information. Current pulses injected in the nanotrack displace the domain structure. A magnetoresistive device, such as a magnetic tunnel junction, measures the domain state traveling in the track (c), and new domains are injected with a local field, generated by the presence of a DW in an adjacent track (d). *Adapted from Parkin* et al. *[42].*

4.3 Digital Information and the Domain Wall Racetrack

The DW racetrack is a data storage device proposed by Parkin *et al.* in 2004 [107], illustrated in Fig. 13. As in other DW digital systems, digital information is encoded by the domains of a magnetic nanotrack. The domain structure of these tracks consists of a variable number of domains along the track that occupy the whole cross-section of the track, with their magnetization parallel to the track axis, alternating between rightwards and leftward domains (red and blue in Fig. 13). Identifying the leftward domains with 0 and the rightwards with 1, it is possible to encode naturally binary information. However, in a perfect straight track, the length of the domain can vary continuously, which introduces ambiguity in the encoding of digital information. It is necessary then to discretize the length of the domains. In the DW racetrack, this is done using artificial defects, such as lateral notches, which stabilize the position of the domain walls at regular intervals. This is not the only way of encoding digital information using nanotracks. Indeed, discrete domain lengths can also be obtained using more complex track shapes, as described later, and information can also be encoded in other discrete magnetic degrees of freedom, such as the chirality of the domain walls separating the domains.

The working principle of the DW racetrack is schematized in Fig. 13. A planar or a vertical track with regularly spaced artificial DW pinning sites stores the information in its magnetic domains. Two geometries were proposed, one with vertical nanotracks (Fig. 13a) for a high areal information density device, and one with planar nanotracks for easier fabrication (Fig. 13b). In order to write information into the device, the local Oersted field of a current line is used to nucleate a new domain (Fig. 13d). Using current pulses, the domains and DWs are shifted along the track in discrete steps (Fig. 13a), and the writing procedure can be repeated. To read the information, the domains are likewise shifted in the track, passing under a magnetoresistive sensor (e.g., an MTJ), which reads serially the orientation of the domains (Fig. 13c). This device is analogous to an electronic shift register, where information is fed and read at the ends of a series of information-bearing simpler devices—the flip-flops in the electronic shift register, and the nanotrack segments in the DW racetrack.

The bit size, i.e., the dimensions of the domains and of the track, can be relatively small, typically a few tens of nanometers [42]. The minimum bit size is limited by the size of the domain wall, by the size of the structure needed to pin the domain wall, and by thermal stability considerations. The important advantage in information density of the nanotrack is brought not by the bit size but by the possibility of fabricating 3D structures [2]. Two properties make the nanotrack especially interesting in this domain, in comparison to other technologies. One is the extremely low intrinsic heat dissipation associated with a magnetic switching, while the switching heat dissipation of flash or complementary metal-oxide semiconductor (CMOS) memory is an important limitation for 3D structures. The other is the simplicity of the physical structure—just a single material magnetic track—which can be relatively easily fabricated vertically or in stacks of a large number of horizontal tracks.

Since its proposal in 2004, the DW racetrack has motivated much of the research on DW propagation in nanotracks, and consequently many of its required elements have been demonstrated: DW propagation by current pulses, local domain reversal with Oersted fields, and the magnetoresistive reading of the nanotrack magnetization (see [42] and references therein). However, many issues have to be solved before the DW racetrack is a reality. The operation of the racetrack depends critically on the controlled and reliable DW propagation by current pulses, as any change to the DW spacing (or their annihilation) would corrupt the encoded information. However, experiments so far show that the distance traveled by the DW, its velocity, and the depinning current can vary stochastically, and depend strongly on hard to control parameters, such as the DW structure and border defects (see, e.g., [30,124]). Reliable and deterministic DW propagation requires very high current densities, which is finally the main obstacle to the DW racetrack. The fabrication of vertical nanotracks with good DW conduit properties, and its integration in planar CMOS electronics, is also to be demonstrated.

4.4 Domain Wall Logic Gates and Circuits

DW logic gates and circuits take the ideas introduced earlier, the propagation and encoding of digital information in bi-stable domains, and exploit the fact that the interaction of DWs with artificial structures and the shape of the nano-track can be used to manipulate this information and implement Boolean functions. A DW logic circuit joins data storage and data processing in the same nonvolatile, low-dissipation and radiation-hard system, of great interest in several applications.

Cowburn *et al.* have demonstrated several Boolean functions in field-driven DW circuits, with both the input and the output encoded as domain and DW sequences [2]. In Fig. 14, some key elements of a Boolean circuit are shown: the unary NOT gate, the binary AND gate, fan-out (i.e., signal amplification), and the crossing of independent tracks. These elements can be integrated in complex circuits, as for example in the circuit shown in Fig. 14a–c containing

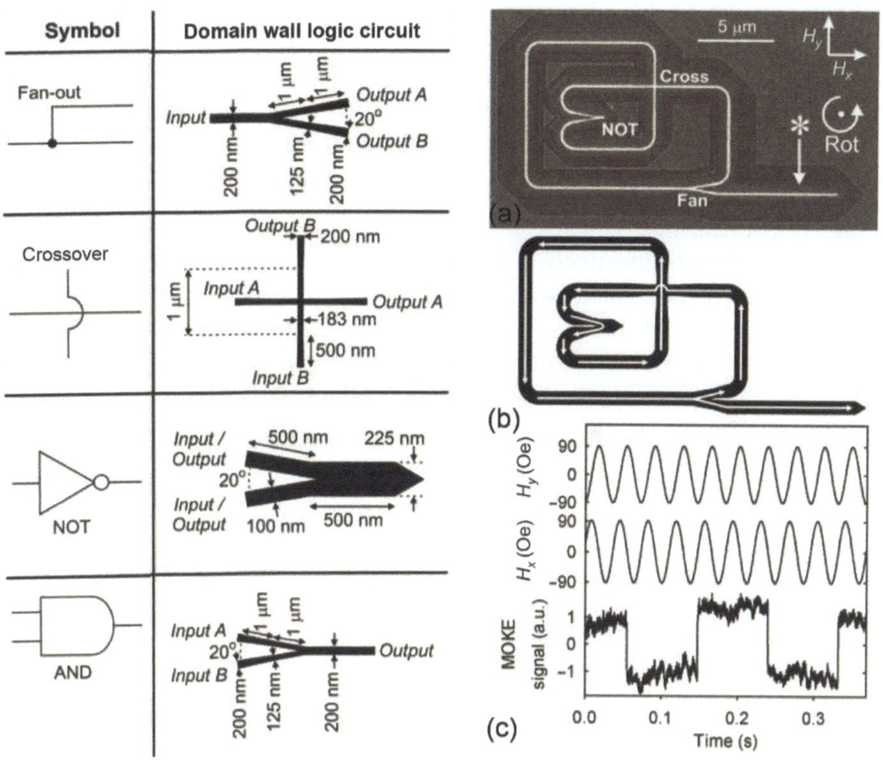

FIGURE 14

Domain wall logic. Left: examples of domain wall logic gates and elements. Right: example of a DW logic circuit containing a NOT gate, a crossover, and a fan-out. (a) SEM image. (b) Schematic of a possible magnetic state. (c) Applied fields and measured MOKE signal in the region marked by an asterisk (*) in (a). *Adapted from Allwood* et al. *[2].*

a NOT gate, a wire crossover, and a wire fan-out. To illustrate how complex functions can be implemented using artificial shapes, we will look into the details of two of these structures: the T gate and the NOT gate.

4.5 The T Gate

The versatility of the interaction of DWs with artificial structures is well illustrated by the T gate [125]. The T gate enables the control of DW propagation with a very high efficiency, either blocking completely the passage of DWs or allowing its propagation unhindered. It is effectively a DW valve. It consists of a track stub joined at 90° with the main track, shown in Fig. 15. The magnetization of the stub points either inwards or outwards from the main track, as indicated by the white or black arrows. The nanotrack dimensions are chosen so that the stable DW structure is the transverse DWs, which are a central magnetization transverse to the track. The blocking behavior, i.e., the closed DW valve state, occurs when the magnetization of the stub and of the DW center are anti-parallel. In this case, a strong exchange interaction stops the DW from entering the gate region. This blocking effect is so strong that if a larger torque is applied a new domain is nucleated before the DW can be pushed through. In the open state, when the magnetization of the stub is parallel to the central DW magnetization, there is no exchange cost of having a DW in the T gate, and the DW passes through the gate with a depinning field close to the typical propagation field.

The T gate is then effectively a binary DW valve, with a magnetically controllable gate. It serves also as a discriminator for transverse DWs. As we have seen, this type of DW has two possible central magnetizations, or chiralities, and only one type of DW can go through the gate. As the DW chirality is conserved under propagation with weak torques, it is possible to envisage

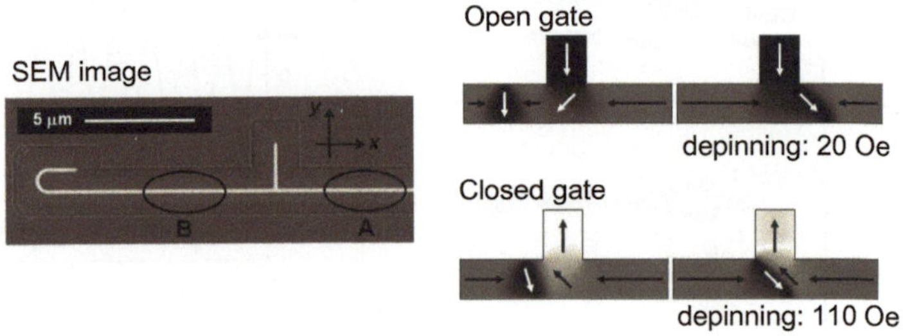

FIGURE 15

DW T gate. SEM image of a T gate (left) and micromagnetic simulation of the internal states (right), for the *open* configuration (with low DW pinning) and for the *closed* configuration (hard DW pinning). *Adapted from* Petit et al. *[125].*

storing information in this binary DW state that could then be detected using the discriminating action of the T gate.

4.6 NOT Gate

The NOT gate is a unary logic gate that transforms "1s" into "0s" and "0s" into "1s." In a magnetic DW logic circuit where we associate the logic bits with the orientation of the domains, the NOT gate is a structure that reverses the domain orientation and, likewise, the polarity of the DWs (head-to-head DWs into tail-to-tail and vice versa). The field-driven NOT gate introduced by Cowburn and colleagues [126] shows how the shape of the nanotrack can be used to implement Boolean functions. The structure consists of a cusp, schematized in Fig. 16a1, which stabilizes the magnetization of both its arms in a symmetric configuration. Under a rotating magnetic field, a DW present in the left arm (the *input arm*) will be pushed into the cusp (Fig. 16a2). As the DW reverses the apex of the cusp (Fig. 16a3) it is annihilated and a new DW, of opposite polarity, is created (Fig. 16a3). As the field continues to rotate, the new DW,

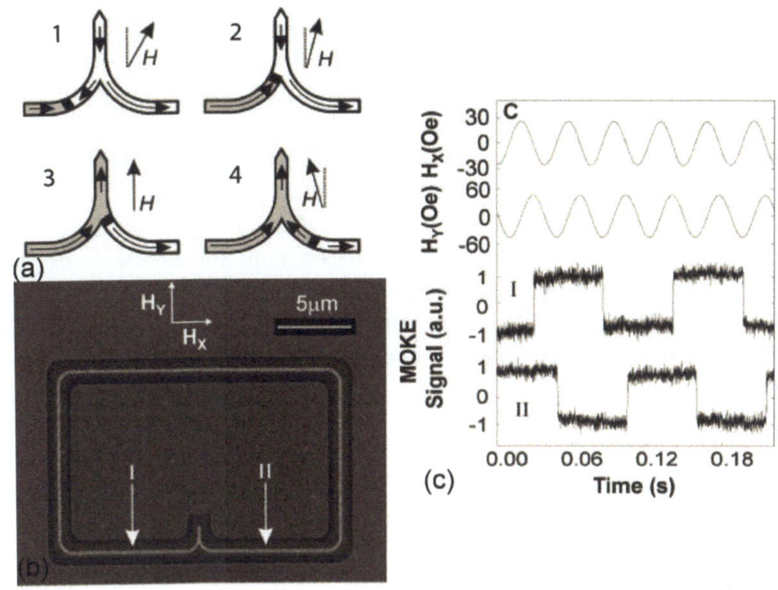

FIGURE 16

Domain wall logic NOT gate. (a) Schematic of the operation of the gate, showing a field-driven head-to-head DW entering the gate (a1 and a2), reversing the cusp and generating a tail-to-tail DW (a3), and the tail-to-tail DW going out (a4). (b) SEM image of a NOT gate integrated in a loop circuit. (c) Applied rotating field (top) and MOKE signal (bottom) measured in regions I and II indicated in (b), showing the 1½ turn period of the NOT gate loop and the ½ turn delay between the gate input and output. *Adapted from Allwood* et al. *[126].*

of opposite polarity to the one that went in the gate, propagates outwards through the right arm (the output arm) (Fig. 16a4). The operation of the NOT gate has two consequences: the domain pattern coming out from the output arm is the reverse of that coming in the input arm (the NOT operation itself) and the propagation of the DWs is delayed by half a field turn (as the field has to rotate 180° before the right arm can reverse). The NOT gate can be used to make a digital oscillator by connecting the output arm back to the input arm. The oscillator period will be the duration of 1.5 field turns: the gate delay (0.5 field turns) plus the delay of the loop of nanotrack connecting the input and output arms (a full field turn). This structure was studied by Allwood *et al.* [126], and an SEM image of their structure is shown in Fig. 16b. Using spatially resolved MOKE, they measured the magnetic state at the input and output arms, which is plotted in Fig. 16c. The recorded signals clearly show that the track magnetization reverses every 1.5 field rotations, and that there is a 0.5 field turn delay between the output and the input.

The NOT gate follows the clock defined by the rotation of the external field, i.e., its output changes in synchrony with the rotation of the field, exactly half a turn after the change of its input. This clocked behavior makes it analogous to the electronic flip-flop, a memory storage element. As with the flip-flop, a chain of NOT gates connected in series make a clocked shift register, with a bit storage capacity equal to the number of NOT gates. Motivated in part by the possibility of making high-density memory storage devices, there has been an effort to explore the scalability in size of these DW NOT gates. While the first proposed device (Fig. 16b) occupies a surface of 4 $\mu m^2 \approx 100F^2$, F being the nanotrack width [126], the optimized design shown in Fig. 17a has a footprint of only $15F^2$ and 0.054 μm^2 [108]. Furthermore, this design allows the dense packing of NOT gates for implementing high-density shift-register memories. A demonstrator is shown in Fig. 17b, with 15 NOT gates organized in three parallel shift registers. Current lines are used to inject DWs on the left arm, while the output arm (on the right) is read using MOKE. In a real device a magnetoresistive device would be used for the read-out. The plot of Fig. 17c shows the current pulses used to input the information in parallel in the three shift registers, and the same pattern appearing on the output five half field turns later (one half turn per gate in each shift register).

In comparison to the racetrack memory presented earlier, the main advantages of the NOT gate shift register arise from its naturally clocked behavior. It reliably isolates the data bits and impedes the annihilation of DWs that can occur in a racetrack. It also enables the propagation of the DWs by a global homogeneous magnetic field, impossible to achieve in a racetrack. The main disadvantages come from the need to produce the global field and from the larger footprint of the NOT gate compared to the artificial pinning site of the racetrack.

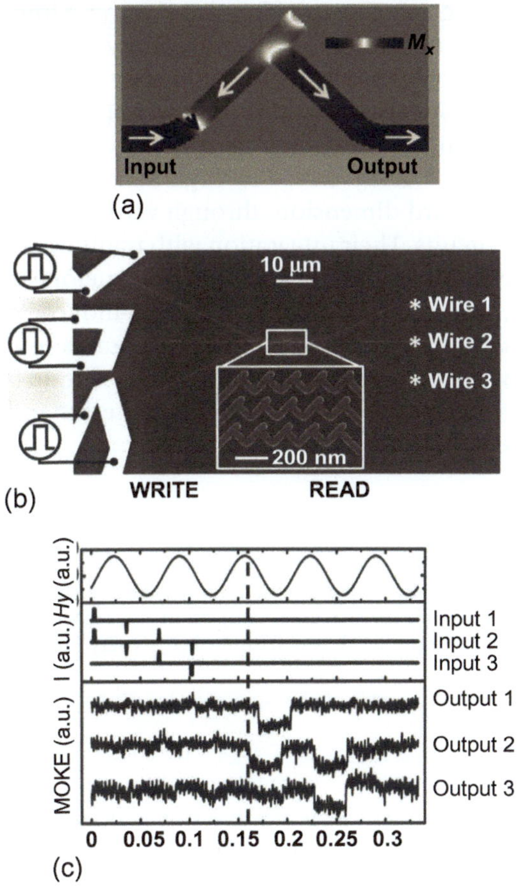

FIGURE 17
Shift register with compact NOT gates. (a) Shape of a compact NOT gate. (b) SEM image of a shift register composed of three (horizontal) magnetic nanotracks with five NOT gates each and three current lines (left) to inject domain walls. (c) Applied rotating field (top), applied current pulses to the current lines (middle), and MOKE signal at the right end of the tracks (bottom). *Adapted from Zeng* et al. *[108].*

5 CONCLUSION

Control over the dynamic magnetic properties of domains and DWs in soft ferromagnetic nanotracks offers interesting phenomena for the implementation of complex digital functions, the binary stable domains offering a natural encoding of digital information. The large ratio between the DW propagation field and the nucleation field makes the nanotrack an effective conductor of DWs, even in complex shaped tracks. Using the track shape, applied external fields or injected currents, nanotracks containing DWs can be used to realize complex digital devices, from digital sensors, memory storage elements and even Boolean logic circuits. These devices present some important advantages relative to the current technologies. They are nonvolatile, a requirement for low-power consuming devices. Being metallic, they are also radiation hard, a requirement for aviation, spatial, and industrial applications, and one that is less and less satisfied by modern high-density CMOS products. Their simple physical structure lends itself easily to high-density fabrication processes, and to the use of the third dimension, through vertical nanotracks or stacking of multiple planar circuits. Their integration with traditional charge-based electronics is possible through giant or tunneling magnetoresistance and related spintronic effects, and the demonstrated current-induced DW propagation. The coming years are sure to offer significant advances towards competitive, efficient and truly nanoscale domain wall devices due to a combination of state-of-the-art lithographic [127] and growth processes and newly identified physical phenomena [128].

Acknowledgments

The authors acknowledge financial support from the European Research Council (Starting Independent Researcher Grant No. ERC 2010 Stg 259068), from the French ANR (Grant ESPERADO 11-BS10-008), and from the Australian Research Council (Discovery Early Career Researcher Award DE120100155).

References

[1] Nagaosa N, Tokura Y. Topological properties and dynamics of magnetic skyrmions. Nat Nanotechnol 2013;8(12):899–911. http://dx.doi.org/10.1038/nnano.2013.243, Nature Publishing Group.

[2] Allwood DA, Xiong G, Faulkner CC, Atkinson D, Petit D, Cowburn RP. Magnetic domain-wall logic. Science 2005;309(5741):1688–92. http://dx.doi.org/10.1126/science.1108813.

[3] Grollier J, Cros V, Nguyen Van Dau F. Dispositif Memristor a Resistance Ajustable Grace Au Deplacement D'une Paroi Magnetique Par Transfert de Spin et Utilisation Dudit Memristor Dans Un Reseau de Neurones. WO Pat, 2010.

[4] He J, Zhang S. Localized steady-state domain wall oscillators. Appl Phys Lett 2007;90 (14):142508. http://dx.doi.org/10.1063/1.2719646.

[5] Vieira G, Henighan T, Chen A, Hauser A, Yang F, Chalmers J, et al. Magnetic wire traps and programmable manipulation of biological cells. Phys Rev Lett 2009;103(12):128101. http://dx.doi.org/10.1103/PhysRevLett.103.128101, American Physical Society.

[6] Malinowski G, Boulle O, Kläui M. Current-induced domain wall motion in nanoscale ferromagnetic elements. J Phys D Appl Phys 2011;44(38):384005. http://dx.doi.org/10.1088/0022-3727/44/38/384005.

[7] Kläui M, Vaz CAF, Bland JAC, Heyderman LJ, Nolting F, Pavlovska A, et al. Head-to-head domain-wall phase diagram in mesoscopic ring magnets. Appl Phys Lett 2004;85 (23):5637. http://dx.doi.org/10.1063/1.1829800.

[8] Koyama T, Chiba D, Ueda K, Kondou K, Tanigawa H, Fukami S, et al. Observation of the intrinsic pinning of a magnetic domain wall in a ferromagnetic nanowire. Nat Mater 2011;10(February):194–7. http://dx.doi.org/10.1038/nmat2961.

[9] McMichael R, Donahue MJ. Head to head domain wall structures in thin magnetic strips. IEEE Trans Magn 1997;33(5):4167–9. http://dx.doi.org/10.1109/20.619698.

[10] Nakatani Y, Thiaville A, Miltat J. Head-to-head domain walls in soft nano-strips: a refined phase diagram. J Magn Magn Mater 2005;290–291:750–3. http://dx.doi.org/10.1016/j.jmmm.2004.11.355.

[11] Mougin A, Cormier M, Adam JP, Metaxas PJ, Ferré J. Domain wall mobility, stability and Walker breakdown in magnetic nanowires. Europhys Lett 2007;78(5):57007. http://dx.doi.org/10.1209/0295-5075/78/57007.

[12] Aharoni A. Introduction to the theory of ferromagnetism. 2nd ed. New York, NY: Oxford University Press; 2000.

[13] Malozemoff AP, Slonczewski JC. Magnetic domain walls in bubble materials. New York, NY: Academic Press; 1979.

[14] Cowburn RP, Koltsov DK, Adeyeye AO, Welland ME. Single-domain circular nanomagnets. Phys Rev Lett 1999;83(5):1042–5. http://dx.doi.org/10.1103/PhysRevLett.83.1042.

[15] Bryan MT, Bance S, Dean J, Schrefl T, Allwood DA. Transverse and vortex domain wall structure in magnetic nanowires with uniaxial in-plane anisotropy. J Phys Condens Matter 2012;24(2):024205. http://dx.doi.org/10.1088/0953-8984/24/2/024205.

[16] Bellec A, Rohart S, Labrune M, Miltat J, Thiaville A. Domain wall structure in magnetic bilayers with perpendicular anisotropy. Europhy Lett 2010;91(1):17009. http://dx.doi.org/10.1209/0295-5075/91/17009.

[17] Rougemaille N, Uhlíř V, Fruchart O, Pizzini S, Vogel J, Toussaint JC. Phase diagram of magnetic domain walls in spin valve nano-stripes. Appl Phys Lett 2012;100(17):172404. http://dx.doi.org/10.1063/1.4704665.

[18] Ndjaka JMB, Thiaville A, Miltat J. Transverse wall dynamics in a spin valve nanostrip. J Appl Phys 2009;105(2):023905. http://dx.doi.org/10.1063/1.3067757.

[19] Thiaville A, Nakatani Y, Piéchon F, Miltat J, Ono T. Transient domain wall displacement under spin-polarized current pulses. Eur Phys J B 2007;60(1):15–27. http://dx.doi.org/10.1140/epjb/e2007-00320-3.

[20] Albert M, Franchin M, Fischbacher T, Meier G, Fangohr H. Domain wall motion in perpendicular anisotropy nanowires with edge roughness. J Phys Condens Matter 2012;24(2):024219. http://dx.doi.org/10.1088/0953-8984/24/2/024219.

[21] Jung S-W, Kim W, Lee T-D, Lee K-J, Lee H-W. Current-induced domain wall motion in a nanowire with perpendicular magnetic anisotropy. Appl Phys Lett 2008;92(20):202508. http://dx.doi.org/10.1063/1.2926664.

[22] Martinez E, Torres L, Lopez-Diaz L. Oscillator based on pinned domain walls driven by direct current. Phys Rev B 2011;83(17):174444. http://dx.doi.org/10.1103/PhysRevB.83.174444.

[23] Shibata J, Tatara G, Kohno H. A brief review of field- and current-driven domain-wall motion. J Phys D Appl Phys 2011;44(38):384004. http://dx.doi.org/10.1088/0022-3727/44/38/384004.

[24] Janutka A. Resonance of spin waves and domain-wall excitations in ferromagnetic stripes. IEEE Magn Lett 2013;4:4000104. http://dx.doi.org/10.1109/LMAG.2013.2261286.

[25] Chauleau J-Y, Weil R, Thiaville A, Miltat J. Magnetic domain walls displacement: automotion versus spin-transfer torque. Phys Rev B 2010;82(21):214414. http://dx.doi.org/10.1103/PhysRevB.82.214414.

[26] Thomas L, Moriya R, Rettner C, Parkin SSP. Dynamics of magnetic domain walls under their own inertia. Science 2010;330(6012):1810–3. http://dx.doi.org/10.1126/science.1197468.

[27] Vogel J, Bonfim M, Rougemaille N, Boulle O, Miron IM, Auffret S, et al. Direct observation of massless domain wall dynamics in nanostripes with perpendicular magnetic anisotropy. Phys Rev Lett 2012;108(24):247202. http://dx.doi.org/10.1103/PhysRevLett.108.247202.

[28] Schryer NL, Walker LR. The motion of 180° domain walls in uniform DC magnetic fields. J Appl Phys 1974;45(12):5406. http://dx.doi.org/10.1063/1.1663252.

[29] Lee J-Y, Lee K-S, Choi S, Guslienko KY, Kim S-K. Dynamic transformations of the internal structure of a moving domain wall in magnetic nanostripes. Phys Rev B 2007;76(18):1–8. http://dx.doi.org/10.1103/PhysRevB.76.184408.

[30] Nakatani Y, Thiaville A, Miltat J. Faster magnetic walls in rough wires. Nat Mater 2003;2 (8):521–3. http://dx.doi.org/10.1038/nmat931.

[31] Beach GSD, Nistor C, Knutson C, Tsoi M, Erskine JL. Dynamics of field-driven domain-wall propagation in ferromagnetic nanowires. Nat Mater 2005;4(10):741–4. http://dx.doi.org/10.1038/nmat1477.

[32] Dourlat A, Jeudy V, Lemaître A, Gourdon C. Field-driven domain-wall dynamics in (Ga, Mn) As films with perpendicular anisotropy. Phys Rev B 2008;78(16):161303. http://dx.doi.org/10.1103/PhysRevB.78.161303.

[33] Glathe S, Mattheis R, Berkov DV. Direct observation and control of the Walker breakdown process during a field driven domain wall motion. Appl Phys Lett 2008;93(7):072508. http://dx.doi.org/10.1063/1.2975181, American Institute of Physics.

[34] Hayashi M, Thomas L, Rettner C, Moriya R, Bazaliy YB, Parkin SSP. Current driven domain wall velocities exceeding the spin angular momentum transfer rate in permalloy nanowires. Phys Rev Lett 2007;98(3):037204. http://dx.doi.org/10.1103/PhysRevLett.98.037204.

[35] Koyama T, Ueda K, Kim K-J, Yoshimura Y, Chiba D, Yamada K, et al. Current-induced magnetic domain wall motion below intrinsic threshold triggered by Walker breakdown. Nat Nanotechnol 2012;7(10):635–9. http://dx.doi.org/10.1038/nnano.2012.151, Nature Publishing Group.

[36] Metaxas PJ, Jamet J-P, Mougin A, Cormier M, Ferré J, Baltz V, et al. Creep and flow regimes of magnetic domain-wall motion in ultrathin Pt/Co/Pt films with perpendicular anisotropy. Phys Rev Lett 2007;99(21):217208. http://dx.doi.org/10.1103/PhysRevLett.99.217208.

[37] Miron IM, Moore T, Szambolics H, Buda-Prejbeanu LD, Auffret S, Rodmacq B, et al. Fast current-induced domain-wall motion controlled by the Rashba effect. Nat Mater 2011;10 (6):419–23. http://dx.doi.org/10.1038/nmat3020, Nature Publishing Group.

[38] Bryan MT, Schrefl T, Atkinson D, Allwood DA. Magnetic domain wall propagation in nanowires under transverse magnetic fields. J Appl Phys 2008;103(7):073906. http://dx.doi.org/10.1063/1.2887918.

[39] Lewis ER, Petit D, O'Brien L, Fernandez-Pacheco A, Sampaio J, Jausovec A-V, et al. Fast domain wall motion in magnetic comb structures. Nat Mater 2010;9(12):980–3. http://dx.doi.org/10.1038/nmat2857.

[40] Sampaio J, Lequeux S, Metaxas PJ, Chanthbouala A, Matsumoto R, Yakushiji K, et al. Time-resolved observation of fast domain-walls driven by vertical spin currents in short tracks. Appl Phys Lett 2013;103(24):242415. http://dx.doi.org/10.1063/1.4848795, Materials Science; Mesoscale and Nanoscale Physics.

[41] Biziere N, Gatel C, Lassalle-Balier R, Clochard MC, Wegrowe JE, Snoeck E. Imaging the fine structure of a magnetic domain wall in a Ni nanocylinder. Nano Lett 2013;13(5):2053–7. http://dx.doi.org/10.1021/nl400317j.

[42] Parkin SSP, Hayashi M, Thomas L. Magnetic domain-wall racetrack memory. Science 2008;320(5873):190–4. http://dx.doi.org/10.1126/science.1145799.

[43] Fukami S, Suzuki T, Nagahara K, Ohshima N, Ishiwata N. Large thermal stability independent of critical current of domain wall motion in Co/Ni nanowires with step pinning sites. J Appl Phys 2010;108(11):113914. http://dx.doi.org/10.1063/1.3518046.

[44] Allwood DA, Xiong G, Cowburn RP. Domain wall diodes in ferromagnetic planar nanowires. Appl Phys Lett 2004;85(14):2848. http://dx.doi.org/10.1063/1.1802388, American Institute of Physics.

[45] Kim J-S, Mawass M-A, Bisig A, Krüger B, Reeve RM, Schulz T, et al. Synchronous precessional motion of multiple domain walls in a ferromagnetic nanowire by perpendicular field pulses. Nat Commun 2014;5(March):3429. http://dx.doi.org/10.1038/ncomms4429.

[46] Petit D, Jausovec A-V, Read DE, Cowburn RP. Domain wall pinning and potential landscapes created by constrictions and protrusions in ferromagnetic nanowires. J Appl Phys 2008;103 (11):114307. http://dx.doi.org/10.1063/1.2936981.

[47] Rodríguez-Rodríguez G, Menéndez JL, Hierro-Rodriguez A, Pérez-Junquera A, Montenegro N, Ravelosona D, et al. Interplay between collective pinning and artificial defects on domain wall propagation in Co/Pt multilayers. J Phys D Appl Phys 2010;43(30):305002. http://dx.doi.org/10.1088/0022-3727/43/30/305002.

[48] Franken JH, Swagten HJM, Koopmans B. Shift registers based on magnetic domain wall ratchets with perpendicular anisotropy. Nat Nanotechnol 2012;7(8):499–503. http://dx.doi.org/10.1038/nnano.2012.111.

[49] Miron I, Zermatten P-J, Gaudin G, Auffret S, Rodmacq B, Schuhl A. Domain wall spin torquemeter. Phys Rev Lett 2009;102(13):137202. http://dx.doi.org/10.1103/PhysRevLett.102.137202.

[50] Bogart L, Atkinson D, O'Shea K, McGrouther D, McVitie S. Dependence of domain wall pinning potential landscapes on domain wall chirality and pinning site geometry in planar nanowires. Phys Rev B 2009;79(5):054414. http://dx.doi.org/10.1103/PhysRevB.79.054414.

[51] Huang S-H, Lai C-H. Domain-wall depinning by controlling its configuration at notch. Appl Phys Lett 2009;95(3):032505. http://dx.doi.org/10.1063/1.3187530.

[52] Kläui M, Ehrke H, Rüdiger U, Kasama T, Dunin-Borkowski RE, Backes D, et al. Direct observation of domain-wall pinning at nanoscale constrictions. Appl Phys Lett 2005;87 (10):102509. http://dx.doi.org/10.1063/1.2042542.

[53] Lepadatu S, Vanhaverbeke A, Atkinson D, Allenspach R, Marrows CH. Dependence of domain-wall depinning threshold current on pinning profile. Phys Rev Lett 2009;102 (12):127203. http://dx.doi.org/10.1103/PhysRevLett.102.127203.

[54] Wuth C, Lendecke P, Meier G. Temperature-dependent dynamics of stochastic domain-wall depinning in nanowires. J Phys Condens Matter 2012;24(2):024207. http://dx.doi.org/10.1088/0953-8984/24/2/024207.

[55] Kunz A, Priem JD. Dynamic notch pinning fields for domain walls in ferromagnetic nanowires. IEEE Trans Magn 2010;46:1559.

[56] Pi U-H, Cho Y-J, Bae J-Y, Lee S-C, Seo S, Kim W, et al. Static and dynamic depinning processes of a magnetic domain wall from a pinning potential. Phys Rev B 2011;84(2):024426. http://dx.doi.org/10.1103/PhysRevB.84.024426.

[57] Bedau D, Kläui M, Krzyk S, Rüdiger U, Faini G, Vila L. Detection of current-induced resonance of geometrically confined domain walls. Phys Rev Lett 2007;99(14):5–8. http://dx.doi.org/10.1103/PhysRevLett.99.146601.

[58] Boone CT, Katine JA, Carey M, Childress JR, Cheng X, Krivorotov IN. Rapid domain wall motion in permalloy nanowires excited by a spin-polarized current applied perpendicular to the nanowire. Phys Rev Lett 2010;104(9):097203. http://dx.doi.org/10.1103/PhysRevLett.104.097203.

[59] Moriya R, Thomas L, Hayashi M, Bazaliy YB, Rettner C, Parkin SSP. Probing vortex-core dynamics using current-induced resonant excitation of a trapped domain wall. Nat Phys 2008;4(5):368–72. http://dx.doi.org/10.1038/nphys936.

[60] O'Brien L, Lewis ER, Fernández-Pacheco A, Petit D, Cowburn RP, Sampaio J, et al. Dynamic oscillations of coupled domain walls. Phys Rev Lett 2012;108(18):187202. http://dx.doi.org/10.1103/PhysRevLett.108.187202.

[61] Saitoh E, Miyajima H, Yamaoka T, Tatara G. Current-induced resonance and mass determination of a single magnetic domain wall. Nature 2004;432(7014):203–6. http://dx.doi.org/10.1038/nature03009.

[62] Martinez E, Lopez-Diaz L, Alejos O, Torres L. Resonant domain wall depinning induced by oscillating spin-polarized currents in thin ferromagnetic strips. Phys Rev B 2008;77(14):144417. http://dx.doi.org/10.1103/PhysRevB.77.144417.

[63] Nozaki T, Maekawa H, Mizuguchi M, Shiraishi M, Shinjo T, Suzuki Y, et al. Substantial reduction in the depinning field of vortex domain walls triggered by spin-transfer induced resonance. Appl Phys Lett 2007;91(8):082502. http://dx.doi.org/10.1063/1.2772174.

[64] Thomas L, Hayashi M, Xin J, Moriya R, Rettner C, Parkin SSP. Resonant amplification of magnetic domain-wall motion by a train of current pulses. Science 2007;315(5818):1553–6. http://dx.doi.org/10.1126/science.1137662.

[65] Lepadatu S, Wessely O, Vanhaverbeke A, Allenspach R, Potenza A, Marchetto H, et al. Domain-wall spin-torque resonators for frequency-selective operation. Phys Rev B 2010;81(6):6–9. http://dx.doi.org/10.1103/PhysRevB.81.060402.

[66] Franken JH, Hoeijmakers M, Lavrijsen R, Swagten HJM. Domain-wall pinning by local control of anisotropy in Pt/Co/Pt strips. J Phys Condens Matter 2012;24(2):024216. http://dx.doi.org/10.1088/0953-8984/24/2/024216.

[67] Lewis ER, Petit D, Thevenard L, Jausovec A-V, O'Brien L, Read DE, et al. Magnetic domain wall pinning by a curved conduit. Appl Phys Lett 2009;95(15):152505. http://dx.doi.org/10.1063/1.3246154.

[68] Franken JH, van der Heijden MAJ, Ellis TH, Lavrijsen R, Daniels C, McGrouther D, et al. Beam-induced Fe nanopillars as tunable domain-wall pinning sites. Adv Funct Mater 2014;24(23):3508–14. http://dx.doi.org/10.1002/adfm.201303540.

[69] Hiramatsu R, Koyama T, Hata H, Ono T, Chiba D, Fukami S, et al. Domain wall pinning by a stray field from NiFe on a Co/Ni nanowire. J Korean Phys Soc 2013;63(3):608–11. http://dx.doi.org/10.3938/jkps.63.608.

[70] Metaxas PJ, Zermatten P-J, Jamet J-P, Ferré J, Gaudin G, Rodmacq B, et al. Periodic magnetic domain wall pinning in an ultrathin film with perpendicular anisotropy generated by the stray magnetic field of a ferromagnetic nanodot array. Appl Phys Lett 2009;94(13):132504. http://dx.doi.org/10.1063/1.3105988.

[71] O'Brien L, Petit D, Lewis ER, Cowburn RP, Read DE, Sampaio J, et al. Tunable remote pinning of domain walls in magnetic nanowires. Phys Rev Lett 2011;106(8):087204. http://dx.doi.org/10.1103/PhysRevLett.106.087204.

[72] Metaxas PJ, Stamps RL, Jamet J-P, Ferré J, Baltz V, Rodmacq B. Expansion and relaxation of magnetic mirror domains in a Pt/Co/Pt/Co/Pt multilayer with antiferromagnetic interlayer coupling. J Phys Condens Matter 2012;24(2):024212. http://dx.doi.org/10.1088/0953-8984/24/2/024212.

[73] O'Brien L, Petit D, Zeng H, Lewis E, Sampaio J, Jausovec A-V, et al. Near-field interaction between domain walls in adjacent Permalloy nanowires. Phys Rev Lett 2009;103(7):077206. http://dx.doi.org/10.1103/PhysRevLett.103.077206.

[74] Bauer U, Emori S, Beach GSD. Voltage-gated modulation of domain wall creep dynamics in an ultrathin metallic ferromagnet. Appl Phys Lett 2012;101(17):172403. http://dx.doi.org/10.1063/1.4764071.

[75] Bauer U, Emori S, Beach GSD. Voltage-controlled domain wall traps in ferromagnetic nanowires. Nat Nanotechnol 2013;8(6):411–6. http://dx.doi.org/10.1038/nnano.2013.96, Nature Publishing Group.

[76] Bernand-Mantel A, Herrera-Diez L, Ranno L, Pizzini S, Vogel J, Givord D, et al. Electric-field control of domain wall nucleation and pinning in a metallic ferromagnet. Appl Phys Lett 2013;102(12):122406. http://dx.doi.org/10.1063/1.4798506.

[77] Franken JH, Yin Y, Schellekens AJ, van den Brink A, Swagten HJM, Koopmans B. Voltage-gated pinning in a magnetic domain-wall conduit. Appl Phys Lett 2013;103(10):102411. http://dx.doi.org/10.1063/1.4819771.

[78] Kakizakai H, Yamada K, Kawaguchi M, Shimamura K, Fukami S, Ishiwata N, et al. Direct observation of domain wall motion in Co/Pt wire under gate electric field. Jpn J Appl Phys 2013;52(7R):070206. http://dx.doi.org/10.7567/JJAP.52.070206.

[79] Schellekens AJ, van den Brink A, Franken JH, Swagten HJM, Koopmans B. Electric-field control of domain wall motion in perpendicularly magnetized materials. Nat Commun 2012;3(January):847. http://dx.doi.org/10.1038/ncomms1848.

[80] Huang Z, Stolichnov I, Bernand-Mantel A, Borrel J, Auffret S, Gaudin G, et al. Ferroelectric control of magnetic domains in ultra-thin cobalt layers. Appl Phys Lett 2013;103(22):222902. http://dx.doi.org/10.1063/1.4833495.

[81] Ferré J, Metaxas PJ, Mougin A, Jamet J-P, Gorchon J, Jeudy V, et al. Universal magnetic domain wall dynamics in the presence of weak disorder. C R Phys 2013;14(8):651. http://dx.doi.org/10.1016/j.crhy.2013.08.001, Elsevier Masson SAS.

[82] Kolton A, Rosso A, Giamarchi T, Krauth W. Creep dynamics of elastic manifolds via exact transition pathways. Phys Rev B 2009;79(18):184207. http://dx.doi.org/10.1103/PhysRevB.79.184207.

[83] Metaxas PJ. Creep and flow dynamics of magnetic domain walls: weak disorder, wall binding and periodic pinning. In: Camley R, Stamps RL, editors. Solid state physics, vol. 62: San Diego, California, USA: Academic Press; 2011.

[84] Bustingorry S, Kolton AB, Giamarchi T. Thermal rounding of the depinning transition in ultrathin Pt/Co/Pt films. Phys Rev B 2012;85(21):214416. http://dx.doi.org/10.1103/PhysRevB.85.214416.

[85] Gorchon J, Bustingorry S, Ferré J, Jeudy V, Kolton AB, Giamarchi T. Pinning-dependent field-driven domain wall dynamics and thermal scaling in an ultrathin Pt/Co/Pt magnetic film. Phys Rev Lett 2014;113:027205. http://dx.doi.org/10.1103/PhysRevLett.113.027205.

[86] Lemerle S, Ferré J, Chappert C, Mathet V, Giamarchi T, Le Doussal P. Domain wall creep in an Ising ultrathin magnetic film. Phys Rev Lett 1998;80(4):849–52. http://dx.doi.org/10.1103/PhysRevLett.80.849.

[87] Miron IM, Garello K, Gaudin G, Zermatten P-J, Costache MV, Auffret S, et al. Perpendicular switching of a single ferromagnetic layer induced by in-plane current injection. Nature 2011;476(7359):189–93. http://dx.doi.org/10.1038/nature10309.

[88] Kim K-J, Lee J-C, Ahn S-M, Lee K-S, Lee C-W, Cho YJ, et al. Interdimensional universality of dynamic interfaces. Nature 2009;458(7239):740–2. http://dx.doi.org/10.1038/nature07874.

[89] Laurson L, Mughal A, Durin G, Zapperi S. Modeling domain wall dynamics in thin magnetic strips with disorder. IEEE Trans Magn 2010;46(2):262–5. http://dx.doi.org/10.1109/TMAG.2009.2033819.

[90] Min H, McMichael RD, Donahue MJ, Miltat J, Stiles MD. Effects of disorder and internal dynamics on vortex wall propagation. Phys Rev Lett 2010;104(21):217201. http://dx.doi.org/10.1103/PhysRevLett.104.217201.

[91] Van de Wiele B, Laurson L, Durin G. Effect of disorder on transverse domain wall dynamics in magnetic nanostrips. Phys Rev B 2012;86(14):144415. http://dx.doi.org/10.1103/PhysRevB.86.144415.

[92] Berger L. Exchange interaction between ferromagnetic domain wall and electric current in very thin metallic films. J Appl Phys 1984;55(6):1954. http://dx.doi.org/10.1063/1.333530.

[93] Freitas PP, Berger L. Observation of S-D exchange force between domain walls and electric current in very thin Permalloy films. J Appl Phys 1985;57(4):1266. http://dx.doi.org/10.1063/1.334524.

[94] Thiaville A, Nakatani Y, Miltat J, Suzuki Y. Micromagnetic understanding of current-driven domain wall motion in patterned nanowires. Europhys Lett 2005;69(6):990–6. http://dx.doi.org/10.1209/epl/i2004-10452-6.

[95] Zhang S, Li Z. Roles of nonequilibrium conduction electrons on the magnetization dynamics of ferromagnets. Phys Rev Lett 2004;93(12):127204. http://dx.doi.org/10.1103/PhysRevLett.93.127204.

[96] Burrowes C, Mihai AP, Ravelosona D, Kim J-V, Chappert C, Vila L, et al. Non-adiabatic spin-torques in narrow magnetic domain walls. Nat Phys 2009;6(1):17. http://dx.doi.org/10.1038/nphys1436, Nature Publishing Group.

[97] Fukami S, Suzuki T, Ohshima N, Nagahara K, Ishiwata N. Micromagnetic analysis of current driven domain wall motion in nanostrips with perpendicular magnetic anisotropy. J Appl Phys 2008;103(7):07E718. http://dx.doi.org/10.1063/1.2830964.

[98] Khvalkovskiy AV, Zvezdin KA, Gorbunov Y, Cros V, Grollier J, Fert A. High domain wall velocities due to spin currents perpendicular to the plane. Phys Rev Lett 2009;102(6):067206. http://dx.doi.org/10.1103/PhysRevLett.102.067206.

[99] Slonczewski JC. Current-driven excitation of magnetic multilayers. J Magn Magn Mater 1996;159(1–2):L1–L7.

[100] Chanthbouala A, Matsumoto R, Grollier J, Cros V, Anane A, Fert A, et al. Vertical-current-induced domain-wall motion in MgO-based magnetic tunnel junctions with low current densities. Nat Phys 2011;7(Apri):626–30, Nature Publishing Group.

[101] Metaxas PJ, Sampaio J, Chanthbouala A, Matsumoto R, Anane A, Fert A, et al. High domain wall velocities via spin transfer torque using vertical current injection. Sci Rep 2013;3(May):1829. http://dx.doi.org/10.1038/srep01829, Nature Publishing Group.

[102] Khvalkovskiy AV, Cros V, Apalkov D, Nikitin V, Krounbi M, Zvezdin KA, et al. Matching domain-wall configuration and spin-orbit torques for efficient domain-wall motion. Phys Rev B 2013;87(2):020402. http://dx.doi.org/10.1103/PhysRevB.87.020402.

[103] Miron IM, Gaudin G, Auffret S, Rodmacq B, Schuhl A, Pizzini S, et al. Current-driven spin torque induced by the Rashba effect in a ferromagnetic metal layer. Nat Mater 2010;9(3):230–4. http://dx.doi.org/10.1038/nmat2613.

[104] Haazen PPJ, Murè E, Franken JH, Lavrijsen R, Swagten HJM, Koopmans B. Domain wall depinning governed by the spin Hall effect. Materials Science, Nat Mater 2013;12(4):299–303. http://dx.doi.org/10.1038/nmat3553, Nature Publishing Group.

[105] Ryu K-S, Yang S-H, Thomas L, Parkin SSP. Chiral spin torque arising from proximity-induced magnetization. Nat Commun 2014;5:3910. http://dx.doi.org/10.1038/ncomms4910, Nature Publishing Group.

[106] Thiaville A, Rohart S, Jué É, Cros V, Fert A. Dynamics of Dzyaloshinskii domain walls in ultrathin magnetic films. Europhys Lett 2012;100(5):57002. http://dx.doi.org/10.1209/0295-5075/100/57002.

[107] Parkin SSP. Shiftable magnetic shift register and method of using the same. US Pat. 6834005, 2004.

[108] Zeng HT, Read DE, O'Brien L, Sampaio J, Lewis ER, Petit D, et al. Asymmetric magnetic NOT gate and shift registers for high density data storage. Appl Phys Lett 2010;96(26):262510. http://dx.doi.org/10.1063/1.3457998.

[109] Locatelli N, Cros V, Grollier J. Spin-torque building blocks. Nat Mater 2014;13:11–20. http://dx.doi.org/10.1038/nmat3823.

[110] Diegel M, Mattheis R. Sensor element for a revolution counter. European Pat. no. EP1740909, 2005.

[111] Donolato M, Torti A, Kostesha N, Deryabina M, Sogne E, Vavassori P, et al. Magnetic domain wall conduits for single cell applications. Lab Chip 2011;11(17):2976–83. http://dx.doi.org/10.1039/c1lc20300b.

[112] Rapoport E, Montana D, Beach GSD. Integrated capture, transport, and magneto-mechanical resonant sensing of superparamagnetic microbeads using magnetic domain walls. Lab Chip 2012;12(21):4433–40. http://dx.doi.org/10.1039/c2lc40715a.

[113] Bisig A, Heyne L, Boulle O, Kläui M. Tunable steady-state domain wall oscillator with perpendicular magnetic anisotropy. Appl Phys Lett 2009;95(16):162504. http://dx.doi.org/10.1063/1.3238314.

[114] Ono T, Nakatani Y. Magnetic domain wall oscillator. Appl Phys Express 2008;1:061301.

[115] Jausovec A-V. Using shape to control the magnetization reversal in ferromagnetic nanostructures. London: Imperial College London; 2008.

[116] Sampaio J, O'Brien L, Petit D, Read DE, Lewis ER, Zeng HT, et al. Coupling and induced depinning of magnetic domain walls in adjacent spin valve nanotracks. J Appl Phys 2013;113(13):133901. http://dx.doi.org/10.1063/1.4798943.

[117] Pignard S, Goglio G, Radulescu A, Piraux L, Dubois S, Declémy A, et al. Study of the magnetization reversal in individual nickel nanowires. J Appl Phys 2000;87(2):824. http://dx.doi.org/10.1063/1.371947.

[118] Khvalkovskiy AV, Apalkov D, Watts S, Chepulskii R, Beach RS, Ong A, et al. Basic principles of STT-MRAM cell operation in memory arrays. J Phys D Appl Phys 2013;46(7):074001. http://dx.doi.org/10.1088/0022-3727/46/7/074001.

[119] Tehrani S, Slaughter JM, Chen E, Durlam M, Shi J, Deherrera M. Progress and outlook for MRAM technology. IEEE Trans Magn 1999;35:2814–9. http://dx.doi.org/10.1109/20.800991.

[120] Suzuki T, Ohshima N, Fukami S, Nagahara K, Ishiwata N. Magnetic domain wall random access memory. US Pat, 2011.

[121] Wang X, Chen Y, Xi H, Dimitrov D, Li H. Spintronic memristor through spin-torque-induced magnetization motion. IEEE Electron Device Lett 2009;30(3):294–7. http://dx.doi.org/10.1109/LED.2008.2012270.

[122] Diegel M, Glathe S, Mattheis R, Scherzinger M, Halder E. A new four bit magnetic domain wall based multiturn counter. IEEE Trans Magn 2009;45(10):3792–5. http://dx.doi.org/10.1109/TMAG.2009.2024426.

[123] Mattheis R, Diegel M, Hübner U, Halder E. Multiturn counter using the movement and storage of 180 magnetic domain walls. IEEE Trans Magn 2006;42(10):3297–9. http://dx.doi.org/10.1109/TMAG.2006.879750.

[124] Kläui M, Jubert P-O, Allenspach R, Bischof A, Bland J, Giancarlo F, et al. Direct observation of domain-wall configurations transformed by spin currents. Phys Rev Lett 2005;95(2):5. http://dx.doi.org/10.1103/PhysRevLett.95.026601.

[125] Petit D, Jausovec A-V, Zeng HT, Lewis ER, O'Brien L, Read DE, et al. High efficiency domain wall gate in ferromagnetic nanowires. Appl Phys Lett 2008;93(16):163108. http://dx.doi.org/10.1063/1.3005586.

[126] Allwood DA, Xiong G, Cooke MD, Faulkner CC, Atkinson D, Vernier N, et al. Submicrometer ferromagnetic NOT gate and shift register. Science 2002;296(5575):2003–6. http://dx.doi.org/10.1126/science.1070595.

[127] Currivan JA, Siddiqui S, Ahn S, Tryputen L, Beach GSD, Baldo MA, et al. Polymethyl methacrylate/hydrogen silsesquioxane bilayer resist electron beam lithography process for etching 25 Nm wide magnetic wires. J Vac Sci Technol B: Microelectron Nanometer Struct 2014;32:021601. http://dx.doi.org/10.1116/1.4867753.

[128] Parkin S, Yang S-H. Memory on the racetrack. Nat Nanotechnol 2015;10(3):195–8. http://dx.doi.org/10.1038/nnano.2015.41, Nature Publishing Group.

CHAPTER 9

Giant Magnetoresistance and Applications

Józef Barnaś*,†, Vitalii K. Dugaev‡,§
*Faculty of Physics, Adam Mickiewicz University, Poznań, Poland
†Institute of Molecular Physics, Polish Academy of Sciences, Poznań, Poland
‡Department of Physics, Rzeszów University of Technology, Rzeszów, Poland
§Departamento de Física and CFIF, Insituto Superior Técnico, Universidade de Lisboa, Lisbon, Portugal

1 INTRODUCTION

Artificially layered magnetic structures have attracted much attention since the early 1980s. This interest was stimulated by the desire to fabricate systems that do not exist naturally but have potentially new properties. Especially promising was the idea of tailoring the properties of metallic magnetic materials with those of nonmagnetic metals. The main difficulty was of a technological nature, because growing metallic systems with good structural properties was a big challenge. But tremendous progress in evaporation techniques, and especially in molecular beam epitaxy (MBE) methods, have enabled fabrication of multilayered structures of high-quality and well-defined interfaces between different layers. Moreover, present-day techniques of structural characterization now allow control of the growing process with atomic monolayer accuracy. This considerable progress in the technology of magnetic multilayered structures and in experimental techniques for their characterization and investigation has been made due to the potential applications in microelectronics devices and in data storage technology [1,2]. This aspect of the problem, however, will not be considered here and we rather focus on the physical properties of magnetic multilayers.

The efforts to fabricate high-quality layered structures consisting of a magnetic transition metal and a nonmagnetic metal resulted in the discovery of several physical phenomena. The most spectacular was the discovery of antiferromagnetic exchange coupling between two Fe films across a thin Cr layer [3], and the subsequent discovery of giant magnetoresistance (GMR) in Fe/Cr multilayers [4] and enhanced magnetoresistance in Fe/Cr/Fe trilayers (also referred to as

Magnetism of Surfaces, Interfaces, and Nanoscale Materials. http://dx.doi.org/10.1016/B978-0-444-62634-9.00009-6

sandwich structures or spin valves) [5]. From the physical point of view, both GMR and enhanced magnetoresistance (MR) are equivalent, and the only difference is in the magnitude of the effect. The GMR effect is also referred to as *spin-valve* MR, which seems to be more appropriate, as it clearly indicates the physical mechanism of the effect without referring to its magnitude. Therefore, the terms *spin-valve* MR and GMR will be used interchangeably in the following.

To understand the physical origin of the *spin-valve* MR effect, it is instructive to look first at the fundamental transport properties of magnetic transition (3d) metals. The basic concept dates back to the work of Mott [6], who proposed that charge transport in magnetic transition metals takes place via two different and almost independent spin channels. This idea is now known as the two-current Mott model. Another important property, from the point of view of GMR, concerns layered structures consisting of thin films of 3d magnetic metals and nonmagnetic metals. It has been shown that in such structures there is a remarkable interlayer exchange coupling between magnetic layers across the nonmagnetic layer (also referred to as a spacer layer or simply spacer). Both the two-current model and interlayer exchange coupling will be described in more detail before presenting the physical origin of the GMR effect and its further implications.

The aim of this chapter is to present some general physics underlying the GMR phenomenon, without going into experimental and theoretical details. Though there are several review articles and book chapters on this topic [7–13], the authors' intention is to provide a simple and concise description of the GMR effect and its influence on the development of the physics of nanomagnetic systems, and especially on the physics of spin-related phenomena.

2 TWO-CURRENT MODEL OF TRANSPORT IN MAGNETIC 3D TRANSITION METALS

It is well known that electron correlations in transition metals lead to a Stoner splitting of the conduction electron band and to the creation of a magnetic moment (itinerant ferromagnetic state). Since the magnetic moment of an electron is opposite to its spin orientation, the majority of electrons have spins opposite to the magnetization orientation, while a minority of electrons have spins aligned with the magnetization. The internal exchange field is relatively strong, so the spin polarization is maintained at high temperatures (well above room temperature).

When an electron propagates through the metal, it is scattered by impurities and other structural defects. Most of scattering processes, especially at low temperatures, conserve the electron spin orientation, while scattering processes with simultaneous reversal of electron spin are relatively rare. Thus, electrons propagating through the material may be ascribed to a particular spin channel—either to the spin-majority or to spin-minority channels. Scattering processes in a particular spin channel are characterized by the corresponding mean free path (MFP), i.e., the average distance between two successive scattering processes of any kind. These MFPs are denoted as λ_M and λ_m for the spin-majority and spin-minority channels respectively. Since the spin-conserving scattering processes are dominant, they determine the MFPs. From time to time, however, electron spin orientation can be changed in a scattering process with spin reversal. Such processes are characterized by the spin diffusion length (SDL), i.e., the average distance between two scattering processes which reverse spin orientation. Electron motion between these two scattering events is diffusive due to more frequent spin-conserving scattering processes. Accordingly, an electron propagates in a particular spin channel until it is switched to the second spin channel by a spin-reversal process.

A fundamental property of magnetic transition metals is the nonequivalence of the two spin channels, i.e., their transport characteristics are significantly different. In other words, the electron MFP depends on the electron spin orientation. This nonequivalence appears mainly due to Stoner splitting of the conduction band, which leads to different densities of electron states at the Fermi level for different spin orientations. Since the scattering probability depends on the density of states available for electrons to be scattered into, this leads, in turn, to the spin dependence of the electron MFP. An additional source of this spin dependence is the spin-dependent scattering potential of impurities.

Assume that the two spin channels are independent, i.e., the role of spin-flip scattering is negligible. At the macroscopic level, the difference in transport properties of the two spin channels can then be described quantitatively by the spin asymmetry parameter β, defined by the following relations [2,14,15]:

$$\rho_M = 2\rho_F^*(1-\beta), \tag{1a}$$

$$\rho_m = 2\rho_F^*(1+\beta), \tag{1b}$$

where ρ_M is the resistivity in the spin-majority channel, ρ_m is the resistivity of the spin-minority channel, and $\rho_F^* = \rho_F/(1-\beta^2)$, with ρ_F denoting the total bulk resistivity of the ferromagnetic metal. In real 3d ferromagnetic metals and their alloys, the parameter β is usually between 0.3 and 0.6, but in some materials, known as Heusler alloys, this parameter can be significantly larger and can exceed 0.9. Generally, the parameter β can be positive as well as negative.

The two-current model was used to account for some transport properties of transition metals [14]. However, the presence of two different spin channels for charge transport in bulk transition metals is of no use from the practical point of view, and also does not lead to physically interesting phenomena. This is because in the bulk limit, the electron transitions from one spin channel to the other one occur randomly and are beyond external control. External control of spin channels is possible in artificially layered systems, when the thickness of the layers is smaller than the electron SDL, as described below. Say, hypothetically, that we split a bulk magnetic transition metal into a sequence of thin films, as shown schematically in Fig. 1. We suppose that we can suppress the internal interactions between the films, so the magnetizations of the films

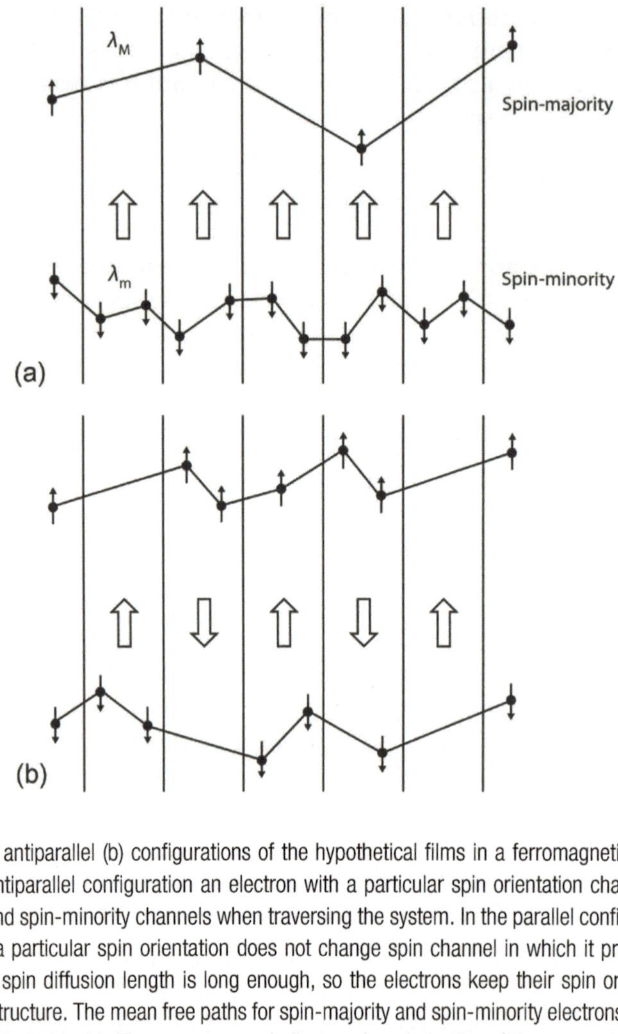

FIGURE 1

Parallel (a) and antiparallel (b) configurations of the hypothetical films in a ferromagnetic 3d transition metal. In the antiparallel configuration an electron with a particular spin orientation changes between spin-majority and spin-minority channels when traversing the system. In the parallel configuration, in turn, an electron of a particular spin orientation does not change spin channel in which it propagates. In both cases the spin diffusion length is long enough, so the electrons keep their spin orientations when traversing the structure. The mean free paths for spin-majority and spin-minority electrons, λ_M and λ_m, are different, as indicated in (a). The open arrows indicate spin polarization of the magnetic layers.

can be rotated independently. Later, we will see how this can be done experimentally.

When the magnetizations of all the layers are parallel, we have a situation like that in the corresponding bulk material. If, however, magnetic moments in adjacent films are antiparallel, then electrons (keeping their spin orientation) change between the spin-majority and spin-minority channels when traversing the structure. Strictly speaking, if in a particular film an electron propagates in the spin-majority channel, then in the adjacent film it propagates in the spin-minority channel. Thus, when changing the magnetic configuration from parallel to antiparallel (or vice versa), one can effectively change the spin channels in which the electrons propagate. This offers a unique way of manipulating the spin channels, which is absent in bulk materials.

To make this channel manipulation effective, one needs to obey the following two conditions. First, one has to find a way to achieve alternating magnetic moments in adjacent layers. Second, an electron should conserve its spin orientation when passing from one magnetic layer to the others, otherwise the spin is reversed in a spin-flip scattering process and the control of spin channels becomes ineffective. In fact, it is sufficient for electrons to maintain their spin orientation when propagating from a particular magnetic layer to the adjacent magnetic ones. To achieve these objectives, one has to suppress or substantially weaken the strong internal coupling responsible for ferromagnetic ordering. This has been done in artificial magnetic layered structures, where a thin layer of a nonmagnetic metal between two ferromagnetic films suppresses the internal coupling and allows manipulation of the magnetic configuration via an external magnetic field [3]. Below, we describe the basic features of these magnetic layered structures.

3 ARTIFICIAL MAGNETIC MULTILAYERS: INTERLAYER EXCHANGE COUPLING

Using sputtering techniques and/or MBE methods, one can grow artificial layered structures, in which thin layers of a magnetic 3d transition metal alternate with thin layers of a nonmagnetic metal. Importantly, the thickness of both magnetic and nonmagnetic layers can be controlled with atomic level accuracy. Indeed, just a few atomic planes of a nonmagnetic metal between two ferromagnetic layers turned out to be sufficient to suppress the strong internal exchange coupling. Thus, by intentionally introducing some (up to a few tens) atomic monolayers of a nonmagnetic metal during the growth process, one can fabricate structures which obey the conditions required for external control of spin channels, as described in Section 2. Moreover, such artificial materials exhibit novel properties which cannot be found in naturally existing materials.

Later in this section, we describe one of those properties, i.e., interlayer exchange coupling. We now focus on the aspects of the coupling phenomenon, which are important from the point of view of the GMR effect. More information on interlayer exchange coupling can be found in relevant review articles [16,17].

Though a thin nonmagnetic metallic layer breaks down the strong internal exchange coupling, some residual interaction between the two magnetic films across the nonmagnetic spacer layer remains when the spacer is up to a few tens of atomic monolayers thick. By analogy to Heisenberg coupling, the leading term in this interaction can be written in the form:

$$E_{ex} = -J\mathbf{m}_1 \cdot \mathbf{m}_2, \tag{2}$$

where E_{ex} is the coupling energy per unit area, J is the interlayer exchange parameter, while $\mathbf{m}_1$ and $\mathbf{m}_2$ are the unit vectors along the magnetic moments of the two coupled magnetic layers. A positive (negative) J corresponds to ferromagnetic (antiferromagnetic) interlayer coupling. The nature of this interaction is different from that in the corresponding bulk ferromagnet. Antiferromagnetic interlayer exchange coupling was discovered in artificially layered Fe/Cr/Fe sandwich structures, in which two 100-nm-thick layers of Fe were separated by a thin layer of Cr [3]. This coupling was ferromagnetic for very thin spacer layers, and then changed to antiferromagnetic for thicker spacers. The presence of antiferromagnetic interlayer coupling was deduced from spin-wave spectra observed using the Brillouin light scattering method. The spin-wave modes of two magnetic layers separated by a nonmagnetic layer depend on the relative orientation of the layers' magnetizations, and the spin-wave frequencies for antiparallel magnetic moments of the two layers are different from those in their parallel alignment. This difference originates from the presence of long-rage magnetic dipolar coupling between the magnetic layers, and the presence of short-range interlayer exchange coupling. When the nonmagnetic layer is thick enough, then only the dipolar coupling survives, while the interlayer exchange coupling vanishes. Both couplings influence the spin-wave frequencies in the case of thin spacing layers. When only dipolar coupling between the magnetic layers is relevant, variation of the corresponding spin-wave modes with the thickness of the nonmagnetic layer is remarkably different from that when there is additionally an exchange interaction between the magnetic layers [18,19].

Detailed evaluation of the spin-wave modes in magnetic sandwich structures, and their comparison with the modes observed experimentally using the Brillouin light scattering technique, allows a quantitative determination of the interlayer coupling parameter. Indeed, in the first experiments on Fe/Cr/Fe structures, this parameter was found to be ferromagnetic at small spacer thickness, but changed sign to antiferromagnetic at spacer thicknesses of the order of 7 Å [19]. Then, it remained antiferromagnetic for larger spacer thicknesses, though its magnitude decayed rather quickly with increasing thickness, and the coupling disappeared for spacer thicknesses larger than 2 or 3 nm.

Further improvement of the structural quality of magnetic layered systems based on 3d transition metals resulted in the discovery of additional features of the phenomenon of interlayer exchange coupling. The most interesting ones are as follows:

- Interlayer exchange coupling is an intrinsic property of all layered structures consisting of ferromagnetic transition-metal films separated by nonmagnetic metallic layers, and it exists in many other systems apart from Fe/Cr.
- The coupling parameter oscillates with increasing spacer thickness, i.e., it changes sign with a period characteristic of a nonmagnetic material, but the amplitude of the coupling parameter decays with increasing spacer thickness d approximately as $1/d^2$.
- In many cases there are two different oscillation periods. One of these is short (usually of the order of two atomic planes), while the other one is longer (about 10 atomic planes).
- The number of oscillation periods as well as the period lengths depend on the crystallographic orientation of the nonmagnetic spacer with respect to the interfaces.
- The magnitude of the interlayer coupling has a component that weakly oscillates with the thickness of magnetic layers.
- The interlayer coupling also has a component that weakly oscillates with the thickness of the nonmagnetic cap layer.
- There are higher order terms in the coupling energy, especially of biquadratic form. The latter coupling favors a perpendicular orientation of the magnetic moments.

Indeed, a long list of various structures displaying oscillatory interlayer coupling has been reported in the relevant literature. This includes, apart from the Fe/Cr system, layered structures such as Co/Ru and Co/Cr [20], Fe/Au [21], Fe/Ag [22], Co/Cu [23], and many others. Thus, oscillatory interlayer coupling is a general feature of magnetic multilayers. In fact, it not only appears in multilayers including magnetic transition metals, but it has also been observed in multilayers of rare earth metals [24]. To observe the oscillations—especially the short-period ones—the interfaces must be of sufficiently high structural quality. The components of the coupling parameter, which oscillate with the thickness of magnetic layers [25] and nonmagnetic cap layers [26], are rather small and their observation requires samples of high structural quality too. Both these effects demonstrate that interlayer exchange coupling is not an interfacial effect, but rather depends on the structure as a whole.

Experimentally, interlayer exchange coupling can be observed using various methods. The method already mentioned above relies on some properties of spin-wave spectra of layered structures, and specifically on their dependence on interlayer exchange coupling. This method enables evaluation of the magnitude of the interlayer exchange parameter, including its sign. There are also other methods that enable determination of the coupling parameter. One such method is based on the magneto-optic Kerr effect (MOKE). When the two magnetic layers of a sandwich structure are coupled antiferromagnetically, the magnetic moments of the layers are antiparallel in the ground-state configuration. An external magnetic field, however, can force their parallel alignment. Since the MOKE signal is sensitive to the total magnetization, it allows one to determine the field at which the magnetization of the system saturates. From this, one can evaluate the interlayer exchange parameter. In its simple version, this method works only for antiferromagnetic coupling, as in the case of ferromagnetic coupling magnetic moments of both films rotate simultaneously in an external magnetic field. However, this difficulty can be easily overcome by pinning the magnetic moment of one layer to a substrate (e.g., due to exchange anisotropy), so only the magnetic moment of the second magnetic layer can rotate freely in the magnetic field. The interlayer coupling in the ferromagnetic regime can then be determined from the field in which the antiparallel configuration is achieved, and this field can be determined from the MOKE spectra as well. Some other techniques, such as neutron scattering or scanning electron microscopy with polarization analysis [22], allow one to observe the magnetization orientation only, and thus allow determination of the sign of the coupling parameter, but not the coupling magnitude.

The physical origin of the interlayer exchange interaction is the same as that of the indirect exchange coupling between magnetic impurities in an electron gas, where coupling between the impurities is mediated by electrons. In other words, a single magnetic impurity polarizes the electron gas in its vicinity due to local exchange coupling of the impurity to mobile electrons. At the same time, each impurity is located in the electron gas, which is spin-polarized by the other impurity, and thus the two magnetic moments *see each other*. This mechanism is known as the Ruderman–Kittel–Kasuya–Yosida (RKKY) interaction. The RKKY interaction was derived through second-order perturbation theory. It is interesting that the RKKY coupling parameter oscillates with increasing distance between the impurities, and the oscillation period Λ is determined by the corresponding Fermi wavelength λ_F of the electron gas, $\Lambda = \lambda_F/2$. The amplitude of the coupling decreases with the distance r between the impurities as $1/r^3$. In layered magnetic structures, the simplest approach assumes RKKY interaction between each magnetic moment of the interfacial atomic plane in one magnetic layer with each magnetic moment of the interfacial atomic plane of the second magnetic layer [27]. Upon summing contributions from all pairs, one gets interlayer coupling that depends on the electronic structure of the nonmagnetic spacing layer between the two magnetic films. Taking into account the exact shape of the Fermi surface, as well as the crystallographic orientation of the layers, one finds the relation between the coupling parameter and the spacer thickness, which is in qualitative agreement with the experimental observations described above. In particular, the coupling parameter oscillates with the spacer thickness with one or two different oscillation periods, while the amplitude of the coupling decreases with the spacer thickness d as $1/d^2$.

The RKKY approach, however, is not able to reproduce correctly all the features of the coupling parameter, because the number of magnetic atoms in the case of layered systems is relatively large and the perturbative approach is then not sufficient. Moreover, magnetic atoms inside magnetic layers and those at interfaces are in different surroundings. More accurate results can be obtained from total energy calculations in the two magnetic configurations, and then defining the coupling parameter as $J = (E_{AP} - E_P)/2$ [28], where E_{AP} and E_P denote the total energy per unit area in the antiparallel and parallel magnetic configurations respectively. There are many papers where the coupling parameter was determined from *ab initio* calculations of the corresponding electronic band structure and total energy in both magnetic configurations (see, e.g., Ref. [29]).

4 EXPERIMENTAL RESULTS ON CURRENT-IN-PLANE GMR

From Section 3, it follows that a few atomic planes of a nonmagnetic metal suppress the strong internal exchange coupling in a bulk ferromagnetic material. When the net interlayer coupling is antiferromagnetic, the ground-state configuration of the layered system is like that presented in Fig. 1b. The situation corresponding to that shown in Fig. 1a can then be easily achieved by applying an external magnetic field, which forces parallel alignment of the magnetic moments of all the magnetic layers. Assume now that an external electric field (or equivalently an external voltage) is applied in the plane of the layers. It is found that the current flowing in the plane of the structure significantly increases at the transition from the antiparallel to the parallel alignment. Alternatively, the electrical resistance of the structure decreases at this transition.

The above-mentioned experiment shows that the electrical resistance of a layered structure consisting of magnetic 3d metallic layers separated by nanomagnetic metallic spacers depends on its magnetic configuration. This dependence is generally called the GMR effect—mainly because the change in resistance associated with the transition from antiparallel to parallel alignment is significantly larger than a similar change in the anisotropic magnetoresistance (AMR), where the resistance depends on the relative orientation of current and magnetization. The GMR effect is usually described quantitatively by the relative resistance change:

$$\mathrm{GMR} = \frac{\Delta R}{R_{\mathrm{P}}} \equiv \frac{R_{\mathrm{AP}} - R_{\mathrm{P}}}{R_{\mathrm{P}}}, \tag{3}$$

where R_{P} and R_{AP} are the resistances in the parallel and antiparallel magnetic configurations respectively. Alternatively, the effect is sometimes quantified as $\mathrm{GMR} = (R_{\mathrm{AP}} - R_{\mathrm{P}})/R_{\mathrm{AP}}$. Because the current flows in the plane of the layered structure, the effect is also called *current-in-plane* GMR (CIP-GMR) or equivalently CIP spin-valve MR. The GMR effect was discovered independently in Fe(100 Å)/Cr(15 Å)/Fe(100 Å) sandwich structures [5], where $\Delta R/R_{\mathrm{P}}$ was about 0.04 at room temperature, and in $[\mathrm{Fe}(30\ \text{Å})/\mathrm{Cr}(9\ \text{Å})]_{40}$ superlattices [4], where $\Delta R/R_{\mathrm{P}}$ was much larger, $\Delta R/R_{\mathrm{P}} \approx 0.8$ at 4.2 K. Even larger values of relative resistance change were observed by Schad *et al.* [30] in $[\mathrm{Fe}(4.5\ \text{Å})/\mathrm{Cr}(12\ \text{Å})]_{50}$ superlattices, where $\Delta R/R_{\mathrm{P}} \approx 2.2$ at 1.5 K.

To verify (or exclude) some of the known physical origins of the MR effects in thin metallic films, for instance the classical size effects, the GMR effect was investigated experimentally in three fundamental configurations: (i) for an external magnetic field (necessary to get the parallel configuration) applied perpendicularly to the layered structure; (ii) for an external magnetic field in the layer plane, but perpendicular to the electric current; and (iii) in the layer plane and along the flowing current. Experimental results obtained by Baibich *et al.* [4] in Fe/Cr superlattices are reproduced in Fig. 2, where different curves correspond to the different experimental geometries mentioned above. Ignoring the difference due to different saturation fields in the cases where the magnetic field is applied in the film plane and normal to this plane, one can say that the relative resistance change is almost independent of the experimental geometry. A small difference between the cases, where the in-plane magnetic field is applied respectively parallel and perpendicular to the electric current, can be attributed to the AMR. These results definitely exclude the AMR and the classical size effect as physical origins of the observed GMR effect. Moreover, the fact that the main contribution to GMR is the same in all three experimental configurations indicates that the amplitude of the effect is independent of the magnetization orientation with respect to the layer plane, and depends only on the relative orientation of the film magnetizations. This, in turn, indicates implicitly that the effect is rather due to spin-channel switching in the antiparallel configuration, as will be described later.

It is also interesting to check whether the physics of the GMR effect has anything to do with the mechanism of interlayer exchange coupling. From the general

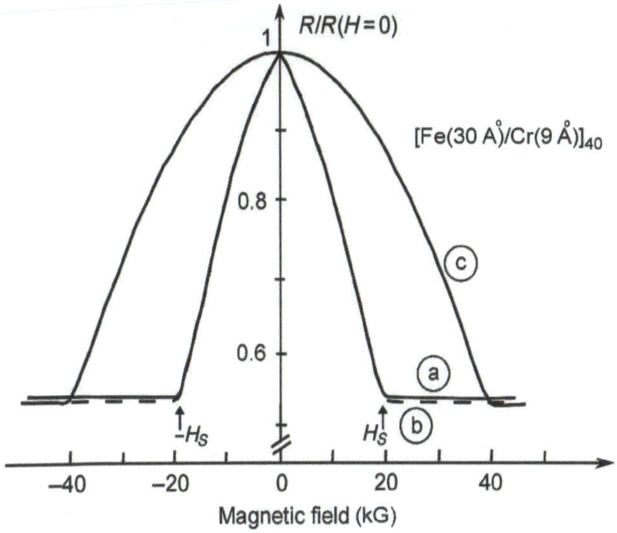

FIGURE 2

Field dependence of the relative resistance of a $[Fe(30\ Å)/Cr(9\ Å)]_{40}$ superlattice for a magnetic field applied in the film plane and parallel to the current (curve (a)), in the film plane and perpendicular to the current (curve (b)), and normal to the film (curve (c)). The resistance R, measured at 4.2 K, is normalized here to the resistance in the antiparallel configuration, $R(H=0)$. *After Ref. [4].*

idea of spin-channel switching (see Fig. 1) and from the experimental data shown in Fig. 2, it follows that to achieve an effective switching of the spin channels one needs only antiparallel magnetic configurations, while the way this configuration is achieved is irrelevant. In the experimental situations described above, the antiparallel alignment was obtained owing to the antiferromagnetic exchange coupling between the magnetic films, which exists for certain spacer thicknesses. However, the antiparallel alignment can also be obtained by other means. One can use, for instance, a sandwich structure without interlayer exchange coupling, but with different coercive fields in the two magnetic films. This problem was studied in detail in Ref. [31]. Results obtained on a Co/Au/Co sample with a sufficiently thick Au spacer to exclude interlayer exchange coupling are shown in Fig. 3. The first Co layer was evaporated on (110)-oriented GaAs substrate, while the second Co layer was grown on the Au spacer. Apart from this, both Co layers had different thicknesses. This was sufficient to produce different coercive fields for both Co layers. Figure 3a shows the MOKE hysteresis curve, which clearly indicates the existence of two regions of the magnetic field where the magnetizations of both films are antiparallel. Figure 3b, in turn, shows the resistance when the field is scanned through the hysteresis loop. Although the GMR effect is not large, the resistance increases remarkably when the antiparallel alignment is achieved. This proves

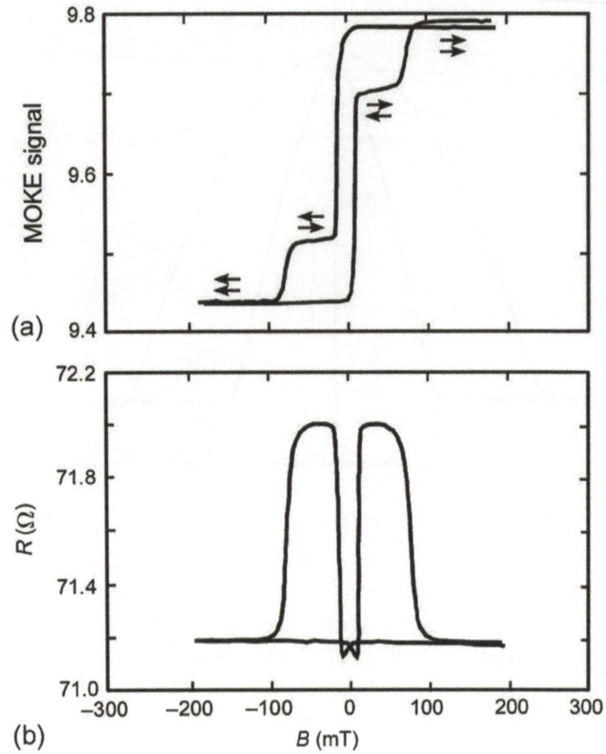

FIGURE 3

Hysteresis curves for the MOKE signal (a) and for the in-plane electrical resistance R (b), obtained on a Co(100 Å)/Au(60 Å)/Co(180 Å) sandwich structure. *After Ref. [31].*

that the GMR effect also appears in the limit of vanishing interlayer exchange coupling between magnetic layers. Thus, the mechanisms of GMR and of interlayer exchange coupling are different. This is rather clear as the interlayer exchange coupling has a quantum origin, while the GMR is of classical origin (disregarding the quantum nature of electron spin), as will be seen below. However, the GMR may be modified in the transport regime, where the quantum nature of electrons becomes relevant. One can find, then, some correlation between the GMR amplitude and the interlayer coupling. As will be shown later, the GMR amplitude includes a contribution that oscillates with the spacer thickness, similarly to interlayer exchange coupling. However, the quantum term in GMR is rather small and is usually suppressed in real structures due to electron-scattering processes.

Layered structures with no interlayer exchange coupling have some advantages over systems in which such coupling exists. In the first experiments on sandwich structures [5] and multilayers [4], the saturation field, i.e., the magnetic field at which the parallel magnetic configuration was achieved, was relatively large, since the magnetic field had to overcome the relatively strong interlayer exchange coupling. This field was especially large in the case of superlattices [4]. Thus, the GMR sensitivity, i.e., the relative resistance change per unit magnetic field, was rather low in those experiments. This essentially makes such systems useless for practical applications. The situation is different in spin valves without interlayer exchange coupling, where the saturation field is low, and therefore the field sensitivity is much higher [31].

If the coercive field of one magnetic layer is high enough, then the corresponding magnetic moment is fixed in external fields below this coercive field [32]. Such a situation can be achieved, for instance, by pinning the magnetic moment of this layer to an antiferromagnetic substrate, which is possible due to a relatively large exchange anisotropy at the interface between ferromagnetic and antiferromagnetic materials [32,33]. If the second magnetic film exhibits no remarkable magnetocrystalline anisotropy, and the dipolar coupling between the layers is weak, then its magnetization can be almost aligned along the external in-plane magnetic field. When rotating the external field in the film plane, one can change the relative orientation of the film magnetizations in a continuous way from parallel to antiparallel, or vice versa. A key feature of such spin valves is the very low saturation field (for one direction only) and therefore very high field sensitivity. It is found that in some spin valves the saturation field is comparable to the magnetic field generated by memory cells in hard drives. The most important practical application of spin valves in the read heads of hard drives exploits this feature of GMR and spin valves, as will be described later.

The CIP-GMR was observed in various artificially layered transition-metal magnetic structures. Though its magnitude varies over a relatively broad range, some general tendencies and features of the effect are rather common to all the systems investigated. The most interesting characteristics of the GMR effect include:

- temperature dependence;
- dependence on the number of magnetic films in a multilayer;
- dependence on the thickness of magnetic and nonmagnetic layers;
- shape of the resistance hysteresis curve in an external magnetic field and/or the corresponding angular dependence of GMR.

The magnitude of the GMR effect decreases with increasing temperature, and the rate of this decrease depends on the multilayer composition and layer thicknesses. In the case of Fe/Cr multilayers, the effect decreases relatively quickly with increasing temperature. For example, in $[Fe(4.5\ Å)/Cr(12\ Å)]_{50}$ superlattices, $\Delta R/R_P = 2.2$ was observed at 1.5 K, and $\Delta R/R_P = 0.42$ at room temperature [30]. The situation is different in Co/Cu multilayers, where $\Delta R/R_P$

decreases rather slowly with increasing temperature. Consequently, a large room-temperature value of the effect, $\Delta R/R_P = 0.5$, was reported in [Co(14 Å)/Cu(60 Å)]$_{120}$ multilayers [34], even though the largest low-temperature value of $\Delta R/R_P$ was observed in the Fe/Cr system. The temperature dependence of $\Delta R/R_P$ in the [Fe/Cr]$_n$/Fe epitaxial multilayers with $n = 1$, 2, and 4 is shown in Fig. 4. For each n, the ratio $\Delta R/R_P$ decreases with increasing temperature. A similar temperature dependence was also observed in other multilayers [35,36]. From Fig. 4, it also follows that the effect increases remarkably with the number n of nonmagnetic layers in the structure (thus also with the number of magnetic layers). This is particularly evident for small numbers n, like those in Fig. 4. However, the effect saturates as n increases above a certain limiting number (or when the total structure thickness exceeds a certain value).

The GMR effect generally decreases with increasing spacer thickness d. This decrease is approximately exponential, with a characteristic decay length of the order of the bulk electron MFP in the spacer material [33]. The dependence on the thickness D of magnetic films is more complex. The effect usually increases with increasing D for small values of D, then reaches a maximum for a certain thickness D, and eventually decreases with a further increase in D [37]. Experimental data on the dependence of the GMR effect on the thickness D of magnetic layers, obtained in Ref. [37] on structures M(D)/Cu(22 Å)/NiFe(50 Å) for M = Co, Ni, and NiFe, are reproduced in Fig. 5.

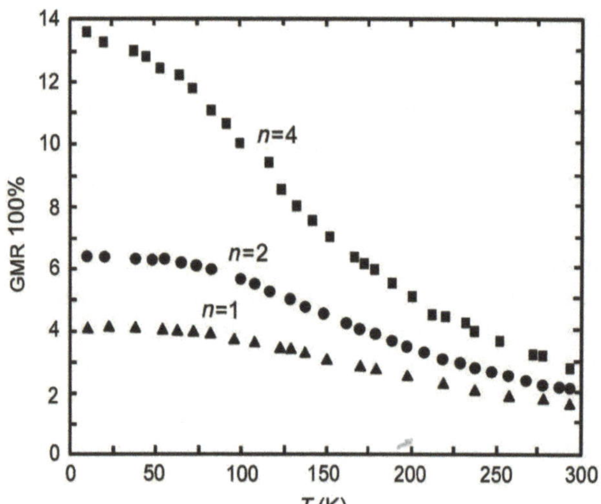

FIGURE 4

Temperature dependence of the GMR effect in [Fe(120 Å)/Cr(10 Å)]$_n$/Fe(120 Å) structures, with the number n as indicated. *After Ref. [31].*

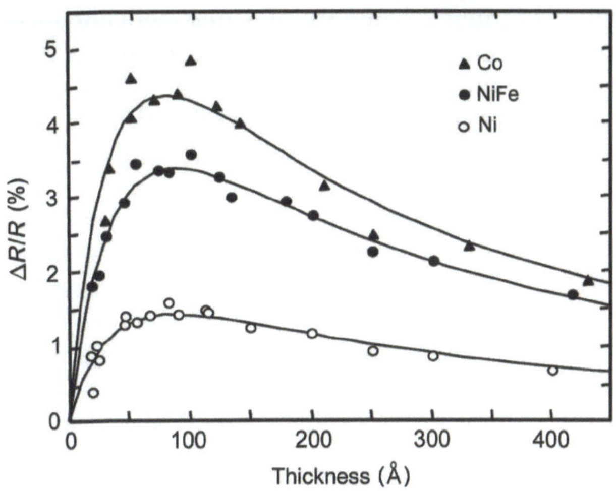

FIGURE 5

Dependence of the CIP-GMR on the thickness *D* of the magnetic layer M in M/Cu/NiFe spin valves for M = Co, Ni, NiFe. Here $\Delta R/R \equiv \Delta R/R_P$. *After Ref. [37].*

The GMR also depends considerably on the sample structural quality, which in turn depends on growth conditions such as the growth temperature, the type of substrate and buffer layer, the deposition technique, etc. The dependence on structural quality is not surprising, because transport properties are determined by the distribution of structural imperfections (scattering centers), and this is related to the growth conditions. However, the role of different scattering centers in GMR is rather complex and has been a topic of extensive investigations and discussion in the relevant literature. Since the resistance of a system is determined by scattering processes from impurities and other defects, the GMR effect also depends on the distribution and type of scattering centers, but it is not clear which scattering processes make the dominant contribution to the experimentally observed GMR. Much experimental work has been done in order to clarify the role of interface roughness and to optimize the growth conditions. The experimental results, however, are rather inconclusive and only show how complex the problem is, and how many different factors have to be taken into account. What follows from the available experimental data is that both bulk and interface structural defects contribute to the GMR. Though it is not possible to formulate some general rules concerning contributions from various scattering centers, the experimental data seem to support the following observations:

- Inserting Au, Ag, Al, Ge, or Ir at the interface between the Fe and Cr layers strongly reduces the GMR effect. If a thin layer of Mn or V is introduced, then the effect is almost the same as in Fe/Cr structures with the same spacer thickness [38].

- Roughening the interfaces by annealing treatment enhances the GMR for annealing temperatures up to 310 °C, and reduces the GMR for annealing at higher temperatures, as observed by Petroff *et al.* [35] on epitaxial Fe(001)/Cr(001) multilayers. This shows that there is an optimum interface roughness in the Fe/Cr system, for which the GMR achieves a maximum value [39].
- A moderate density of steps at the interfaces enhances the GMR effect, whereas a significant interface roughness reduces the effect, as concluded by Belien *et al.* [40] from results obtained on polycrystalline Fe/Cr superlattices grown at different substrate temperatures with and without a Cr seed layer. This proves the existence of an optimum interface roughness for the GMR effect too.
- Alloying the Fe films with Cr enhances the GMR effect [41].

We will come back to this problem in the section on the theoretical description of the CIP-GMR (Section 6), where it will become clear that the GMR effect is due to spin-dependent scattering processes, and that scattering at interfaces, surfaces, and inside the ferromagnetic films contributes to the effect.

The other important parameters which characterize the GMR effect are the saturation field B_s and the field sensitivity, $(\Delta R/R_P)/B_s$. The former quantity indicates how strong a magnetic field has to be applied to obtain the maximum resistance change, and the latter indicates how the electrical resistance is sensitive to a change in magnetic field. These two characteristics are important mainly from the point of view of possible applications in microelectronics, and especially in magnetic field sensors. In fact, these two parameters are determined by the corresponding resistance hysteresis curve in an external magnetic field. In turn, this hysteresis curve is related to the magnetization hysteresis curve, and reflects the magnetization processes when a system is switched from one magnetic configuration to the other. In the simplest case of a sandwich structure with antiferromagnetic interlayer exchange coupling, the resistance hysteresis curve reflects the rotation of the moments from one magnetic configuration to the other, and is generally different from that when the switching between the parallel and antiparallel configurations takes place via domain wall motion.

The GMR ratio defined in Eq. (3) provides information on the change in resistance, which is associated with the transition from antiparallel to parallel magnetic configurations. However, this ratio gives no information on how the resistance changes with increasing angle between the magnetic moments. This angular dependence of GMR is closely related to the shape of the resistance hysteresis curve. To describe the angular dependence of the GMR effect more quantitatively, one generalizes the resistance ratio, Eq. (3), as follows:

$$\text{GMR}(\varphi) = \frac{R(\varphi) - R_\text{P}}{R_\text{P}} \equiv \frac{R(\varphi) - R(\varphi = 0)}{R(\varphi = 0)}, \tag{4}$$

where φ is the angle between magnetic moments of the two magnetic layers in a sandwich structure. Obviously, R_P is equivalent to $R(\varphi = 0)$, while R_AP corresponds to $R(\varphi = \pi)$. The angular variation of GMR was investigated experimentally in Refs. [42–44], where a roughly linear dependence of the effect on $\sin^2(\varphi)$ was observed. However, remarkable deviations from this linear behavior have been found in other systems. The key experimental difficulty is the precise determination of the angle between magnetic moments. This was done in systems in which the orientation of the magnetic moment of one layer was fixed. The orientation of the magnetic moment of the second layer can then be evaluated from the corresponding magnetic hysteresis loops, and this is sufficient to determine the angle between magnetic moments.

In the experimental situations presented above, the magnetic films in a sandwich structure or in a multilayer were either made from the same material or the corresponding effective spin asymmetries of both magnetic layers were of the same sign. The corresponding GMR ratio was then positive in all considered situations. Such an effect is called *normal* GMR. We note that the effective spin asymmetry of a particular magnetic layer is a result of the interplay of scattering by impurities inside the layers (described by the parameter β) and by the interface roughness. Obviously, for zero interface roughness, the effective spin asymmetry is described by the parameter β only. In general, the effective spin asymmetry is positive (negative) when the spin-majority channel in a magnetic layer is more (less) conducting than the spin-minority one.

If the two magnetic films in a sandwich structure are made of different materials [45], the situation may be qualitatively different, providing the corresponding effective spin asymmetries of the two magnetic layers have opposite signs. This problem was investigated experimentally by George *et al.* [46] and later by Renard *et al.* [47]. In the former case, by intercalating thin Cr layers inside every second magnetic layer of a Fe/Cu multilayer, the authors found an *inverse* GMR effect, i.e., an increase in the resistance at the transition from the antiparallel to the parallel magnetic configuration. In the latter case, the authors showed that the positive GMR ratio (*normal* GMR effect) in Fe/Au/Co trilayers changes sign when the Fe layer is doped with vanadium. This also shows that different magnetic materials are not sufficient to produce an inverse effect, and that the spin asymmetry in the doped Fe layer in the $Fe_{1-x}V_x$/Au/Co trilayers changes sign. Interestingly, the magnetic moments of the films in the latter case were oriented perpendicularly to the films. All this supports the idea that spin-channel switching in the antiparallel configuration, and its suppression during the magnetization reorientation to the parallel configuration, is the origin of the GMR effect, as will be discussed in more detail in Section 5.

5 PHYSICAL ORIGIN OF CURRENT-IN-PLANE GMR

Physically, the GMR effect is a consequence of the switching between the spin-majority and spin-minority channels in the antiparallel configuration, when an electron propagates through the layered structure and keeps its spin orientation unchanged, and the absence of such switching in the parallel configuration [31,44]. It is important that switching between the spin channels is a consequence of the antiparallel magnetic configuration, and has nothing to do with spin reversal, which is absent on the considered length scale. Let us assume that a magnetic layered structure is in the antiparallel magnetic state, as in Fig. 1b. It is rather obvious that Fig. 1 is also applicable to real magnetic multilayers, provided the artificial border between neighboring magnetic films in Fig. 1 is replaced by a spacing layer of finite thickness. This spacing layer generally contributes to electron scattering independent of electron spin, as well as to spin-dependent scattering via interface roughness. Consider for simplicity a symmetrical sandwich structure, with only two magnetic films separated by a nonmagnetic spacer. When the thickness of the nonmagnetic spacing layer is significantly smaller than the SDL, an electron keeps its spin orientation when propagating across the spacer, say from the left magnetic layer to the right magnetic layer. If the electron propagates in the spin-majority channel in the left magnetic layer, it becomes a spin-minority electron in the right magnetic layer. In turn, an electron propagating in the spin-minority channel in the left layer is transmitted to the spin-majority channel in the right layer. Thus, the electrons switch between the spin-majority and spin-minority channels, when they are transmitted from one magnetic layer to the other. In other words, an electron propagating in a highly conducting channel of one magnetic layer is transferred to the weakly conducting channel of the second magnetic layer, and vice versa, an electron propagating in a weakly conducting channel is transferred to the highly conducting channel. Thus, electrons of both spin orientations are strongly scattered in the antiparallel configuration, either in the left or in the right magnetic layer.

The situation changes when the magnetic moments of the films become parallel, e.g., in an external magnetic field. An electron propagating in a particular spin channel (spin-majority or spin-minority) in one magnetic layer belongs to the same spin channel when it is transmitted to the other magnetic layer. Thus, electrons of one spin orientation propagate in the weakly conducting channel while electrons of the opposite spin orientation propagate in the highly conducting channel. Since the conductivities of the spin-majority and spin-minority channels are significantly different, the transition from the antiparallel to the parallel magnetic configuration is associated with a drop in electrical resistance, and consequently with an increase in the electric current when the voltage is kept constant. The transition from the antiparallel configuration to the parallel one is reminiscent of a short-cut in an electrical circuit with two resistors (one large and the other small) connected in parallel (see also Fig. 6).

The above picture also qualitatively explains the origin of the *inverse* CIP-GMR. The inverse effect takes place when the left and right magnetic layers of the sandwich structure have opposite effective spin asymmetries [45]. Then, the spin-majority channel in one magnetic layer is a good conductor, while in the second magnetic layer it is weakly conducting. The same also applies to the spin-minority channel. Thus, in the parallel magnetic configuration electrons of both spin orientations are strongly scattered, either in the left or in the right layer, while in the antiparallel configuration electrons of one spin orientation propagate in the strongly conducting channel in both magnetic layers, while electrons of the opposite spin orientation propagate in the weakly conducting channel in both layers. The situation is similar to that in the *normal* GMR effect, except that the parallel and antiparalell configurations are now interchanged. Accordingly, the electrical resistance in the parallel configuration is larger than in the antiparallel one, which leads to a negative value of the GMR ratio given by Eq. (3).

The qualitative picture presented above for a sandwich structure is also applicable to layered systems including more magnetic layers, as well as to superlattices. The addition of a new magnetic (and also spacing) layer enhances the effect, and this continues until the total thickness of the structure becomes

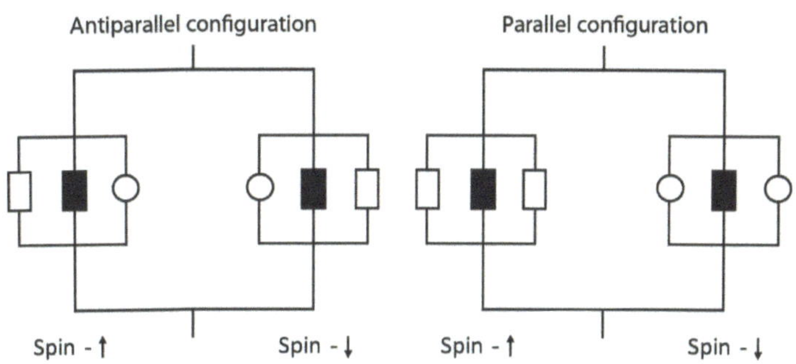

FIGURE 6

Resistor network model for the antiparallel and parallel magnetic configurations in the macroscopic limit. The black rectangles correspond to the resistance (per spin) of the spacer, whereas the open rectangles and circles correspond to the resistance of the ferromagnetic films respectively for the spin-majority and spin-minority channels. In the CIP geometry, the resistance in the macroscopic limit is the same in both parallel and antiparallel magnetic configurations.

comparable with the spin-diffusion length. Beyond this thickness, the addition of new layers does not change the GMR ratio. Thus, the key parameter for the effect to occur seems to be the spin-diffusion length, which assures coherent spin contact between the layers. However, this condition does not tell us anything about the magnitude of the effect. In fact, the effect saturates earlier, as the characteristic length for the CIP-GMR is MFP rather than SDL. This is because electrons which ensure *spin contact* between the magnetic layers separated by a distance equal to SDL propagate perpendicularly to the layers, and thus do not contribute to current flowing in the layer plane (and thus also to GMR). In fact, when the spacer thickness is larger than the MFP, the effect almost disappears. This limit corresponds to the so-called macroscopic limit, in which all layers can be treated as independent resistors. The corresponding resistor network model depicted in Fig. 6 shows that the resistance of the system at the macroscopic limit is independent of the magnetic configuration, and therefore the CIP-GMR vanishes in this limit. This will become clearer later when considering the problem theoretically.

The physical picture presented above is oversimplified, though it captures the basic features of the GMR effect. In Section 6, we present a more accurate theoretical description of the phenomenon. We also distinguish between bulk, interface, and surface scattering processes, which all contribute to the GMR effect.

6 THEORETICAL DESCRIPTION OF CURRENT-IN-PLANE GMR

To describe quantitatively all the features of the GMR effect, one should calculate the transport properties of the layered system taking into account its specific properties, such as the exact electronic structure of the multilayer, the distribution of structural defects and impurities, the shape of the impurity scattering potential, the geometrical interface roughness, and others. The whole problem seems to be quite complex and requires extensive numerical calculations. However, the basic features of the CIP-GMR effect can be accounted for qualitatively by assuming simplified models. The simplifications concern (i) the description of the electronic structure of the layered system, (ii) the description of the scattering centers, especially of the interfacial roughness, and (iii) the theoretical model of electronic transport. All these simplifications, however, should capture the basic features of the layered systems and transport properties. Apart from this, the description should also include the basic idea of channel switching, as described qualitatively in Section 5.

6.1 Collinear Configurations: Quasiclassical Description

Bearing in mind the basic features of the two-current model, we now briefly present the simplest theoretical description of the CIP-GMR. First, we simplify the electronic structure by assuming a free electron gas model for all layers. This assumption may be reasonable for s electrons, but generally it is a large over-simplification of the real electronic band structure. Spin asymmetry in the transport channels due to scattering on impurities distributed inside magnetic layers is included via spin-dependent MFPs. Second, diffuse scattering by surface and interface roughness will be described in terms of Fuchs–Sondheimer specular reflection and transmission parameters. These parameters are spin dependent and describe the interface contribution to the spin asymmetry of the spin channels. Third, we use the Boltzmann kinetic equation to describe the transport characteristics, assuming independent spin channels. In the following we consider a symmetric spin valve, in which two ferromagnetic films of thicknesses D are separated by a nonmagnetic layer of thickness d. However, by imposing perfectly reflecting boundary conditions at the external surfaces, the obtained results are also applicable to an infinite superlattice with doubled thickness of the magnetic layers [31,48]. Let the axis y be perpendicular to the layers, and the left and right surfaces be at $y=y_l=0$ and $y=y_r=d+2D$ respectively, while the left and right internal interfaces are at $y=D$ and $y=D+d$ (see Fig. 7). The magnetic moments of the two ferromagnetic films are assumed to be collinear (either parallel or antiparallel).

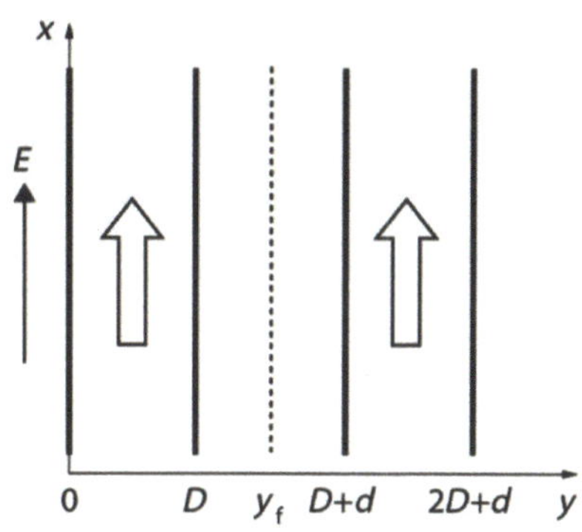

FIGURE 7

Schematic of the layered structure and the coordinate system used in the theoretical description. The magnetic layers have thickness D while the thickness of the nonmagnetic spacer is d. Electric field is along the axis x, while the axis y is perpendicular to the layers. In this figure magnetic moments of the layers are parallel. In the antiparallel configuration the magnetic moment of the right layer is reversed. The dotted line in the middle of the spacer is a fictitious plane assumed when considering noncollinear configuration (Section 6.2).

Conduction electrons at the Fermi level are scattered by impurities distributed inside the films and by rough interfaces and surfaces. Electron scattering inside the films is taken into account by the appropriate MFPs or equivalently by the relaxation times. In the magnetic films, the relaxation time is spin dependent, and is denoted as τ_{M} for the spin-majority electrons and τ_{m} for the spin-minority ones, while in the spacer layer it is independent of the electron spin orientation and is equal to τ. The corresponding MFPs are then λ_{M} and λ_{m} for the spin-majority and spin-minority electrons in the ferromagnetic layers, and λ in the nonmagnetic spacer. The electron distribution functions inside the magnetic and nonmagnetic layers can be calculated by solving the corresponding Boltzmann kinetic equations. These solutions have to be matched with the appropriate boundary conditions at all interfaces and surfaces.

Assume an external electric field E is oriented in the film plane and along the axis x. This field drives the system out of equilibrium. The distribution function is generally spin dependent, $f_s(y,\mathbf{v})$, where $s=\uparrow$ and $s=\downarrow$ for the two spin orientations. We distinguish between the local and global quantization axes (by a local axis we mean the axis opposite to the local magnetization orientation). Spin projection on the global axis is denoted as $s=\uparrow$ (spin-up) and $s=\downarrow$ (spin-down). Obviously, when the global quantization axis coincides with the local one, then spin-$\uparrow$ coincides with the spin-majority and spin-$\downarrow$ with the spin-minority orientations. As the global quantization axis in collinear configurations, we take the local axis of one of the two layers. The distribution function depends on the electron velocity $\mathbf{v}$ and also on the coordinate y. The latter dependence is a result of the lack of translational symmetry along the axis y. Generally, one can decompose the distribution function into two parts, $f_s(y,\mathbf{v})=f_0(\mathbf{v})+g_s(y,\mathbf{v})$, where $f_0(\mathbf{v})$ is the equilibrium distribution function corresponding to the Fermi energy E_{F}, while $g_s(y,\mathbf{v})$ is a field-induced deviation from the equilibrium distribution. The Boltzmann kinetic equation can be then written, for each layer and each spin orientation, in the form [31,48,49]:

$$\frac{\partial g_s(y,\mathbf{v})}{\partial y}+\frac{g_s(y,\mathbf{v})}{\tau_s v_y}=\frac{eE}{m v_y}\frac{\partial f_0(\mathbf{v})}{\partial v_x}, \qquad (5)$$

where e and m denote the electron charge ($e>0$) and electron effective mass respectively, while τ_s is the corresponding relaxation time, $\tau_s=\tau$ for the nonmagnetic layer and $\tau_s=\tau_{\mathrm{M}}$ or $\tau_s=\tau_{\mathrm{m}}$ inside the magnetic layers.

To find the distribution function $g_s(y,\mathbf{v})$, one needs to solve the Boltzmann equation in each layer separately, and then to match the solutions in different layers using appropriate boundary conditions. It is convenient to decompose $g_s(y,\mathbf{v})$ into two terms, $g_s^+(y)$ and $g_s^-(y)$, respectively, for $v_y>0$ and $v_y<0$. The general solution of Eq. (5) can then be written in the form [31,48,49]:

$$g_s^{\pm}(y) = \frac{e\tau_s E}{m}\frac{\partial f_0(\mathbf{v})}{\partial v_x}\left[1 + F_s^{\pm}(\mathbf{v})\exp\left(\frac{\mp y}{\tau_s|v_y|}\right)\right], \tag{6}$$

where the functions $F_s^{\pm}(\mathbf{v})$ are generally different in different layers and can be determined from the relevant boundary conditions. These boundary conditions for an interface located at y_0 ($y_0 = D$ or $y_0 = D + d$) can be written in the form [31,48,49]:

$$g_s^{\mp}(y = y_0 \mp 0) = T_s g_s^{\mp}(y = y_0 \pm 0), \tag{7}$$

where $y_0 - 0$ and $y_0 + 0$ mean that the distribution function is to be taken on the left ($y < y_0$) and right ($y > y_0$) side of the interface at y_0 respectively. In turn, T_s is a spin-dependent coefficient of specular transmission through the interfaces (assumed the same for both interfaces). This coefficient will be denoted as T_M when s corresponds to the spin-majority electrons in the adjacent ferromagnetic layer, and $T_s = T_m$ when s corresponds to the spin-minority electrons. The coefficients T_M and T_m can be written as $T_M = T(1+\chi)$ and $T_m = T(1-\chi)$, where χ describes the spin dependence of transmission probability. Since there is no band mismatch at the interfaces, there is also no specular reflection. Accordingly, the probability D_s of diffuse electron scattering at the interfaces is given by $D_s = 1 - T_s$. In turn, the boundary conditions for the outer surfaces have the forms $g_s^+(y_l + 0) = p_s g_s^-(y_l + 0)$ for the left surface (located at $y_l = 0$) and $g_s^-(y_r - 0) = p_s g_s^+(y_r - 0)$ for the right surface (at $y_r = 2D + d$). The phenomenological coefficient p_s describes the probability of specular reflection from the left and right surfaces, and can be written as $p_M = p(1+\chi')$ and $p_m = p(1-\chi')$ for the spin-majority and spin-minority channels respectively. For simplicity we assume that the parameters of specular reflection are the same at both surfaces, though they can be different in general. Note that the effective spin asymmetry of the two spin channels in magnetic layers is described by three parameters: β, χ, and χ'. The boundary conditions allow us to determine the unknown functions $F_s^{\pm}(\mathbf{v})$ and thus to determine the distribution function for the whole structure.

Writing the distribution function $g_s(y, \mathbf{v})$ as $g_s(y, \mathbf{v}) = E\hat{g}_s(y, \mathbf{v})$, one can express the local spin-dependent in-plane conductivity $\sigma_s(y)$ in the form:

$$\sigma_s(y) = -e(m/h^3)\int d^3\mathbf{v}v_x\hat{g}_s(y, \mathbf{v}), \tag{8}$$

and the global conductivity σ as

$$\sigma = \frac{1}{L}\sum_s \int \mathrm{d}y\sigma_s(y), \tag{9}$$

where L is the total thickness of the spin valve, $L=d+2D$. Thus, having found the distribution functions from the set of boundary equations, these formulas allow one to find the electrical resistance in both parallel and antiparallel magnetic configurations, and thus the magnitude of the CIP-GMR. Such an approach was widely used in numerical interpretations of the experimental results on GMR in sandwich structures, multilayers, and in superlattices [31,48,50–52]. Importantly, all the basic features of the effect have been reproduced, with good qualitative agreement with the experimental results described in Section 4. The approach, however, is too simple to describe the effect quantitatively.

Though the above description of the CIP-GMR is for a sandwich structure, the corresponding results can also be used to simulate GMR in superlattices with doubled thickness of the magnetic layers, by assuming perfectly reflecting boundary conditions at the external surfaces, $p_M=p_m=1$. This description has also been generalized to multilayers with the number of nonmagnetic layers larger than 1 [21]. What follows from the description is that the CIP-GMR effect appears to be due to the *spin dependence of electron scattering processes*. Some contribution to the spin-dependent scattering comes from scattering processes on impurities inside the magnetic layers, which are described by the spin-dependent relaxation time and/or spin-dependent MFP. Scattering on impurities inside the nonmagnetic layers is generally independent of the spin orientation, and therefore does not contribute to GMR, but contributes to the resistance that is not sensitive to the magnetic configuration, and thus diminishes the GMR effect. An important contribution to the CIP-GMR effect originates from scattering due to interface roughness, which is rather unavoidable in the growth process, and which is spin dependent in a general case. Scattering processes at the outer surfaces are spin dependent, and thus they contribute to the GMR effect as well. Moreover, the contribution to GMR due to interface roughness can be even larger than that due to scattering on impurities inside the magnetic layers, especially when the layers are sufficiently thin. Separation of these contributions is rather difficult due to the interplay of the interface and bulk effects, and also due to scattering processes that are independent of spin orientation. However, some information on the relative contributions from bulk and interface scattering can be found from the variation of these contributions with certain external parameters, especially with the layer thicknesses. Since this variation is different for bulk and interface contributions, it may be used to draw some conclusions on the role of particular scattering processes. There has been some discussion in the relevant literature on whether the CIP-GMR is due to scattering at the interfaces or inside magnetic layers. There is, however, no consensus on this, as it depends on several factors, including the quality of the structures, their composition, layer thicknesses, spin asymmetries, etc. Nonetheless, most of the available data for the Fe/Cr system suggest that the corresponding GMR is due to scattering at the interfaces.

The model presented above is based on several simplifications; in particular, the assumption of a free electron gas with no potential steps at the interfaces is a large oversimplification of the real situation, even though quantum calculations within this model improve transport characteristics [53]. This simplification is also a reason why above there was only specular transmission (or some diffuse scattering) and there was no specular reflection at the interfaces. Moreover, in the Sondheimer formulations of the boundary conditions, the specular transmission is independent of the angle of electron incidence at the interface, which is also a great simplification. The quasiclassical model was improved and extended by Hood *et al.* [54,55], who assumed different band edges in different layers, which were additionally spin dependent in the magnetic films. The corresponding calculations were performed with more realistic parameters of electron scattering at the interfaces, taking into account their dependence on the angle of incidence at the interface. In this description, electrons traversing an interface can be either specularly reflected, or specularly transmitted, or diffusively scattered. Generally, this description improves quantitatively the transport characteristics of the layered systems. We also note that there are other approaches to the boundary conditions for scattering at interfaces [56,57].

6.2 Noncollinear Configuration: Quasiclassical Description

In the collinear magnetic configurations (parallel and antiparallel) considered above, the electron distribution function was diagonal in the spin space, when the global quantization axis was collinear with the magnetizations. The situation is different in magnetic multilayers with noncollinear magnetic moments. First, there is no global quantization axis, which would be collinear with both local quantization axes. Second, exchange of electrons between the spacer and both magnetic layers in a sandwich structure leads to nondiagonal components of the distribution function, even in the local coordinate systems. In the free electron model, like that assumed in the case of collinear configurations, the nondiagonal components of the distribution function extend throughout the magnetic layers. This, however, is a considerable simplification, as the exchange field in ferromagnetic films usually suppresses the component of the spin density perpendicular to the magnetization, and this suppression takes place within a few interfacial atomic planes in strong ferromagnets. Thus, the nondiagonal components of the distribution function in magnetic films also disappear at this distance from the interfaces. The opposite situation corresponds to the limit of a strong exchange field, which suppresses to zero the nondiagonal components of the distribution function in magnetic layers.

Of course, in real systems, the situation is somewhere between these two limits, i.e., the spin density perpendicular to the layer's magnetization penetrates the magnetic film up to a certain distance from the interfaces. Since the angular dependence is rather weakly sensitive to the penetration depth, and in order to be consistent with the considerations in the case of a collinear configuration, we will analyze now the angular dependence of the GMR in the free electron gas model, which neglects the influence of exchange field on the distribution function.

The electron distribution function, $\mathbf{f}(y,\mathbf{v})$, is a matrix in the spin space, and generally diagonal components, $f_{\uparrow\uparrow}$ and $f_{\downarrow\downarrow}$, and nondiagonal ones, $f_{\uparrow\downarrow}$ and $f_{\downarrow\uparrow}$, are nonzero [58]. It is convenient to solve the Boltzmann equation for the magnetic layers in the corresponding local coordinate systems. In turn, for the nonmagnetic layer one can assume the local quantization axis of one of the magnetic layers. Instead of this, we introduce a fictitious plane in the middle of the spacer layer, i.e., at $y_{\mathrm{f}}=D+d/2$ (see also Fig. 7), and for each part of the spacer, we assume the local quantization axis of the adjacent magnetic layer. When an external electric field is applied to the system, the distribution function $\mathbf{f}(y,\mathbf{v})$ can be written as $\mathbf{f}(y,\mathbf{v})=\mathbf{f}_0(\mathbf{v})+g(y,\mathbf{v})$, where $\mathbf{f}_0(\mathbf{v})$ is the equilibrium distribution function while $g(y,\mathbf{v})$ is a term driven by the electric field. The Boltzmann kinetic equation for $g(y,\mathbf{v})$ can be then written in the form [59]:

$$\frac{\partial \mathbf{g}(y,\mathbf{v})}{\partial y}+\frac{1}{v_y}\boldsymbol{\tau}^{-1}\mathbf{g}(y,\mathbf{v})=\frac{eE}{mv_y}\frac{\partial f_0}{\partial v_x}\mathbf{1}, \tag{10}$$

where $\mathbf{1}$ is the 2×2 unit matrix in the spin space, and $\boldsymbol{\tau}$ is a diagonal matrix with the following diagonal elements: $\boldsymbol{\tau}_{\uparrow\uparrow}=\tau_{\mathrm{M}}$ and $\boldsymbol{\tau}_{\downarrow\downarrow}=\tau_{\mathrm{m}}$ in the magnetic layers, and $\boldsymbol{\tau}_{\uparrow\uparrow}=\tau_{\downarrow\downarrow}=\tau$ in the nonmagnetic spacer.

As in the collinear situations, the solutions to the Boltzmann equation have to be matched with the assistance of the appropriate boundary conditions. In the local coordinate systems, the boundary conditions for transmission across the two interfaces include specularity transmission coefficients, T_{M} and T_{m}. In turn, the boundary conditions at the external surfaces include specularity reflection coefficients, p_{M} and p_{m}. The solutions also have to be matched at the internal fictitious plane. Since the two quantization axes used for the left and right parts of the spacer layer are different, the solutions on one side (say the left one) have to be rotated in the spin space to the quantization axis on the other side, and then they have to be set equal. Thus, the boundary condition at $y=y_{\mathrm{f}}$ can be written as

$$\mathbf{g}^{\pm}(y=y_{\mathrm{f}}+0)=\mathbf{R}^{-1}(\varphi)\mathbf{g}^{\pm}(y=y_{\mathrm{f}}-0)\mathbf{R}(\varphi), \tag{11}$$

where $\mathbf{g}^+$ and $\mathbf{g}^-$ are the left and right propagating solutions to the Boltzmann kinetic equation, which are taken on both sides of the fictitious plane and written in the corresponding local quantization axes. In turn, the matrix $\mathbf{R}(\varphi)$,

$$\mathbf{R}(\varphi) = \begin{bmatrix} \cos(\varphi/2) & -\sin(\varphi/2) \\ \sin(\varphi/2) & \cos(\varphi/2) \end{bmatrix}, \tag{12}$$

describes the rotation of the spin quantization axis by the angle φ between the magnetic moments of the two layers.

Figure 8 shows the angular dependence of the normalized MR due to spin-dependent electron scattering at the interfaces. This figure shows that the deviation from the simple linear dependence of GMR on $\sin^2(\varphi/2)$ is rather small for the assumed parameters. The different curves in this figure correspond to different values of the MFP λ in the nonmagnetic layer, while the MFP in the magnetic layers is assumed to be independent of spin orientation and equal to λ. The specularity factors at the external surfaces are assumed to be equal to 1 (no diffuse scattering at the surfaces), so the result also describes a superlattice with doubled thickness of the magnetic layers. Spin-dependent diffuse scattering occurs only at the internal interfaces, where the specular transmission is equal to 1 (no diffusive scattering) for the spin-minority channel, $T_m = 1$,

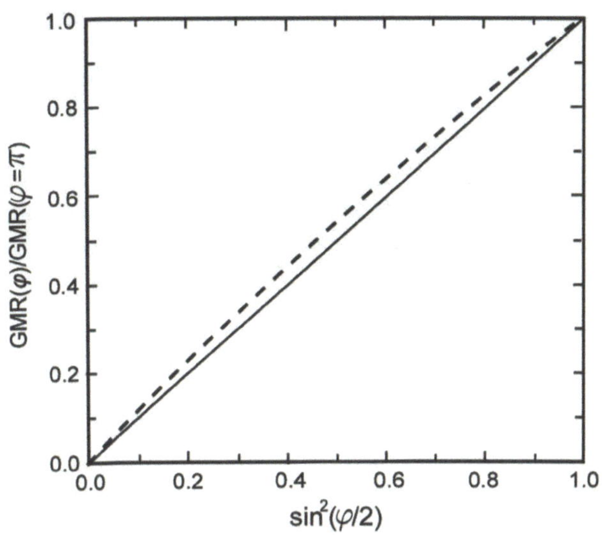

FIGURE 8
Angular dependence of GMR for a sandwich structure with specular surfaces for $\lambda = 100$ Å (solid line) and $\lambda = 1000$ Å (dashed line). In each case, $\lambda_M = \lambda_m = \lambda$ is assumed. The other parameters are: $E_F = 4$ eV, $T_M = 0.5$, $T_m = 1$, $d = 20$ Å, and $D = 100$ Å. The GMR effect is generated here by spin-dependent scattering at the interfaces. *After Ref. [59].*

while for the spin-majority channel the specular transmission is equal to 0.5, which also means that the probability of diffuse scattering of spin-majority electrons at the interfaces is equal to 0.5.

The angular dependence shown in Fig. 8 reflects the trends observed experimentally. It is surprising that such a simple model describes the general features rather well, at least qualitatively. For a more accurate description of the angular dependence of GMR one should relax some of the assumptions, especially the one concerning the influence of exchange field on the nondiagonal elements of the distribution function in magnetic films. Assuming that this exchange field is strong, one can neglect the nondiagonal components of the distribution function in the magnetic layers, and calculate the angular dependence of GMR taking into account the full distribution function in the spacer. However, the difference between these two limiting situations is rather weak for the parameters used in Fig. 8.

6.3 Role of Spin-Mixing Processes

One of the key assumptions underlying the models described above was the lack of mixing of the two spin channels. In real situations, however, the two spin channels are not independent as there are, albeit rare, electron scattering processes which reverse the electron spin orientation. The spin reversal processes which leave the electron momentum almost unchanged are of particular importance. Such processes do not contribute to the electrical resistance in nonmagnetic metals. However, the situation is different in ferromagnetic metals, where such spin-mixing processes transfer the electron momentum between two nonequivalent spin channels, and therefore can contribute to the electrical resistance [60–62]. The contribution due to the spin-mixing processes can be taken into account by an additional term in the Boltzmann equation, as shown by Fert [63] for bulk ferromagnetic metals. For layered structures with collinear magnetizations, this equation takes the form [63]:

$$\frac{\partial g_s(y,\mathbf{v})}{\partial y}+\frac{g_s(y,\mathbf{v})}{\tau_s v_y}+\frac{g_s(y,\mathbf{v})-g_{-s}(y,\mathbf{v})}{\tau_{sf} v_y}=\frac{eE}{m v_y}\frac{\partial f_0(\mathbf{v})}{\partial v_x} \tag{13}$$

for $s=\uparrow$ and $s=\downarrow$, where τ_{sf} is the spin-mixing relaxation time [63]. The third term in this equation describes the spin-mixing processes, which tend to reduce the difference between distribution functions for electrons with opposite spin orientations. It is worth noting that the spin-flip scattering processes contribute generally to the spin-dependent momentum relaxation time τ_s as well as to the electron-momentum-conserving spin-mixing relaxation time τ_{sf}. The dominant contribution to τ_{sf} originates from electron scattering from long-wavelength spin waves [63]. The energy and quasimomentum conservation rules make the scattering predominantly forward.

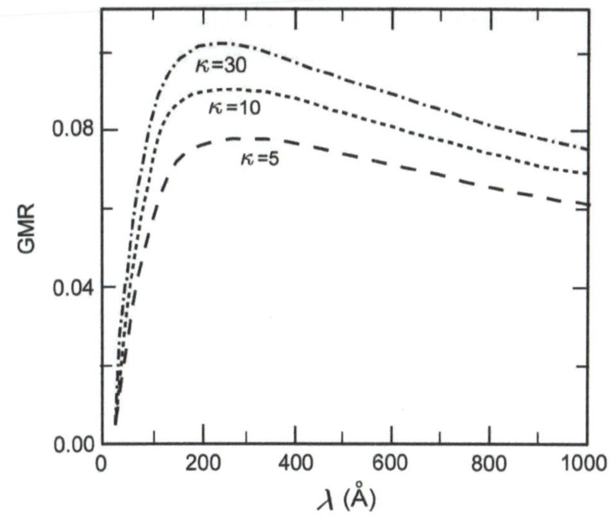

FIGURE 9
Reduction of GMR by spin-mixing scattering processes. The GMR ratio is shown as a function of mean free path in the nonmagnetic spacer for indicated values of the parameter $\kappa = \lambda_{sf}/\lambda$. The other parameters are $d = D = 20$ Å, $p_M = p_m = T_M = T_m = 0.9$, $p_{sf} = T_{sf} = 0$, $(\lambda_M + \lambda_m)/2 = \lambda$, and $\lambda_M/\lambda_m = 5$.

The above generalized Boltzmann kinetic equation can be solved similarly as the Boltzmann equation (5), with the boundary conditions at the surfaces and interfaces supplemented by additional parameters p_{sf} and T_{sf}, which describe the specular reflection from the surfaces and specular transmission through the interfaces with simultaneous spin reversal respectively. As a result one finds that the spin reversal processes reduce the GMR, as one might expect from a simple argument that spin-mixing processes reduce the spin asymmetry of the two spin channels. This is shown explicitly in Fig. 9, where GMR is plotted as a function of MFP in the spacer material, $\lambda = v_F \tau$ (with v_F denoting the Fermi velocity), for different values of the parameter $\kappa = \tau_{sf}/\tau = \lambda_{sf}/\lambda$, where τ_{sf} is the spin-mixing relaxation time and $\lambda_{sf} = v_F \tau_{sf}$ is the spin-mixing MFP in the nonmagnetic spacer layer. It is assumed in Fig. 9 that λ_{sf} in the magnetic layers is the same as in the spacer layer, and λ is equal to average MFPs in the magnetic layers, $\lambda = (\lambda_M + \lambda_m)/2$. The limit of independent spin channels corresponds to $\kappa \to \infty$, where the GMR ratio is maximal. When τ_{sf} is finite, the factor κ is also finite and the GMR is reduced.

Additionally, Fig. 9 also shows that GMR is low in the short MFP regime, which reflects the fact that GMR disappears in the macroscopic limit. Moreover, GMR has a maximum at a certain value of MFP and becomes suppressed with a further increase in MFP. This suppression appears because the resistance for large values of MFP is dominated by interface scattering, which is independent of electron spin for the parameters assumed in Fig. 9.

6.4 Influence of Quantum Size Effect

The quasiclassical approaches describe qualitatively the main features of the GMR effect. The question is if the quantum effects modify the GMR or not. The key correction of quantum origin follows from the quantum size effect, i.e., from size quantization of electron states along the orientation normal to the layers. This quantization is due to an external confining potential and potential steps (band mismatch) at the interfaces. Another weak point of the quasiclassical approach used above is the fact that electron scattering of impurities inside the layers and the electron scattering from rough interfaces are not taken into account on an equal footing. Despite this inconsistency, the quasiclassical approach gives satisfactory results, except in the limit of very clean layers, when the bulk electron MFPs are much longer than the layer thicknesses. The role of interface scattering is then rather underestimated. This is a well-known problem in the theory of electronic transport in single metallic films, where the quasiclassical approach breaks down in the limit of MFP much longer than the layer thickness. This breakdown appears because the electrons propagating inside the layers and parallel to the interfaces do not *sample the interface roughness* in the classical description. To describe such a situation properly, one has to employ a quantum-mechanical method [53,60,64–67].

Below, we describe briefly the quantum corrections to GMR in a sandwich structure that was described earlier within the quasiclassical approach. We also relax the assumption of no band mismatch at the interfaces. The corresponding Hamiltonian of the system can then be written in the general form:

$$H = -\frac{\hbar^2}{2m}\nabla^2 + U_s(y) + V_s(\mathbf{r}), \tag{14}$$

where the first term is the usual kinetic term, while $U_s(y)$ is the confining potential which effectively describes the position of the bottom band edges inside the layers. This potential is assumed to be constant within a particular layer, but is generally different in different layers and additionally spin dependent in the magnetic layers. Thus, its shape throughout the structure changes when the magnetic configuration is changed from antiparallel to parallel. The last term, $V_s(\mathbf{r})$, is the spin-dependent scattering potential of impurities and interface roughness,

$$V_s(\mathbf{r}) = \sum_i v_{is}\delta(\mathbf{r}-\mathbf{r}_i) + \sum_\mu \xi_\mu(\boldsymbol{\rho}) V_{\mu s}^{\mathrm{eff}} \delta\left(y - y_\mu^c(\boldsymbol{\rho})\right). \tag{15}$$

The first term represents point-like impurities, with v_{is} being the spin-dependent scattering potential of the ith impurity located at $\mathbf{r}_i$. The scattering potential of the impurities as well as their concentration may be generally different in different layers. The second term in the above Hamiltonian is the scattering potential of rough interfaces, where $\xi_\mu(\boldsymbol{\rho})$ describes the geometrical roughness of the μth interface, $V_{\mu s}^{\text{eff}}$ is the effective spin-dependent potential step at the μth interface, while $y_\mu^c(\boldsymbol{\rho})$ describes position of the *center* of the interface scattering potential, $y_\mu^c(\boldsymbol{\rho}) = y_\mu + (1/2)\xi_\mu(\boldsymbol{\rho})$ [67]. Here, $\boldsymbol{\rho}$ is the two-dimensional in-plane position vector, $\boldsymbol{\rho} = (x,z)$, y_μ is the average position of the μth interface, and the sum is over all interfaces. The Hamiltonian (14) does not include spin-mixing processes, so it describes effectively two independent spin channels.

One of the quantum-mechanical approaches to electronic transport is based on the Kubo formalism. Let us neglect for a while scattering from interface roughness. Following the Kubo approach, one can write the nonlocal spin-dependent conductivity $\sigma_s(y, y')$ in the form [53]:

$$\sigma_s\left(y, y'\right) = -\frac{e^2}{(2m)^2}\frac{\hbar}{(2\pi)^2}\int \mathrm{d}q q^3 \left[\mathrm{G}_{qs}^{\mathrm{r}}\left(y, y'\right) - \mathrm{G}_{qs}^{\mathrm{a}}\left(y, y'\right)\right]\left[\mathrm{G}_{qs}^{\mathrm{r}}\left(y', y\right) - \mathrm{G}_{qs}^{\mathrm{a}}\left(y', y\right)\right], \quad (16)$$

where $G_{qs}^{\mathrm{r(a)}}(y, y')$ is the retarded (advanced) Green function for the in-plane wavevector q ($q = |\mathbf{q}|$) and spin s. The Green functions can be obtained from the corresponding differential equation, in which scattering by impurities is effectively taken into account by appropriate relaxation times (related to self-energy). This equation can be solved separately for each layer, assuming a constant relaxation time, which can be different in different layers and also spin dependent inside the magnetic layers. Solutions have to be matched at all interfaces with the relevant boundary conditions [65]. Having found the Green functions, one can calculate the global conductivity as

$$\sigma = \frac{1}{L}\sum_s \int \mathrm{d}y' \int \mathrm{d}y \sigma_s\left(y, y'\right). \quad (17)$$

Detailed numerical calculations based on the formula (17) show that GMR has a component that indeed oscillates with the spacer thickness [65].

The interface roughness can also be taken into account within the Kubo formalism. The intermixing or alloying effects at an interface can be included in the above formalism by assuming an additional thin layer of reduced relaxation time at this interface. A more accurate description of the role of geometrical interface roughness requires formulation of the Kubo formula in reciprocal space and calculation of the appropriate self-energy. Instead of this one can use a conceptually simpler approach based on the Boltzmann kinetic equation, in which the probabilities of electron scattering on structural defects are calculated with quantum electronic states of the corresponding defect-free system (which include the size effects). Such quantum states in the case of a layered structure form a set of two-dimensional subbands, and for each subband η one can write the following Boltzmann equation for the corresponding distribution function [66–68]:

$$\frac{-e}{\hbar}\mathbf{E}\cdot\nabla_{\mathbf{q}}f_{\eta s}(\mathbf{q})=\sum_{\eta' \mathbf{q}'}\left\{W^{\eta s}_{\eta' s}\left(\mathbf{q}',\mathbf{q}\right)f_{\eta' s}\left(\mathbf{q}'\right)\left(1-f_{\eta s}(\mathbf{q})\right)-W^{\eta' s}_{\eta s}\left(\mathbf{q},\mathbf{q}'\right)f_{\eta s}(\mathbf{q})\left(1-f_{\eta' s}\left(\mathbf{q}'\right)\right)\right\}, \tag{18}$$

where $W^{\eta s}_{\eta' s}$ is the probability per unit time of electron transitions from the state $|\eta \mathbf{q} s\rangle$ to the state $|\eta' \mathbf{q}' s\rangle$, while $f_{\eta s}(\mathbf{q})$ is the spin-dependent distribution function for the subband η. The transition rates $W^{\eta s}_{\eta' s}$ are determined by matrix elements of the scattering potential between the states corresponding to the subbands η and η'. Solution of the above set of Boltzmann equations allows one to find the global conductivity in the form [67,68]:

$$\sigma=\frac{2e^2}{\hbar L}\sum_s\sum_{\eta\eta'}\left(E_F-\varepsilon_{\eta s}\right)\left(E_F-\varepsilon_{\eta' s}\right)\left[\mathbf{C}_s^{-1}(E_F)\right]_{\eta\eta'}, \tag{19}$$

where $\varepsilon_{\eta s}$ is the band-edge energy of the ηth subband for spin s, while $\mathbf{C}$ is a certain matrix determined by transition rates due to scattering potentials (for details see Ref. [67]). The summation over η and η' is over all occupied subbands.

The above description, similar to the Kubo approach, leads to a contribution to GMR that oscillates with the thickness of the spacer layer, as shown in Fig. 10. This figure also shows that GMR decreases with increasing spacer thickness, and

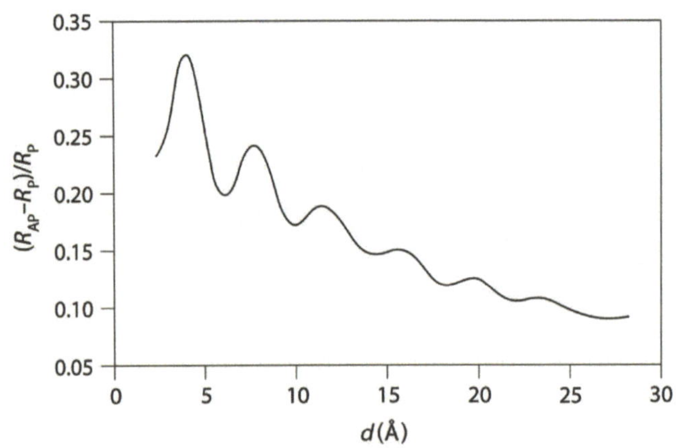

FIGURE 10

Quantum size oscillations in the dependence of the GMR effect on the spacer thickness for a magnetic sandwich structure with $D=20$. *After Ref. [67], where other parameters can also be found.*

this decrease on average is exponential-like. More importantly, there are damped oscillations in GMR with increasing spacer thickness. These oscillations have a similar origin to the oscillations in RKKY interlayer exchange coupling. Accordingly, some correlation between the quantum size effects in GMR and the interlayer exchange coupling can appear [66]. The quantum oscillations in GMR may occur in very clean structures and at low temperatures. Such oscillations, however, have not yet been observed experimentally in real structures, since they are rather strongly damped by scattering processes. Apart from this, electron–electron interaction was neglected in all descriptions presented above, which also may diminish the amplitude of the oscillations due to the size effect.

Though the quantum size effects in GMR may be apparent in the dependence on the layer thicknesses, they have no significant influence on the angular dependence, which is very similar to that presented in Fig. 8, i.e., in the absence of quantum size effects. This angular dependence, i.e., almost linear variation with $\sin^2(\theta/2)$, is thus a rather general feature of CIP-GMR in the diffusive transport regime, independently of how this limit is treated, classically or quantum mechanically. However, quite remarkable deviations from this linear dependence can occur in the limit of weak scattering, when quantum features in transport become crucial [62].

6.5 Other Theoretical Aspects

The free-electron-like models, corresponding to parabolic electron bands—spin-polarized or not—describe qualitatively the basic features of the GMR effect in a satisfactory way. However, they are not able to reproduce satisfactorily all the features of transport properties, including the GMR effect. The further improvements rely on taking into account the discrete atomic structure as well as the exact electronic band structure of magnetic multilayers. This has been done in terms of parameterized tight binding models and *ab initio* methods. Scattering potentials due to various impurities can also be included in these models. The advantage of such calculations is a reliable quantitative description of the effect and its correlation with the distribution of impurities in the multilayers [69–73]. Such calculations, however, require extensive computational work. There are already some review articles on this issue, so this problem will not be addressed in any more detail here.

7 CURRENT-PERPENDICULAR-TO-PLANE GMR

In the experimental situation described in Section 4 and in the corresponding theoretical part presented in Sections 5 and 6, the net charge current was flowing in the plane of a layered structure. However, the GMR effect can also appear in magnetic multilayers for current flowing along the growth direction, i.e., perpendicular to the layers. The effect is then known as *current-perpendicular-to-plane* GMR (CPP-GMR). But even a more general geometry, with current flowing along a direction at an arbitrary angle to the growth direction, has been realized experimentally. The latter case will not be considered here, as it essentially does not contain new physics which would be absent either in the CIP or CPP geometries.

Since the layered structures used in typical CIP experiments are very thin, their resistance in the CPP geometry is very small. This leads to the main experimental difficulty concerning resistance measurements in the CPP geometry, as one has to extract the sample resistance from the total resistance. The latter resistance includes a contribution from external electrodes, which is dominant. Thus, in order to measure the resistance change in the CPP geometry, one has to increase the relative contribution of the magnetic layered structure to the experimentally measured resistance. This can be done either by reducing the resistance of external contacts, or by augmenting the resistance of the measured sample. In fact, both options have been realized experimentally. To measure GMR in Co/Cu and Co/Ag multilayers in the perpendicular geometry, Pratt *et al.* [74] used superconducting electrodes, which reduced the resistance of external contacts, and therefore enabled extraction of the resistance of the multilayer from the total resistance measured experimentally. Alternatively, one can increase the resistance of a multilayer by reducing its lateral dimensions. This can be done using patterning techniques, which allow one to fabricate nanopillars from a typical multilayer grown, for instance, by the MBE method. This technique was used by Gijs *et al.* [75] to fabricate Fe/Cr nanopillars, and is now a commonly used technique in nanopillar fabrication. There is also another method, which allows fabrication of multilayered nanowires. This method relies on electrochemical deposition into pores created in membranes by heavy particle bombardment [76,77]. The latter technique was used to fabricate, for instance, Co/Cu nanowires. Typical diameters of the nanowire cross-sections are of an order of a few tens up to a few hundreds of nanometers, and nanopillar cross-sections are of comparable size. The multilayered structures described above, used for measurements in the CPP geometry, are shown schematically in Fig. 11.

Experimental results show that $\Delta R/R_P$ in the CPP geometry is usually larger than the corresponding ratio obtained in the CIP geometry. Moreover, the characteristic spacer thickness at which the CPP-GMR disappears is larger than in the case of CIP-GMR. On the other hand, the main qualitative arguments which explain the physical origin of the CIP-GMR also hold for the CPP geometry. However, there is an important difference between these two configurations. In the CPP geometry, the current density is the same in all layers, while in the CIP geometry the current is not equally distributed, and the current density is larger in layers

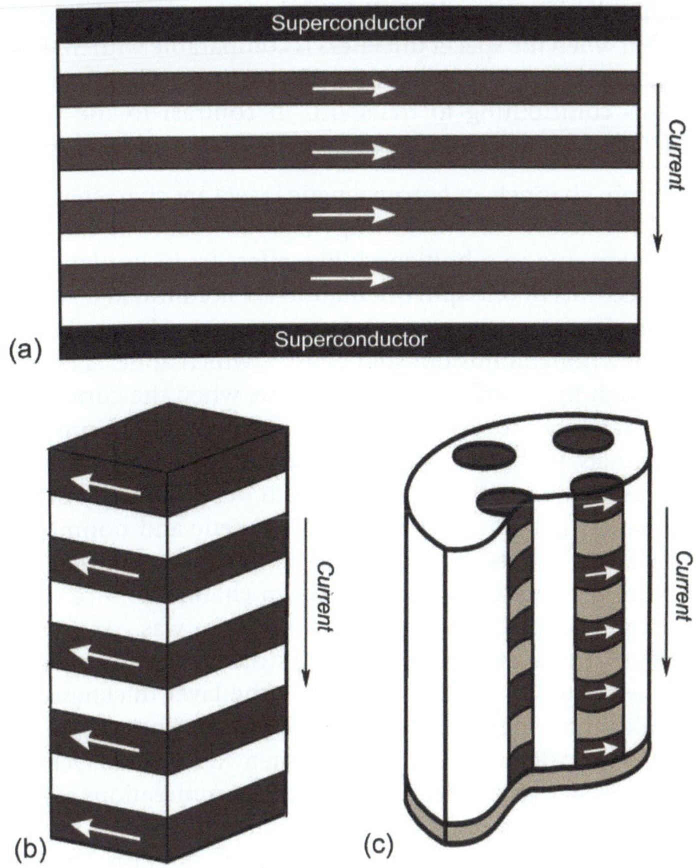

FIGURE 11
Schematic picture of systems fabricated to measure GMR in the CPP geometry: (a) multilayer with superconducting electrodes, (b) nanopillar obtained from a multilayer by patterning techniques, and (c) nanowire grown by electrochemical deposition into nanopores in a membrane.

of higher conductivity. Another important difference is that the current crosses the interfaces between the layers, which leads to further consequences. One of these consequences is the presence of interface resistance [15], which is associated with a band mismatch at the interfaces [78]. The point is that when electrons are reflected specularly from the interface, their velocity parallel to the interface is not changed, and such reflections have no influence on in-plane current, while the velocity component normal to the interface changes sign, therefore such reflections have an important impact on transport perpendicular to the interfaces. The concept of interface resistance has been used extensively in the theoretical description of CPP-GMR [79–82]. Moreover, the relevant scale length in the CPP geometry is not the MFP but rather the SDL, i.e. the CPP-GMR also exists in the limit, where the CIP-GMR has already vanished. To explain this we note that when the spacer thickness is comparable with SDL, *spin contact* between magnetic layers is mediated by electrons traveling normally to the layers and thus contributing to transport, in contrast to the CIP geometry (see Sections 5 and 6), where such electrons do not contribute to the current.

Since the two spin channels in ferromagnetic layers are characterized by different transport parameters, current flowing perpendicular to an interface leads to spin accumulation due to a bottleneck-like effect (spin bottleneck effect). In other words, electrons of one spin orientation are accumulated at the interface, while electrons of the opposite spin orientation are then depopulated. This spin accumulation is a nonequilibrium spin density, which appears only when current flows through the interface, and disappears when the current is switched off. Spin-flip scattering processes tend to restore the equilibrium between different spin channels, so the spin accumulation vanishes from the interface in an exponential-like fashion with the decay length determined by SDL. The spin accumulation at a single interface between magnetic and nonmagnetic semi-infinite systems was studied by van Son *et al.* [83], who showed that current conversion from the spin-up to the spin-down channel or vice versa leads to an interfacial spin-coupled resistance. This problem was considered later by Valet and Fert [15] in the case of multilayered structures, where successive interfaces cannot be treated as independent when the layer thicknesses are of the order or smaller than the SDL. When the layer thicknesses are larger than the SDL, the CPP-GMR vanishes. In turn, when the spacer layer thickness is of the order of the SDL or smaller, then the spin accumulations at neighboring interfaces interfere and the resultant spin accumulation in the parallel configuration is different from that in the antiparallel configuration. The current conversion cannot then be described by spin-coupled resistances of individual interfaces, as discussed in detail by Valet and Fert [15]. A specific situation appears in the limit of layer thicknesses much smaller than the SDL. In such a case electrons traversing the structure keep their spin orientation, and the current in each spin channel must be constant across the structure. This requires some redistribution of electrons between the spin channels (spin accumulation), which is uniform in individual layers. In this limit one can present the CPP-GMR as a simple resistor network model, which is shown for a superlattice in Fig. 12. The condition of constant current in each spin channel in the antiparallel configuration requires equal resistances of both spin channels in the ferromagnetic layers [15], as shown in Fig. 12. In the parallel magnetic configuration, the currents flowing in opposite spin channels can be different. From this resistor network model one can easily find that the transition from the antiparallel to the parallel configuration leads to the CPP-GMR, which in the limit of an SDL much longer than the layer thicknesses is given by the formula:

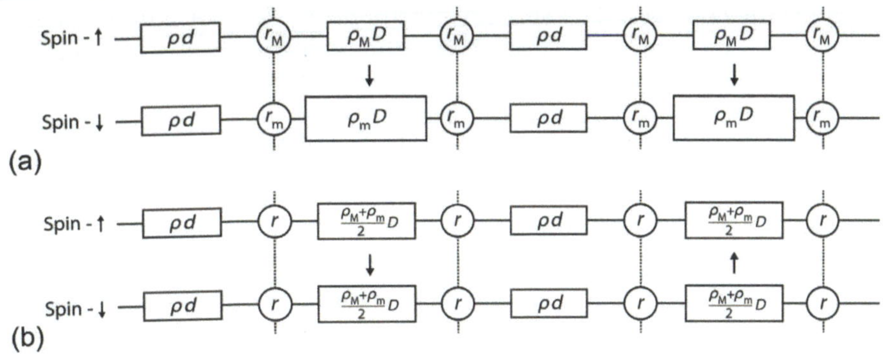

FIGURE 12
The resistor network model for a superlattice in the CPP geometry for the parallel (a) and antiparallel (b) configurations in the limit of layer thicknesses much smaller than the spin-diffusion length. Here ρ, ρ_M, and ρ_m are the bulk resistivities of the nonmagnetic metal (per spin channel) and ferromagnetic metal for spin-majority and spin-minority channels respectively, while r_M and r_m denote the interface resistance per unit area in the spin-majority and spin-minority channels. In the antiparallel configuration (b), r is the average value of r_M and r_m, $r=(r_M+r_m)/2$. The thicknesses of the nonmagnetic and magnetic layers are denoted by d and D respectively. The dotted vertical lines represent the interfaces.

$$\frac{\Delta R}{R_P}=\frac{[(\rho_M-\rho_m)D+2(r_M-r_m)]^2}{4(\rho_M D+\rho d+2r_M)(\rho_m D+\rho d+2r_m)}, \tag{20}$$

where ρ_M and ρ_m denote the bulk resistivity of the magnetic material in the spin-majority and spin-minority channels respectively, ρ is the bulk resistivity of the nonmagnetic spacer (per spin channel), while r_M and r_m are the interface resistances of a unit square for the spin-majority and spin-minority channels. These two resistances are usually written in a form analogous to Eqs. (1a and 1b) [15]:

$$r_M=2r_F^*(1-\gamma), \tag{21a}$$

$$r_m=2r_F^*(1+\gamma), \tag{21b}$$

where γ is the corresponding interface spin asymmetry parameter, while $r_F^*=r_F/(1-\gamma^2)$, with r_F denoting the total interface resistance of a unit square. In general, the spin asymmetry parameter γ is different from the corresponding spin asymmetry factor β of the bulk resistivity. Note that in a structure of finite thickness, the surface resistance can correspond to a spin asymmetry parameter which is different from γ.

The formula (20) for CPP-GMR in the limit of thin layers is valid for a superlattice. If ρ_M is different from ρ_m and/or r_M is different from r_m, the CPP-GMR does not vanish. The resistor network model for the CPP geometry was used to account for many experimental data obtained for multilayers [84], and to determine the corresponding spin asymmetry parameters for bulk resistivities as well as for interface resistances. When the layer thicknesses increase, and become comparable to SDL, the current in a particular spin channel is not conserved (though the total current is conserved), and the CPP-GMR decreases. Finally, it disappears in the limit of layer thicknesses larger than SDL.

The spin accumulation changes significantly with the angle between magnetic moments, and may lead to some interesting angular dependence of the CPP-GMR. Let us consider now this general situation in more detail. The spin accumulation is equivalent to a spin splitting of the electrochemical potential, and appears only near the interfaces. Consider a spin valve with noncollinear magnetic moments. The diffusion equation for both magnetic and nonmagnetic films takes the general form [85–87]:

$$\mathbf{D}\frac{\partial^2 \mathbf{f}}{\partial y^2}=\frac{1}{\tau_{sf}}\left[\mathbf{f}-\mathbf{1}\frac{Tr\{\mathbf{f}\}}{2}\right], \tag{22}$$

where $\mathbf{D}$ is a matrix of diffusion parameters in spin space, and τ_{sf} is the spin-flip relaxation time. Since the internal exchange field inside ferromagnetic 3d metals is strong enough, the component of the distribution function perpendicular to the local magnetization vanishes at short distances from an interface (see also Section 6.2). Thus, in the limit of a strong exchange field, the distribution function in magnetic films is diagonal, when the spin quantization axis is parallel to the local magnetization. In the following we assume this is the case. Equation (22) determines the behavior of the spin accumulation, from which one can derive the following equations for the electrochemical potentials in the spin-majority, μ_M, and spin-minority, μ_m, channels:

$$\frac{\partial^2(\mu_M-\mu_m)}{\partial y^2}=\frac{1}{l_{sf}^2}(\mu_M-\mu_m), \tag{23a}$$

$$\frac{\partial^2(\mu_M+\mu_m)}{\partial y^2}=\zeta\frac{\partial^2(\mu_M-\mu_m)}{\partial y^2}, \tag{23b}$$

where l_{sf} is the spin diffusion length, defined as $1/l_{sf}^2=\left(1/l_M^2+1/l_m^2\right)/2$ with $l_{M(m)}^2=D_{M(m)}\tau_{sf}$, whereas $\zeta=(D_m-D_M)/(D_m+D_M)$, with D_m and D_M denoting the diffusion constants for the spin-minority and spin-majority electrons.

The electrochemical potentials in the ferromagnetic layers can be written as a matrix in the spin space, $\boldsymbol{\mu}=\mu_0\mathbf{1}+g\sigma_z$, where g is the spin accumulation, $g=(\mu_M-\mu_m)/2$, μ_0 is the average electrochemical potential, $\mu_0=(\mu_M+\mu_m)/2$, while σ_z is the zth component of the Pauli matrices, $\boldsymbol{\sigma}=(\sigma_x,\sigma_y,\sigma_z)$. The explicit forms of μ_M and μ_m can be obtained from the solution of Eqs. (23a and 23b). In turn, the current matrix $\mathbf{J}$, defined by the formula:

$$\mathbf{J}=-\rho(E_{\mathrm{F}})\mathbf{D}\frac{\partial\boldsymbol{\mu}}{\partial y}, \tag{24}$$

can be written as $\mathbf{J}=(j_0\mathbf{1}+j_z\sigma_z)/2$, where j_0 is the particle current, while j_z is the zth component of the spin current (in the units of $\hbar/2$). In the magnetic layers, only the zth components of the spin accumulation and of the spin current are nonzero. The situation is different in the nonmagnetic spacer layer, where the nondiagonal components of the spin accumulation $\boldsymbol{\mu}$ and of the spin current $\mathbf{j}$ do not vanish in the general case. The spin accumulation can be then written as $\boldsymbol{\mu}=\mu_0\mathbf{1}+\mathbf{g}\cdot\boldsymbol{\sigma}$, while the current matrix is $\mathbf{J}=(j_0\mathbf{1}+\mathbf{j}\cdot\boldsymbol{\sigma})/2$. Both $\boldsymbol{\mu}$ and $\mathbf{J}$ can be found from the corresponding equations (23a and 23b) and (24), with $D_{\mathrm{m}}=D_{\mathrm{M}}=D$.

The solutions for the spin accumulation $\mathbf{g}$, spin current $\mathbf{j}$, electrochemical potential μ_0, and particle current j_0 in different layers have to be joined with some interface boundary conditions [85]:

$$e^2 j_0=(G_{\mathrm{M}}+G_{\mathrm{m}})\left(\mu_0^{\mathrm{F}}-\mu_0^{\mathrm{N}}\right)+(G_{\mathrm{M}}-G_{\mathrm{m}})\left(g_z^{\mathrm{F}}-g_z^{\mathrm{N}}\right), \tag{25a}$$

$$e^2 j_z=(G_{\mathrm{M}}-G_{\mathrm{m}})\left(\mu_0^{\mathrm{F}}-\mu_0^{\mathrm{N}}\right)+(G_{\mathrm{M}}+G_{\mathrm{m}})\left(g_z^{\mathrm{F}}-g_z^{\mathrm{N}}\right), \tag{25b}$$

$$e^2 j_x=-2\mathrm{Re}\{G_{\mathrm{mix}}\}g_x^{\mathrm{N}}+2\mathrm{Im}\{G_{\mathrm{mix}}\}g_y^{\mathrm{N}}, \tag{25c}$$

$$e^2 j_y=-2\mathrm{Re}\{G_{\mathrm{mix}}\}g_y^{\mathrm{N}}-2\mathrm{Im}\{G_{\mathrm{mix}}\}g_x^{\mathrm{N}}, \tag{25d}$$

where the superscripts F and N refer to the ferromagnetic and normal metal sides of the interface. Here G_{M} and G_{m} denote the interface conductance for the spin-majority and spin-minority channels ($G_{\mathrm{M(m)}}=1/r_{\mathrm{M(m)}}$) respectively, while G_{mix} is the mixing conductance.

Applying the boundary conditions to all interfaces, one can determine the charge current as well as all components of the spin current. In addition, one can find the spin accumulation and the electrochemical potential μ_0. From this one can calculate the resistance and thus also GMR and its angular dependence. This angular variation is generally given by the following formula [87]:

$$\frac{\mathrm{GMR}(\varphi)}{\mathrm{GMR}(\varphi=\pi)}=\frac{\sin^2(\varphi)}{1+\psi\cos^2(\varphi/2)}, \tag{26}$$

as derived from magneto-circuit theory [87] and from the diffusive approach [88]. In the measurements on Py/Cu/Py spin valves [89], the parameter ψ was treated as a fitting parameter, and a relatively good agreement with the formula (26) was achieved.

In symmetrical structures, the angular dependence of CPP-GMR is similar to that in the CIP geometry (see Fig. 8). In asymmetrical structures, however, a minimum in resistance (and also in GMR) may occur in a noncollinear configuration. The angular dependence of the normalized GMR in a Cu/Co(d_1)/Cu(10 nm)/Co(8 nm)/Cu spin valve, calculated in the diffusive approach described above, is shown in Fig. 13 for two different values of d_1. For $d_1 = 60$ nm, the normalized GMR exhibits a minimum for $\theta \approx \pi/3$. In turn, for $d_1 = 16$ nm, the GMR varies monotonically with the angle between magnetic moments. The global minimum arises as a result of the spatial depletion of the electrical field in the thick magnetic layer, which is a consequence of the spin accumulation discontinuity at the normal-metal/ferromagnet interface, controlled by the mixing conductance [86,90].

8 GMR IN OTHER SYSTEMS

The GMR effect may also occur in other systems, provided there are two remarkably different spin channels for transport. Indeed, to observe this phenomenon one does not need hybrid systems in the form of layered structures, but it is enough to have a granular material, in which small particles (grains) of a magnetic transition metal are distributed in a nonmagnetic metallic matrix. Due to the random shape anisotropy of the grains, their magnetic moments are oriented randomly, so the net magnetic moment of the material vanishes in the absence of an external field. Magnetic moments of the particles become aligned in an external magnetic field, and the transition from the random distribution

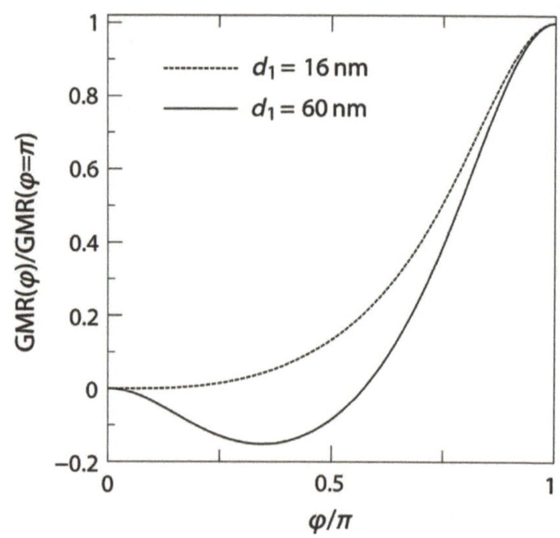

FIGURE 13
Angular dependence of the normalized magnetoresistance for the spin valve Cu/Co(d_1)/Cu(10)/Co(8)/Cu (thickness in nanometers) and for d_1 as indicated. *After Ref. [90].*

of the magnetic moments to the state with the moments aligned along the magnetic field is associated with a decrease in resistance. Such resistance behavior has indeed been observed in some granular systems [91,92]. One advantage of the granular systems is the relatively simple technology needed for their preparation. But they usually have too large a saturation field to be useful for applications. The physical mechanism of the GMR effect in granular structures is qualitatively similar to that of GMR in layered structures. However, the detailed description is more complex due to a nonuniform current distribution and random distribution of the magnetic nanoparticles.

The phenomenon of GMR is characteristic of all magnetic systems with a significant spin asymmetry in transport characteristics. Successful applications of GMR initiated extensive experimental and technological research for other magnetic systems, especially semiconductor spin valves. This is because integration of spintronic devices based on semiconducting ferromagnets with micro- and nanoelectronics is easier than integration of devices based on metallic 3d ferromagnets, especially when the conventional electronics is based mainly on semiconductors. This research aimed to find ferromagnetic semiconductors by doping some conventional semiconductors with magnetic atoms. One such system is GaAs doped with Mn atoms, which becomes ferromagnetic at some doping level. However, the key point is to find the conditions and the doping levels where the induced ferromagnetism is maintained at room temperature [93]. The main obstacle is the phase segregation which occurs for doping above a certain level. Magnetic atoms then concentrate into ferromagnetic nanoparticles, and instead of a uniform ferromagnetic system one gets a system similar to a granular one. There are a few review articles on this and related issues, so this problem will not be addressed in any more detail here. We only note that spin valves based on magnetic semiconductors have been constructed, and show the typical GMR phenomenon [94].

Another road in the search for novel materials for spintronic devices, and especially for spin valves, includes molecular systems connected to ferromagnetic electrodes. A typical example of such a hybrid system is a spin valve based on a carbon nanotube (CNT) contacted to metallic ferromagnetic electrodes. The main technological difficulty is to achieve good contacts between the electrodes and CNTs – contacts which ensure high efficiency of spin injection from a ferromagnetic metallic (or semiconducting) electrode to the CNT. When the contacts between metallic electrodes and CNTs are good, transport through the CNT is metallic. Moreover, when the spin relaxation time is long enough, the electrons keep their spin orientation when traversing the CNT from one electrode to the other; the conductance depends on the relative orientation of electrodes' magnetic moments and reveals features typical of the GMR effect. This behavior was observed, for instance, in a spin valve based on multiwall CNTs connected to two cobalt electrodes [95].

Instead of CNTs, one can use another carbon nanostructure, for instance a narrow ribbon of graphene, a two-dimensional array of carbon atoms. Indeed, the GMR effect in graphene-based spin valves was observed in Ref. [96], where transport through a graphene ribbon connected to two leads of permalloy was measured at room temperature. The resistance change associated with magnetization rotation was of the order of 10% at room temperature, which is comparable to typical values observed in metallic spin valves. The subsequent experiments have revealed Fabry–Perot-like patterns and remarkable oscillations of GMR [97]. The magnitude of GMR is very sensitive in this case to the quality of the contacts between leads and graphene nanoribbon, similarly as in the case of CNT-based spin valves. This problem was also investigated theoretically in Ref. [98].

In the case of graphene and other graphene-like nanoribbons (like silicene, for example), there is also a GMR effect of another type, which is different from that discussed above. As the GMR studied up to now was due to rotation of the electrodes' magnetic moments from antiparallel to parallel alignment, the new kind of GMR is based on the rotation of the edge magnetic moments in the nanoribbons. In the ground state these moments at one edge are opposite to the moments at the other edge. In a magnetic field, however, the moments at both edges become oriented in parallel, and this is associated with a resistance change as in the case of the normal (or inverse) GMR [99,100]. An important point is that now the electrodes can be nonmagnetic.

9 PRACTICAL APPLICATIONS OF GMR AND CONCLUDING REMARKS

The GMR effect has proved to be a phenomenon that in a short time has found practical applications [2,101]. The key point underlying this successful application was the high sensitivity of the GMR effect, i.e., a large resistance change in a very weak magnetic field. Shortly after its discovery, GMR was supposed to be an alternative for magnetic sensors based on AMR in thin magnetic layers. Such AMR sensors utilize the dependence of current (and thus resistance) on the relative orientation of the current and magnetic moment. When an external magnetic field switches magnetic moment of a layer between parallel and perpendicular to the current orientations, the resistance changes by a factor of an order of 1–4%. This change, though not too high, was sufficient to use field sensors based on AMR in read heads in hard drives [2,101]. Magnetic fields from a small region on the hard disk have been sufficient to reorient the magnetic moments of the sensor, and thus to read the stored information. Spin valves based on the CIP-GMR effect have much better characteristics, from the point of view of applications in magnetic field sensors, than field sensors based on AMR. A condition for this application is that the magnetic configuration should be switched between parallel and antiparallel configurations in a very weak magnetic field. Because of this, layered structures with large exchange interlayer coupling are of no practical use. It was necessary to exclude this coupling by a sufficiently thick nonmagnetic spacer, and also reduce the dipolar coupling between the magnetically pinned layer and a magnetically free layer. Moreover, the magnetic anisotropy in the free layer also has to be reduced. Such spin valves exhibit a high field sensitivity.

GMR-based sensors have other advantages compared to sensors based on AMR. The magnetic sensors based on the GMR effect can have remarkably smaller dimensions than those based on AMR sensors. As a result, magnetic sensors based on the CIP-GMR spin valves replaced AMR-based sensors in read heads of hard drives. All of this lead to an increase in storage density on hard drives of roughly two orders of magnitude.

Spin valves have also been used to construct nonvolatile magnetic random access memories (MRAMs) [2,101]. Parallel and antiparallel configurations of the magnetic moments in a spin valve can then be considered as two logic states, "0" and "1." Writing information consists of changing this configuration, which can be achieved by passing a current pulse through word and sensing lines in the MRAMs. To read the information, in turn, one can use the GMR effect. Recently, it has been found that spin valves have further interesting properties, i.e. their magnetic configuration can be switched by a current pulse passing directly through the spin valve. This is the so-called current-induced magnetic switching (CIMS) phenomenon [2,101]. This phenomenon can be used to write the information in memory elements based on spin valves. This shows that spin valves are multifunctional elements. It is also interesting to note here that there are several other well-known mechanisms of MR in metals and semiconductors (like, e.g., classical MR, negative MR due to weak localization, etc.), which have never been used in practical applications. But even more important is the fact that the discovery of GMR has initiated a new area of research, where spin-resolved effects and spin manipulation play a crucial role.

In summary, in this chapter, we have reviewed the basic properties and characteristics of the GMR effect in layered structures. The main focus was on the physical origin of GMR. Some applications of the phenomenon have also been briefly discussed. It is worth mentioning that a similar MR effect appears when the metallic layer separating the ferromagnetic layers in a conventional spin valve is replaced with a thin insulating layer (tunneling barrier). Transport then occurs due to quantum tunneling through the barrier and also depends on the magnetic configuration of the junction. This effect is known as tunnel magnetoresistance (TMR). As in conventional GMR, the tunnel resistance is usually larger when the electrodes have antiparallel magnetic moments and the resistance is decreased when this orientation changes to parallel (normal TMR) [102], but inverse TMR may also appear in some systems [103]. There are a number of differences between TMR and GMR. TMR occurs only in the current-perpendicular-to-plane geometry, and the magnitude of the current is lower than in the case of metallic spin valves. Importantly, the TMR can be even larger than the typical GMR in metallic structures. A particularly large TMR was observed in tunnel junctions based on MgO barriers [102]. There are also many other systems, including for instance single molecules and semiconductor quantum dots coupled to ferromagnetic metallic electrodes, which also exhibit a typical GMR/TMR effect. Moreover, such systems reveal single-electron transport features, which are especially interesting for low-energy electronics and information processing technology.

Acknowledgments

The authors acknowledge support from the National Science Center in Poland under Project No. DEC-2012/04/A/ST3/00372 and the National Center of Research and Development in Poland under Grant No. NCBR/ERA.NET RUS/01/2012.

References

[1] Grünberg PA. Nobel lecture: from spin waves to giant magnetoresistance and beyond. Rev Mod Phys 2008;80:1531.

[2] Fert A. Nobel lecture: origin, development, and future of spintronics. Rev Mod Phys 2008;80:1517.

[3] Grünberg P, Schreiber R, Pang Y, Brodsky MB, Sowers H. Layered magnetic structures: evidence for antiferromagnetic coupling of Fe layers across Cr interlayer. Phys Rev Lett 1986;57:2442.

[4] Baibich MN, Broto JM, Fert A, Van Dau FN, Petroff F, Eitenne P, et al. Giant magnetoresistance of (001)Fe/(001)Cr magnetic superlattices. Phys Rev Lett 1988;61:2472.

[5] Binasch G, Grünberg P, Saurenbach F, Zinn W. Enhanced magnetoresistance in layered magnetic structures with antiferromagnetic interlayer exchange. Phys Rev B 1989;39:4828.

[6] Mott NF. Electrons in transition metals. Adv Phys 1964;13:325.

[7] Shinjo T. Experiments of giant magnetoresistance. In: Maekawa S, Shinjo T, editors. Spin dependent transport in magnetic nanostructures. London: Taylor & Francis; 2002. p. 1–46.

[8] Levy PM, Mertig I. Theory of giant magnetoresistance. In: Maekawa S, Shinjo T, editors. Spin dependent transport in magnetic nanostructures. London: Taylor & Francis; 2002. p. 47–112.

[9] Parkin SSP. Applications of magnetic nanostructures. In: Maekawa S, Shinjo T, editors. Spin dependent transport in magnetic nanostructures. London: Taylor & Francis; 2002. p. 237–78.

[10] Bass J. Giant magnetoresistance: experiment. In: Tsymbal EY, Žutić I, editors. Handbook of spin transport and magnetism. New York: CRC Press; 2012. p. 69–94.

[11] Tsymbal YY, Pettifor DG, Maekawa S. Giant magnetoresistance: theory. In: Tsymbal EY, Žutić I, editors. Handbook of spin transport and magnetism. New York: CRC Press; 2012. p. 95–114.

[12] Hall DA, Gaster RS, Wang SX. GMR spin-valve biosensors. In: Tsymbal EY, Žutić I, editors. Handbook of spin transport and magnetism. New York: CRC Press; 2012. p. 715–30.

[13] Gijs MAM, Bauer GEW. Perpendicular giant magnetoresistance of magnetic multilayers. Adv Phys 1997;46:285.

[14] Fert A, Campbell IA. Two-current conduction in nickel. Phys Rev Lett 1968;21:1190.

[15] Valet T, Fert A. Theory of perpendicular magnetoresistance in magnetic multilayers. Phys Rev B 1993;48:7099.

[16] Fert A, Bruno P. Interlayer exchange interactions in magnetic multilayers. In: Heinrich B, Bland JAC, editors. Ultrathin magnetic structures. vol. 2. Berlin: Springer Verlag; 1994. p. 82–117 [chapter 2.2].

[17] Edwards DM, Umerski A. Exchange coupling in magnetic multilayers. In: Kronműller H, Parkin S, editors. Handbook of magnetism and advanced magnetic materials. Fundamentals and theory, vol. 1. New York: John Wiley; 2007.

[18] Grünberg P. Light scattering from magnetic surface excitations. Prog Surf Sci 1985;18:1. Fassbender J, Nortemann F, Stamps RL, Camley RE, Hillebrands B, Giintherodt G, et al. Oscillatory interlayer exchange coupling of Co/Ru multilayers investigated by Brillouin light scattering. Phys Rev B 1992;46:5810.

[19] Barnaś J, Grünberg P. Spin waves in exchange coupled epitaxial double-layers. J Magn Magn Mater 1989;82:186.

[20] Parkin SSP, More N, Roche KP. Oscillations in exchange coupling and magnetoresistance in metallic superlattice structures: Co/Ru, Co/Cr, and Fe/Cr. Phys Rev Lett 1990;64:2304.

[21] Grünberg P, Demokritov S, Fuss A, Schreiber R, Wolf JA, Purcell ST. Interlayer exchange, magnetic transport and magnetic domains in Fe/Cr layered structures. J Magn Magn Mater 1992;104–107:1734.

[22] Unguris J, Celotta RJ, Pierce DT. Oscillatory magnetic coupling in Fe/Ag/Fe(100) sandwich structures. J Magn Magn Mater 1993;127:205.

[23] de Miguel JJ, Cebollada A, Gallego JM, Miranda R, Schneider CM, Schuster P, et al. Influence of growing conditions on the magnetic properties of fcc cobalt films: from monolayers to superlattices. J Magn Magn Mater 1991;93:1.

[24] Majkrzak CF, Cable JW, Kwo J, Hong M, McWhan DB, Yafet Y, et al. Observation of a magnetic antiphase domain structure with long-range order in a synthetic Gd-Y superlattice. Phys Rev Lett 1986;56:2700.

[25] Bloemen PJH, Johnson MT, van de Vorst MTH, Coehoorn R, de Vries JJ, Jungblut R, et al. Magnetic layer thickness dependence of the interlayer exchange coupling in (001) Co/Cu/Co. Phys Rev Lett 1994;72:764.

[26] de Vries JJ, Schudelaro AAP, Jungblut R, Bloemen PJH, Reinders A, Kohlhepp J, et al. Oscillatory interlayer exchange coupling with the Cu cap layer thickness in Co/Cu/Co/Cu(100). Phys Rev Lett 1995;75:4306.

[27] Bruno P, Chappert C. Oscillatory coupling between ferromagnetic layers separated by a nonmagnetic metal spacer. Phys Rev Lett 1991;67:1602.

[28] Barnaś J. Coupling between two ferromagnetic films through a non-magnetic metallic layer. J Magn Magn Mater 1992;111:L215.

[29] See, e.g.,Opitz J, Zahn P, Binder J, Mertig I. *Ab initio* calculation of the interlayer exchange coupling in Fe/Au multilayers: the role of impurities at the interface. Phys Rev B 2001;63:094418.

[30] Schad R, Potter CD, Belien P, Verbanck G, Moshchalkov VV, Bruynseraede Y. Giant magnetoresistance in Fe/Cr superlattices with very thin Fe layers. Appl Phys Lett 1994;64:3500.

[31] Barnaś J, Fuss A, Camley RE, Grünberg P, Zinn W. Novel magnetoresistance effect in layered magnetic structures: theory and experiment. Phys Rev B 1990;42:8110.

[32] Barnaś J, Fuss A, Camley RE, Walz U, Grünberg P, Zinn W. Layered magnetic structures: magnetoresistance due to antiparallel alignment. Vacuum 1990;41:1243.

[33] Dieny B, Speriosu VS, Metin S, Parkin SSP, Gurney BA, Baumgart P, et al. Magnetotransport properties of magnetically soft spin-valve structures. J Appl Phys 1991;69:4774.

[34] Kano H, Kagawa K, Suzuki A, Okabe A, Hagashi K, Aso K. Substrate temperature effect on giant magnetoresistance of sputtered Co/Cu multilayers. Appl Phys Lett 1993;63:2839.

[35] Petroff F, Barthelemy A, Hamzic A, Fert A, Etienne P, Lequien S, et al. Magnetoresistance of Fe/Cr superlattices. J Magn Magn Mater 1991;93:95.

[36] Mattson JE, Brubaker ME, Sowers CH, Conover M, Qiu Z, Bader SD. Temperature dependence of the magnetoresistance of sputtered Fe/Cr superlattices. Phys Rev B 1991;44:9378.

[37] Dieny B, Humbert P, Speriosu VS, Metin S, Gurney BA, Baumgart P, et al. Giant magnetoresistance of magnetically soft sandwiches: dependence on temperature and on layer thickness. Phys Rev B 1992;45:806.

[38] Takanashi K, Obi Y, Mitani Y, Fujimori H. Relationship between magnetoresistance and lattice uncertainty at the interface of sputtered Fe/Cr multilayer films. J Phys Soc Jpn 1992;61:1169.

[39] Schad R, Belien P, Barnaś J, Verbanck G, Potter CD, Gladyszewski G, et al. On the Fe thickness dependence of the giant magnetoresistance in epitaxial Fe/Cr superlattices. J Magn Magn Mater 1996;156:341.

[40] Belien P, Schad R, Potter CD, Verbanck G, Moshchalkov VV, Bruynseraede Y. Relation between interface roughness and giant magnetoresistance in MBE-grown polycrystalline Fe/Cr superlattices. Phys Rev B 1994;50:9957.

[41] Chen LH, Tiefel TH, Jin S, van Dover RB, Gyorgy EM, Fleming RM. Enhanced giant magnetoresistance in Fe/Cr multilayer films by Cr alloying of the Fe layers. Appl Phys Lett 1993;63:1279.

[42] Chaiken A, Prinz GA, Krebs JJ. Magnetotransport sudy of Fe-Cr-Fe sandwiches grown on ZnSe(100). J Appl Phys 1990;67:4892.

[43] Dieny B, Speriosu VS, Parkin SSP, Gurney A, Wilhoit DR, Mauri D. Giant magnetoresistance in soft ferromagnetic multilayers. Phys Rev B 1991;43:1297.

[44] Steren LB, Barthelemy A, Duvail JL, Fert A, Morel R, Petroff F, et al. Angular dependence of the giant magnetoresistance effect. Phys Rev B 1995;51:292.

[45] Barnaś J. Transport in magnetic layered structures: giant magnetoresistance. Acta Phys Pol A 1994;85:165.

[46] George JM, Pereira LG, Barthelemy A, Petroff F, Steren L, Duvail JL, et al. Inverse spin-valve-type magnetoresistance in spin engineered multilayer structures. Phys Rev Lett 1994;72:408.

[47] Renard J-P, Bruno P, Megy R, Bartenlian B, Beauvillain P, Chappert C, et al. Inverse magnetoresistance in the simple spin-valve system $Fe_{1-x}V_x$/Au/Co. Phys Rev B 1995;51:12821.

[48] Camley RE, Barnaś J. Theory of giant magnetoresistance effects in magnetic layered structures with antiferromagnetic coupling. Phys Rev Lett 1989;63:664.

[49] Sondheimer EH. The mean free path of electrons in metals. Adv Phys 1952;1:1.

[50] Trigui F, Velu E, Dupas C. Fuchs-Sondheimer theory as explanation of the magnetoresistance effects in multilayers Au/Co. J Magn Magn Mater 1991;93:421.

[51] Barthelemy A, Fert A. Theory of the magnetoresistance in magnetic multilayers: analytical expressions from a semiclassical approach. Phys Rev B 1991;43:13124.

[52] Johnson BL, Camley RE. Theory of giant magnetoresistance effects in Fe/Co multilayers: spin-dependent scattering from impurities. Phys Rev B 1991;44:9997.

[53] Levy PM, Zhang S, Fert A. Electrical conductivity of magnetic multilayered structures. Phys Rev Lett 1990;65:1643.

[54] Hood RQ, Falicov LM. Boltzmann-equation approach to the negative magnetoresistance of ferromagnetic-normal-metal multilayers. Phys Rev B 1992;46:8287.

[55] Hood RQ, Falicov LM, Penn DR. Effects of interfacial roughness on the magnetoresistance of magnetic metallic multilayers. Phys Rev B 1994;49:368.

[56] Dugaev VK, Litvinov VI, Petrov PP. Electric-current transmission through the contact of two metals. Phys Rev B 1995;52:5306.

[57] Litvinov VI, Dugaev VK, Willikens MMH, Swagten HJM. Current-in-plane magnetoresistance: an approach to boundary conditions. Phys Rev B 1997;55:8374.

[58] Sheng L, Xing DY, Wang ZD, Dong J. Quasiclassical approach to magnetotransport in magnetic inhomogeneous systems. Phys Rev B 1997;55:5908.

[59] Barnaś J, Baksalary O, Fert A. Angular dependence of giant magnetoresistance in magnetic multilayers. Phys Rev B 1997;56:6079.

[60] Zhang S, Levy PM. Enhanced temperature-dependent magnetoresistivity of Fe/Cr superlattices. Phys Rev B 1991;43:11048.

[61] Duvail JL, Fert A, Pereira LG, Lottis DK. Calculation of the temperature dependence of the giant MR and application to Co/Cu multilayers. J Appl Phys 1994;75:7070.

[62] Barnaś J, Baksalary O, Bruynseraede Y. Effects of interchannel transitions in the current-in-plane giant magnetoresistance. J Phys Condens Matter 1995;7:6437.

[63] Fert A. Two-current conduction in ferromagnetic metals and spin wave – electron collisions. J Phys Solid State Phys 1969;2:1784.

[64] Vedyayev A, Dieny B, Ryzhanova N. Quantum theory of giant magnetoresistance of spin-valve sandwiches. Europhys Lett 1992;19:329.

[65] Bulka B, Barnaś J. Parallel magnetoresistance in magnetic multilayers. Phys Rev B 1995;51:6348.

[66] Barnaś J, Bruynseraede Y. Influence of quantum size effect and interface roughness on the giant magnetoresistance in ultrathin magnetic layered structures. Europhys Lett 1995;32:167.

[67] Barnaś J, Bruynseraede Y. Electronic transport in ultrathin magnetic multilayers. Phys Rev B 1996;53:5449.

[68] Fishman G, Calecki D. Surface-induced resistivity of ultrathin metallic films: a limit law. Phys Rev Lett 1989;62:1302. Fishman G, Calecki D. Influence of surface roughness on the conductivity of metallic and semiconducting quasi-two-dimensional structures. Phys Rev B 1991;43:11581.

[69] Zahn P, Binder J, Mertig I. Impurity scattering and quantum confinement in giant magnetoresistance systems: a comparative *ab initio* study. Phys Rev B 2005;72:174425.

[70] Herper HC, Szunyogh L, Entel P, Weinberger P. Influence of the interlayer exchange coupling on the electric transport in Fe/Cr/Fe and Fe/Cr/T/Fe (T=Mn, V): an *ab initio* study. Phys Rev B 2003;68:134421.

[71] Erler F, Zahn P, Mertig I. Finite-size effects in giant magnetoresistance: an *ab initio* calculation. Phys Rev B 2001;64:094408.

[72] Zahn P, Mertig I, Richter M, Eschrig H. *Ab initio* calculations of the giant magnetoresistance. Phys Rev Lett 1995;75:2996.

[73] Tsymbal EY, Pettifor DG. Effects of band structure and spin-independent disorder on conductivity and giant magnetoresistance in Co/Cu and Fe/Cr multilayers. Phys Rev B 1996;54:15314.

[74] Pratt Jr. WP, Lee SF, Slaughter JM, Loloee R, Schroeder PA, Bass J. Perpendicular giant magnetoresistance of Ag/Co multilayers. Phys Rev Lett 1991;66:3060.

[75] Gijs MAM, Lenczowski SKJ, Giesbers JB. Perpendicular giant magnetoresistance of microstructured Fe/Cr magnetic multilayers from 4.2 to 300 K. Phys Rev Lett 1993;70:3343.

[76] Piraux L, Fert A, Schroeder PA, Laloee R, Etienne P. Large magnetothermoelectric power in Co/Cu, Fe/Cu and Fe/Cr multilayers. J Magn Magn Mater 1992;110:L247.

[77] Blondel A, Meier JP, Doudin B, Ansermet J-P. Giant magnetoresistance of nanowires of multilayers. Appl Phys Lett 1994;65:3019.

[78] Barnaś J, Fert A. Interface resistance for perpendicular transport in layered magnetic structures. Phys Rev B 1994;49:12835.

[79] Barnaś J, Fert A. Interfacial scattering and interface resistance for perpendicular transport in magnetic multilayers. J Magn Magn Mater 1994;136:260.

[80] Bauer GEW, Tserkovnyak Y, Huertas-Hernando D, Brataas A. Universal angular magnetoresistance and spin torque in ferromagnetic/normal metal hybrids. Phys Rev B 2003;67:094421.

[81] Brataas A, Bauer GEW, Kelly PJ. Non-collinear magnetoelectronics. Phys Rep 2006;427:157.

[82] Kovalev AA, Bauer GEW, Brataas A. Perpendicular spin valves with ultrathin ferromagnetic layers: magnetoelectronic circuit investigation of finite-size effects. Phys Rev B 2006;73:054407.

[83] van Son PC, van Kempen H, Wyder P. Boundary resistance of the ferromagnetic-nonferromagnetic metal interface. Phys Rev Lett 1987;58:2271.

[84] Bass J, Pratt Jr WP. Spin-diffusion lengths in metals and alloys, and spin-flipping at metal/metal interfaces: an experimentalist's critical review. J Phys Condens Matter 2007;19:183201.

[85] Brataas A, Nazarov YV, Bauer GEW. Spin transport in multi-terminal normal metal-ferromagnet systems with non-collinear magnetizations. Eur Phys J B 2001;22:99.

[86] Barnaś J, Fert A, Gmitra M, Weymann I, Dugaev VK. From giant magnetoresistance to current-induced switching by spin transfer. Phys Rev B 2005;72:024426.

[87] Kovalev AA, Brataas A, Bauer GEW. Spin transfer in diffusive ferromagnet-normal metal systems with spin-flip scattering. Phys Rev B 2002;66:224424.

[88] Shpiro A, Levy PM, Zhang S. Self-consistent treatment of nonequilibrium spin torques in magnetic multilayers. Phys Rev B 2003;67:104430.

[89] Urazhdin S, Loloee R, Pratt Jr WP. Noncollinear spin transport in magnetic multilayers. Phys Rev B 2005;71:100401(R).

[90] Gmitra M, Barnaś J. Correlation of the angular dependence of spin-transfer torque and giant magnetoresistance in the limit of diffusive transport in spin valves. Phys Rev B 2009;79:012403.

[91] Xiao JQ, Jiang JS, Chien CL. Giant magnetoresistance in multilayer magnetic systems. Phys Rev Lett 1992;68:3749.

[92] Berkowitz AE, Mitchell JR, Carey MJ, Young AP, Zhang S, Spada FE, et al. Giant magnetoresistance in heterogeneous Cu-Co alloys. Phys Rev Lett 1992;68:3745.

[93] Dietl T. A ten-year perspective on dilute magnetic semiconductors and oxides. Nat Mater 2010;9:965.

[94] Mattana R, Elsen M, George JM, Jaffres H, Nguyen Van Dau F, Fert A, et al. Chemical profile and magnetoresistance of $Ga_{1-x}Mn_xAs/GaAs/Al/GaAs/Ga_{1-x}Mn_xAs$ tunnel junctions. Phys Rev B 2005;71:075206.

[95] Tsukagoshi K, Alphenaar BW, Ago H. Coherent transport of electron spin in a ferromagnetically contacted carbon nanotubes. Nature 1999;401:572.

[96] Hill E, Geim AK, Novoselov K, Schedin F, Blake P. Graphene spin valve devices. IEEE Trans Magn 2006;42:2694.

[97] Cho S, Chen Y-F, Fuhrer MS. Gate-tunable graphene spin valve. Appl Phys Lett 2007;91:123105.

[98] Krompiewski S. Theoretical studies of spin-dependent electronic transport in ferromagnetically contacted graphene flakes. Phys Rev B 2009;80:075433.

[99] Yazyev OV. Emergence of magnetism in graphene materials and nanostructures. Rep Prog Phys 2010;73:056501.

[100] Zberecki K, Wierzbicki M, Barnaś J, Swirkowicz R. Thermoelectric effects in silicene nanoribbons. Phys Rev B 2013;88:115404.

[101] Chappert C, Fert A, Nguyen van Dau F. The emergence of spin electronics in data storage. Nat Mater 2007;6:813.

[102] Yuasa S, Nagahama T, Fukushima A, Suzuki Y, Ando K. Giant room-temperature magnetoresistance in single-crystal Fe/MgO/Fe magnetic tunnel junctions. Nat Mater 2004;3:868.

[103] Kado T. Large room-temperature inverse magnetoresistance in tunnel junctions with a Fe_3O_4 electrode. Appl Phys Lett 2008;92:092502.

CHAPTER 10

Planar Magnetic Devices for Signal Processing in the Microwave and Millimeter Wave Frequency Range

Zbigniew Celinski, I.R. Harward, N.R. Anderson, Robert E. Camley
Center for Magnetism and Magnetic Nanostructures, University of Colorado at Colorado Springs, Colorado Springs, CO, USA

1 INTRODUCTION

Magnetic materials have long been used as elements of signal processing in the microwave range. These devices include band-pass filters, notch filters, phase shifters, and signal-to-noise enhancers. Microwave devices are widely used in radar systems and in both military and civilian communications systems. In this chapter, we explore some recent work on ultra-small tunable signal processing devices which are planar in structure and can be integrated with high-frequency electronics. As we shall see, the small size of these structures allows for some new implementations of current devices.

There are several reasons why it is important to reexamine microwave devices based on magnetic materials. First, there is an increasing demand for devices which will operate at higher frequencies (10–120 GHz) and this is not a range that is appropriate for the ferrite materials typically used today because it would require very large magnetic fields that must be produced by heavy permanent magnets. Second, new growth methods have allowed the creation of high-quality thin films and structured magnetic materials. In particular, metallic magnets such as Fe or Permalloy can be made small enough that eddy currents do not play a large role. As we will see, this allows the use of higher magnetization materials, with a corresponding increase in operational frequency. Third, it is important to reduce the size and weight of devices, particularly for use in satellite systems, and this requires on-wafer devices which can be directly integrated with the electronic circuits. Finally, higher frequency operation also inspires one to look at new materials such as hexagonal ferrites or antiferromagnets, where the typical resonance frequencies are much higher, 40–80 GHz for the hexagonal ferrites and in the terahertz range for the antiferromagnets. This range of new materials also encourages one to investigate new device structures where the electromagnetic fields are strongly contained within

Magnetism of Surfaces, Interfaces, and Nanoscale Materials. http://dx.doi.org/10.1016/B978-0-444-62634-9.00010-2

a very small region, which can lead to high powers and nonlinear effects. The new materials also include multilayer magnetic films which have very high susceptibilities (compared to ferrite materials in the gigahertz range) and can therefore be used in inductors at moderate frequencies.

Tunable filters based on the ferrimagnetic dielectric Yttrium Iron Garnet (YIG) [1] are a well-established technology that works well at lower frequencies in the 100 MHz to the low gigahertz range. Band-stop or notch filters, for example, rely on ferromagnetic resonance (FMR) to absorb microwave power at the FMR frequency. This frequency, f, is set by material properties, such as saturation magnetization, M_s, anisotropy fields, H_a, the gyromagnetic ratio, γ, and the magnitude of an applied field, H. If, for example, the applied field is parallel to the surface of a ferromagnetic thin film and along the easy axis, the resonance frequency is given by

$$f = \gamma\sqrt{(H + H_a)(H + H_a + 4\pi M_s)} \tag{1}$$

and therefore the resonance frequency can be adjusted or tuned with an electromagnet. The maximum field produced by the electromagnet determines the upper limit for the band-stop frequency. It is instructive to compare FMR frequencies for different materials. For YIG with a low $4\pi M_s = 1.75$ kG and no anisotropy, one finds that the operating frequency is only 4.8 GHz at an applied field of 1 kOe. An applied field of over 11 kOe is necessary to reach frequencies of about 35 GHz. Such large fields are incompatible with devices of a limited size since substantial electromagnets as well as large power consumption are required.

An alternative is to use a material with a high M_s such as Fe [2–4]. We compare the operating frequencies of several different materials in Table 1. While Fe has a much higher resonance frequency for the same applied field, its conductivity can lead to eddy current losses at microwave frequencies. However, structures utilizing thin Fe films—typical thicknesses are below 200 nm—minimize conduction loss while still producing significant attenuation at the band-stop frequency, as discussed below.

A schematic of a typical microstrip structure is shown in Fig. 1a. The device is bounded by conductors on both sides, with a dielectric film and a magnetic material in between. The thickness of the dielectric, D, is typically in the range of 3–6 μm. The thickness of the ferromagnet, d, is about 0.2 μm. The width of the microstrip is usually in the range of 5–15 μm. For structures with thin dielectric films the only propagating mode is a transverse magnetic (TM) mode with oscillating E and H fields, as seen in Fig. 1a. Because of the small dimensions, the fields in both the magnetic material and the dielectric can be quite large and the magnetic film, even if very thin, can have a significant impact

Table 1 Saturation Magnetization, Anisotropy Fields, and Operational Frequency of Notch Filters for YIG, Permalloy, Fe, and BaM

Material	$4\pi M_s$ (kG)	H_a (kOe)	Frequency (GHz) at $H = 1$ kOe	Frequency (GHz) at $H = 5$ kOe
YIG	1.75	0	4.8	16.9
Permalloy	10.0	0	9.7	25.2
Fe	21.5	0.5	17.2	35.6
BaM *c*-axis out-of-plane	4.2	16.4	37.7	48.8
BaM *c*-axis in-plane	4.2	16.4	54.3	65.9

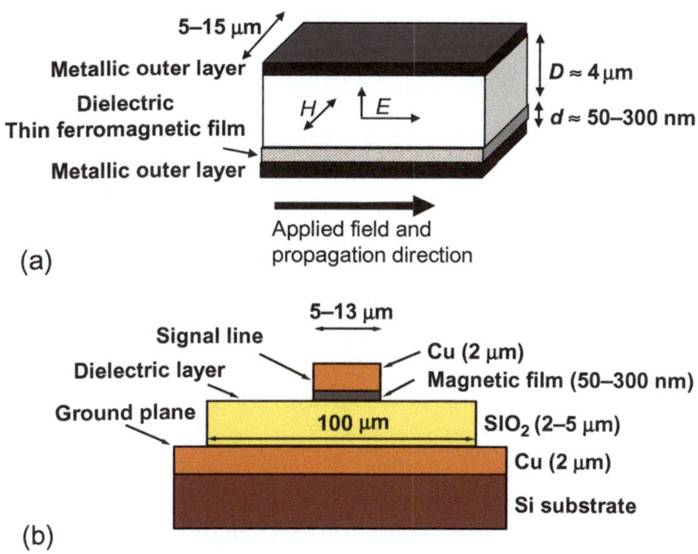

FIGURE 1

(a) Schematic diagram of microstrip structure showing the dimensions, the oscillatory *E* and *H* field directions, and the applied field. The propagation in this case is parallel to the applied field. The small cross-section can lead to a large energy density in the microstrip and thus to large oscillating *E* and *H* fields. (b) Example of a cross-section of the experimental structure.

on the propagation characteristics. Figure 1b shows the cross-section of a device as typically constructed.

Early attempts at producing Fe-film-based structures succeeded in making filters with high band-stop frequencies and low broadband loss [3,4]. However, because the thickness of the dielectric was large, several hundred micrometers, the maximum attenuation only reached about 4–5 dB/cm. This can be understood through a simple theoretical calculation [5,6], which shows that the power attenuation for the notch filter in Fig. 1 decays exponentially with distance, i.e., $P(z) = P_o e^{-\alpha z}$ The attenuation factor is given by [7]

$$\alpha = 8\pi\sqrt{\varepsilon}\left(\frac{d}{D}\right)\left(\frac{\omega}{c}\right)\mathrm{Im}(\chi_{xx}). \tag{2}$$

Here, χ_{xx} is the dynamic magnetic susceptibility of the magnetic material which includes the contribution of the demagnetizing field. χ_{xx} is a function of frequency and applied field, and its imaginary portion generally has a peak at the position of the FMR. The permittivity of the dielectric is ε, c is the speed of light, and ω is the angular frequency of the propagating electromagnetic wave. From Eq. (2), the attenuation is inversely proportional to the thickness of the dielectric. Based on this work, we have constructed microstrip band-stop filters with a much higher attenuation [8–10].

We note that there are other approaches to signal processing in the microwave range. Noise suppression, for example, can be accomplished by using low-conductivity metallic materials with lower M_s values [11–13]. These materials are primarily useful in the 1–10 GHz range. A higher operational frequency with low insertion loss can also be achieved by using a composite of metallic ferromagnetic nanowires in a polycarbonate matrix [14–16]. High-frequency circulators which can operate at zero DC fields have also been fabricated using ferromagnetic nanowires [17]. Magnetic nanowires separated by nonmagnetic material are used to exploit the shape anisotropy of the nanowires to enhance the bandwidth of a microwave filter [18]. The use of magnetic patterned materials [19,20] or materials with an internal domain structure [21,22] and layered materials [23] has also been explored. An interesting variation in the layered structures is the construction of a negative index material using YIG and copper films [24].

Of course, there are also some nonmagnetic approaches to signal processing at high frequencies. Devices using ferroelectrics [25,26] and MEMS devices [27,28] are promising candidates. A review of some of these devices is given in Ref. [29]. Ferrite–ferroelectric heterostructures are used to build electrically tunable band-pass filters in the gigahertz frequency range [30].

Here, we present a review of experimental and theoretical results for microscopic, on-wafer signal processing devices that operate in the 10–70 GHz range. This chapter is organized as follows. Section 2 gives information on the growth and fabrication of the devices and the dynamic characterization. Section 3 describes several devices based on ferromagnetic metals, including notch filters, local band-pass filters, phase shifters, and a signal-to-noise enhancer that is based on the nonlinear dynamic motion of the magnetization. Section 4 examines devices that use a planar geometry but which use materials other than metallic ferromagnets. These materials include hexagonal ferrites, antiferromagnets, and liquid crystals. As we will see, each of these materials has advantages and disadvantages compared to the ferromagnetic metals. A summary is presented in Section 5.

2 FABRICATION AND CHARACTERIZATION METHODS

We used GaAs(001) and Si substrates and sputtering techniques to construct our devices. The microstrips were grown using the following procedure. After cleaning the GaAs substrate in an ultrasonic bath, we annealed it at 200 °C inside a vacuum chamber. All the depositions were done at room temperature and at an Ar pressure of 2–3 mTorr during sputtering. The chamber background pressure was maintained around $\sim 2 \times 10^{-7}$ Torr.

A typical growth scheme is as follows. First, a 5-nm-thick Ti layer is deposited for good adhesion to the substrate. This is followed by a 2-μm-thick Ag (or Cu) layer, which is used as a ground plane for our device. We dust the Ag (Cu) surface with Ti to increase adhesion of the dielectric layer. The next sequence of deposition is made through a shadow mask. We typically deposit a 3- to 4-μm-thick SiO_2 film as a dielectric layer by employing an e-gun source and/or by rf sputtering. We determined that films grown by rf sputtering generally have much higher quality than films grown by e-beam. Then we deposit ~100 nm of Fe or Permalloy as the magnetic layer. At the end, we deposit a 2-μm-thick Ag (or Cu) layer, which is used as a signal line for the device. At this point, a portion of the sample is removed for FMR measurements. We pattern the films by photolithography and then dry etch to obtain the required strip widths and lengths for our devices.

The device characterization was done using a vector network analyzer along with a microprobe station. Noise, delay due to uncompensated transmission lines connectors, its frequency dependence, and cross talk which occurred in measurement data have been taken into account by performing through-reflect-line calibration using NIST Multical® software [31]. A DC bias magnetic field is applied along the length of the microstrip line or perpendicular to the plane of the sample. Either geometry ensures the FMR condition, because the rf magnetic field and the DC magnetic field are perpendicular to each other. The widths of the signal lines are 5–24 μm and the length of the device is 2–6 mm. The microstrip band-stop filters were designed for a 50 Ω characteristic impedance.

3 PLANAR DEVICES USING FERROMAGNETIC METALS

3.1 Notch Filters

As a first example of devices based on metallic ferromagnets, we present results for a tunable notch filter using a single Permalloy film as the active part of the device [32]. Figure 2 shows a plot of transmission as a function of frequency for the Permalloy (Py)-based microstrips. The frequency of the notch is clearly tunable by adjusting the external magnetic field. Furthermore, the tuning range is quite large, spanning the range 5–22 GHz within a single device by adjusting the external magnetic field from 0 to 4 kOe. The insertion loss over most of the region is of the order of 2–3 dB, while the power attenuation is close to a value of 60 dB/cm. The width of the attenuation dip, measured at 3 dB above the minimum (i.e., half-maximum) becomes distinctly narrower at higher frequencies. For example, it is about 0.4 GHz for the dip at 20 GHz compared to a width of 0.82 GHz at 4.3 GHz. This narrowing of the width of the attenuation peak is consistent with our theoretical results, as seen in the inset. In fact, this behavior of the frequency linewidth decreasing as the frequency increases is a common feature in these systems. The reasons for this are given elsewhere [33].

Figure 3 shows the results for a notch filter using an Fe/Cu multilayer instead of Permalloy. In comparison to the results for a filter based on Py, one notices here that the typical frequency range is substantially higher, 12–29 GHz, using the same fields as for the Py device, i.e., up to 4 kOe. This is to be expected based on

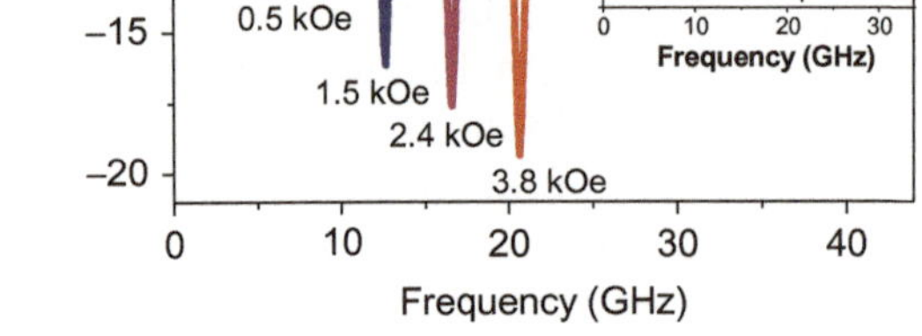

FIGURE 2

Transmission parameter S_{12} in the Permalloy-based notch filter as a function of frequency for different applied magnetic fields. Inset shows theoretical results. The thickness of the Permalloy film is 350 nm, the length is 3 mm, and the width is 18 μm. The SiO_2 thickness is 4.5 μm.

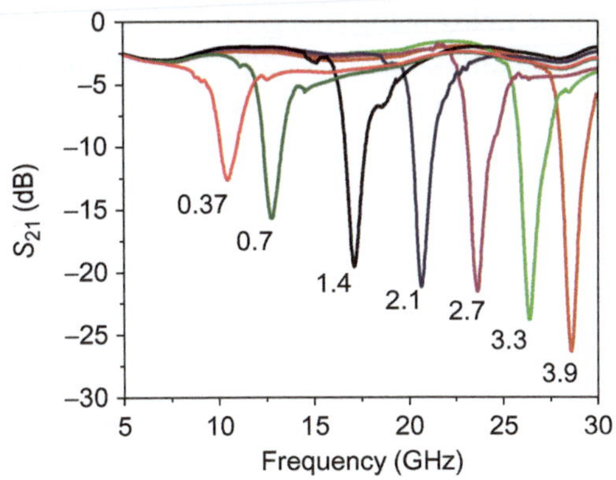

FIGURE 3

Transmission response of a 3.3-mm-long and 9-μm-wide microstrip band-stop filter for different applied fields. The magnetic film is [Fe(5 nm)/Cu(0.8 nm)]$_{20}$ (total Fe thickness = 100 nm). The SiO_2 film has a thickness of 3 μm. The numbers in the figure show the applied field in kOe and the motion of the notch with field. The Fe/Cu multilayer has a considerably narrower linewidth than a pure Fe film grown by sputtering. This leads to relatively narrow transmission dips.

the higher value of M_s found in Fe, as discussed earlier. Of course, a single sputtered film of Fe also would have this increase in the operational frequency; however, the Fe/Cu multilayer can have a significantly lower FMR linewidth than a pure Fe film, resulting in narrower absorption notches [34].

There are a number of methods that can be used to increase the operational frequency of these devices. These include:

(1) Shaping the magnetic element to produce internal fields [32]. As the magnetization precesses, dynamic magnetic poles are generated at the surfaces and sides of the ferromagnetic material. This leads to dynamic demagnetizing fields which can influence the precession frequency. The theoretical resonance frequency for a shaped magnetic element is calculated from the following resonance condition [35]:

$$f = \gamma\sqrt{(H + H_a + (N_y - N_z)M_s)(H + H_a + (N_x - N_z)M_s)}, \tag{3}$$

where N_x, N_y, and N_z are the demagnetization factors which depend on the geometry of the magnetic material. For a long bar with a square cross-section, Eq. (3) reduces to $f = 2\pi\gamma M_s$ at very low applied fields if the anisotropy field, H_a, is negligible. In principle, one can achieve operational frequencies above 30 GHz in Fe-based devices for very small magnetic fields using this geometry. Nonetheless, there may be additional difficulties, such as eddy current damping, in large structures with square cross-sections. In real devices we have obtained values of about 20 GHz near zero applied field [36].

(2) Exchange coupling one magnetic material to another produces internal fields, which can raise the operational frequency. For example, the resonance frequency in very thin Fe films coupled to SmCo films can be increased to 70–85 GHz, for external magnetic fields in the 0–6 kOe range [37]. Our theoretical calculations [38] have shown that a multilayer made up of repeated units of Fe/SmCo could be used in notch filters and band-pass filters which would operate in the 50–100 GHz range. Initial experimental data on a $[NiFe/SmCo]_{15}$ multilayer shows an upshift of about 10–15 GHz in the operational frequency compared to the Permalloy alone [39].

(3) The introduction of an effective anisotropy in the magnetic material. For example, a miscut GaAs substrate (2.5° off from a (001) direction) can create a strong uniaxial anisotropy (0.47 kOe) even for sputtered Fe films with thicknesses of the order of 1 μm [10]. We note that this result is for a coplanar waveguide geometry and it might be difficult to induce this effect in a microstrip geometry because the large thickness of the ground plane could minimize the influence of the miscut substrate.

3.2 Local Band-Pass Filter

A local band-pass filter can be created with two ferromagnetic metals. In this case one uses a structure similar to that in Fig. 1, but with an Fe film on the bottom of the dielectric and a Py film directly above the dielectric as part of the signal line. The different values of saturation magnetization in Fe ($4\pi M_s = 21.5$ kG) and Py ($4\pi M_s = 10$ kG) result in different resonance frequencies. This leads to two frequency regions where propagation is reduced. The frequency between the two transmission dips creates a local band-pass region with adjacent frequencies strongly attenuated [40].

The transmission as a function of frequency for a local band-pass filter is shown in Fig. 4. This device used a 2-μm-thick Ag layer as a ground plane. In Fig. 4a and b, the first magnetic layer is 250 nm of Py followed by a 4-μm-thick SiO_2 film, and 170 nm of Fe. At the end, a 2-μm-thick Ag layer was deposited, which was again used as the signal line for the structure. As discussed above, there is a local band-pass region with a distinct attenuation region on each side. The position of the notches on either side of the pass band occurs at the frequencies given by the FMR condition and is tunable with the external field. Figure 4b shows the experimental S_{21} response for the same structure at an applied field of 4 kOe. Clearly,

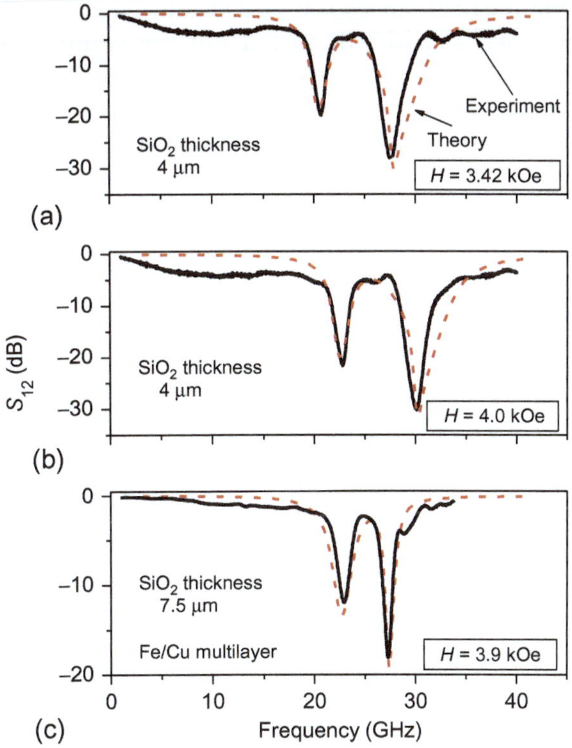

FIGURE 4

Transmission response of a local band-pass filter for different applied fields. In (a) and (b), the structure has 240 nm of Py, a 4-μm-thick SiO_2 film, and 170 nm of Fe. (c) Results for a filter with a Py thickness of 300 nm, an SiO_2 thickness of 7.5 μm, and with an Fe/Cu magnetic element [Fe 5 nm/Cu 0.8 nm]$_{12}$. The experiment is shown as a solid line and the theory is the dashed curve. In all cases, the structure is 3.3-mm long and the microstrip has a width of 18 μm.

the band-pass region has moved, almost as a single unit, to higher frequencies. We emphasize that these are initial results on such structures. In real devices, one would want lower insertion loss in the band-pass region and a sharper transition from the band-pass region to the rejection regions.

There are a number of variations which could be introduced into the local band-pass filter. For example, the width of the band-pass region may be adjusted by choosing different magnetic materials. This can be seen in Fig. 4c, which shows results for a local band-pass filter constructed using a 300-nm Py film and an Fe/Cu multilayer. The effective magnetization in the Fe/Cu multilayer is lower than that of pure Fe, and as a result the band-pass region is narrower.

A nonuniform biasing field can also be used to adjust the properties of this local band-pass filter. Such a field will spread out the width of the absorption dips because the different fields will cause the FMR to occur at slightly different frequencies. This spreading will reduce the width of the band-pass region and increase the width of the band-reject regions neighboring the band-pass region.

3.3 Phase Shifters

The same microstrip devices can also produce a phase shift because the magnetic material influences both the real part (wavelength) and the imaginary part (attenuation) of the propagation wavevector. We have characterized the phase shift as a function of the magnetic field for a microstrip with an Fe layer that is grown by molecular beam expitaxy [9]. Figure 5 shows the performance for the Fe-based device at two different frequencies. The attenuation of the signal in these devices, for each frequency, is below 3 dB. The results measured in the Fe-based microstrip are of the order of 360 deg/cm with an applied field of 1 kOe. The observed phase shift for the Permalloy-based devices was smaller, of the order of 100 deg/cm, again with a magnetic field of 1 kOe [9]. Phase shifters with other materials are also possible, as will be discussed later in this chapter.

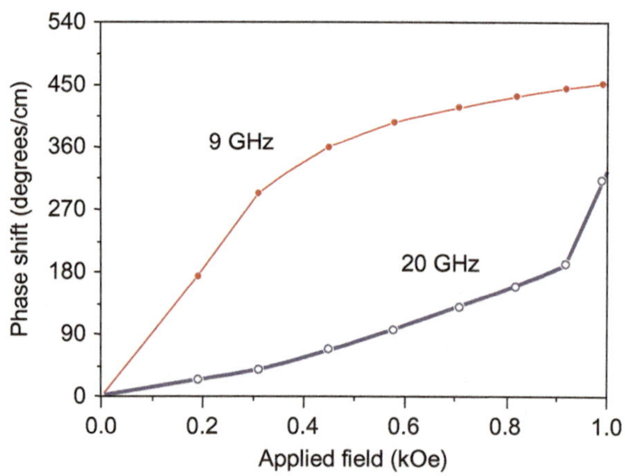

FIGURE 5

Relative changes in phase of transmitted microwaves through an Fe-based microstrip versus applied magnetic field for two frequencies. The Fe film (200 nm) was grown with molecular beam epitaxy in this case, and the SiO_2 film (4 μm) was grown with e-beam evaporation. The width of the microstrip is quite large, about 100 μm. Additional details are given in Ref. [9].

3.4 Nonlinear Devices

Up to this point all the discussion has been about devices that work in the linear regime. In terms of the theory, this means that the dynamic magnetization components (typically these are the transverse components $M_x(t)$ and $M_y(t)$) occur only in the linear form in the Landau–Lifshitz equation of motion or in Maxwell's equations. However, the Landau–Lifshitz equation is inherently nonlinear and is linearized only in the limit of small precession angles. The nonlinear magnetic behavior has a number of typical features:

(1) The frequency of a mode now depends on the amplitude of that mode
(2) A driving field at one frequency can generate modes at multiple frequencies
(3) At strong enough driving frequencies, one can get a chaotic response.

The physics of nonlinear magnetization dynamics has been covered elsewhere in this book (see the chapter by K. Livesey), and in this section, we concentrate on possible device applications. Typical devices [41,42] include the following:

(1) Signal-to-noise enhancer
(2) Power limiter
(3) Upshifting and downshifting frequency converters
(4) Nonlinear phase shifter.

In fact, many of the nonlinear devices listed above have been investigated over the past 30 years using YIG. As noted earlier, YIG has the disadvantage of requiring large bias fields for high-frequency operation. In addition, the lack of thermal stability with YIG has hindered practical applications. The use of ferromagnetic metal films, with higher operational frequencies and better thermal stability, may allow more widespread implementation of these devices. We do note, however, that a signal-to-noise enhancer based on YIG film technology has recently been investigated and evaluated as practical for noise reduction in direct broadcast satellite systems [43].

In order to have a nonlinear response, one needs large precession angles for the magnetization. This can be accomplished in two ways: In a material with low damping, such as YIG, one can drive the system near the resonance frequency. Because the damping is small, the response near resonance will be large, leading to large values for the transverse magnetization components. In materials with larger damping, such as ferromagnetic metals, one must use large oscillating magnetic fields to produce a large response.

As pointed out in the introduction, the small size of a microstrip waveguide concentrates both the oscillating E and H fields. One can make a simple estimation of the strength of the fields in the following way. Imagine we put a

1 W signal through a microstrip structure with a cross-sectional area A. The time-averaged Poynting vector is given by

$$\langle S\rangle = \frac{\text{Power}}{\text{Area}} = \frac{c}{8\pi}\vec{E}\times\vec{H}^{*}. \tag{4}$$

If we assume the wave is concentrated in the cross-sectional area and is approximately given by a plane wave in this region, one has the relation

$$|E| = |H|\sqrt{\frac{\mu}{\varepsilon}}. \tag{5}$$

Using the last two relations and solving for the applied field, one has

$$H = \sqrt{\frac{8\pi}{c}\left(\frac{\text{Power}}{\text{Area}}\right)\sqrt{\frac{\varepsilon}{\mu}}}. \tag{6}$$

Assuming a small cross-section of 4 μm × 4 μm and a 1 W input signal, along with typical values of $\varepsilon = 3$ and $\mu = 1$, one gets an oscillating H field of about 300 G, which is easily large enough to induce large-angle precession dynamics in metals. One can get similarly large values in a coplanar waveguide with a small spacing between the signal line and the ground plane.

We now turn to studies of individual nonlinear devices. Our first example is that of a signal-to-noise enhancer [44]. In Fig. 6, we show the FMR absorption

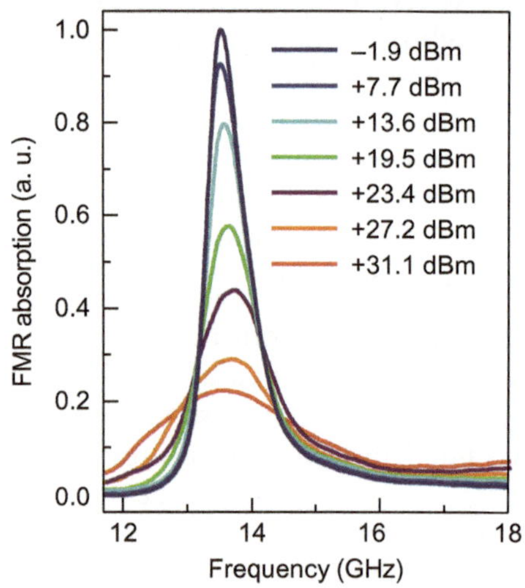

FIGURE 6

Normalized power ratio for a microstrip waveguide with a 70-nm Permalloy film. As the power is increased, the relative absorption decreases, thus creating a signal-to-noise enhancer, which allows strong signals to pass and weak signals to be absorbed. The SiO_2 thickness is 2.8 μm. The applied field is 2 kOe.

in a microstrip device as a function of frequency for various power levels. The structure is the same as for a notch filter. The SiO_2 thickness is 2.8 μm and the ferromagnet is a 70-nm-thick Py film. The applied field is 2 kOe.

The different power levels correspond to different strengths of the oscillating rf field within the magnetic film. As we have emphasized earlier, the small structure concentrates the electric and magnetic fields in a very small region. This results in some astonishing values for the oscillating magnetic field, h_{rf}. We estimate that at the lower power levels the magnitude of h_{rf} is about 0.1 Oe, while at the highest levels the value can be as high as 500 Oe. (The precise value depends on the details of the structure, particularly the width of the microstrip line and the thickness of the dielectric below it.) With such a large value for h_{rf} the motion of the magnetization becomes highly nonlinear, and this is reflected in the power absorption shown in Fig. 6, where we plot the power ratio, absorbed power/input power, for different input power levels. At low powers one sees a strong absorption at the FMR frequency. As the power is increased, the power ratio decreases substantially. (This occurs because the absorbed power is proportional to $\vec{m} \cdot \vec{h}$ while the input power is proportional to h^2. Here the use of lower case indicates the transverse components. At low power, m is proportional to h, and the ratio is independent of the input power. But at high powers the transverse component of the magnetization reaches a maximum and the ratio decreases as $1/h$.) Thus, if the input to the device consists of low-power noise and a high-power signal, the high-power signal will be preferentially transmitted, effectively creating a signal-to-noise enhancer. We note that similar devices have been made previously with YIG as the active element [45], but these generally operate at lower frequencies, as noted earlier.

As a second example of nonlinear magnetic devices, we will consider frequency converters. An interesting early paper [46] using a thin film of Py showed that a single input signal could generate a variety of output signals. In this case, a 25-nm Py film was grown on an Si substrate. The Py film was covered by a 500-nm SiO_2 film and an asymmetric coplanar antenna carrying a microwave current was used to generate the magnetostatic wave in the Py. A second antenna was used as a receiver. The two antennas were separated by about 4 μm. First with an input frequency of 0.5 GHz, multiple output harmonics were seen at 1.0, 1.5, 2.0, and 2.5 GHz. In addition, a wave at half the input frequency could also be observed and amplified.

Somewhat later it was shown both theoretically and experimentally [47], using the microstrip waveguide structure, that one could mix two input frequencies at f_1 and f_2 to get a variety of output frequencies. The output frequencies included the combinations $2f_1 \pm f_2$, $2f_2 \pm f_1$, $3f_1$, $3f_2$, etc. At higher powers a frequency comb was generated. Some of the results are shown in Fig. 7. These

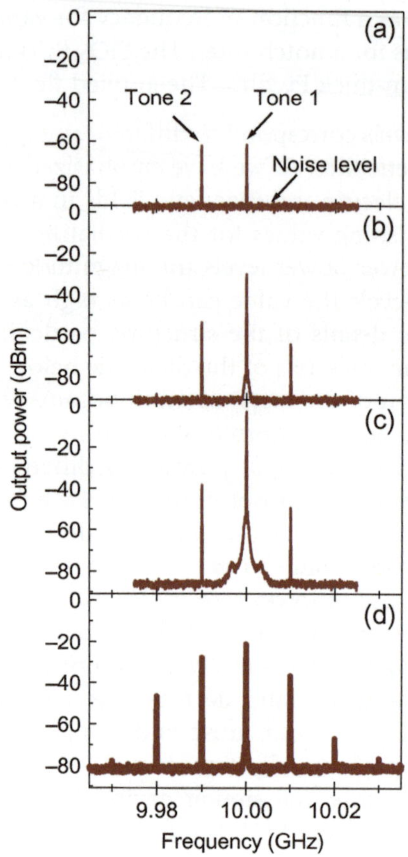

FIGURE 7

(a) Output power measured by spectrum analyzer when the input is two small signals, both at −29 dBm, at frequencies $f_2 = 9.99$ and $f_1 = 10.0$ GHz. (b) Output power when the input of f_1 has been increased to −2 dBm and f_2 has an input of power of −29 dBm. Note the appearance of a new signal at $f_3 = 10.01$ GHz and amplification of the f_2 signal. (c) Output power when the input of f_1 is +10 dBm and f_2 has an input power of −29 dBm. Note the further amplification of the f_2 and f_3 signals and the appearance of a frequency comb. From Ref. [47].

combinations were shown to arise because of a cubic nonlinear term in the magnetic equations of motion [48].

Based on the cubic nonlinearity, it was somewhat surprising that second harmonic waves could be generated. It was subsequently shown [49] that a quadratic nonlinearity would arise if there was even a small misalignment between the applied field and the direction of propagation (the long axis of the microstrip waveguide). This lead to a significant enhancement of all the even harmonics. We illustrate this in Fig. 8, showing the theoretical results

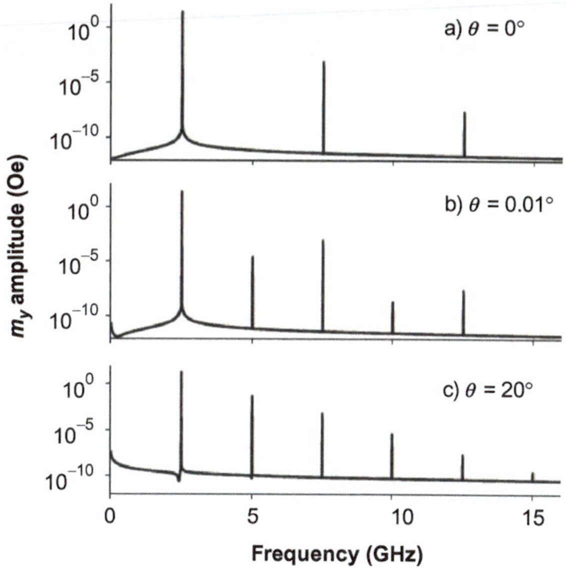

FIGURE 8
Results of a macrospin calculation for the amplitude of m_y as a function of frequency for different angles between the waveguide and the magnetic field. The input wave is at a frequency $f_1 = 2.5$ GHz and is driven by an oscillating field with amplitude $h_y = 10$ Oe. The static field is 220 Oe. Note that even a very small change in angle away from $\theta = 0$ produces a significant signal at the second and fourth harmonics. From Ref. [49].

for how the amplitude of a macrospin depends on the angle of the waveguide with respect to the magnetization direction.

This idea was used in a later study [50] showing one could efficiently generate second harmonic signals using a coplanar waveguide with the magnetization in a 30-nm Py film directed at an angle with respect to the waveguide. In that work it was found that the ratio of the second harmonic power to the first harmonic power was about three orders of magnitude higher than that demonstrated in ferrites!

3.5 Nonreciprocal Devices

A nonreciprocal device is one where reversing the direction of propagation results in a wave with different properties. For example, the frequency could be different, i.e., $\omega\left(+\vec{k}\right) \neq \omega\left(-\vec{k}\right)$, where ω is the frequency and $\vec{k}$ is the propagation wavevector. Similarly, the attenuation can depend on the direction of propagation. The two most common nonreciprocal devices are isolators and circulators. An isolator allows a wave to pass through a guide in one direction, but blocks the wave passing through in the reverse direction. Circulators are generally three-port devices, where a wave that enters at port 1 is directed to port 2 and the wave entering at port 2 is directed to port 3, etc.

Isolators and circulators require either magnetic or superconducting components. However, superconductor-based devices require energy for their operation, and operate only at low power, not exceeding 1 mW, and at low temperatures. In addition, magnetic nonreciprocal devices have desirable properties such as being tunable via an applied magnetic field.

Nonreciprocal [51] devices based on magnetic components use the gyrotropic properties of magnetic materials. Magnetization precession has a natural gyrotropy, i.e., a definite sense of precession with respect to the direction of the internal magnetic field. As a result, the magnetic susceptibility of a magnetic medium is substantially different for left-handed and right-handed polarizations. This property is related to the axial nature of the magnetization and not present in any nonmagnetic materials or materials without an applied magnetic field. Thus, the properties of a magnetic medium can be different for electromagnetic waves propagating in opposite directions and, therefore, it is possible to realize nonreciprocal devices.

Both the Y-junction circulator and the isolator operate by breaking the symmetry of the waveguide. For the circulator an external magnetic field is oriented perpendicular to the three ports. The gyrotropic property of the magnetic material forces an electromagnetic wave to travel in a preferred direction around the circulator. By blocking off one of the three ports with a matched termination the wave propagating in one direction is damped while the other direction is not. In the isolator a number of different geometries are possible as long as symmetry is broken by the magnetic material. There are a number of designs of both circulators [52,53] and isolators [54–56].

Early development of nonreciprocal devices was mostly limited to ferrites rather than metallic, magnetic materials [57,58]. Metallic magnets allow eddy currents which produce insertion losses in both directions. These losses, typically above 5 dB, are too large for modern devices. Ferrites drastically reduce insertion losses and are therefore more desirable materials to work with. However, as pointed out earlier, ferrite devices generally require large magnetic fields for operation at higher frequencies. One solution to these issues is to use nanostructured metallic devices.

Recent advancements in fabrication techniques have allowed the production of on-wafer isolators. These are significantly smaller than previous devices and allow fields to be confined to regions very near the signal line, allowing the use of smaller fields while still producing significant effects. One example of an isolator using metallic magnetic elements [59] has a series of ferromagnetic

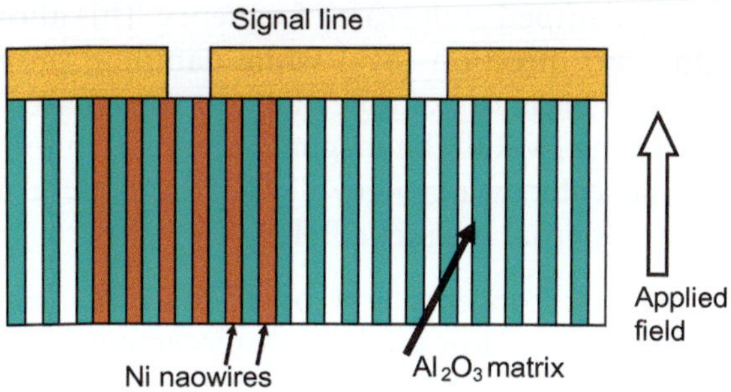

FIGURE 9
Cross-section of a coplanar isolator with metallic ferromagnetic nanowires. The nanowires are used to decrease losses to eddy currents. The magnetic nanowires are placed off-center in the waveguide to allow isolation.

nanorods positioned to one side of the signal line in a coplanar waveguide, as shown in Fig. 9. For this device, there is a porous Al_2O_3 structure which contains ferromagnetic (FM) nanowires on one side. Above this layer lies a Cu signal line as the topmost layer. By using the nanowires for the ferromagnetic material eddy currents are reduced. In addition, using Ni as the ferromagnetic material increases the saturation magnetization (compared to typical ferrites) and therefore increases the operational frequency of the device.

Theoretical and experimental results are shown for this device in Fig. 10. The S_{21} transmission is significantly damped at around 23 GHz, whereas the S_{12}

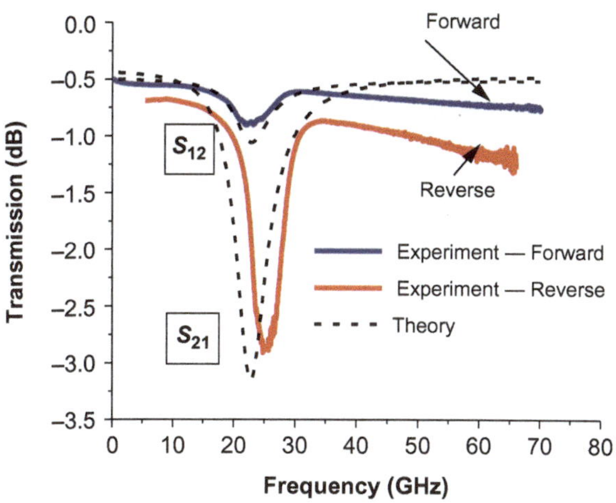

FIGURE 10
Theoretical and experimental signal transmission for forward and reverse (S_{12} and S_{21}) directions for the isolator shown in Fig. 9. After Ref. [59].

transmission is lightly damped at the same frequency. This allows propagation of waves in the forward direction (S_{12}) while damping propagation in the reverse (S_{21}) direction. This is an example of a resonance-based isolator, where the damping is due to the FMR in the magnetic element. In this case, the operation frequency can be set by the magnetization of the material used or it can be tuned by applying an external magnetic field.

Isolators based on other concepts and other materials are also possible. A second type of common isolator is a field-displacement (or edge-mode) isolator, where the forward and reverse directions of propagation localize the wave energy in different regions. Edge-mode isolators have also been fabricated with magnetic nanowires [60]. Another novel isolator design [61] uses an inverted L-shaped transducer with a YIG slab above. When the YIG slab is rotated at 45° with respect to the transducer, nonreciprocal coupling takes place, producing isolation as high as 22 dB in the C-band (0.5–1 GHz) with <3 dB of insertion loss.

Isolators based on ferrites have also been proposed. Theoretical calculations for isolators show a substantial increase in isolation for increased thicknesses of ferrite layers [62]. This is particularly interesting for M-type barium hexagonal ferrite (BaM) as this has a high anisotropy field (16.4 kOe) which allows for self-biased devices, meaning no external applied magnetic field is required for operation. An interesting calculation showed that for a hexagonal ferrite, with anisotropy axis in the plane of the film, one could get large isolation for a structure where the axis of the waveguide was at an angle with respect to the anisotropy direction [63]. With the increasing ability [64,65] to create thick ferrite films we see a number of new device designs and implementations [66]. However, with the necessity for perpendicular magnetization and small linewidths, experimental results have not yet achieved the results predicted by theory. Other hexagonal ferrite devices will be discussed later in this chapter.

Current research in circulators has focused on reducing the size of the devices to typical stripline and microstrip geometries. In this pursuit a number of interesting circulators have been designed. One design involves the integration of magnetic nanowires [67] into a polycarbonate and requires no bias field. Miniaturized circulators have also been produced [68] using 15-μm-thick YIG films in a coplanar waveguide. This device showed an isolation of 8 dB at 10 GHz but has a large insertion loss of around 18 dB.

3.6 On-Wafer Inductors

The integration of rf electronics components into complete monolithic solutions requires the design and fabrication of on-wafer inductors. These passive devices are essential components of integrated circuits such as voltage-controlled oscillators, low-noise amplifiers, mixers, DC–DC power converters, or intermediate frequency filters, for example. For applications at frequencies below 5 GHz, on-wafer air-core inductors (with typical values of inductance in the range of 1–10 nH) can occupy a significant portion of the surface area of a wafer, up to 1 mm^2 [69]. In order to reduce the physical size of the inductors while increasing the value of inductance, magnetic materials with high permeability are employed. Below we discuss the most common geometries used for on-wafer inductors, the magnetic materials used in the construction of the inductors, and the benefits and challenges resulting from the use of these magnetic materials.

The geometry of an on-wafer inductor is determined by many factors, including the following: the required value of the inductance, the required quality factor Q, the available surface area, the degree of complications that implementation of a particular design can afford, etc. Among the possible geometries, the strip inductor is the easiest to implement and build. It is simply a strip of highly conductive material (typically Cu) of width w and length l surrounded by the magnetic material of significantly lower conductivity and high relative permeability μ_r. The inductance L of such an inductor is given approximately by

$$L = \mu_o \mu_r \frac{t_m l}{2w}, \tag{7}$$

where t_m is the thickness of the magnetic material and μ_o is the permeability of free space [70,71]. (This equation is in SI units, which is convenient because the inductance is calculated directly in henrys.) This is only correct, however, if the skin depth in the magnetic material is significantly larger than t_m, indicating the importance of low conductivity for the magnetic layer. This expression also assumes magnetic flux closure surrounding the conductive element. It is rather difficult to make perfectly vertical walls surrounding the conductive element due to built-in stress during construction. Therefore, so-called magnetic "flanges" (see Fig. 11) or magnetic vias are used [72,73]. Magnetic vias, especially, improve the overall performance of the inductor by reducing eddy currents in metallic magnetic materials.

In order to increase the value of the inductance in a limited space, meander lines (where the long line is made into a bunch of parallel lines connected by curved segments) are sometimes used [74,75]. The increased inductance of meander lines is simply associated with the increased length of the devices. However, the increase is partly reduced by the negative mutual inductance between neighboring conductor strips in which rf currents flow in opposing directions. The most effective design, with respect to minimizing the surface area (footprint) used and maximizing the inductance, are planar spiral

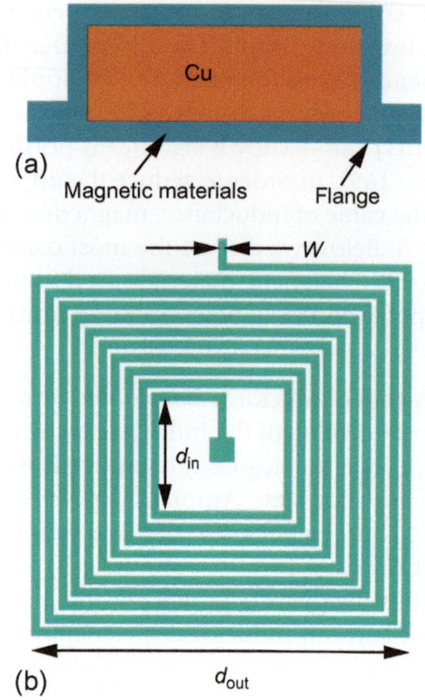

FIGURE 11

(a) Schematic cross-section of the strip inductor with flanges. (b) Schematic layout of the spiral inductor.

inductors. Among those, square spiral inductors are the most popular choice due to their ease of design and fabrication. These advantages outweigh the fact that other planar inductor geometries, such as hexagonal, octagonal or circular, have slightly better performance [76]. The inductance is proportional to the square of the number of turns and depends on other geometrical factors such as the width of the conducting element, the spacing between elements, and the inner and outer diameter of the spiral (see Fig. 11b) [76,77]. To further increase the inductance, 3D stacked spiral inductors have also been proposed. However, this requires a significant increase in complexity in terms of design and fabrication [78].

The use of magnetic materials significantly increases inductance, but may also result in undesired losses. As seen in Eq. (7), an ideal magnetic material would exhibit the highest possible value of the relative permeability μ_r. The relative permeability is typically proportional to the saturation magnetization M_s, and inversely proportional to the anisotropy field H_k. Materials with a very low anisotropy field will therefore exhibit high values of μ_r. At the same time, however, the utility of such materials is limited to the frequency range below the FMR frequency, so one wants this to be as high as possible. For magnetic materials with the magnetization in the plane of the film at zero applied field, the FMR frequency, f_{FMR} (see chapter on magnetic dynamics), is given by

$$f_{FMR} = \gamma' \sqrt{4\pi M_s H_k},$$

where γ' is the gyromagnetic constant. The FMR frequency is proportional to the square root of H_k, and therefore inversely proportional to the square root of μ_r [79,80]. This clearly indicates the need for material optimization in terms of the desired operational frequency range and the specified value of inductance. Additionally, in the vicinity of the FMR, the permeability is mostly imaginary and there are significant losses. It is also important, therefore, to choose materials with small FMR linewidth.

Another important physical property of the magnetic material used for inductors is the electrical conductivity. Insulating materials ensure small losses due to eddy currents. The electrical conductivity also determines the value of the skin depth, important for determining the required thickness of the magnetic materials in inductors operating at high frequencies. Finally, it is important that the magnetic material used remains in a single-domain state and exhibits small magnetostriction to avoid additional losses.

In the literature, there are three main groups of magnetic materials (oxides, amorphous and crystalline metallic alloys) used to make on-chip planar inductors. Below we discuss their properties and the advantages they provide. Among oxides, NiZn ferrites are most commonly used [81–83]. Typically these are substoichiometric compounds such as $Ni_{0.20}Cu_{0.20}Zn_{0.62}Fe_{1.98}O_{3.99}$, $Mg_{0.20+x-y}Cu_yZn_{0.62-x}Fe_{1.98}O_{3.99}$ or $Mn_{0.2}Ni_{0.49}Zn_{0.49}Co_{0.02}Fe_{1.90}O_x$ [84,85]. These compounds may have a relative permeability as high as 900 and also have a very high value of electrical resistance, which allows one to use relatively thick materials in the inductor structures. Typically these ferrites are used in powder form with an appropriate binder material. The constituent particles have a grain size of the order of 5 μm, a size that ensures a single-domain state. The very good properties of NiZn ferrites are limited, however, to frequencies below 100 MHz due to strong absorption associated with FMR. Feng Liu *et al.* demonstrated an example of a planar inductor structure using NiZn ferrites [86].

Soft metallic amorphous films, such as FeCoB [87], CoNbZr [88], and CoNbTaB [89], or nanocrystalline grains dispersed in an amorphous matrix ($Fe_{55}Hf_{17}O_{28}$) [90], are very interesting for inductor applications. Typically, the DC magnetron sputtering growth method is used, while sometimes reactive sputtering is used in the case of materials containing oxygen or nitrogen [91]. These materials exhibit reasonably high values of electric resistivity ($\rho > 1$ mΩ cm) and high relative permeability (up to 1600). With the saturation magnetization, $4\pi M_s$, of the order of 10 kG and anisotropy field of approximately 10 Oe, the FMR observed in these materials is at frequencies above 1 GHz. The small value of conductivity can be further decreased by proper lamination of these compounds with 1- to 2-nm-thick layers of oxides such as CoO or SiO_2. Such procedures significantly reduce eddy current losses in the inductor. The observed losses are thus dominated by FMR losses associated with the FMR linewidth.

Among crystalline metallic magnetic materials Permalloy, a NiFe alloy, takes a special place. It has a high $4\pi M_s = 10$ kG ($Ni_{80}Fe_{20}$ composition), low anisotropy, and nearly zero magnetostriction. When properly prepared, its FMR linewidth can be quite narrow and the coercive field can be below 1 Oe (see Fig. 12). As a result, Permalloy exhibits a large relative permeability and is widely used in inductor structures. Permalloy is typically grown using two techniques, magnetron sputtering [92,93] and electroplating [94]. In order to decrease the conductivity of Permalloy films, one can make a laminated film with Permalloy interspersed with layers of SiO_2 or Cr [73]. The typical operational frequency range for inductors with Permalloy is below 1 GHz [92]; however, there have been efforts to significantly extend the upper operational frequency up to 5 GHz. This can be achieved by creating exchange coupled structures of NiFe/FeCoB [95] or by proper patterning of the Permalloy elements [96].

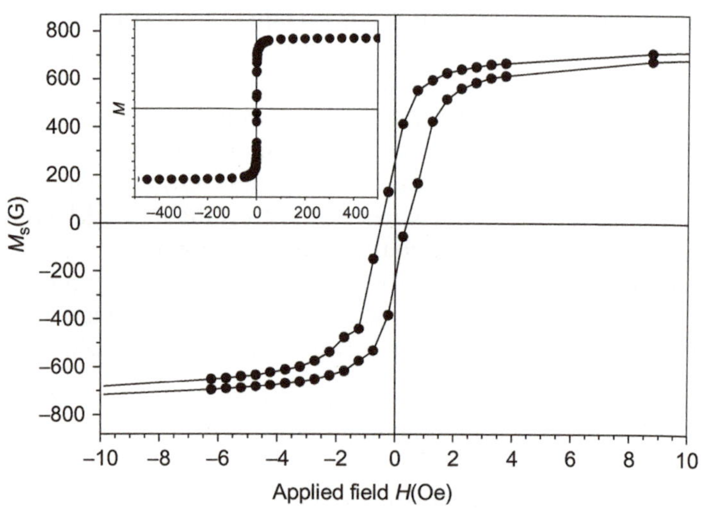

FIGURE 12

Hysteresis loops of a (30 nm NiFe/5 nm SiO_2)$_{50}$ multilayer. The measured coercive field is 0.5 Oe and value of relative permeability $\mu_r = 528$. The inset shows the hysteresis curve over a larger field range.

FeCo alloys are another interesting group of crystalline ferromagnetic metallic materials. Theses alloys may have saturation magnetization values as high as 24 kG and anisotropy fields H_k up to 200 Oe [75,97]. As a result, the relative permeability can reach values of the order of 100–200 and FMR is observed in these films above 4 GHz. The damping parameters was also relatively low ($\alpha \sim 0.025$), as seen in FMR experiments. In order to reduce the conductivity of FeCo films, they are frequently laminated with SiO_2 layers. For example, Namikawa *et al.* reported resistivity as high as 1 mΩ cm [97]. The exchange bias between FeCo layers and antiferromagnets such as IrMn, FeMn or NiO is used to further extend the operational frequency range of inductors. Various groups have reported inductors operating in the 5–7 GHz range [75,98,99].

4 PLANAR DEVICES WITH OTHER MATERIALS

4.1 Hexagonal Ferrites

A key point of much of the earlier material is that the confined geometry presents opportunities to explore parameters and devices that are not as easily produced in large-scale bulk versions. One example of this concerns the use of hexagonal ferrites. Hexagonal ferrites have been investigated for many years for use in high-frequency signal processing because they contain large internal anisotropy fields which, according to Eq. (1), will result in high-frequency resonance without large external fields [100,101]. Furthermore, these materials are insulators, eliminating eddy current losses, which can be a problem in the metallic ferromagnets. There have been a number of nice reviews on the growth and use of hexagonal ferrite films recently [102,103].

Despite the advantages of hexagonal ferrites, there are also problems. One such problem is associated with the size of the samples required for useful signal processing. One often sees in the literature phrases such as [104]: "These films are potentially useful for thin film millimeter-wave devices such as circulators, isolators, and phase shifters, provided that thick films (e.g., 20–100 μm) … can be grown." This has lead to a significant number of explorations of how to grow high-quality thick hexaferrite films. In fact, our theoretical calculations and initial experimental results show that in the small geometries explored here, one may be able to make useful devices with M-type BaM films which have a thickness of less than 1 μm.

Figure 13 shows a theoretical calculation for transmission as a function of frequency for a microstrip device with two materials between the metallic guiding plates [105]. The geometry is similar to that shown in Fig. 1 except that the magnetic field is oriented perpendicular to the interfaces. The thickness of the hexagonal ferrite is 500 nm and the thickness of the SiO_2 spacer is 4.5 μm. The assumed linewidth is 400 Oe at 40 GHz. The results show a substantial

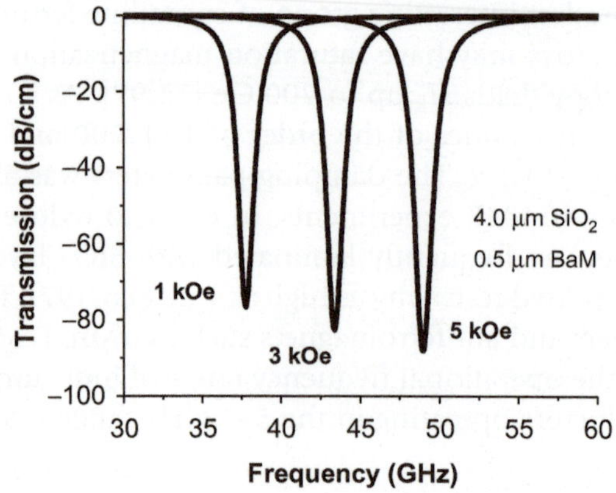

FIGURE 13

Theoretical calculations for transmission as a function of frequency for a microstrip waveguide constructed with SiO_2 and BaM films, for three different values of the applied field. The applied field is along the *c*-axis and perpendicular to the plane of the film; however, propagation of the electromagnetic wave is still in the plane of the sample. From Ref. [105].

symmetric dip or notch in the transmission, with the position of the notch tunable with an external field. For this geometry the maximum transmission loss is about 85 dB/cm. The theory also indicated that such a device could be used as a phase shifter.

Initial studies of hexagonal ferrites in planar-like geometries were based on a flip-chip geometry where a thin BaM film was placed on top of a coplanar waveguide. The Wu group demonstrated a number of important results in a series of papers [106–108]. These included the following:

(1) A self-biased notch filter that operated in the 53 GHz range at zero field [106]
(2) Tunable notch filters that operated in the 52–56 GHz range with biasing fields in the range of about 1 kOe [107]
(3) Tunable millimeter-wave phase shifters [107].

The experimental results were found to be in good agreement with the general theory obtained earlier.

The flip-chip geometry is convenient for testing materials; however, for practical devices it is often necessary to have integrated devices such as a microstrip waveguide. To accomplish this, it was necessary to grow BaM on a metallic substrate. This is a substantial technical problem since BaM growth often requires high-temperature processing steps. One solution for this was to use a metallo-organic decomposition technique with a Pt substrate [109]. The Pt substrate

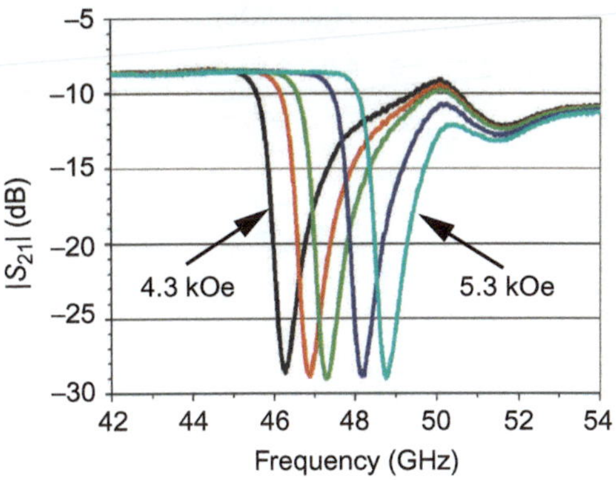

FIGURE 14
Experimental results for transmission of microwaves through a microstrip containing a 300-nm-thick BaM film and a 3.9-μm-thick SiO_2 film. The applied field is perpendicular to the film plane. The curves are for different magnetic fields, incremented by 0.25 kOe. The length of the structure was 3.3 mm. After Ref. [109].

was used because it could withstand the high temperatures required for growth of the BaM.

Figure 14 shows some experimental results for a notch filter constructed with a BaM film in a microstrip geometry [110]. The microstrip has a Pt ground plane (300 nm thick), a BaM film (320 nm thick), an SiO_2 film (3.9 μm thick) and a Cu signal line (740 nm thick), and the device is 3.3 mm long. The key result seen in this figure is that the attenuation at the notch is indeed quite substantial, about 60 dB/cm, as expected from the theoretical calculation shown earlier. The background attenuation of 8 dB/cm is higher than desired, but this is attributed, in part, to the thin Pt film used in the structure, which allows radiation to leak away from the device.

As can be seen above, the natural frequency range signal processing with BaM is typically in the 45–55 GHz range. In fact, this range can be widely extended by doping the BaM with Al [111]. This is illustrated in Fig. 15, which shows the FMR frequency as a function of applied field for $BaAl_xFe_{12-x}O_{19}$ as a function of the doping parameter x. One sees that, even at zero field, the resonance frequency can vary from about 33 to about 70 GHz.

4.2 Antiferromagnets

It is, in principle, also possible to use similar devices to operate in a very different frequency range. Signal processing at higher frequencies, 100 GHz–10 THz, is also of interest for radar and satellite communication. For these very

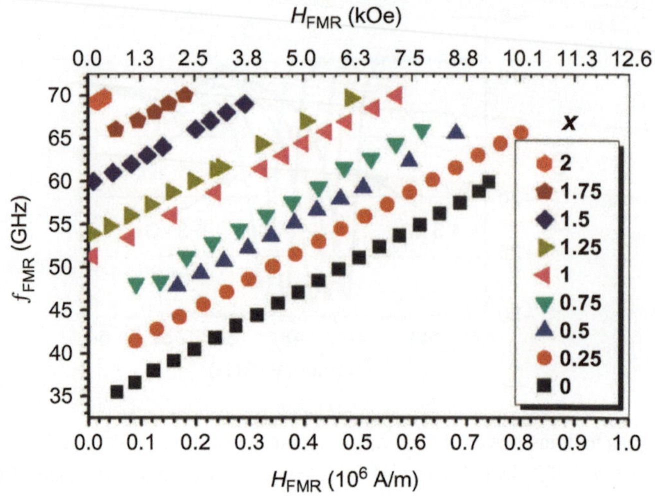

FIGURE 15

FMR frequency versus field for different doping levels, x, of Al in BaM, $BaAl_xFe_{12-x}O_{19}$. From Ref. [111].

high frequencies one can employ antiferromagnetic materials, which have resonance frequencies in the far-infrared region of the spectrum rather than in the microwave region as for ferromagnets. Also, many of the antiferromagnets are insulators and eddy current losses are minimal as for the hexagonal ferrites.

It is well known that the resonance frequency on a uniaxial antiferromagnet is given by $f=\gamma\sqrt{H_a(2H_e+H_a)}$ in the absence of demagnetizing effects and an external field [112]. Here H_a is the anisotropy field, H_e is the exchange field, γ is the gyromagnetic ratio, and f is the frequency.

A theoretical calculation [113] for a tunable notch filter using an antiferromagnet as the active medium shows the following results:

(1) One can obtain significant attenuation (above 35 dB/cm) so that ultrasmall devices with thicknesses of the order of 1–10 μm can be made in planar geometry. The attenuation of the electromagnetic wave is strongly influenced by the anisotropy field, with larger H_a values leading to larger losses at resonance. The absorption can, in fact, be quite large for FeF_2, as seen in Fig. 16.

(2) In contrast to the results for a ferromagnet, the filters with a single uniaxial antiferromagnet can also be used to provide a local band-pass region. The width of the band-pass region is proportional to the applied magnetic field, as is seen in Fig. 16.

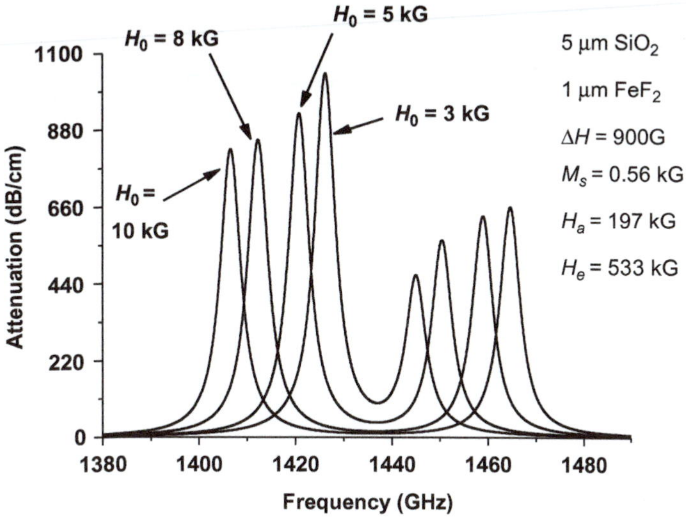

FIGURE 16

Theoretical calculations of the attenuation as a function of frequency for a filter constructed with an antiferromagnet for different applied fields. The SiO_2 has a thickness of 5 μm and the FeF_2 film has a thickness of 1 μm. As the external field is increased, a local band-pass region develops. From Ref. [113].

4.3 Liquid Crystals

We point out that other geometries and other classes of materials may also be of interest for high-frequency signal processing. One such material is the group of liquid crystals [114,115]. Some microstrip phase shifters have already been investigated in the range 10–20 GHz [116]. Liquid-crystal-based phase shifters have a number of advantages:

(1) They have a small footprint—on-wafer integration is possible
(2) They are inexpensive to produce
(3) Large phase shifts are possible, over 300 deg/cm at 110 GHz
(4) Because resonances in liquid crystals are typically in the infrared, broadband operation (10 GHz to hundreds of GHz) is possible
(5) They have low voltage operation and low power consumption
(6) They have a linear dependence of phase shift on frequency
(7) Because the change in dielectric constant is based on a reorientation and not a resonance, there is a constant value of the group delay—no dispersion
(8) They are expected to have good power handling capabilities
(9) Liquid crystal materials have relatively low losses at microwave frequencies.

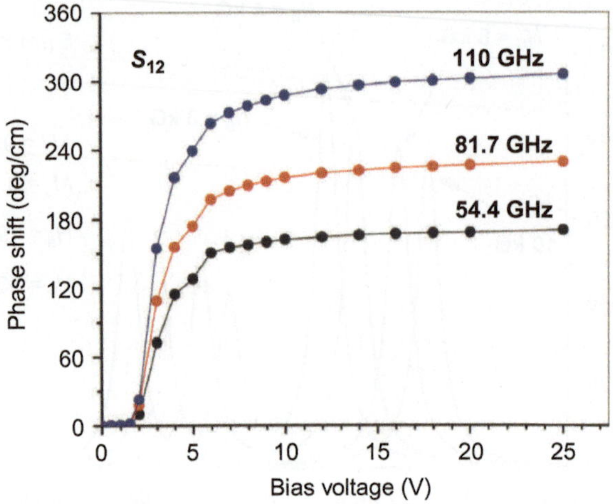

FIGURE 17

Phase shift as a function of bias voltage for different frequencies for a liquid-crystal mixture W-1825 in an inverted microstrip waveguide. From Ref. [117].

There are also some disadvantages; liquid crystal switching times are relatively slow, often in the hundreds of milliseconds range.

We have studied the phase-shifting properties of these materials in the range 0–110 GHz using an inverted microstrip structure where the thickness of the liquid crystal layer was about 25 μm [117]. Figure 17 presents the results for phase shift as a function of biasing voltages between the central conductor and the ground planes. We see a tunable phase shift of about 300 deg/cm at 110 GHz.

4.4 Voltage-Controlled Devices

We have discussed several examples of magnetically tunable microwave devices above. These devices take advantage of FMR in the magnetic material, and therefore an operational frequency shift can be achieved by changing the strength of an applied DC magnetic field (see Eqs. 1 and 3). There are, however, drawbacks to using a magnetic field for tuning, which is why there has been interest in devices that use electric field tuning. The reasons for this are that voltage control is generally faster and requires far less power than applying a changing magnetic field, where you must drive current through an electromagnet.

Electric tuning results from coupling the ferromagnet to a ferroelectric or piezoelectric material, often in a bilayer or multilayer film heterostructure. Several kinds of magnetoelectric (ME) coupling have been identified, including interface-charge-mediated [118,119], exchange-bias-mediated [120], and mechanical strain-induced coupling [121–123]. Interface-charge- and exchange-bias-mediated ME coupling are surface effects, and have been studied primarily in epitaxial ultrathin ferromagnetic films/interfaces with an eye towards spintronics applications. Work on electrically tuned microwave devices so far has primarily exploited strain-induced coupling.

Strain-induced coupling was demonstrated, for example, by Srinivasan *et al.* using a ferrite/ferroelectric bilayer of YIG/PMN-PT [30]. In a bilayer with 4.9 μm YIG and 500 μm PMN-PT, FMR measurements show an upshift of 5.4 Oe in the resonance field at 9.3 GHz for every kV/cm of applied electric field across the structure. This change is an order of magnitude higher than was measured in a bulk YIG/PZT heterostructure [124]. They proposed a band-pass filter design using the bilayer that would operate at 9.3 GHz, and their calculations show an upshift in the band-pass frequency of roughly 700 MHz for every 50 kV/cm applied, with only a 2.5 dB insertion loss at the band-pass frequency. This YIG-based design still requires a permanent biasing magnet to operate at 9.3 GHz. The results from a prototype of their proposed device were published a year later, and exhibited a 125-MHz tuning range for $E=0$–3 kV/cm and an insertion loss of 5 dB at 6.5 GHz [125]. They also measured the electrically induced differential phase shift in a similar bilayer structure, obtaining a 40° differential phase shift between 0 and 10 kV/cm over the 6–9 GHz band [126].

Over the last decade, many groups have also developed electrically tuned phase shifter technology operating between 5 and 10 GHz [127–131]. Using a YIG/PZT bilayer, Tatarenko *et al.* measured a 180° phase shift with $E=8$ kV/cm at 9.6 GHz [127]. In this case, the FMR frequency shift with $E=8$ kV/cm was only 6 MHz, but nevertheless a substantial differential phase shift was demonstrated. Yang *et al.* used YIG/PMN-PT heterostructures and achieved a 119° differential phase shift under an electric field of 11 kV/cm [131]. A very interesting feature of this particular YIG-based design, however, is that the device operates from 5 to 6 GHz without the need for a biasing magnetic field. Geiler *et al.* developed a very interesting approach, using magnetic fringe fields emanating from a multiferroic transducer to tune the YIG-based device, rather than direct mechanical contact with an electrically activated strain element [128]. The device also operates above and far away from the FMR frequency. In this way, they were able to achieve phase shifts up to 65° over a large bandwidth of 500 MHz, centered at 5.95 GHz, using electric fields of only 6 kV/cm and a 200 Oe biasing magnetic field. In principle, their approach could lead to a phase shifter with low insertion loss, suitable for applications, due to the fact that the operating frequency is far away from the FMR frequency, and thus additional losses associated with the magnetic resonance are avoided.

5 SUMMARY AND DISCUSSION

In this chapter, we have examined the properties of a set of microwave devices which use metallic ferromagnets such as Fe and Permalloy as the tunable magnetic element. These devices operate in the 10–40 GHz frequency range and include tunable notch filters, band-pass filters, phase shifters, and a signal-to-noise enhancer. They are grown using techniques such as magnetron sputtering, which are compatible with industrial standards. The devices are planar, very small, and can be integrated with microwave electronics on a single chip. All of them exhibit a small insertion loss (2–3 dB) away from the FMR. The notch filters have high attenuation (60–100 dB/cm) at a frequency that is tunable by an external magnetic field.

In addition, we have explored the use of other materials such as hexagonal ferrites, antiferromagnets, and liquid crystals in similar geometries. Because of large internal fields, the hexagonal ferrites and antiferromagnets can operate at much higher frequencies ranging from 50 GHz to 1.5 THz. Our theoretical calculations and initial experimental work indicate that it is possible to construct useful devices with films that are of the order of 1 μm in thickness. The liquid-crystal devices show promise as phase shifters.

We emphasize that the parameters and devices shown here can be modified by use of different materials or different physical geometries. For example, we showed that the operational frequency of the devices can be significantly increased by shape anisotropy or by exchange coupling of one magnetic film to another. The linewidth of the device can also be altered by using a multilayer structure, as shown earlier.

There are several issues which can impact the implementation of these devices:

(1) While the magnetic materials exhibit significant operational frequency range, they must be tuned by a magnetic field. As we have pointed out, the use of ferromagnetic metals and other materials reduces the size of the required field, but it is not eliminated. We estimate, for example, that a notch filter which could be tuned over a 10 GHz range would require a small electromagnet. This could be done with a weight of about 50 g and 5 W of power. Ideally, one would like a tunable, planar magnet which is on the same chip as the filtering device.

(2) There are some losses associated with the metallic guides at high frequencies. For example, our data shows that the losses within a fixed structure increase substantially as the frequency increases above 40 GHz. These losses can be reduced by increasing the thickness of the dielectric layer or by increasing the thicknesses of the conducting layers [132]. However, this may reduce some of the advantages of the small-scale geometry discussed earlier.

In summary, new materials and new growth methods offer the opportunity to significantly change many microwave signal processing devices. There are significant reductions in size and weight that are possible, but in addition to these practical issues, the new devices also invite exploration into new physics, particularly in nonlinear physics. We anticipate that future work in this area will continue to involve new materials, such as ferrimagnets which have high-frequency resonances, and planar forms of YIG.

Acknowledgments

The authors acknowledge support from DOA Grant No. W911NF-04-1-0247. The authors would also like to thank Karen Livesey for a careful reading of the chapter and for helpful comments. Portions of this chapter were taken from a previous article, Ref. [133].

References

[1] Adam JD, Davis LE, Dionne GF, Schloemann EF, Steven N. Stitzer, ferrite devices and materials. IEEE Trans Microwave Theory Tech 2002;50:721.

[2] Schloemann E, Tustison R, Weissman J, Jerrold Van Hook H, Varitimos T. Epitaxial Fe films on GaAs for hybrid semiconductor-magnetic devices. J Appl Phys 1988;63:3140.

[3] Liau VS, Wong T, Stacey W, Ali S, Schloemann E. Tunable band-stop filter based on epitaxial Fe film on GaAs. IEEE MTT-S 1991;3:957.

[4] Tsai CS, Su J, Lee CC. Wideband electronically tunable microwave bandstop filters using iron film-gallium arsenide waveguide structures. IEEE Trans Magn 1999;35:3178.

[5] Camley RE, Mills DL. Theory of microwave propagation in dielectric/magnetic film multilayer structures. J Appl Phys 1997;82:3058.

[6] Astalos RJ, Camley RE. Theory of a high frequency magnetic tunable filter and phase shifter. J Appl Phys 1998;83:3744.

[7] Kuanr B, Harward IR, Marvin DL, Fal T, Camley RE, Mills DL, et al. High-frequency signal processing using ferromagnetic metals. IEEE Trans Magn 2005;41:3538.

[8] Cramer N, Lucic D, Camley RE, Celinski Z. High attenuation tunable microwave notch filters utilizing ferromagnetic resonance. J Appl Phys 2000;87:6911.

[9] Cramer N, Lucic D, Walker DK, Camley RE, Celinski Z. Incorporation of ferromagnetic metallic films in planar transmission lines for microwave device applications. IEEE Trans Magn 2001;37:2392.

[10] Kuanr B, Malkinski L, Camley RE, Celinski Z, Kabos P. Iron and permalloy based magnetic monolithic tunable microwave devices. J Appl Phys 2003;93:8591.

[11] Hyeon Kim K, Yamaguchi M, Arai K-I, Nagura H, Ohnuma S. Effect of radio-frequency noise suppression on the coplanar transmission line using soft magnetic thin films. J Appl Phys 2003;93:8002.

[12] Hyeon Kim K, Ohnuma S, Yamaguchi M. RF integrated noise suppressor using soft magnetic films. IEEE Trans Magn 2004;40:2838.

[13] Li L, Lee DW, Mao M, Schneider T, Bubber R, Hwang KP, et al. High-frequency responses of granular CoFeHfO and amorphous CoZrTa magnetic materials. J Appl Phys 2007;101:123912.

[14] Goglio G, Pignard S, Radulescu A, Piraux L, Huynen I, Janvier DV, et al. Microwave properties of metallic nanowires. Appl Phys Lett 1999;75:1769.

[15] Encinas A, Demand M, Vila L, Piraux L, Huynen I. Tunable remanent state resonance frequency in arrays of magnetic nanowires. Appl Phys Lett 2002;81:2031.

[16] Huynen I, Piraux L, Janvier DV, Vorst AV. In: Proceedings of PIERS 2000 symposium, July 2000; 2000. p. 323.

[17] Saib A, Darques M, Piraux L, Janvier DV, Huynen I. Unbiased microwave circulator based on ferromagnetic nanowire arrays of tunable magnetization state. J Phys D Appl Phys 2005;38:2759.

[18] Ye B, Li F, Cimpoesu D, Wiley JB, Jung J-S, Stancu A, et al. Passive high-frequency devices based on superlattice ferromagnetic nanowires. J Magn Magn Mater 2007;316:e56.

[19] Zhuang Y, Rejaei B, Boellaard E, Vroubel M, Burghartz JN. GHz bandstop microstrip filter using patterned $Ni_{78}Fe_{22}$ ferromagnetic film. IEEE Microwave Wireless Compon Lett 2002;12:473.

[20] Amiri PK, Rejaei B, Vroubel M, Zhuang Y, Burghartz JN. Experimental determination of the nonuniform shape-induced anisotropy field in thin Ni–Fe films. IEEE Trans Magn 2007;43:1880.

[21] Schoenstein F, Aublanc P, Pagès H, Queste S, Barentin V, Adenot A-L, et al. Influence of the domain structure on the microwave permeability of soft magnetic films and multilayers. J Magn Magn Mater 2005;292:201.

[22] Vukadinovic N, Vacus O, Labrune M, Acher O, Pain D. Magnetic excitations in a weak-stripe-domain structure: a 2D dynamic micromagnetic approach. Phys Rev Lett 2000;85:2817–20.

[23] Salahun E, Quéffélec P, Tanné G, Adenot A-L, Acher O. Correlation between magnetic properties of layered ferromagnetic/dielectric material and tunable microwave device applications. J Appl Phys 2002;91:5449.

[24] He Y, He P, Yoon SD, Parimi PV, Rachford FJ, Harris VG, et al. Tunable negative index metamaterial using yttrium iron garnet. J Magn Magn Mater 2007;313(1):187.

[25] Matters-Kammerer M, Mackens U, Reimann K, Pietig R, Hennings D, Schreinemacher B, et al. Material properties and RF applications of high k and ferrite LTCC ceramics. Microelectron Reliab 2006;46:134.

[26] Lee S-J, Eon Moon S, Kwak M-H, Ryu H-C, Kim Y-T, Kang K-Y. Enhanced dielectric properties of (Ba, Sr) TiO_3 thin films for high-performance microwave phase shifters. Integr Ferroelectr 2005;72:39.

[27] Rebeiz GM. RFMEMS theory, design, and technology. New York: Wiley; 2003.

[28] De los Santos HJ. RF MEMS circuit design. Boston, MA: Artech House; 2002.

[29] Marsan E, Gauthier J, Chaker M, Wu K. Tunable microwave devices, In: The 3rd international IEEE-NEWCAS conference, 19–22 June 2005; 2005. p. 279.

[30] Srinivasan G, Tatarenko AS, Bichurin MI. Electrically tunable microwave filters based on ferromagnetic resonance in ferrite-ferroelectric bilayers. Electron Lett 2005;41:596.

[31] Marks RB. A multiline method of network analyzer calibration. IEEE Trans Microwave Theory Tech 1991;39:1205.

[32] Kuanr B, Camley RE, Celinski Z. Tunable high-frequency band-stop magnetic filters. Appl Phys Lett 2003;83:3969.

[33] Kuanr B, Camley RE, Celinski Z. Narrowing of the frequency-linewidth in structured magnetic strips: experiment and theory. Appl Phys Lett 2005;87:012502.

[34] Kuanr BK, Harward IR, Deiotte RT, Camley RE, Celinski Z. Magnetically tunable micro-strip band-stop filter: design optimization and characterization. J Appl Phys 2005;97:10Q103.

[35] Kittel C. On the theory of ferromagnetic resonance absorption. Phys Rev 1948;73:155.

[36] Khivintsev YV, Kuanr BK, Harward I, Camley RE, Celinski Z. Iron-based microstrip band-stop filters at higher microwave frequency range: design optimization using shape anisotropy. J Appl Phys 2006;99:08P512.

[37] Grimsditch M, Camley R, Fullerton EE, Jiang S, Bader SD, Sowers CH. Exchange-spring systems: coupling of hard and soft ferromagnets as measured by magnetization and Brillouin light scattering. J Appl Phys 1999;85:5901.

[38] Astalos RJ, Camley RE. Magnetic permeability for exchange-spring magnets: application to Fe/Sm-Co. Phys Rev B 1998;58:8646.

[39] Kuanr BK, Khivintsev YV, Harward I, Camley RE, Celinski ZJ, Bedenbecker M, et al. Increasing operational frequency in microwave devices by using [SmCo/NiFe] multilayered structures. IEEE Trans Magn 2007;43:2648.

[40] Kuanr BK, Marvin DL, Christensen TM, Camley RE, Celinski Z. High-frequency magnetic microstrip local bandpass filters. Appl Phys Lett 2005;87:222506.

[41] Stitzer SN, Emtage PR. Nonlinear microwave signal-processing devices using thin ferrimagnetic films. Circuits Syst Signal Process 1985;4:227.

[42] Adam JD. Analog signal processing with microwave magnetic. Proc IEEE 1988;76:159.

[43] Nomoto T, Kuki T. MSW signal-to-noise enhancers for noise reduction in DBS reception. J Phys IV France 1997;07:387.

[44] Khivintsev YV, Zagorodnii VV, Hutchison AJ, Camley RE, Celinski ZJ. On-wafer microwave signal-to-noise enhancer using NiFe films. Appl Phys Lett 2009;92:022512; Khivintsev Y, Kuanr B, Fal TJ, Haftel M, Camley RE, Celinski Z, et al. Nonlinear ferromagnetic resonance in permalloy films: a nonmonotonic power-dependent frequency shift. Phys Rev B 2010;81:054436.

[45] Adam JD, Stitzer SN. A magnetostatic wave signal-to-noise enhancer. Appl Phys Lett 1980;36:485.

[46] Bao M, Khitun A, Wu Y, Lee J-Y, Wang KL, Jacob AP. Coplanar waveguide radio frequency ferromagnetic parametric amplifier. Appl Phys Lett 2008;93:072509.

[47] Khivintsev Y, Marsh J, Zagorodnii V, Harward I, Lovejoy J, Krivosik P, et al. Nonlinear amplification and mixing of spin waves in a microstrip geometry with metallic ferromagnets. Appl Phys Lett 2011;98:042505.

[48] Marsh J, Camley RE. Two-wave mixing in nonlinear magnetization dynamics: a perturbation expansion of the Landau-Lifshitz-Gilbert equation. Phys Rev B 2012;86:224405.

[49] Marsh J, Zagorodnii V, Celinski Z, Camley RE. Nonlinearly generated harmonic signals in ultra-small waveguides with magnetic films: tunable enhancements of 2nd and 4th harmonics. Appl Phys Lett 2012;100:102404.

[50] Cheng C, Bailey WE. High-efficiency gigahertz frequency doubling without power threshold in thin-film $Ni_{81}Fe_{19}$. Appl Phys Lett 2013;103:242402.

[51] Camley RE. Nonreciprocal surface waves. Surf Sci Rep 1987;7:103.

[52] Bosma H. A general model for junction circulators; choice of magnetization and bias field. IEEE Trans Magn 1968;4:587.

[53] Schloemann E, Blight RE. Broad-band stripline circulators based on YIG and Li-ferrite single crystals. IEEE Trans MTT 1986;34:1394.

[54] Wen CP. Coplanar waveguide: a surface strip transmission line suitable for nonreciprocal gyromagnetic device applications. IEEE Trans MTT 1969;17:1087.

[55] Forterre G, Chiron B, Courtois L. A survey of broad band stripline ferrite isolators. IEEE Trans Magn 1975;11:1279.

[56] Beguhn S, Yang X, Sun NX. Wideband ferrite substrate integrated waveguide isolator using shape anisotropy. J Appl Phys 2014;115:17E503.

[57] Pozar DM. Microwave engineering. 2nd ed. New York: Wiley; 1998.

[58] Huynen I, Goglio G, Vanhoenacker D, Vander Vorst A. A novel nanostructured microstrip device for tunable stopband filtering applications at microwaves. IEEE Microw Guid Wave Lett 1999;9:401.

[59] Kuanr BK, Veerakumar V, Marson RL, Mishra SR, Camley RE, Celinski Z. Nonreciprocal microwave devices based on magnetic nanowires. Appl Phys Lett 2009;94:202505.

[60] Carignan LP, Caloz C, Menard D. Ferromagnetic nanowire metamaterials: theory and applications, In: International microwave symposium IEEE MTT-S, Anaheim, USA. vol. 1336. ; 2010.

[61] Wu J, Yang X, Lou J, Beguhn S, Sun NX. Non-reciprocal tunable low-loss bandpass filters with ultra-wideband isolation based on magnetostatic surface wave. IEEE Trans MTT 2012;60:3959.

[62] Capraro S, Chatelon J-P, Le Berre M, Joisten H, Rouiller T, Bayard B, et al. Barium ferrite thick films for microwave applications. J Magn Magn Mater 2004;272:E1805.

[63] Anderson NR, Camley RE. Multilayer magnetic waveguides: optimizing nonreciprocal propagation in the 50–70 GHz range. J Appl Phys 2014;116:023903.

[64] Oliver SA, Yoon SD, Kozulin I, Chen ML, Vittoria C. Growth and characterization of thick oriented barium hexaferrite films on MgO (111) substrates. Appl Phys Lett 2000;76:3612.

[65] Hu B, Su Z, Bennett S, Chen Y, Harris VG. Epitaxial growth of 100-μm thick M-type hexaferrite crystals on wide bandgap semiconductor GaN/Al_2O_3 substrates. J Appl Phys 2014;115:17a513.

[66] Bayard B, Vincent D, Simovski CR, Noyel G. Electromagnetic study of a ferrite coplanar isolator suitable for integration. IEEE Trans MTT 2003;51:1809.

[67] Saib A, Darques M, Piraux L, Vanhoenacker-Janvier D, Huynen I. An unbiased integrated microstrip circulator based on magnetic nanowired substrate. IEEE Trans MTT 2005;53:2043.

[68] Zahwe O, Abdel Samad B, Sauviac B, Chatelon JP, Blanc Mignon MF, Rousseau JJ. YIG thin film used to miniaturize a coplanar junction circulator. J Electromagn Waves Appl 2010;24:25.

[69] Burghartz JN, Rejaei B. On the design of RF spiral inductors on silicon. IEEE Trans Electron Devices 2003;50:718.

[70] Korenivski V, van Dover RB. Magnetic film inductors for radio frequency applications. J Appl Phys 1997;82:5247.

[71] Korenivski V. GHz magnetic film inductors. J Magn Magn Mater 2000;215:800.

[72] Gardner DS, Schrom G, Hazucha P, Paillet F, Karnik T, Borkar S, et al. Integrated on-chip inductors using magnetic material. J Appl Phys 2008;103:07E927.

[73] Wu H, Gardner DS, Zhao S, Huang H, Yu H. Control of magnetic flux and eddy currents in magnetic films for on-chip radio frequency inductors: role of the magnetic vias. J Appl Phys 2014;115:17E719.

[74] Piekarz I, Sorocki J, Wincza K, Gruszczynski S. Modeling and performance improvement of folded coupled lines in miniaturized quasi-lumped directional couplers. Int J RF Microwave Comput Aided Eng 2015;25:1.

[75] Michel JP, Lamy Y, Royet AS, Viala B. New RF magnetic stripe inductors with flanges based on exchange-coupled magnetic films. IEEE Trans Magn 2006;42:3368.

[76] Mohan SS, del Mar Hershenson M, Boyd SP, Lee TH. Simple accurate expressions for planar spiral inductances. IEEE J Solid-State Circuits 1999;34:1419.

[77] Musunuri S, Chapman PL, Zou J, Liu C. Design issues for monolithic DC-DC converters. IEEE Trans Power Electron 2005;20:639.

[78] Chen J, Liou JJ. On-chip spiral inductors for RF applications: an overview. J Semicond Tech Sci 2004;4:149.

[79] Snoek JL. Dispersion and absorption in magnetic ferrites at frequencies above one Mc/s. Physica (Amsterdam) 1948;14:207.

[80] Acher O, Dubourg S. Generalization of Snoek's law to ferromagnetic films and composites. Phys Rev B 2008;77:104440.

[81] Su H, Zhang H, Tang X, Jing Y, Liu Y. Effects of composition and sintering temperature on properties of NiZn and NiCuZn ferrites. J Magn Magn Mater 2007;310:17.

[82] Kawano K, Hachiya M, Iijima Y, Sato N, Mizuno Y. The grain size effect on the magnetic properties in NiZn ferrite and the quality factor of the inductor. J Magn Magn Mater 2009;321:2488.

[83] Lečić N, Mukkattu Kuniyil NC, Stojanović G. Characterization of ferrite materials used as a core for multi-phase coupled inductors. In: 2014 International conference and exposition on electrical and power engineering (EPE 2014), December 1st, 2014, Iasi, Romania.

[84] Mürbe J, Töpfer J. High permeability Ni–Cu–Zn ferrites through additive-free low-temperature sintering of nanocrystalline powders. J Eur Ceram Soc 2012;32:1091.

[85] Mürbe J, Töpfer J. Mg–Cu–Zn ferrites for multilayer inductors. Int J Appl Ceram Technol 2007;4:415.

[86] Liu F, Ren T, Yang C, Liu L, Wang AZ, Yu J. NiCuZn ferrite thin films for RF integrated inductors. Mater Lett 2006;60:1403.

[87] Klemmer TJ, Ellis KA, Chen LH, van Dover B, Jin S. Ultrahigh frequency permeability of sputtered Fe–Co–B thin films. J Appl Phys 2000;87:830.

[88] Kima KH, Yamaguchi M. The physics of metals and metallography, vol. 102. Moscow, Russia: Pleiades Publishing, Inc; 2006. p. S74–S79 (Suppl. 1).

[89] Wu H, Gardner DS, Zhao S, Huang H, Yu H. Control of magnetic flux and eddy currents in magnetic films for on-chip radio frequency inductors: role of the magnetic vias. J Appl Phys 2014;115:17E719.

[90] Huijbregtse J, Roozeboom F, Sietsma J, Donkers J, Kuiper T, van de Riet E. High-frequency permeability of soft-magnetic Fe–Hf–O films with high resistivity. J Appl Phys 1998;83:1569.

[91] van de Riet E, Roozeboom F. Ferromagnetic resonance and eddy currents in high-permeable thin films. J Appl Phys 1997;81:350.

[92] Xu W, Sinha S, Dastagir T, Wu H, Bakkaloglu B, Gardner DS, et al. Performance enhancement of on-chip inductors with permalloy magnetic rings. IEEE Elec Dev Lett 2011;32:69.

[93] Dastagir T, Xu W, Shinha S, Wu H, Cao Y, Yu H. Tuning the permeability of permalloy films for on-chip inductor applications. Appl Phys Lett 2010;97:162506.

[94] Wang N, O'Sullivan EJ, Herget P, Rajendran B, Krupp LE, Romankiw LT, et al. Integrated on-chip inductors with electroplated magnetic yokes. J Appl Phys 2012;111:07E732.

[95] Hashimoto A, Nakagawa S, Yamaguchi M. Improvement of soft magnetic properties of Si/NiFe/FeCoB thin films at gigahertz band frequency range by multilayer configuration. IEEE Trans Magn 2007;43:2627.

[96] Ni W, Kim J, Kan EC. Permalloy patterning effects on RF inductors. IEEE Trans Magn 2006;42:2827.

[97] Namikawa M, Munakata M, Yagi M, Shimada Y, Miyazawa Y, Yamaguchi M. Magnetic properties and GHz permeability of (FeCO)-(SiO_2) films with high saturation magnetization of bcc-FeCo phase. J Magn Soc Japan 2004;28:873.

[98] Jiang R-F, Shams N, Rahman M, Lai C-H. Exchange-coupled IrMn/CoFe mulitlayers for RF-integrated inductors. IEEE Trans Magn 2007;43:3930.

[99] Lamy Y, Viala B. Combination of ultimate magnetization and ultrahigh uniaxial anisotropy in CoFe exchange-coupled multilayers. J Appl Phys 2005;97:10F910.

[100] Song Y-Y, Kalarickal S, Patton CE. Optimized pulsed laser deposited barium ferrite thin films with narrow ferromagnetic resonance linewidths. J Appl Phys 2003;94:5103.

[101] Chen Y, Geiler AL, Chen T, Sakai T, Vittoria C, Harris VG. Low-loss barium ferrite quasi-single-crystals for microwave application. J Appl Phys 2007;101:09M501.

[102] Harris V, Geiler A, Chen Y, Dae Yoon S, Wu M, Yang A, et al. Recent advances in processing and applications of microwave ferrites. J Magn Magn Mater 2009;321:2035.

[103] Wu M. M-type barium hexagonal ferrite films. In: Malkinski L, editor. Advanced magnetic materials: Rijeka (Croatia): InTech; 2012. ISBN 978-953-51-0637-1 DOI: 10.5772/39103.

[104] Dorsey PC, Chrisey DB, Horwitz JS, Lubitz P, Auyeung RCY. Oriented barium hexaferrite thick films grown on c-plane and m-plane sapphire substrates. IEEE Trans Magn 1994;30:4512.

[105] Fal T, Camley RE. Hexagonal ferrites for use in microwave notch filters and phase shifters. J Appl Phys 2008;104:023910.

[106] Song YY, Sun Y, Lu L, Bevivino J, Wu M. Self-biased planar millimeter wave notch filters based on magnetostatic wave excitation in barium hexagonal ferrite thin films. Appl Phys Lett 2010;97:173502.

[107] Wang Z, Song YY, Sun Y, Bevivino J, Wu M, Veerakumar V, et al. Millimeter wave phase shifter based on ferromagnetic resonance in a hexagonal barium ferrite thin film. Appl Phys Lett 2010;97:072509.

[108] Lu L, Song YY, Bevivino J, Wu M. Planar millimeter wave band-stop filters based on the excitation of confined magnetostatic waves in barium hexagonal ferrite thin film strips. Appl Phys Lett 2011;98:212502.

[109] Nie Y, Harward I, Kalin K, Beaubien A, Celinski Z. Preparation and characterization of barium hexagonal ferrite thin films on a Pt template. J Appl Phys 2010;107:073903.

[110] Harward I, Camley RE, Celinski Z. On-wafer magnetically tunable millimeter wave notch filter using M-phase Ba hexagonal ferrite/Pt thin films on Si. Appl Phys Lett 2014;105:173503.

[111] Harward I, Nie Y, Gardner A, Reisman L, Celinski Z. Al doped Ba hexaferrite ($BaAl_xFe_{12-x}O_{19}$) thin films on Pt using metallo-organic decomposition. J Appl Phys 2012;111:07A514.

[112] Kittel C. Introduction to solid state physics. 4th ed. New York: John Wiley; 1971. pp. 603.

[113] Veerakumar V, Camley RE. High frequency tunable band stop filters using antiferromagnetic films. J Appl Phys 2007;101:093906.

[114] Weil C, Müller S, Scheele P, Best P, Lüssem G, Jakoby R. Highly-anisotropic liquid-crystal mixtures for tunable microwave devices. Electron Lett 2003;39:1732.

[115] Dolfi D, Labeyrie M, Joffre P, Huignard JP. Liquid crystal microwave phase shifter. Electron Lett 1993;29:926.

[116] Weil C, Luessem G, Jakoby R. Tunable inverted-microstrip phase shifter device using nematic liquid crystals. IEEE MTT-S International Microwave Symposium Digest 2002;1:367.

[117] Garbovskiy Y, Zagorodnii V, Krivosik P, Lovejoy J, Camley RE, Celinski Z, et al. Liquid crystal phase shifters at millimeter wave frequencies. J Appl Phys 2012;111:054504.

[118] Nozaki T, Yakushiji K, Tamaru S, Sekine M, Matsumoto R, Konoto M, et al. Voltage-induced magnetic anisotropy changes in an ultrathin FeB layer sandwiched between two MgO layers. Appl Phys Express 2013;6:073005.

[119] Zhang H, Richter M, Koepernik K, Opahle I, Tasnadi F, Eschrig H. Electric-field control of surface magnetic anisotropy: a density functional approach. New J Phys 2009;11:043007.

[120] Heron JT, Trassin M, Ashraf K, Gajek M, He Q, Yang SY, et al. Electric-field-induced magnetization reversal in a ferromagnet-multiferroic heterostructure. Phys Rev Lett 2011;107:217202.

[121] Liu M, Obi O, Lou J, Chen Y, Cai Z, Stoute S, et al. Giant electric field tuning of magnetic properties in multiferroic ferrite/ferroelectric heterostructures. Adv Funct Mater 2009;19:1826.

[122] Wu T, Bur A, Zhao P, Mohanchandra KP, Wong K, Wang KL, et al. Giant electric-field-induced reversible and permanent magnetization reorientation on magnetoelectric Ni/(011) $[Pb(Mg_{1/3}Nb_{2/3})O_3]_{(1-x)}$–$[PbTiO_3]_x$ heterostructure. Appl Phys Lett 2011;98:012504.

[123] Jin T, Hao L, Cao J, Liu M, Dang H, Wang Y, et al. Electric field control of anisotropy and magnetization switching in CoFe and CoNi thin films for magnetoelectric memory devices. Appl Phys Express 2014;7:043002.

[124] Bichurin MI, Petrov RV, Kiliba YV, Srinivasan G. Magnetic and magnetoelectric susceptibilities of a ferroelectric/ferromagnetic composite at microwave frequencies. Phys Rev B 2002;66:134404.

[125] Tatarenko AS, Gheevarughese V, Srinivasan G. Magnetoelectric microwave bandpass filter. Electron Lett 2006;9:540.

[126] Tatarenko AS, Srinivasan G, Bichurin MI. Electrically tunable microwave phase shifter based on ferrite piezoelectric layered structure, In: 18th Int. crimean conference "microwave & telecommunication technology" (CriMiCo'2008); 2008. p. 508.

[127] Tatarenko AS, Srinivasan G, Bichurin MI. Magnetoelectric microwave phase shifter. Appl Phys Lett 2006;88:183507.

[128] Geiler AL, Gillette SM, Chen Y, Wang J, Chen Z, Yoon SD, et al. Multiferroic heterostructure fringe field tuning of meander line microstrip ferrite phase shifter. Appl Phys Lett 2010;96:053508.

[129] Leach JH, Liu H, Avrutin V, Rowe E, Ozgur U, Morkoc H, et al. Electrically and magnetically tunable phase shifters based on a barium strontium titanate-yttrium iron garnet layered structure. J Appl Phys 2010;108:064106.

[130] Zhu Y, Qiu G, Tsai CS. A magnetically- and electrically-tunable microwave phase shifter using yttrium iron garnet/gadolinium gallium garnet thin film. J Appl Phys 2012;111:07A502.

[131] Yang X, Gao Y, Wu J, Zhou Z, Beguhn S, Nan T, et al. Voltage tunable multiferroic phase shifter with YIG/PMN-PT heterostructure. IEEE Microwave Wireless Compon Lett 2014;24:191.

[132] Pucel RA, Massé DJ, Hartwig CP. Losses in microstrip. IEEE Trans Microwave Theory Tech 1968;16:342.

[133] Camley RE, Celinski Z, Fal T, Glushchenko AV, Hutchison AJ, Khivintsev Y, et al. High-frequency signal processing using magnetic layered structures. J Magn Magn Mater 2009;321:2048.

Index

Note: Page numbers followed by "*f*" indicate figures and "*t*" indicate tables.

M

N

Kolibri 360 GmbH
Ferdinand-Jühlke-Straße 7
99095 Erfurt, Deutschland
produktsicherheit@kolibri360.de